Bacteriophages in the Control of Food- and Waterborne Pathogens

Bacteriophages in the Control of Food- and Waterborne Pathogens

EDITED BY

Parviz M. Sabour
Guelph Food Research Centre, Research Branch
Agriculture & Agri-Food Canada
Guelph, Ontario, Canada

Mansel W. Griffiths
Canadian Research Institute for Food Safety
Department of Food Science
University of Guelph
Guelph, Ontario, Canada

ASM PRESS WASHINGTON, DC

Cover image: Ed Atkeson, Berg Design

ASM Press
American Society for Microbiology
1752 N St., N.W.
Washington, DC 20036-2904

Library of Congress Cataloging-in-Publication Data

Bacteriophages in the control of food- and waterborne pathogens / edited by Parviz M. Sabour and Mansel W. Griffiths.
p. ; cm.
Includes bibliographical references and index.
ISBN-13: 978-1-55581-502-8 (hardcover : alk. paper)
ISBN-10: 1-55581-502-2 (hardcover : alk. paper)
1. Bacteriophages—Therapeutic use. 2. Bacteriophages—Diagnostic use. 3. Food contamination. 4. Water—Pollution. I. Sabour, Parviz M. II. Griffiths, Mansel. III. American Society for Microbiology.
[DNLM: 1. Bacteriophages. 2. Food Contamination—prevention & control. 3. Food Microbiology. 4. Water Microbiology. QW 161 B1318 2010]

QR342.B3385 2010
579.2′6—dc22

2010014688

10 9 8 7 6 5 4 3 2 1

Printed in the United States of America

Address editorial correspondence to ASM Press, 1752 N St., N.W., Washington, DC 20036-2904, USA

Send orders to ASM Press, P.O. Box 605, Herndon, VA 20172, USA
Phone: 800-546-2416; 703-661-1593
Fax: 703-661-1501
E-mail: books@asmusa.org
Online: http://estore.asm.org

CONTENTS

CONTRIBUTORS

Craig Billington
Food Safety Programme
Institute of Environmental Science and Research Ltd.
Christchurch 8540, New Zealand

Patrick Boerlin
Department of Pathobiology
Ontario Veterinary College
University of Guelph
Guelph, Ontario N1G 2W1, Canada

Richard P. Bonocora
Laboratory of Molecular and Cellular Biology
National Institute of Diabetes and Digestive and Kidney Diseases
National Institutes of Health
Bethesda, MD 20892

Harald Brüssow
Nutrition and Health Department
Nestlé Research Center
Vers-chez-les-Blanc
CH-1000 Lausanne 26, Switzerland

Alan J. Castle
Department of Biological Sciences
Brock University
500 Glenridge Avenue
St. Catharines, Ontario L2S 3A1, Canada

Pieter-Jan Ceyssens
Department of Biosystems
Division of Gene Technology
Kasteelpark Arenberg 21 bus 2462
B-3001 Leuven, Belgium

Gail E. Christie
Department of Microbiology & Immunology
Virginia Commonwealth University
P.O. Box 980678, 1101 East Marshall Street, 6-034 Sanger Hall
Richmond, VA 23298-0678

Jason J. Gill
Department of Biochemistry and Biophysics and Center for Phage Technology
Texas A&M University
2128 TAMU
College Station, TX 77843

Lawrence D. Goodridge
Department of Animal Sciences
Center for Meat Safety and Quality
Colorado State University
Fort Collins, CO 80523

Mansel W. Griffiths
Department of Food Science and Canadian Research Institute for Food Safety
University of Guelph
43 McGilvray Street
Guelph, Ontario N1G 2W1, Canada

James T. Hoopes
Center for Advanced Research in Biotechnology
University of Maryland Biotechnology Institute
Rockville, MD 20850

J. Andrew Hudson
Food Safety Programme
Institute of Environmental Science and Research Ltd.
Christchurch 8540, New Zealand

Juan Jofre
Department of Microbiology
Faculty of Biology
University of Barcelona
Avinguda Diagonal 645
08028 Barcelona, Spain

Andrew M. Kropinski
Laboratory for Foodborne Zoonoses
Public Health Agency of Canada
110 Stone Road West
Guelph, Ontario N1G 3W4, Canada

John Kuzio
Department of Microbiology and Immunology
Queen's University
Kingston, Ontario K7L 3N6, Canada

Simon Labrie
Département de biochimie et de microbiologie
Faculté des sciences et de génie
Groupe de recherche en écologie buccale
Faculté de médecine dentaire
Félix d'Hérelle Reference Center for Bacterial Viruses
Université Laval
Québec City, Québec G1V 0A6, Canada

Rob Lavigne
Department of Biosystems
Division of Gene Technology
Kasteelpark Arenberg 21 bus 2462
B-3001 Leuven, Belgium

Susan M. Lehman
Centers for Disease Control and Prevention
1600 Clifton Road NE, Mail Stop C-16
Atlanta, GA 30333

Marcin Łoś
Department of Molecular Biology
University of Gdansk
Kladki 24
80-822 Gdansk, Poland

Francisco Lucena
Department of Microbiology
Faculty of Biology
University of Barcelona
Avinguda Diagonal 645
08028 Barcelona, Spain

Lynn McIntyre
Food Safety Programme
Institute of Environmental Science and Research Ltd.
Christchurch 8540, New Zealand

Michael R. McConnell
Department of Biology
Point Loma Nazarene University
3900 Lomaland Drive
San Diego, CA 92106

Sylvain Moineau
Département de biochimie et de microbiologie
Faculté des sciences et de génie
Groupe de recherche en écologie buccale
Faculté de médecine dentaire
Félix d'Hérelle Reference Center for Bacterial Viruses
Université Laval
Québec City, Québec G1V 0A6, Canada

Toshihiro Nakai
Laboratory of Fish Pathology
Graduate School of Biosphere Sciences
Hiroshima University
Higashi-Hiroshima 739-8528, Japan

Daniel C. Nelson
Center for Advanced Research in Biotechnology
University of Maryland Biotechnology Institute
Rockville, MD 20850

Gary R. Pasternack
Intralytix, Inc.
The Columbus Centre
701 East Pratt Street
Baltimore, MD 21202

Parviz M. Sabour
Guelph Food Research Centre
Research Branch
Agriculture and Agri-Food Canada
93 Stone Road West
Guelph, Ontario N1G 5C9, Canada

Caren J. Stark
Center for Advanced Research in Biotechnology
University of Maryland Biotechnology Institute
Rockville, MD 20850

Alexander Sulakvelidze
Intralytix, Inc.
The Columbus Centre
701 East Pratt Street
Baltimore, MD 21202

Antonet M. Svircev
Research Branch
Agriculture and Agri-Food Canada
4902 Victoria Avenue North, P.O. Box 6000
Vineland Station, Ontario L0R 2E0, Canada

Qi Wang
Guelph Food Research Centre
Research Branch
Agriculture and Agri-Food Canada
93 Stone Road West
Guelph, Ontario N1G 5C9, Canada

Grzegorz Węgrzyn
Department of Molecular Biology
University of Gdansk
Kladki 24
80-822 Gdansk, Poland

FOREWORD

This very timely book effectively addresses a number of important issues: the growing concerns about food safety in light of widespread outbreaks of pathogens such as *Salmonella* and *Escherichia coli,* leading to massive food recalls; the rise in resistance to current antimicrobials, exacerbated by their extensive use in agriculture; the challenges of quickly and specifically identifying pathogenic bacterial strains; and the frustrations of phage invasion of food-fermentation vats. As antibiotic resistance becomes more widespread, it makes sense to aggressively explore the therapeutic potential of phages, taking advantage of modern biological tools to select the most effective, characterize them thoroughly under relevant physiological conditions, and carry out well-controlled studies. Gaining research funding and approval for human therapy implementation has proven very challenging in the Western world for a variety of regulatory and financial reasons. As a result, many of those drawn to such work have decided to focus at present on nonmedical applications in less regulated key areas such as agriculture and food safety, thus dealing with significant parts of the problem while collecting relevant phage experience and developing expertise at commercialization.

The most significant recent advances toward phage therapy are in fact related to food safety, as discussed in several chapters of this new ASM book. A major milestone was the 2006 FDA approval of an Intralytix, Inc. cocktail of phages targeting *Listeria monocytogenes* for use on ready-to-eat meats and cheeses, including the right to substitute related well-characterized phages for those in the cocktail as initially approved—a crucial factor for many phage applications. In this context chapter 15, by Alexander Sulakvelidze and Gary R. Pasternack of Intralytix, on industrial and regulatory issues related to bacteriophage applications in food production and processing is particularly useful, complementing chapter 7, by the New Zealand team of J. Andrew Hudson et al., on phage control of pathogenic and spoilage bacteria in food processing and distribution.

Much interesting and productive work is already under way. A Utah phage company, OmniLytics, Inc., has won permission to market phage treatments for such applications as managing plant pathogens and preslaughter reduction of *E. coli* O157 levels in cattle, with clear official recognition that this approach is far less potentially hazardous than standard chemical methods. Both the U.S. military and the U.S. Department of Agriculture are supporting sophisticated pathogen detection and food safety work. The research cluster of the Guelph Food Research Centre, the University of Guelph and its Canadian Research Institute for Food Safety, and the Public Health Agency of Canada, all located in Guelph, Ontario, are deeply committed to exploring and applying phage approaches for livestock, crops, and food safety, along with providing badly needed underpinnings in basic research. Significant initiatives are also under way in Australia, New Zealand, Great Britain, Japan, and other countries. *Bacteriophages in the Control of Food- and Waterborne Pathogens* does an excellent job of bringing the resultant academic, governmental, and corporate work together for the first time. It can be expected to help spearhead future research and increasingly rapid progress in this new field, with many possibilities for new applications and patents. (It should be noted here that the very extensive phage research in Eastern Europe was so deeply committed to human therapy that little agricultural or water quality work was carried out there until very recently; the same was true of the work going on in the West over the first half of the last century.)

Historically, the first-ever ASM monograph (Mathews et al., 1983) focused on a phage—bacteriophage T4, the primary tool in the development of the field of molecular biology. There were detailed discussions of T4's structure and initiation of infection; of the enzymes of DNA metabolism that it encoded; of the processes of DNA replication, recombination, and repair; of RNA polymerase modifications and the regulation of the various classes of genes; and of the processes of head, tail, and tail fiber morphogenesis. It also included a restriction map of the whole T4 genome and details of the 133 T4 genes known to date. The next ASM phage book (Karam, 1994) again focused on T4. It elaborated much further on molecular aspects of the infection process and included the first analysis of the complete T4 genomic sequence, a two-dimensional protein gel portrayal of the pattern of synthesis of T4 early proteins, and a chapter on the effects of host physiology on the infection process. It also included a very extensive section on methods of working with phage—the first such compendium since the classic book *Bacteriophages* by Adams (1959). That book, still used worldwide, was based on the Cold Spring Harbor phage courses of the 1940s and 1950s, and was enriched by the phage meetings there (as were the Cold Spring Harbor Laboratory volumes on phages lambda [e.g., Hendrix et al., 1983] and mu and the RNA phages).

Both ASM T4 books grew out of the Evergreen International T4 Meetings, with competing research labs often collaborating on the chapters, reflecting the strong collegiality of the phage community that has been so important to the field since the 1940s. The 1994 book had an amazing 105 contributors from around the world but not a single mention of phage applications. At that time the West was just beginning to be aware of the ongoing Soviet phage therapy work and no one had even begun to consider most of the possibilities discussed in the present book. Who would have dreamed that, just 15 years later, our deep understanding of T4 and its relatives would form the basis for a chapter on safety and efficacy issues in phage control of bacterial diarrhea (chapter 14 by Harald Brüssow)—from a scientist at Nestlé in Switzerland, doing work with infant patients in Bangladesh? In parallel with this evolution, the biennial Evergreen International T4 Meetings broadened into Evergreen International Phage Biology meetings, with the recent 18th meeting drawing 150 scientists from 30 countries to discuss phage molecular biology, ecology, genomics, and a range of applications in biotechnology, agriculture, food safety, and human phage therapy. This was an excellent mix of scientists young and old from academia, industry, and governmental and clinical venues, leading to new collaborations and progress toward a formal International Phage Society oriented toward promoting phage therapy and other implementations.

The third ASM phage book (Waldor et al., 2005) dealt almost exclusively with temperate phages, taking advantage of the explosion of microbial genomic sequencing as it explored their many roles in bacterial pathogenesis and evolution and introduced new applications of phages in biotechnology. How far that subfield has come in the last 5 years is reflected in Marcin Łoś et al.'s very extensive and interesting chapter on seroconversion in the present book (chapter 9). Building on explosive advances in genomics, protein structure, and microbial pathogenesis, it goes systematically through the protein structures and mechanisms of action of the various phage-encoded molecules involved in pathogenesis for the major food-borne pathogens *Shigella dysenteriae, E. coli, Staphylococcus aureus,* and *Clostridium botulinum,* and includes over 300 references in the field.

A range of other new books in recent years complement each other in providing further understanding of the extent and complexity of work related to phage and phage therapy that has gone on over the last 90 years, as well as tools for phage applications. These include Summers' (1999) *Félix d'Hérelle and the Origins of Molecular Biology;* Häusler's (2006) detailed and readable *Viruses vs. Superbugs: a Solution to the Antibiotics Crisis?; Bacteriophages: Biology and Application* (Kutter and Sulakvelidze, 2004); the new edition of *The Bacteriophages* (Calendar, 2005); *Bacteriophage: Genetics and Molecular Biology* (McGrath and van Sinderen, 2007); *Bacteriophage Ecology: Population Growth, Evolution, and Impact of Bacterial Viruses* (Abedon, 2008); and the in-depth methods compendium assembled by Clokie and Kropinski (2009). In addition, new electronic tools have at last made much of the older literature, as well as the mounting stream of current work in the field, highly accessible, and we are coming to appreciate the depth of careful research and understanding reflected in the original books of Félix d'Hérelle (1926).

In the midst of this plethora of resources, *Bacteriophages in the Control of Food- and Waterborne Pathogens* brings a very new and important integrative perspective and road map. Important chapters include detailed discussions by leaders in each field of the use of phages in controlling pathogens in aquaculture and in the production of food animals and plants, environmental factors affecting phage efficacy and enhancement of efficacy through phage encapsulation, and potential applications of bacterial lysins in food safety, in addition to an excellent overview of relevant basic phage biology and the support chapters discussed above. It is to be hoped that many will take inspiration from the ideas and challenges and the wide range of information presented in this book. Anyone who is instructed by this material and inspired to extend this work will be gladly welcomed into the worldwide phage community, made up of investigators committed to making effective use of these powerful allies—the most abundant living entities on earth. The ASM, the editors, and the authors are all to be highly commended for making this excellent resource available to those interested in phage application. It is very appropriate that the assembly of this book has been based in Guelph, Ontario, which for many years has been the world center of work related to phage applications in agriculture and food safety.

REFERENCES

Abedon, S. T. (ed.). 2008. *Bacteriophage Ecology: Population Growth, Evolution, and Impact of Bacterial Viruses.* Cambridge University Press, Cambridge, United Kingdom.

Adams, M. H. 1959. *Bacteriophages.* Interscience Publishers, New York, NY.

Calendar, R. (ed.). 2005. *The Bacteriophages,* 2nd ed. Oxford University Press, New York, NY.

Clokie, M. R. J., and A. M. Kropinski (ed.). 2009. *Bacteriophages: Methods and Protocols,* vol. 1. *Isolation, Characterization, and Interactions.* Vol. 2, *Molecular and Applied Aspects.* Humana Press, Springer. New York, NY.

d'Hérelle, F. 1926. *The Bacteriophage and Its Behavior.* Williams & Wilkins, Baltimore, MD.

Häusler, T. 2006. *Viruses vs. Superbugs: a Solution to the Antibiotics Crisis?* Macmillan, Basingstoke, United Kingdom.

Hendrix, R. W., J. W. Roberts, F. W. Stahl, and R. A. Weisberg (ed.). 1983. *Lambda II.* Cold Spring Harbor Laboratory, Cold Spring Harbor, NY.

Karam, J. D. (ed.). 1994. *Molecular Biology of Bacteriophage T4.* ASM Press, Washington, DC.

Kutter, E., and A. Sulakvelidze (ed.). 2004. *Bacteriophages: Biology and Applications.* CRC Press, Boca Raton, FL.

Mathews, C. K., E. M. Kutter, G. Mosig, and P. B. Berget (ed.). 1983. *Bacteriophage T4.* ASM Press, Washington, DC.

McGrath, S., and D. van Sinderen (ed.). 2007. *Bacteriophage: Genetics and Molecular Biology.* Caister Academic Press, Norfolk, United Kingdom.

Stent, G. S. 1963. *Molecular Biology of Bacterial Viruses.* W. H. Freeman & Co., San Francisco, CA.

Summers, W. C. 1999. *Félix d'Hérelle and the Origins of Molecular Biology.* Yale University Press, New Haven, CT.

Waldor, M. K., D. I. Friedman, and S. L. Adhya (ed.). 2005. *Phages: Their Role in Bacterial Pathogenesis and Biotechnology.* ASM Press, Washington, DC.

ELIZABETH KUTTER
Laboratory of Phage Biology
The Evergreen State College
Olympia, Washington

PREFACE

Bacteriophages are the most abundant life form on our planet, and their importance has been recognized since their discovery almost 100 years ago. Even before that, it was recognized that river waters had the capacity to cure diseases such as leprosy, and in the late 1800s an agent was isolated from the Ganges that was active against cholera. Bacteriophages have been used successfully in Eastern Europe to combat infectious diseases, but with the advent of antibiotics, interest in their use as antibacterial agents diminished. However, the rise of "superbugs" resistant to a wide range of antibiotics has regenerated interest in their use as therapeutic agents. One area where bacteriophages are particularly relevant as both biocontrol agents and as biosensing agents is in food production and processing.

There has been increasing focus on the safety of our food and water supply, especially as there have been numerous large outbreaks of food- and waterborne illness over the last 10 years. These include outbreaks of *Escherichia coli* O157:H7 related to water and spinach; salmonellosis linked to chocolate, nuts, and fresh produce; as well as listeriosis associated with ready-to-eat meat. Indeed, in developed countries it is accepted that about 30% of the population will be subjected to a bout of food-borne illness each year. The situation in developing countries is of even greater concern, with the World Health Organization estimating that in 2005 alone, 1.8 million people died from diarrheal diseases through consumption of contaminated food or water. As well as public health concerns, food-borne illness has significant economic implications, and its full cost in developed countries may amount to 0.5% of GDP. Thus, finding ways to improve the safety and quality of our food and water supplies remains paramount, and bacteriophages offer some unique opportunities as nanotools in our efforts to achieve this goal.

In *Bacteriophages in the Control of Food- and Waterborne Pathogens,* leading researchers provide an insight into the applications of bacteriophages to detect and control bacterial contamination of food and water. Introductory chapters provide the rationale for their use and describe the way they function. There follow chapters discussing opportunities for the use of bacteriophages to detect bacterial contamination of foods and as indicators of water quality. Their application as biocontrol agents during both food production and food processing is an area that is receiving increasing attention, and research carried out to date on the control of pathogens by bacteriophages during animal, fish, and plant production and at food processing plants is reviewed. A chapter is devoted to the antimicrobial activity of bacteriophage lytic enzymes and their use as sensing agents for bacterial detection. As well as being beneficial, bacteriophages can also present a significant problem to the food industry, particularly in the production of fermented products such as cheese. In this instance they can cause failure of the fermentation at a significant economic cost to the company. A chapter discusses the importance of bacteriophage control during the manufacture of fermented foods. Currently, there is limited information on the delivery of bacteriophages for biocontrol applications in food animal production, so two chapters are devoted to this topic. One discusses the contribution of chemical and physical factors to the efficacy of bacteriophage biocontrol strategies, and the other describes encapsulation as a means to deliver bacteriophages to animals. Finally, safety and regulatory issues, which are crucial to the success of bacteriophage use, are covered.

This book is the first to comprehensively address all aspects of the application of bacteriophages for food industry use and will prove useful to food microbiologists, food industry professionals, individuals in government with responsibility for food safety, as well as scientists from other disciplines who are interested in these fascinating microorganisms.

The editors hold a debt of gratitude to all the authors who contributed their valuable time and knowledge to prepare their chapters. They are all leaders in their respective fields of research, and it is a privilege to be associated with them.

PARVIZ M. SABOUR
MANSEL W. GRIFFITHS

ACKNOWLEDGMENTS

We are grateful to all of the authors who graciously contributed to this book. We also thank Peggy A. Pritchard for her assistance in preparation of the application to ASM Press and Kyly Whitfield for her editorial assistance in preparing the chapters for final submission. We thank Eleanor S. Riemer, Senior Consulting Editor, and John Bell, Senior Production Editor, at ASM Press for their encouragement and valuable editorial advice.

P.M.S. acknowledges financial support of Guelph Food Research Centre research grant #1274.

IMPLICATIONS OF ANTIMICROBIAL AGENTS AS THERAPEUTICS AND GROWTH PROMOTERS IN FOOD ANIMAL PRODUCTION

Patrick Boerlin

1

For decades, antimicrobial agents represented wonder tools to fight infectious diseases in human and veterinary medicine. Unfortunately, their intensive use has put such a strong selective pressure on bacterial populations that antimicrobial resistance (AMR) has now reached levels threatening the effectiveness of these agents. This chapter provides an overview of antimicrobial use in animals and of the rise of AMR in commensal bacteria and zoonotic agents. It demonstrates the need for alternatives to antimicrobial agents, such as bacteriophages, to preserve the safety of human food and to control the spread of zoonotic agents from animals to humans.

USE OF ANTIMICROBIAL AGENTS IN FARM ANIMALS AND FOOD PRODUCTION

The high concentrations of food animals reared under industrial conditions to fulfill the needs of the continuously increasing human population represent a prime substrate for the development and spread of infectious diseases. This predisposition is accentuated by several factors: animals' physiologies being pushed to their limits through extreme growth and production rates, mixing of young animals from diverse origins with naïve immune systems, social and physical stress, and exhaustion from long-distance travel. Infectious diseases therefore represent a major challenge for the food animal industry (Silbergeld et al., 2008). In order to minimize the impact of infections and the development of disease outbreaks, increasingly strict biosafety and hygiene protocols are implemented worldwide in the farming industry. Despite all these precautions, antimicrobial agents are still needed as an additional tool to control infectious diseases in the swine, poultry, cattle, and fish farming industries (Sarmah et al., 2006; Cabello, 2006).

As in human medicine, antimicrobial agents are extensively used in veterinary medicine and agriculture (both animal and plant) for the treatment of infectious disease. Unlike in human medicine, they are used not only for the treatment of individual cases but also for metaphylaxis, for evident practical reasons (Aarestrup, 2005). Although this has been a controversial subject for a long time and local regulations may vary significantly on this last

Patrick Boerlin, Department of Pathobiology, Ontario Veterinary College, University of Guelph, Guelph, Ontario N1G 2W1, Canada.

Bacteriophages in the Control of Food- and Waterborne Pathogens
Edited by Parviz M. Sabour and Mansel W. Griffiths, © 2010 ASM Press, Washington, DC

issue, antimicrobial agents are also widely used in many countries at low concentrations as feed additives for the purpose of growth promotion (Aarestrup, 2005; Prescott, 2008).

The families of antimicrobial agents and many of the actual active substances used in farm animals and human medicine are the same. These include older antimicrobial agents such as penicillins, aminopenicillins, and narrow-spectrum cephalosporins, tetracyclines, sulfonamides, macrolides, lincosamides, and some aminoglycosides including streptomycin, neomycin, and gentamicin. However, newer antimicrobials or antimicrobial families of great value for human medicine such as expanded-spectrum cephalosporins, fluoroquinolones, and streptogramins are also frequently used in veterinary medicine and agriculture (Silbergeld et al., 2008).

Figures on the amounts of antimicrobial agents used in food animals are not easy to come by. Most frequently, only overall consumption or sales numbers expressed in weight units are available. Separate information by target species is usually not available, and frequently all antimicrobials, including ionophores, are pooled together in the available data. In addition, the exact biological significance of these numbers is difficult to evaluate because of the diversity of molecules used and their variable potency per unit of weight, their mode of administration, and the usual lack of information on animal species distribution (Grave et al., 2006). Nevertheless, the numbers are staggering and range in the millions of kilograms per year, according to the Union of Concerned Scientists (http://www.ucsusa.org/food_and_agriculture/science_and_impacts/impacts_industrial_agriculture/hogging-it-estimates-of.html). For instance, an estimated 1.3 million kg of antimicrobial agents were distributed in Canada for use in animals in 2006 (Government of Canada, 2009), excluding 456 tons of ionophores, coccidiostats, and arsenicals, and 121 tons were distributed in Denmark in 2007 (80% of it in pigs). Older estimates suggest that more than 9 million kg of antimicrobials (including ionophores and other nontherapeutic antimicrobials) were used in animals each year in the United States at the turn of the 21st century (Sarmah et al., 2006). A comparison with human medicine remains very difficult, but some estimates suggest that veterinary medicine and agriculture use at least half of the antimicrobial agents sold on the planet. This assumption is supported by a few numbers such as in Denmark and New Zealand, where 70% and 57% of antimicrobials (including ionophores) used in these countries, respectively, were for animals (Aarestrup 2005; Sarmah et al., 2006).

AMR IN BACTERIA FROM ANIMALS AND ITS LINKS TO HUMAN INFECTIONS

By the pure virtue of Darwinian selection and the immense reproduction rates and adaptability of prokaryotes to new environments, the use of antimicrobial agents in both humans and animals inevitably leads to the emergence of AMR. Since most antimicrobial agents have a relatively broad activity spectrum not restricted to their major target, both pathogenic bacteria and nontarget commensal microorganisms are the simultaneous object of selection by antimicrobial agents (van den Bogaard and Stobberingh, 2000). The emergence of resistant bacteria in these populations is the result not only of mutations in preexisting genes, but also of the acquisition of AMR genes through horizontal gene transfer (Boerlin and Reid-Smith, 2008). The latter phenomenon is of major relevance, since it speeds up the spread of AMR beyond the "simple" transmission of resistant bacterial clones within and between hosts. Bacteria have evolved mechanisms which seem to boost mutation rates and the ability to acquire and retain new DNA through horizontal gene transfer, both at the individual and population level, when under stressful conditions (Chopra et al., 2003; Foster, 2007). Therefore, they seem to have developed the ability to adapt promptly to survive changing conditions and environmental challenges such as exposure to antimicrobial agents. Recent developments suggest that pathogens have acquired new resistance genes, not only from antibiotic-producing bacteria (Davies, 1994) but also from naturally resistant environmental bacteria, as seems to be the case for very important antimicrobial agents such as fluoroquinolones (Canton, 2009). A large reservoir of resistant bacteria exists in the environment, which represents an inexhaustible source of new resistance genes to new and older antimicrobial agents (D'Costa et al., 2007).

A single resistance determinant provides resistance not to a single antimicrobial agent but usually to multiple agents of the same family of molecules (a phenomenon called cross-resistance). In addition, multiple AMR genes are frequently physically linked and can move together on the same mobile genetic element (Bennett, 2008). Consequently, the use of a microbial agent may lead to the coselection of resistance to antimicrobial agents of different and unrelated antimicrobial families. Coselection has been suggested to be the reason for the persistence of resistance to antimicrobial agents despite the discontinuation of their use (Aarestrup et al., 2001), sometimes over extended periods of time (Travis et al., 2006). Thus, the recommended use of different molecules in veterinary and human medicine is not a complete guarantee that no resistance of significance for humans would be selected by antimicrobial use in animals. In addition, AMR genes are frequently linked to virulence genes and there is a potential for the use of antimicrobial agents to coselect for virulence genes and therefore for more-virulent pathogens (Martínez and Baquero, 2002).

The epidemiology of AMR is further complicated by the role that commensal bacteria from animals and humans may play as a reservoir of AMR genes for pathogens. Many commensal bacteria can be exchanged between humans and animals, which can potentially transfer AMR genes to zoonotic agents and to host-specific pathogens (van den Bogaard and Stobberingh, 2000). However, the dynamics of these exchanges is still poorly understood and the epidemiology of AMR in commensals and pathogens may still differ significantly, even if they are closely related and share the same host compartment (Kozak et al., 2009).

Overall, a very complex and confusing picture of the global epidemiology of AMR emerges, with an intricate network of relationships linking a large variety of ecological niches (Boerlin and Reid-Smith, 2008). This makes the effect of interventions to reduce AMR frequencies difficult to predict, and sometimes counterintuitive.

Despite the complexity of the AMR epidemiology and some controversies on the solidity of supporting evidence, the vast majority of investigations show the presence of links between the use of antimicrobial agents in animals, the occurrence of AMR in bacteria from animals, and AMR in zoonotic pathogens in humans. This evidence relies on a variety of approaches, including outbreak investigations, case studies of sporadic infections, limited field studies on the effect of changes in antimicrobial use in animals, ecological and temporal associations between AMR in bacteria from animals and humans, and comparisons of resistant bacteria and AMR determinants in animals and human populations using molecular methods (for reviews on the subject, see, for instance, Angulo et al., 2004; and Mølbak, 2004).

Among the most striking examples of the transmission of resistant bacteria and AMR determinants between food animals and humans are those of (i) the spread of streptothricin resistance, (ii) the vancomycin-resistant enterococci (VRE) in Europe, (iii) the fluoroquinolone-resistant *Campylobacter,* and (iv) extended-spectrum cephalosporin (ESC) resistance in *Salmonella* in North America. The case of streptothricin resistance illustrates how resistance determinants can spread across bacterial populations in the absence of direct selection. The use of nourseothricin in swine in the former East Germany selected for the emergence of specific resistance determinants that rapidly spread to bacteria from farmers and then to the general population (Hummel et al., 1986; van den Bogaard and Stobberingh, 2000). Despite the absence of streptothricin and nourseothricin use outside of East Germany and the discontinuation of its use in animals, the *sat* genes responsible for resistance to these agents can now be found worldwide in commensals and pathogens from both animals and humans.

The emergence of VRE in food animals in Europe shows how a specific resistance problem can lead to preventive precautionary measures of very large amplitudes. There is little doubt that the use of avoparcin as a growth promoter in animals in Europe led to the selection of VRE in farm animals (Wegener et al., 1999). Several pieces of evidence show that these VRE subsequently reached and colonized human populations (Angulo et al., 2004). This eventually led to the ban of avoparcin and antimicrobial growth promoters in general in Denmark and other European countries. These measures seem to have ultimately achieved, in most instances, a reduction of both VRE prevalence in food animals (Aarestrup et al., 2001; Bager et al., 1999) and of VRE colonization in humans in the general population (van den Bogaard et al., 2000; Klare et al., 1999).

The examples of fluoroquinolone resistance in *Campylobacter* and of ESC resistance in *Salmonella* are more directly related to the effects of antimicrobial use in food animals and foodborne zoonoses in the general population. Fluoroquinolone resistance can be readily selected in *Campylobacter* when these antimicrobial agents are used in chickens (McDermott et al., 2002). Despite the earlier licensing and broad use of fluoroquinolones in human medicine, the emergence of fluoroquinolone resistance in *Campylobacter* infections in humans in the United States began only after the licensing and actual use of fluoroquinolones in food animals (Gupta et al., 2004). Similar trends were seen in other countries in which fluoroquinolones started to be used in food animals, but not in others where they were not used (Angulo et al., 2004). Thus, despite their ecological nature, such studies strongly suggest that the use of fluoroquinolones in food animals (i.e., poultry) is at least partially responsible for the emergence of fluoroquinolone-resistant strains and associated therapeutic challenges in human infections. Finally, molecular studies have demonstrated the spread of ESC resistance determinants (i.e., $bla_{\text{CMY-2}}$ genes and associated plasmids) across bacterial species and host populations (Winokur et al., 2000, 2001; Mulvey et al., 2009). Sporadic case investigations have additionally demonstrated the possibility of transfer of ESC-resistant *Salmonella* from animals to humans (Fey et al., 2000). Other studies further support the link between the use of expanded-spectrum cephalosporins in food animals and the occurrence of ESC-resistant *Salmonella* in human infections. For instance, recent surveillance data show that the use and subsequent voluntary withdrawal of ceftiofur in chickens in some Canadian provinces were paralleled by an increase and a subsequent drop in resistance to expanded-spectrum cephalosporins in *Salmonella enterica* serovar Heidelberg from both chicken and humans (Government of Canada, 2007). Despite their ecological nature, such studies are among the most striking illustrations of the connections between antimicrobial agent use in food animals and AMR in zoonotic agents with direct implications for human health (Webster, 2009).

ECONOMIC COST IN ANIMAL HUSBANDRY AND HUMAN MORBIDITY AND MORTALITY

As can be concluded from the above, antimicrobial agent use and the rise of resistance in bacteria from animals have consequences for both animal and human health. Resistance frequencies in some animal pathogens are at levels which would be deemed unacceptable in human medicine. For instance, resistance to bacitracin (an important antimicrobial agent in the prevention of necrotic enteritis in poultry) can be detected in up to 90 to 100% of *Clostridium perfringens* strains from broiler chickens raised with conventional methods in some regions of Canada (Chalmers et al., 2008a,

2008b). Among pathogenic *Escherichia coli* from swine and poultry in North America, resistance frequencies between 20 and 100% for important antimicrobial agents used in therapy and prevention, such as apramycin, neomycin, gentamicin, tetracycline, and trimethoprim-sulfonamide, can be observed (Maynard et al., 2003; Boerlin et al., 2005; Zhao et al., 2005). Thus, besides strategies to reduce resistance levels for existing antimicrobial agents, truly new molecules, not just modifications of existing molecules with slightly different activity spectra or pharmacokinetics, would be needed to circumvent this resistance problem. However, there is little hope that such molecules will be available soon (Norrby et al., 2005), so other alternatives are urgently needed. In the meanwhile, there are strong trends in the farming industry to use even more-potent classes of antimicrobial agents, with the potential consequence of contributing to eroding the health system's power to treat human infections.

Only a few studies are available on the economic impact of the ban of growth promoters in Europe, and they show variable outcomes (Bengtsson and Wierup, 2006). Some reports on the cost of antimicrobial use in animal husbandry also suggest that this is not always cost-effective (Graham et al., 2007), thus providing further incentives for the development of alternatives to antimicrobial agents. Even fewer data are available on the real direct cost of AMR for the farming industry, and only pleas for serious sound research in this area can be found in the literature (Miller et al., 2006; Grugel and Wallmann, 2004; Coast and Smith, 2003; Cromwell, 2002; Wassenaar, 2005).

The effects of resistance on human health can be differentiated into three major categories (Barza, 2002; Angulo et al., 2004): (i) the occurrence of infections that would otherwise not have taken place (resistant pathogens are more likely to establish themselves in high-risk patients on antimicrobial treatment), (ii) increased frequency of treatment failure associated with resistance to clinically relevant antimicrobial agents, and (iii) increased severity of disease. The last effect may be related to initial therapy failure, but also possibly to increased virulence of some resistant bacteria (Martinez and Baquero, 2002). A number of studies have demonstrated and partially quantified the actual impact of AMR in zoonotic agents in general on human health, both for *Campylobacter* and for *Salmonella* (see, for instance, Barza and Travers, 2002; Travers and Barza, 2002; Mølbak, 2004; and Angulo et al., 2004). These effects are undeniable in terms of number of clinical infections, frequency of hospitalization, duration of disease and hospitalization, frequency of bloodstream infections, and mortality. For instance, some estimates suggest that AMR is responsible for more than 17,000 cases of *Campylobacter* infection and 500,000 days of *Campylobacter*-related diarrhea per year in the United States alone, of which 400,000 days could possibly be caused by resistant isolates originating from animals (Barza and Travers, 2002; Travers and Barza, 2002). The same authors suggest that more than 29,000 infections and 8,000 days of hospitalization per year are attributable to resistance in *Salmonella* in the United States. However, little is known quantitatively about the real impact and role of AMR originating from antimicrobial use in animals in this overall effect, and some controversies on methodological approaches and interpretation for this quantification are still present (Bartholomew et al., 2005; Cox and Popken, 2004; Cox, 2005). The transfer of resistant zoonotic agents from animals to humans nevertheless remains an overwhelming reality, and both the prudent use of antimicrobial agents and the development of alternatives to antimicrobial agents to reduce the frequency of AMR and to treat infections with resistant organisms should be strongly encouraged.

INTERNATIONAL TRADE RESTRICTIONS

As demonstrated by international outbreaks of food-borne infections, the spread of zoonotic agents, including resistant ones, through trade and distribution of food and ingredients occurs not only locally but also across borders. Because of its indirect implications for human health, the transfer and spread of resistant zoonotic agents through international animal trade is less frequently considered. However, its occurrence was clearly demonstrated in a case of a breeding swine exported from Canada to Denmark, which was a carrier of ESC-resistant *Salmonella* serovar Heidelberg (Aarestrup et al., 2004). This raises the potential threat that AMR could be considered in the future as an impediment for trade and be included in international legislations in the same way as zoonotic agents. Although, to the best of our knowledge, no restrictions have yet been implemented on the basis of AMR and antimicrobial use, it is realistic to envisage that international trade restrictions associated with the control and adequate use of antimicrobial agents in food animals and agriculture could become a reality in the future (Fidler, 1998). The Codex Alimentarius Commission of the World Health Organization and Food and Agriculture Organization, and the documents released by its intergovernmental task force on AMR, may for instance represent a likely basis for such trade restrictions (Fidler, 1999). Besides the international intergovernmental regulations and recommendations, the pressure exerted by consumers and the efforts made by some major food distributors to market meat produced without antibiotics puts limits on the use of antimicrobial agents in food animals. This has led to the emergence of numerous "antibiotic-free" meat production programs. Since most of these programs rely on intensive production techniques, they still face the challenges associated with maximum productivity and infectious diseases under such conditions and are therefore in a continuous search for efficient alternatives to antimicrobial agents.

NEED FOR NEW ALTERNATIVES

Alternatives to antimicrobial agents are needed to sustain growth promotion and control infectious diseases. Many strategies have been developed for these two purposes (Lawrence and Fowler, 2002), including mainly the use of prebiotics and probiotics (Callaway et al., 2004), control and optimization of the gut flora by chemical and natural feed additives such as zinc oxide (Quintavalla and Agostini, 2007), the use of plant extracts and essential oils (Benchaar et al., 2008), or modulation of the immune system (Gallois and Oswald, 2008). Although some of these approaches may show encouraging results, to the best of our knowledge none of them has been able to challenge the efficacy of antimicrobial agents with regard to disease prevention and therapy. The most promising alternative approaches to antimicrobial agents remain vaccines and bacteriophages. In contrast to most of the other strategies, including antimicrobial agents, these last two methods target specific pathogens and are not likely to disturb the homeostasis and protective qualities of the normal resident flora (Barza and Travers, 2002). Vaccines rely on long-accumulated knowledge and wide experience and are constantly improving as our understanding of host immunity and pathogen biology progresses. However, vaccines present a major disadvantage in today's intensive production systems: their effect is not immediate and there are gaps in the protection provided by passive and active immunity that leave the door open to pathogens at some crucial times in the short life span of food animals. This contrasts with bacteriophages, which, if delivered in a timely fashion, can both prevent and cure infections. In addition, bacteriophages have the advantage over any other control method that they can survive and multiply as long as their target organism persists. These advantages, and many more, will be discussed and illustrated in the following chapters of this book.

REFERENCES

Aarestrup, F. M. 2005. Veterinary drug usage and antimicrobial resistance in bacteria of animal origin. *Basic Clin. Pharmacol. Toxicol.* **96:**271–281.

Aarestrup, F. M., H. Hasman, I. Olsen, and G. Sørensen. 2004. International spread of bla_{CMY-2}-mediated cephalosporin resistance in a multiresistant *Salmonella enterica* serovar Heidelberg isolate stemming from the importation of a boar by Denmark from Canada. *Antimicrob. Agents Chemother.* **48:**1916–1917.

Aarestrup, F. M., A. M. Seyfarth, H. D. Emborg, K. Pedersen, R. S. Hendriksen, and F. Bager. 2001. Effect of abolishment of the use of antimicrobial agents for growth promotion on occurrence of antimicrobial resistance in fecal enterococci from food animals in Denmark. *Antimicrob. Agents Chemother.* **45:**2054–2059.

Angulo, F. J., V. N. Nargund, and T. C. Chiller. 2004. Evidence of an association between use of anti-microbial agents in food animals and antimicrobial resistance among bacteria isolated from humans and human health consequences of such resistance. *J. Vet. Med. B* **51:**374–379.

Bager, F., F. M. Aarestrup, M. Madsen, and H. C. Wegener. 1999. Glycopeptide resistance in *Enterococcus faecium* from broilers and pigs following discontinued use of avoparcin. *Microb. Drug Resist.* **5:**53–56.

Bartholomew, M. J., D. J. Vose, L. R. Tollefson, and C. C. Travis. 2005. A linear model for managing the risk of antimicrobial resistance originating in food animals. *Risk Anal.* **25:**99–108.

Barza, M. 2002. Potential mechanisms of increased disease in humans from antimicrobial resistance in food animals. *Clin. Infect. Dis.* **34**(Suppl. 3):S123–S125.

Barza, M., and K. Travers. 2002. Excess infections due to antimicrobial resistance: the "attributable fraction." *Clin. Infect. Dis.* **34**(Suppl. 3):S126–S130.

Benchaar, C., S. Calsamiglia, A. V. Chaves, G. R. Fraser, D. Colombatto, T. A. McAllister, and K. A. Beauchemin. 2008. A review of plant-derived essential oils in ruminant nutrition and production. *Anim. Feed Sci. Technol.* **145:**209–228.

Bengtsson, B., and M. Wierup. 2006. Antimicrobial resistance in Scandinavia after a ban of antimicrobial growth promoters. *Anim. Biotechnol.* **17:** 147–156.

Bennett, P. M. 2008. Plasmid encoded antibiotic resistance: acquisition and transfer of antibiotic resistance genes in bacteria. *Br. J. Pharmacol.* **153**(Suppl. 1):S347–S357.

Boerlin, P., and R. J. Reid-Smith. 2008. Antimicrobial resistance: its emergence and transmission. *Anim. Health Res. Rev.* **9:**115–126.

Boerlin, P., R. M. Travis, C. L. Gyles, R. Reid-Smith, N. Janecko, H. Lim, V. Nicholson, S. A. McEwen, R. Friendship, and M. Archambault. 2005. Antimicrobial resistance and virulence genes of *Escherichia coli* from swine in Ontario. *Appl. Environ. Microbiol.* **71:**6753–6761.

Cabello, F. C. 2006. Heavy use of prophylactic antibiotics in aquaculture: a growing problem for human and animal health and for the environment. *Environ. Microbiol.* **8:**1137–1144.

Callaway, T. R., R. C. Anderson, T. S. Edrington, K. J. Genovese, R. B. Harvey, T. L. Poole, and D. J. Nisbet. 2004. Recent preharvest supplementation strategies to reduce carriage and shedding of zoonotic enteric bacterial pathogens in food animals. *Anim. Health Res. Rev.* **5:**35–47.

Canton, R. 2009. Antibiotic resistance genes from the environment: a perspective through newly identified antibiotic resistance mechanisms in the clinical setting. *Clin. Microbiol. Infect.* **15**(Suppl. 1): 20–25.

Chalmers, G., H. L. Bruce, D. B. Hunter, V. R. Parreira, R. R. Kulkarni, Y. F. Jiang, J. F. Prescott, and P. Boerlin. 2008a. Multilocus sequence typing analysis of *Clostridium perfringens* isolates from necrotic enteritis outbreaks in broiler chicken populations. *J. Clin. Microbiol.* **46:**3957–3964.

Chalmers, G., S. W. Martin, D. B. Hunter, J. F. Prescott, L. J. Weber, and P. Boerlin. 2008b. Genetic diversity of *Clostridium perfringens* isolated from healthy broiler chickens at a commercial farm. *Vet. Microbiol.* **127:**116–127.

Chopra, I., A. J. O'Neill, and K. Miller. 2003. The role of mutators in the emergence of antibiotic-resistant bacteria. *Drug Resist. Updates* **6:** 137–145.

Coast, J., and R. D. Smith. 2003. Antimicrobial resistance: cost and containment. *Expert Rev. Anti Infect. Ther.* **1:**241–251.

Cox, L. A. 2005. Some limitations of a proposed linear model for antimicrobial risk management. *Risk Anal.* **25:**1327–1332.

Cox, L. A., and D. A. Popken. 2004. Quantifying human health risks from virginiamycin used in chicken. *Risk Anal.* **24:**271–288.

Cromwell, G. L. 2002. Why and how antibiotics are used in swine production. *Anim. Biotechnol.* **13:** 7–27.

Davies, J. 1994. Inactivation of antibiotics and the dissemination of resistance genes. *Science* **264:**375–382.

D'Costa, V. M., E. Griffiths, and G. D. Wright. 2007. Expanding the soil antibiotic resistome: exploring environmental diversity. *Curr. Opin. Microbiol.* **10:**481–498.

Fey, P. D., T. J Safranek, M. E. Rupp, E. F. Dunne, E. Ribot, P. C. Iwen, P. A. Bradford, F. J. Angulo, and S. H. Hinrichs. 2000. Ceftriaxone-resistant *Salmonella* infection acquired by a child from cattle. *N. Engl. J. Med.* **342:**1242–1249.

Fidler, D. P. 1998. Legal issues associated with antimicrobial drug resistance. *Emerg. Infect. Dis.* **4:** 169–177.

Fidler, D. P. 1999. Legal challenges posed by the use of antimicrobials in food animal production. *Microbes Infect.* **1:**29–38.

Foster, P. L. 2007. Stress-induced mutagenesis in bacteria. *Crit. Rev. Biochem. Mol. Biol.* **42:**373–397.

Gallois, M., and I. P. Oswald. 2008. Immunomodulators as efficient alternatives to in-feed antimicrobial in pig production. *Archiva Zootehnica* **11:**15–32.

Government of Canada. 2007. *Canadian Integrated Program for Antimicrobial Resistance Surveillance (CIPARS) 2005.* Public Health Agency of Canada, Guelph, Ontario, Canada.

Government of Canada. 2009. *Canadian Integrated Program for Antimicrobial Resistance Surveillance (CIPARS) 2006.* Public Health Agency of Canada, Guelph, Ontario, Canada.

Graham, J. P., J. J. Boland, and E. Silbergeld. 2007. Growth promoting antibiotics in food animal production: an economic analysis. *Public Health Rep.* **122:**79–87.

Grave, K., V. F. Jensen, S. McEwen, and H. Kruse. 2006. Monitoring of antimicrobial drug usage in animals: methods and applications, p. 375–395. *In* F. M. Aarestrup (ed.), *Antimicrobial Resistance in Bacteria of Animal Origin.* ASM Press, Washington, DC.

Grugel, C., and J. Wallmann. 2004. Antimicrobial resistance in bacteria from food-producing animals. Risk management tools and strategies. *J. Vet. Med. B* **51:**419–421.

Gupta, A., J. M. Nelson, T. J. Barrett, R. V. Tauxe, S. P. Rossiter, C. R. Friedman, K. W. Joyce, K. E. Smith, T. F. Jones, M. A. Hawkins, B. Shiferaw, J. L. Beebe, D. J. Vugia, T. Rabatsky, J. A. Benson, T. P. Root, and F. J. Angulo. 2004. Antimicrobial resistance among *Campylobacter* strains, United States, 1997–2001. *Emerg. Infect. Dis.* **10:**1102–1109.

Hummel, R., H. Tschäpe, and W. Witte. 1986. Spread of plasmid-mediated nourseothricin resistance due to antibiotic use in animal husbandry. *J. Basic Microbiol.* **26:**461–466.

Klare, I., D. Badstübner, C. Konstabel, G. Böhme, H. Claus, and W. Witte. 1999. Decreased incidence of VanA-type vancomycin-resistant enterococci isolated from poultry meat and from fecal samples of humans in the community after discontinuation of avoparcin usage in animal husbandry. *Microb. Drug Resist.* **5:**45–52.

Kozak, G. K., D. L. Pearl, J. Parkman, R. J. Reid-Smith, A. Deckert, and P. Boerlin. 2009. Distribution of sulfonamide resistance genes in *Escherichia coli* and *Salmonella* isolates from swine and chickens at abattoirs in Ontario and Québec, Canada. *Appl. Environ. Microbiol.* **75:**5999–6001.

Lawrence, T. L. J., and V. R. Fowler. 2002. 'Growth promoters,' performance enhancers, feed additives and alternative approaches, p. 320–329. *In* T. L. J. Lawrence and V. R. Fowler (ed.), *Growth of Farm Animals.* CABI Publishing, Wallingford, United Kingdom.

Martínez, J. L., and F. Baquero. 2002. Interactions among strategies associated with bacterial infection: pathogenicity, epidemicity, and antibiotic resistance. *Clin. Microbiol. Rev.* **15:**647–679.

Maynard, C., J. M. Fairbrother, S. Bekal, F. Sanschagrin, R. C. Levesque, R. Brousseau, L. Masson, S. Larivière, and J. Harel. 2003. Antimicrobial resistance genes in enterotoxigenic *Escherichia coli* O149:K91 isolates obtained over a 23-year period from pigs. *Antimicrob. Agents Chemother.* **47:**3214–3221.

McDermott, P. F., S. M. Bodeis, L. L. English, D. G. White, R. D. Walker, S. Zhao, S. Simjee, and D. D. Wagner. 2002. Ciprofloxacin resistance in *Campylobacter jejuni* evolves rapidly in chickens treated with fluoroquinolones. *J. Infect. Dis.* **185:**837–840.

Miller, G. Y., P. E. McNamara, and R. S. Singer. 2006. Stakeholder position paper: economist's perspectives on antibiotic use in animals. *Prev. Vet. Med.* **73:**163–168.

Mølbak, K. 2004. Spread of resistant bacteria and resistance genes from animals to humans—the public health consequences. *J. Vet. Med. B* **51:** 364–369.

Mulvey, M. R., E. Susky, M. McCracken, D. W. Morck, and R. R. Read. 2009. Similar cefoxitin-resistance plasmids circulating in *Escherichia coli* from human and animal sources. *Vet. Microbiol.* **134:**279–287.

Norrby, S. R., C. E. Nord, and R. Finch. 2005. Lack of development of new antimicrobial drugs: a potential serious threat to public health. *Lancet Infect. Dis.* **5:**115–119.

Prescott, J. F. 2008. Antimicrobial use in food and companion animals. *Anim. Health Res. Rev.* **9:** 127–133.

Quintavalla, F., and A. Agostini. 2007. Zinc oxide: an interesting alternative in swine management. *Large Anim. Rev.* **13:**21–26.

Sarmah, A. K., M. T. Meyer, and A. B. A. Boxall. 2006. A global perspective on the use, sales, exposure pathways, occurrence, fate and effects of veterinary antibiotics (VAs) in the environment. *Chemosphere* **65:**725–759.

Silbergeld, E. K., J. Graham, and L. B. Price. 2008. Industrial food animal production, antimicrobial resistance, and human health. *Annu. Rev. Public Health* **29:**151–169.

Travers, K., and M. Barza. 2002. Morbidity of infections caused by antimicrobial-resistant bacteria. *Clin. Infect. Dis.* **34**(Suppl. 3)**:**S131–S134.

Travis, R. M., C. L. Gyles, R. Reid-Smith, C. Poppe, S. A. McEwen, R. Friendship, N. Janecko, and P. Boerlin. 2006. Chloramphenicol and kanamycin resistance among porcine *Escherichia coli* in Ontario. *J. Antimicrob. Chemother.* **58:** 173–177.

van den Bogaard, A. E., and E. E. Stobberingh. 2000. Epidemiology of resistance to antibiotics. Links between animals and humans. *Int. J. Antimicrob. Agents* **14:**327–335.

van den Bogaard, A. E., N. Bruinsma, and E. E. Stobberingh. 2000. The effect of banning avoparcin on VRE carriage in The Netherlands. *J. Antimicrob. Chemother.* **46:**145–153.

Wassenaar, T. M. 2005. Use of antimicrobial agents in veterinary medicine and implications for human health. *Crit. Rev. Microbiol.* **31:**155–169.

Webster, P. 2009. Poultry, politics, and antibiotic resistance. *Lancet* **374:**773–774.

Wegener, H. C., F. M. Aarestrup, L. B. Jensen, A. M. Hammerum, and F. Bager. 1999. Use of antimicrobial growth promoters in food animals and *Enterococcus faecium* resistance to therapeutic antimicrobial drugs in Europe. *Emerg. Infect. Dis.* **5:** 329–335.

Winokur, P. L., A. Brueggemann, D. L. DeSalvo, L. Hoffmann, M. D. Apley, E. K. Uhlenhopp, M. A. Pfaller, and G. V. Doern. 2000. Animal and human multidrug-resistant, cephalosporin-resistant *Salmonella* isolates expressing a plasmid-mediated CMY-2 AmpC β-lactamase. *Antimicrob. Agents Chemother.* **44:**2777–2783.

Winokur, P. L., D. L. Vonstein, L. J. Hoffman, E. K. Uhlenhopp, and G. V. Doern. 2001. Evidence for transfer of CMY-2 AmpC β-lactamase plasmids between *Escherichia coli* and *Salmonella* isolates from food animals and humans. *Antimicrob. Agents Chemother.* **45:**2716–2722.

Zhao, S., J. J. Maurer, S. Hubert, J. F. De Villena, P. F. McDermott, J. Meng. S. Ayers, L. English, and D. G. White. 2005. Antimicrobial susceptibility and molecular characterization of avian pathogenic *Escherichia coli* isolates. *Vet. Microbiol.* **107:**215–224.

INTRODUCTION TO BACTERIOPHAGE BIOLOGY AND DIVERSITY

Pieter-Jan Ceyssens and Rob Lavigne

2

INTRODUCTION

During the last decades, it has become widely accepted that bacterial viruses, or bacteriophages, are extremely abundant and exert enormous influences on the biosphere. Phages kill between 4 and 50% of the bacteria produced every day, are a driver of global geochemical cycles, and are a reservoir of the greatest genetic diversity on earth (Suttle, 2005). In this light, it is surprising that they were not recognized for almost 40 years after the beginning of serious bacteriological work in European and American laboratories around 1880. Two independent observations made by Frederick W. Twort (1915) and Félix d'Hérelle (1917), who described glassy transformations of micrococci colonies and antagonistic bacterial microbes, respectively, are seen as the onset of modern phage research (Summers, 1999). It was d'Hérelle who introduced the name *bacteriophages,* derived from "bacteria" and the Greek *phagein,* which means "to eat." He also defined the term *plaque* to describe the roundish area of clearing caused by infection of a single phage on double-layered agar plates. Soon after their discovery, d'Hérelle observed that phage titers increased in stool samples from dysentery patients. This finding encouraged scientists almost immediately to explore the therapeutic potential of these bacterial viruses. The first report of the therapeutic use of phage was a note by Bruynoghe and Maisin (1921) from Louvain, who noticed reduction in swelling and pain after injection of staphylococcal phage in the local region of cutaneous boils. d'Hérelle himself conducted trials on humans after he tested his phage preparation by self-administration. The work which probably drew the most attention to phage therapy was his report of four cases of bubonic plague which he treated successfully with antiplague phage (Summers, 1993).

While many early phage therapy trials were reported to be successful, and many of the major pharmaceutical companies sold phage preparations (e.g., L'Oreal in France, Lilly in the United States), there were also failures. Several factors, including an inadequate understanding of phage biology and imperfections in the diagnostic bacteriology techniques available at that time, contributed to the failure of early phage therapy studies and to a decline of interest in phage therapy in the West. The loss of interest in phages was fueled

Pieter-Jan Ceyssens and Rob Lavigne, Department of Biosystems, Division of Gene Technology, Kasteelpark Arenberg 21 bus 2462, B-3001 Leuven, Belgium.

Bacteriophages in the Control of Food- and Waterborne Pathogens
Edited by Parviz M. Sabour and Mansel W. Griffiths, © 2010 ASM Press, Washington, DC

by two negative reports on phage therapy (Eaton and Bayne-Jones, 1934; Krueger and Scribner, 1941), and by the upcoming availability of broad-spectrum antibiotics. Bacteriophage therapy, nonetheless, continued to be offered by the Soviet military and by medical centers in collaboration with the Eliava Institute (Tbilisi, Georgia) and later also by others, like the Hirszfeld Institute of Immunology and Experimental Therapy in Poland (Deresinki, 2009). In the West, phage research was diverted to a more fundamental level. The study of phages played a central role in some of the most significant discoveries in the biological sciences, from the identification of DNA as the genetic material to the deciphering of the genetic code to the development of molecular recombinant technology. Phage-derived proteins are currently being used as molecular machines (Smith et al., 2001), as diagnostic and therapeutic agents (Loeffler et al., 2001; Schuch et al., 2002), and for drug discovery (Liu et al., 2004). The Western skepticism toward phage therapy itself was again followed by renewed interest and reappraisal, mainly due to the emergence of antibiotic-resistant bacteria (Thacker, 2003). In this chapter, we focus on basic aspects of phage biology and phage classification, and on the abundance and diversity of these organisms.

BASIC PHAGE BIOLOGY

Bacteriophages can be defined as obligate intracellular bacterial parasites that lack an independent metabolism. A century of phage research has revealed these viruses as extremely diversified and ubiquitously present in the biosphere, preying on eubacteria and archaea in a wide range of biological niches. Accordingly, the genome sizes of phages vary enormously, from a few thousand base pairs up to 498 kilobases in phage G, the largest phage sequenced to date. Although the size of this genome resembles that of an average bacterium, even phage G lacks genes for essential bacterial machinery like ribosomes, emphasizing the purely parasitic nature of these organisms. In the following section, basic aspects of the morphology and infection cycle of bacteriophages are discussed.

Phage Morphology Forms the Basis for Classification

A bacteriophage particle or virion consists of a single-stranded (ss) or double-stranded (ds) DNA or RNA molecule, encapsulated inside a protein or lipoprotein coat. The International Committee on the Taxonomy of Viruses (ICTV) uses virion morphology and nucleic acid composition as a basis for the classification of phages into 13 families. Over 95% of all phages described in the literature and almost all important phages associated with food-borne pathogens belong to the order of the *Caudovirales* or tailed dsDNA phages. The three main families comprising the *Caudovirales* are distinguished by their very distinct tail morphologies: 60% of the characterized phages are *Siphoviridae,* with long, flexible tails; 25% are *Myoviridae,* with double-layered, contractile tails; and 15% are *Podoviridae,* with short, stubby tails (Fig. 1). Polyhedral, filamentous, and pleomorphic phages represent only 3 to 4% of the studied phages and belong to 10 families, some of which are very small (Ackermann, 2007).

Tailed virions are typically composed of about a 1:1 ratio of proteins and dsDNA. In all dsDNA phages, the chromosome is highly condensed within the virion head, and represents 20 to 50% of its mass (Earnshaw and Harrison, 1977). The mature capsid lattice must be sufficiently robust to maintain this chromosome in its condensed form: unlike many animal viruses, dsDNA phages deliver their genetic material through an injection mechanism rather than an uncoating process. All *Caudovirales* have heads with icosahedral (20 sides/12 vertices) symmetry or elongated derivatives thereof with known triangulation numbers ($T = 4, 7, 13, 16,$ and 52), assembled from many copies of one or two major head proteins. The size of the phage heads is correlated to the genome size it packages, and varies between 45 and 100 nm. The corners

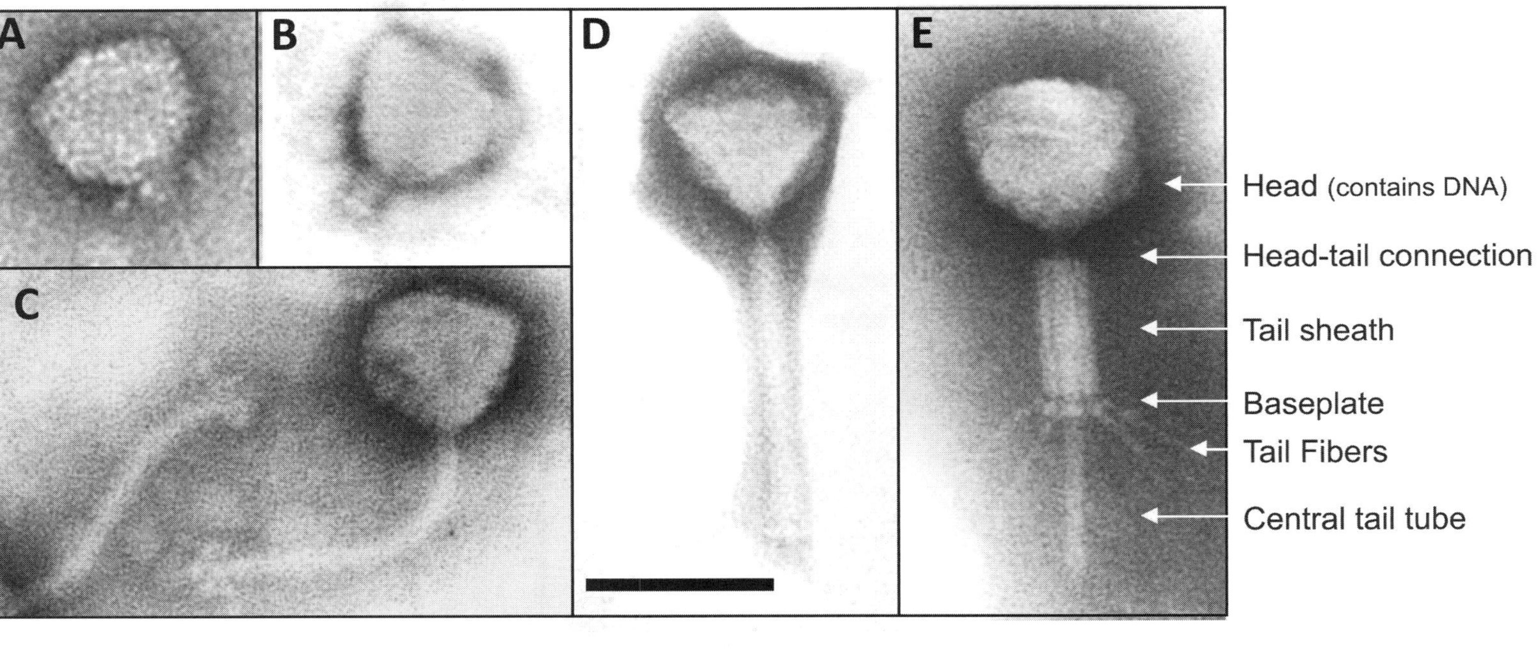

FIGURE 1 Transmission electron microscopic recordings of negatively stained phage particles. Phages A and B are short-tailed *Podoviridae;* phage C has a flexible tail and belongs to the *Siphoviridae*. The myovirus on the right is displayed with noncontracted (D) and contracted (E) tail. The scale bar represents 100 nm; the different parts of the myovirus particle are indicated on the right.

of the icosahederon are generally made up of pentamers of the capsid proteins, while the rest of each side is made up of hexamers of the same (or a similar) protein (Guttman et al., 2005).

The protein connecting the head and tail structures is a key viral structure involved in several steps of the infection cycle. This connector is a homo-oligomeric structure found in a unique vertex of the phage head and controls the entry of the DNA and assembly of the tail to the immature head during morphogenesis of new particles. During infection, it changes its conformation and allows the DNA to exit the virion toward the bacterial cell (Valpuesta et al., 2000).

The tail attaches the phage to the host and serves as conduit for genome insertion. Tail shafts have sixfold or (rarely) threefold symmetry, and are helical or stacks of subunit disks between 3 and 825 nm in length. They usually have base plates, spikes, or terminal fibers at the distal end. As stated before, long phage tails may be contractile, as found in the *Myoviridae* (e.g., phage T4), or noncontractile, as found in the *Siphoviridae* (e.g., phage λ). All long-tailed phages possess a large gene (usually >2 kbp) encoding a tape measure protein, which is responsible for precisely determining tail length. Contractile tails are often very complex structures; for example, the tail of T4 is composed of 22 different proteins (Mesyanzhinov, 2004). The tube of noncontractile tails is composed primarily of multiple copies of one protein, known as the major tail protein. On infection, the overall shape of the noncontractile tail is maintained. In contrast, contractile tails include a central tail tube protein assembly that is surrounded by a tail sheath protein assembly. On infection, the tail sheath contracts and the tail tube protein penetrates the outer cell membrane and cell wall (Fig. 1E) (Leiman et al., 2004). Short-tailed *Podoviridae* may have some key infection proteins enclosed inside the head that can form a sort of extensible tail upon contact with the host, as shown most clearly for coliphage T7 (Molineux, 2001). After the ejection of the genomic DNA, the empty phage particle stays behind on the surface of the target bacterium, and is called a *phage ghost.*

The ICTV classification method as it is described here is currently being reevaluated, since it ignores the vast amount of available genome sequence data which occasionally causes contradictory issues. For example, despite a complete lack of DNA sequence relatedness, the current taxonomic system lists both the *Salmonella* P22 phage and the coliphage T7 as belonging to the *Podoviridae* family based on the shared presence of short tails. However, it has been known for more than 35 years that P22 is so genetically related to the long-tailed phage lambda that recombination between the two genomes forms functional hybrids (Botstein and Herskowitz, 1974). Cases like these encouraged scientists to look for alternative (molecular) classification methods, but this has proven to be difficult because no universal gene, analogous to the 16S rRNA gene used for microbes, exists throughout all phage families (Paul et al., 2002). Comparative approaches based on terminase sequences (Serwer et al., 2004), structural genes (Chibani-Chenouffi et al., 2004), and whole proteomes (Rohwer and Edwards, 2002) were proposed as novel taxonomic tools. In 2008, Lavigne and coworkers delineated new phage subfamilies and genera using a method that integrated BLAST-based tools with a careful review of available literature. This work laid the basis for systematic and genome-based classification of sequenced phages currently undergoing ratification by the ICTV.

Bacteriophage Infection Cycle

In essence, phages carry their genomes from one susceptible bacterial cell to another, in which they direct the production of progeny phages. The target host for each phage is a specific group of bacteria: this group is often a subset of one species, but several related species can sometimes be infected by the same phage. The bacteriophage infection cycle follows a number of tightly programmed steps (Fig. 2), of which the efficiency and timing depend strongly on the metabolic state of the host.

ADSORPTION

Infection by tailed phages starts when specialized adsorption structures, such as tail fibers or spikes, bind to specific surface molecules or capsules on their target bacteria. In gram-negative bacteria, virtually any of the proteins, oligosaccharides, and liposaccharides present in the outer membrane can be used during phage attachment. The more complex murein of gram-positive bacteria offers a very different set of potential binding sites. The receptor-recognizing areas of phage proteins seem to be confined to single regions on the polypeptide chains (Heller, 1992). Many phages require clusters of specific kinds of molecules present in high concentration to properly position the phage tail for surface penetration. However, coliphage N4 manages to use a receptor (NfrA) which has a limited copy number per cell (McPartland and Rothman-Denes, 2009). Adsorption to the host-cell surface is often a multistage process, as illustrated for *Escherichia coli*-infecting bacteriophage T4. Irreversible binding of at least three of the six tail fibers to a glucose residue of the outer core of the lipopolysaccharide (LPS) matrix positions the phage correctly on the cell surface and triggers a structural rearrangement in the tail (Crawford and Goldberg, 1980). This allows the tail-spikes to irreversibly bind to the heptose residue of the LPS inner core (Montag et al., 1987). Phage adsorption of *Podoviridae* is in some cases assisted by the presence of enzymatically active tail proteins, which are able to cleave extracellular capsules or the O-antigen in order to facilitate access to the bacterial receptor (Barbirz et al., 2008).

Speed and efficiency of adsorption are strongly dependent on external parameters and the physiological state of the host. For example, the receptor of phage lambda (LamB) is expressed only in the presence of maltose (Boos and Shuman, 1998), and many phages require divalent cations for stable adsorption to the bacterial surface.

Bacterial resistance against phage attack typically occurs by mutation or loss of the receptor. In response, the phage can acquire a compensating adaptation; for example, through a series of mutations in its tail fiber, phage Ox2 can switch its receptor-recognition site from OmpA to OmpC to OmpP to the *E. coli* B LPS (Henning and Hashemolhosseini, 1994). Other phages like P1 and Mu encode multiple versions of their tail proteins. Not surprisingly, there is much interest in engineering new receptor-recognition elements into the phage tail fibers, enabling infection of taxonomically distant hosts. In addition, phages having a virulence or adhesion factor as receptor are of special interest, since in case resistance to the phage develops due to loss or mutation of the receptor, the virulence of the bacteria is reduced at the same time (see, e.g., Spears et al., 2008).

DNA TRANSFER AND HOST CONVERSION

After irreversible adhesion, the phage genome must cross either two or three major bacterial barriers (the outer membrane, the peptidoglycan layer, and the inner membrane) to obtain access to the transcription/translation metabolism of its gram-positive or gram-negative host. For this, phages employ strategies that vary with phage particle morpholjogy. In general, the phage tail carries a peptidoglycan-degrading enzyme at its tip, and binding of the tail disables a mechanism that blocks the premature extrusion of the genomic material from the capsid. The DNA is then rapidly drawn into the cell by a mechanism which requires metabolic energy and is based on available ATP or membrane potential (Letellier et al., 2004). Phages have many strategies to cope with exonucleases and restriction enzymes present in the bacterial cell. A widely used system is the rapid circulariza-

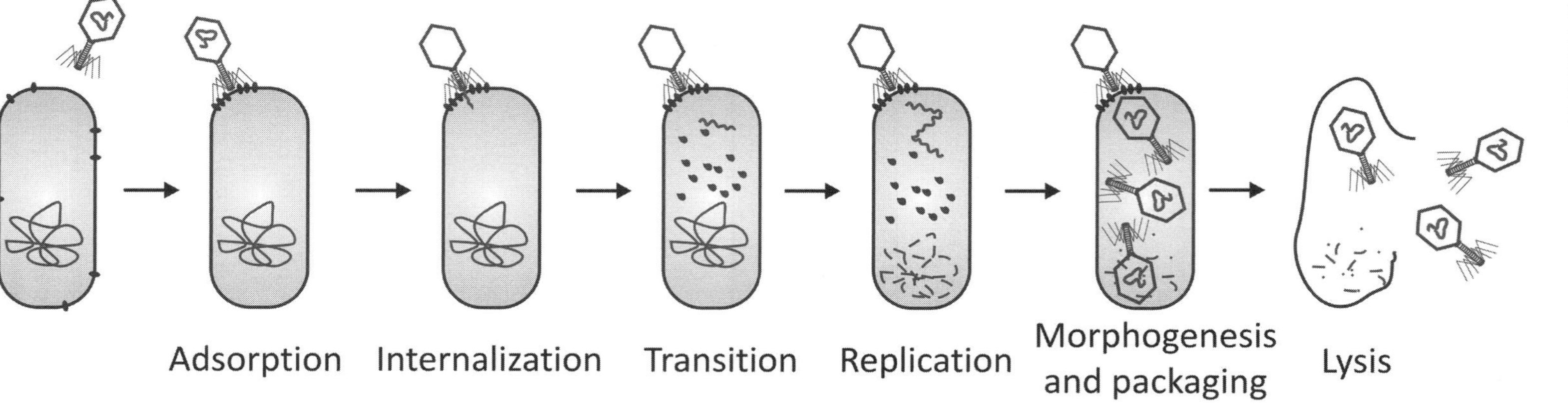

FIGURE 2 Schematic overview of the infection cycle of virulent bacteriophages. The bacteriophage adsorbs to the host cell. Immediately after internalization of the genome, the first phage proteins involved in transition from host-directed to phage-directed metabolism are produced. Next, replication of the bacteriophage genome takes place. Afterwards the new phage particles are assembled and the genome is packed inside. Finally, the new virions are released. The phage shown in this scheme has the characteristics of a member of the *Myoviridae*, phages with an icosahedral head and a contractile tail.

tion of the DNA by means of sticky ends or terminal repeats or the protection of linear genome ends by proteins. Other phages encode unusual nucleotides in their genome like hydroxymethyldeoxyuridine (Schellenberger et al., 1995), or have eliminated throughout evolution those restriction sites that would be recognized by the type II restriction enzymes present in their hosts, as is the case with coliphage N4 (Gutmann et al., 2005). Type I restriction enzymes are sometimes inhibited by their interaction with a phage-encoded protein that structurally mimics B-folded DNA.

After internalization of the bacteriophage genome, the host-encoded RNA polymerase recognizes strong promoters located on the phage genome, leading to transcription of the phage early genes. Alternatively, internal virion proteins may be injected into the host cell together with the phage genome. These early proteins are involved in the takeover of the host-cell metabolism. Generally they provide protection from bacterial defense mechanisms or redirect the host-cell metabolism to establish optimal conditions for the synthesis of new virions. To establish these optimal conditions, various macromolecular processes like transcription, translation, or replication, as well as single enzymatic functions, may be redirected or inhibited (Miller et al., 2003). Early proteins are very phage specific and generally poorly conserved, even between two closely related phages, and typically display no similarity to any known protein in the databases.

DNA REPLICATION AND PARTICLE FORMATION

Once optimal metabolic conditions are established, replication of the bacteriophage genome starts. While most temperate phages are dependent on the host cell for their replication, virulent phages often encode their own replication machinery. Phages use a wide variety of different replication systems, extensively reviewed by Weigel and Seitz (2006).

Before the replicated, concatemerized phage genomes are packed, the new phage heads or capsids have to be assembled in a highly regulated, multistage process. Briefly, this process can be dissected into three major stages for most tailed phages. First, the phage head—at this stage called the procapsid or prohead—is assembled around scaffolding proteins. The pore complex, which forms the attachment point for the tail, is located at one vertex of the prohead. In the next phase, the terminase, which docks to the pore complex, translocates the phage genome into the procapsid, transforming it into a mature capsid. DNA translocation is driven by ATP hydrolysis and coupled to the processing of the concatemerized phage genome (Fujisawa and Morita, 1997). After DNA packaging, the tail is attached to the pore complex. While the phage tail is assembled separately for *Myoviridae* and *Siphoviridae* (Kutter et al., 2005), tail proteins of *Podoviridae* cooperatively assemble on the pore complex (Matsuo-Kato et al., 1981; Tao et al., 1998; Johnson and Chiu, 2007).

PHAGE RELEASE

At the end of the replication cycle, newly formed phage particles exit the host cell in search of new prey cells. For this, bacteriophages developed two basic strategies. During their unique morphogenesis, filamentous phages continuously extrude their progeny across the cell wall without killing their host cells (reviewed by Russel, 1995). In contrast, most dsDNA phages accomplish fatal lysis of the host cell using specific lysis proteins. During phage infection, a soluble and active endolysin accumulates in the host cell. To gain access to the peptidoglycan layer, a small hydrophobic membrane-spanning protein, designated holin, is essential. After accumulation and oligomerization of the holin in the cytoplastic membrane, the holins form a membrane lesion at a genetically predetermined time, permeabilizing the inner membrane for the cognate endolysin. The endolysin, a phage-encoded muralytic enzyme, will degrade the peptidoglycan layer until the cell can no longer withstand the internal osmotic pressure. The cell bursts open and the mature phage particles are released (Young et al., 2000; Wang, 2006).

The holin-endolysin two-component cell lysis system described here was long thought to be universal (see chapter 8 in this volume). However, a second, holin-independent lysin transport system has been described in phages that encode a lysin containing an N-terminal signal sequence, transported to the periplasm by the host Sec machinery, which subsequently induces lysis after cleavage of the signal sequence (Loessner et al., 1997; São-José et al., 2000). Recently, a hybrid between these two systems called the signal arrest and release (SAR) system was identified in phage P1 (Xu et al., 2004). Phages using SAR encode both a holin and a lysin with a noncleavable N-terminal signal (type II signal anchor). Hence, the N terminus of the lysin remains part of the mature lysin and stays embedded in the inner cell membrane in an inactive form (due to the close proximity of the catalytic residues to the membrane or because of rearranged disulfide bonds in the lysine) (Xu et al., 2005). In general, triggering of the lysin is accomplished by the holins, accumulating in clusters in the inner membrane to a critical concentration which punctures the membrane, destroying membrane potential and releasing the lysin to the periplasm. In some gram-negative bacteria-infecting phages, additional proteins (Rz/Rz1) that degrade the outer membrane have also been described (Summer et al., 2007).

Phage-encoded endolysins can be used for the biopreservation of food and feed. Because of their specificity, they can selectively kill gram-positive food-borne pathogens such as *Listeria monocytogenes* without affecting beneficial organisms (Gaeng et al., 2000; Loessner, 2005).

Lytic versus Lysogenic Infection Cycle

Based on their propagation cycle, most phages can be divided into two major groups: virulent and temperate phages. Virulent phages immediately redirect the host metabolism toward the production of new phage virions, which are released upon cell death within minutes to hours after the initial infection event. Temperate phages act more subtly and can replicate either by lysing the host cell in the same way as a virulent phage or by establishing a stable relationship with their host bacteria. In this state, the phage DNA is replicated together with the host's chromosome, and virus genes that are detrimental to the host are not expressed (Fig. 3). Usually, the phage DNA physically integrates at a specific (e.g., λ) or random (e.g., Mu) site within the host genome (Campbell, 1962; Harshey, 1988), or can in some cases reside as a circular (P1) or linear (LE1) plasmid in the bacterial cytoplasm (Ikeda and Tomizowa, 1968; Girons et al., 2000) and is called a *prophage*. These resident prophages are very common: they can constitute as much as 10 to 20% of a bacterium's genome and are major contributors to differences between individuals within species (Casjens, 2005). Some of the genes expressed from the prophages in a lysogen are "lysogenic conversion" genes and alter the properties of the host bacterium. The products of these genes can have very important effects on the host, which range from protection against further phage infection to increasing the virulence of a pathogenic host (reviewed by Boyd and Brüssow, 2002; see also chapter 7 in this volume).

Under stress conditions, bacteriophage virions can be released from cells containing an intact prophage by a process called induction, during which expression of prophage genes required for lytic growth is turned on and progeny virions are produced and released from the cell. Cells carrying a prophage are called *lysogens* because of this potential to induce and lyse. Induction can happen spontaneously and randomly in a small fraction of the bacteria that harbor a given prophage, or specific environmental signals can cause simultaneous induction of a particular prophage in many cells. Generally, prophage induction is associated with physical breakdown of the protein that represses the lytic cycle. This can

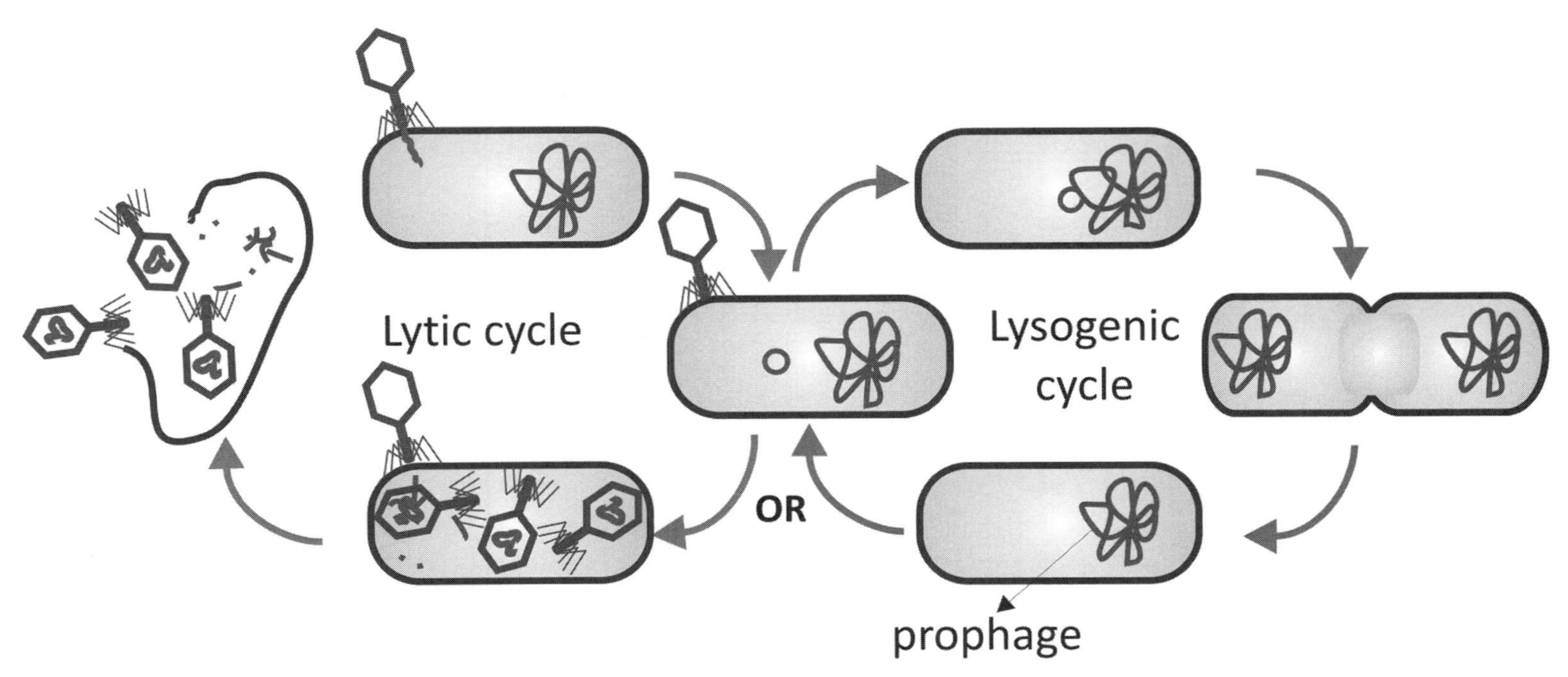

FIGURE 3 After circularization of phage DNA upon infection, temperate phages make a choice between a lytic or a lysogenic cycle. In the lytic cycle, new phage particles are produced and released upon lysis of the host cell. If the phage enters a lysogenic stage, the phage DNA in most cases enters the bacterial chromosome and replicates along with the host DNA, and a stable relationship is established. Occasionally, the prophage may excise from the host genome by another recombination event and a lytic cycle is initiated.

be artificially induced by UV light, mitomycin, and prolonged starvation.

The molecular mechanism behind the switch between the lysogenic and lytic state has been studied in great detail for phage λ (Oppenheim et al., 2005). The key regulation system is the competition between the CI and Cro proteins: while Cro promotes the lytic cycle, CI represses the expression of all genes except its own. The transition toward lysogeny is also promoted by two other proteins, CII and CIII, which bind to critical promoters and stimulate transcription of the *cI* gene and others (Fig. 4). The stability of CII depends on the cell's cyclic AMP concentration: a high level of cyclic AMP (indicative of starvation) stabilizes CII and thus promotes lysogeny. In other words, the phage senses upon infection whether its host carries enough energy to achieve a large burst of new virions, or whether the energy level is low and the best strategy for survival is to go into a prophage state.

Due to imprecise DNA excision during induction of the prophage, bacterial DNA can be packed inside the phage particle and subsequently transferred to another bacterial cell, a process called transduction. When this event occurs with a phage which integrates at its own specific site, this transducing phage can effect a *specialized transduction* by carrying this particular piece of bacterial DNA into other cells followed by recombination and genome change. A temperate phage that randomly integrates in the host genome can incorporate random pieces of bacterial DNA and provokes *generalized transduction*. Transduction can also be caused by (lytic or temperate) phages that mistakenly produce particles which contain only host DNA (Guttman et al., 2005); those are generally rare and cannot reproduce, but may still contribute substantially to bacterial genetic exchange in nature. The importance of temperate phages as "replicons endowed with horizontal transfer capabilities" (Briani et al., 2001) has been widely recognized: at any moment on planet Earth, there are literally tons of bacterial DNA being swapped around by phages (Thiel, 2004).

Differences in Plaque Formation

Usually, phage infection is studied in a layer of soft agar which allows the phage to diffuse rapidly. Upon cell lysis, progeny phage particles are released and can diffuse to and subsequently infect neighboring cells. This process continues, and ultimately results in a circular area of cell lysis in a turbid lawn of cells, called a *plaque*. The morphology of the plaque depends upon the phage, the host, and the growth conditions. The size of the plaque is proportional to the efficiency of adsorption, the length of the *latent period* (time between infection and cell lysis), and the *burst size* (amount of particles released upon cell lysis) of the phage. The last two parameters are studied by single-step growth curves of phage infection (Fig. 5A). Its size is also affected by

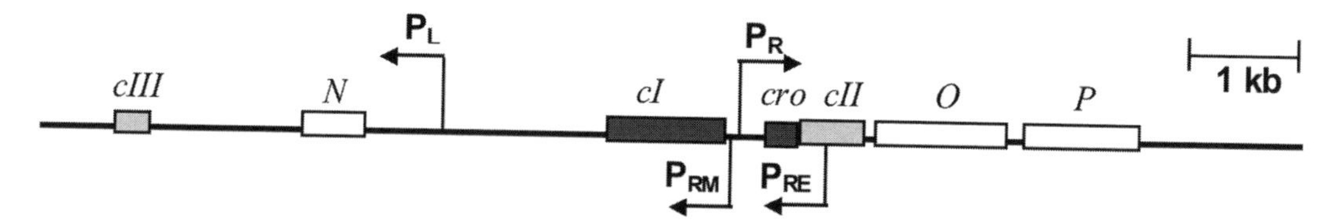

FIGURE 4 Map of the critical regulatory region of phage λ, controlling the switch between lytic and lysogenic life cycles. The proteins CII and CIII bind to P_{RE}, promoting transcription of the *cI* gene. The DNA-binding CI protein represses various operator sites and prevents formation of the N protein, which is required for expression of late genes of the lytic cycle. CI also promotes its own transcription by binding to P_{RM}. These processes are in competition with transcription of genes *cro, O, P,* and *Q* (not depicted) from P_R, which are required for lytic growth. Cro and CI compete for binding sites in the complex promoter/operator site that includes both P_{RM} and P_R, and determine in this way the decision between lytic and lysogenic growth.

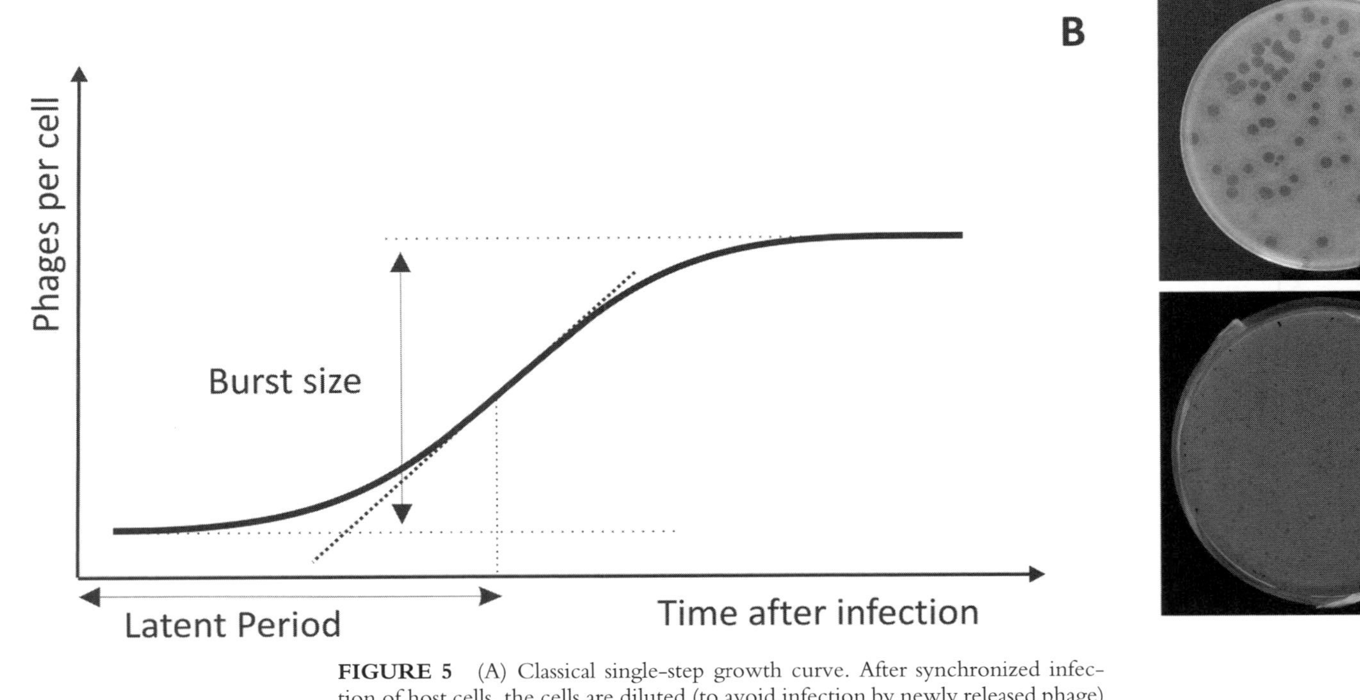

FIGURE 5 (A) Classical single-step growth curve. After synchronized infection of host cells, the cells are diluted (to avoid infection by newly released phage) and samples are plated at various times to determine *infective centers*. An infective center is either a single phage particle or an infected cell that lyses on the plate to produce a plaque. Initially, the number of plaques remains constant for a time (the latent period), rises sharply at a certain moment, and levels off as each host cell has liberated its progeny phage. The ratio between the numbers of infected centers before and after lysis is the burst size. (B) Differences in plaque morphologies between lytic and temperate phages. Shown in this picture are lytic *Pseudomonas putida* phage ε15 (top) and temperate *Pseudomonas aeruginosa* phage LKR3 (bottom).

such factors as the concentration of the soft agar and dryness of the plate. Theoretical aspects of plaque formation have been studied in detail by Abedon and Yin (2008).

Plaque assays can be used to make a tentative discrimination between lytic and temperate phages. In general, lytic phages produce completely clear zones of killing (Fig. 5B), while temperate phages form plaques with a turbid center. The latter characteristic plaque morphology is due to the effect of the varying amount of phage particles and changing cell physiology on the lysis-lysogeny decision. As stated in the previous paragraph, lytic growth is favored when cells are growing rapidly. When the cells are initially infected with phage, the ratio of phage to cells (the *multiplicity of infection*) is relatively low and nutrients are plentiful, so the bacteria grow rapidly and the phages grow lytically. After several lytic cycles the local multiplicity of infection increases and most of the cells are lysed, producing a plaque in the lawn of cells. As the cell lawn becomes saturated, the rate of cell growth slows down and, since lysis requires rapid metabolism, the plaque stops increasing in size in most cases. However, any lysogens that formed in the center of the plaque are immune to lysis and can continue to grow since they do not have to compete with nearby cells for nutrients. Thus, lysogens begin to grow in the center of the plaque, giving the plaque the turbid appearance.

Genomic Architectures and Evolution of Tailed Phages

The linear dsDNA genomes of the *Caudovirales* encode from 27 to over 600 genes that are highly clustered according to function and tend to be arranged in large operons, as schematically drawn in Fig. 6. Gene expression is largely time ordered, and groups of genes are sequentially expressed. "Early genes" are expressed first and are largely involved in host-cell modification and viral DNA replication, while "late genes" specify virion structural proteins and lysis proteins. In addition, temperate phages like phage λ encode a cluster of genes responsible for integration into the host chromosome and maintenance of the lysogenic state. Although this scenario is followed by many phages, larger tailed viruses like T4 have much more complex genomes and gene expression cascades. For more details on the transcriptional control and genome content of these and other phages, we refer to several excellent reviews written on this subject (e.g., Miller et al., 2003; Hendrix and Casjens, 2005; Molineux, 2005).

Evolution of tailed phages is influenced by a combination of horizontal and vertical gene transfer. In specific genera, genomes display a mosaic genetic structure, consisting of gene modules (individual genes or small groups of genes) which are exchanged using host- or phage-encoded recombination machinery. This phenomenon was first recognized in the "lambdoid" phages as a result of DNA-DNA heteroduplex experiments (Simon et al., 1971). This mosaicism most likely results from random, nonhomologous (illegitimate) recombination between (mostly) temperate phages, resulting in a heterogeneous mix of recombinant types which are almost all defective. The few recombinants that survive are the ones in which the functions of essential genes are not disrupted and whose genomes still fit inside the virion capsid (Lawrence et al., 2002).

For other clades like the T4-related phages, limited lateral gene transfer is observed, as a set of core genes is conserved among them (Comeau et al., 2007). For this clade, the cause for this limited horizontal gene transfer lies in the great complexity of these phages, and in their multitude of protein-protein interactions between the constituents of the virion (Leiman et al., 2003) and replication complexes (Karam and Konigsberg, 2000). Despite the occurrence of (sometimes rampant) horizontal gene transfer, the phylogeny of dsDNA phages is tractable, based on complete proteome analysis.

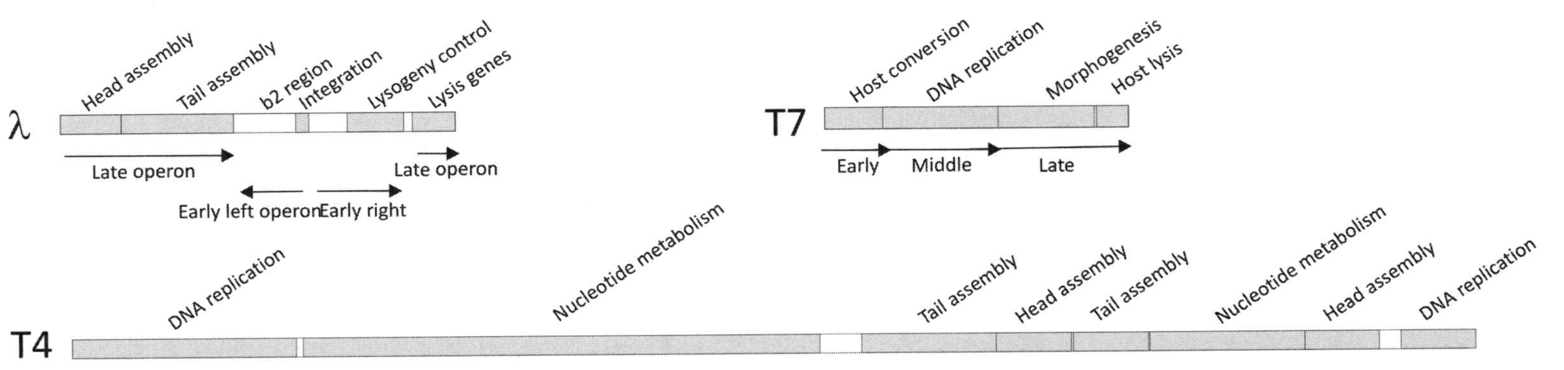

FIGURE 6 Schematic overview of the genome build-up of the *Siphovirus* λ (48,502 bp), *Podovirus* T7 (39,937 bp), and *Myovirus* T4 (168,904 bp), all infecting *E. coli*. The genomes are divided into functional blocks, arranged head to tail for reasons of simplicity. For lambda and T7, the timing of gene expression is indicated with arrows. In contrast to these two phages, the early and late transcription units of T4 (and related phages) are not clearly separated but are interdigitated (e.g., late genes *26* and *25* and the middle gene *uvsY*). The direction of T4 transcription does not unambiguously distinguish between early and late transcription, and in many cases early and late genes are cotranscribed (Mosig and Hall, 1994).

BACTERIOPHAGES IN THE ENVIRONMENT

Phage Abundance

The introduction of direct counts of phage particles in environmental samples using transmission electron microscopy (Bergh et al., 1989) and epifluorescence techniques (Hennes et al., 1995) revealed the vast abundance of phage particles in nature. Most of these counts were performed in aquatic environments, indicating that, typically, there are 10^7 viruses per ml of seawater. By extrapolation, the total population of phages was estimated to be about 10^8 species and 10^{31} particles in the biosphere, making them the most abundant organisms on this planet (Rohwer, 2003). As could have been expected, viral abundance often correlates strongly with microbial abundance, with a fairly constant ratio of 5 to 10 virus-like particles per bacterium (Wommack and Colwell, 2000). In marine sediments, abundances are even higher, with 10^8 to 10^9 viruses cm^{-3} in near-shore surface sediments (Paul et al., 1993). Subsequent application of these counting methods in soil samples has revealed that viruses are also extraordinarily abundant in a diverse range of soil types and locations: in a survey of Delaware soils which included wetlands and agricultural soils, viral abundance was 80 to 390 $\times 10^7$ (Williamson et al., 2007). Surprisingly, despite the vast differences in composition and geographic location of these soil samples, there was only a 16-fold range in viral abundance. By comparison, across marine environments ranging from the deep sea to freshwater marshes, viral abundance changes over 2,000-fold (Srinivasiah et al., 2008).

Aside from water and soil samples, bacteriophages have been isolated from a wide range of food products including ground beef, poultry meat, pork sausage, chicken, sardines, and cheese (see, e.g., Gautier et al., 1995). To give just one example, Kennedy et al. (1984) isolated 3.3 to 4.4 $\times 10^{10}$ PFU/100 g of fresh chicken and up to 2.7 $\times 10^{10}$ PFU/100 g of roast turkey breast samples. Several other studies have suggested that 100% of the ground beef and chicken meat sold at retail contains various levels of various bacteriophages (Hsu et al., 2002). These data indicate that naturally occurring bacteriophages are commonly consumed by humans and animals; the daily ingestion of phages may be an important natural strategy to replenish the phage population in the gastrointestinal tract and to regulate the microbalance of the colon (Sulakvelidze and Barrow, 2005). Phages that infect *E. coli*, *Salmonella* spp., and *Bacteroides fragilis* have been isolated from human fecal samples at concentrations ranging up to 10^5 phages per g of dry feces (Calci et al., 1998; Gantzer et al., 2002). Breitbart et al. (2003) performed a metagenomic analysis of an uncultured viral community from human feces and found that of the recognizable viruses, 81% were *Siphoviridae*. The community contained an estimated 1,200 viral genotypes, and most of the sequences were unrelated to anything previously reported: this issue is tackled further in the next section.

Phage population dynamics have been studied extensively for marine viruses. It has been proposed that phage populations behave in a typical "kill-the-winner" fashion, where the most abundant host population is reduced by its viral predators. As a result, an open niche is created which can enable a new host to become abundant (Breitbart and Rohwer, 2005). Supporting this model, culturing and genome size distribution studies provide evidence that specific viruses become abundant, decay to undetectable levels, and then return on seasonal cycles (Wommack et al., 1999; Marston and Sallee, 2003).

Phage Diversity

The dramatic increase in the number of published completely sequenced phage genomes over the last decade (from 40 in January 1997 to 546 in June 2009) provides evidence that bacteriophage diversity is extraordinarily high. Phages isolated from different bacterial hosts typically have little or no identifiable nucleotide sequence similarity, and even phages capable of infecting a common host may exhibit great diversity (see, e.g., Kwan et al., 2005). It is not uncommon for reports of newly sequenced phage genomes to reveal that only a small proportion (20 to 40%) of the encoded genes have similarity to current database entries (Comeau et al., 2008). Moreover, characterized phages in current databases are clearly biased toward easily cultured phages that infect generally well-known bacteria. For example, only eight (1.5%) have genomes as long as 200 kb. Statistical analysis, however, reveals a significant undersampling of long-genome bacteriophages (Claverie et al., 2006). The strong possibility exists that long-genome bacteriophages (>200-kb genome) are more frequent and are major contributors to microbial ecology, but are undersampled because of the use of classical bacteriophage propagation procedures and possibly also classical processing of environmental samples for microscopy. For example, bacteriophage G was discovered by accident ~40 years ago through electron microscopy of a preparation of another bacteriophage (Donelli, 1968). This well-known plate count anomaly (Staley and Konopka, 1985) has forced microbial ecologists to rely on cultivation-independent approaches to explore natural phage populations.

In an attempt to overcome these constraints and to get a better overview of natural phage populations, various research groups have employed some of the best-conserved phage genes (polymerases, structural genes) as molecular markers for population studies. The resulting data gave first insights into an enormous genetic variation not represented in cultures and showed that very similar sequences were present in distant oceans (Short and Suttle, 2002). Even more remarkable is that nearly identical sequences at the nucleotide level occurred in environments as far-reaching as the Southern Ocean, the Gulf of Mexico, and a meltwater pond on an Arctic ice shelf (Short and Suttle, 2005). Similarly, a study that targeted phage T7-like DNA polymerases found that these sequences were ubiquitous in the environment (Breitbart et al., 2004). These single-locus studies do not completely bypass the above-mentioned problems, since environmental PCR primers are still based on sequences from cultured plates. Nevertheless, these results indicate that closely related hosts and their phages are distributed widely and that horizontal gene transfer has occurred among phage communities from very different environments.

Shotgun sequencing from environmental metagenomes can bypass this dilemma and provide quantitative as well as qualitative information of each genotype (Edwards and Rohwer, 2005). The online bioinformatic tool PHACCS makes it possible to use metagenomic data to mathematically model possible community structures. Using these models, estimates of viral community structures and of the richness, evenness, and abundance of each viral genotype can be made (Angly et al., 2005). Although the metagenomic approach holds drawbacks such as free DNA in the environment, viral genes that kill the host in the cloning process, and unclonable viral DNA (Edwards and Rohwer, 2005), the technique gave valuable insights into the local and global distribution of phage genotypes. In numerous metagenomic studies, an extraordinarily high diversity of both local and global scale was found. For example, in 1 kilogram of marine sediment, at least 10^4 different viral genotypes were found (Breitbart et al., 2004). In an analysis of marine viromes across four oceanic regions, no fewer than 129,000 genotypes were predicted in the viral metagenome of British Columbian coastal waters (Angly et al., 2006). Interestingly, these studies also emphasize that regional diversity can be as high as global diversity due to viral migration. It has been shown that viruses taken from one ecosystem can replicate in a totally different environment, implying that phages can move between biomes (Sano et al., 2004). As an illustration, a phage like *Prochlorococcus* phage P-SSM4 was found at all sampling sites during the Sorcerer II Global Ocean Sampling Expedition (Williamson et al., 2008). In this respect, Angly et al. (2006) suggested that local viral community composition and nucleic acid type (i.e., dsDNA versus ssDNA) is a function of the geographic location and regional selective pressure. As more and more metagenomic data accumulate, deeper insights into global phage will spread, and evolutionary mechanisms and composition of phage communities will come to light.

ACKNOWLEDGMENT

P.-J.C. holds a postdoctoral scholarship (PDM) of the "Bijzonder Onderzoeksfonds" at the K. U. Leuven.

REFERENCES

Abedon, S. T., and J. Yin. 2008. Bacteriophage plaques: theory and analysis. *Methods Mol. Biol.* **501:**161–174.

Ackermann, H. W. 2007. 5500 Phages examined in the electron microscope. *Arch. Virol.* **152:**227–243.

Angly, F. E., B. Felts, M. Breitbart, P. Salamon, P. A. Edwards, C. Carlson, A. M. Chan, M. Haynes, S. Kelley, H. Liu, J. M. Mahaffy, J. E. Mueller, J. Nulton, R. Olson, R. Parsons, S. Rayhawk, C. A. Suttle, and F. Rohwer. 2006. The marine viromes of four oceanic regions. *PLoS Biol.* **4:**e368.

Angly, F., B. Rodriguez-Brito, D. Bangor, P. McNairnie, M. Breitbart, P. Salamon, B. Felts, J. Nulton, J. Mahaffy, and F. Rohwer. 2005. PHACCS, an online tool for estimating the structure and diversity of uncultured viral communities using metagenomic information. *BMC Bioinformatics* **6:**1.

Barbirz, S., J. J. Müller, C. Uetrecht, A. J. Clark, U. Heinemann, and R. Seckler. 2008. Crystal structure of *Escherichia coli* phage HK620 tailspike: podoviral tailspike endoglycosidase modules are evolutionarily related. *Mol. Microbiol.* **69:**303–316.

Bergh, O., K. Y. Børsheim, G. Bratbak, and M. Heldal. 1989. High abundance of viruses found in aquatic environments. *Nature* **340:**467–468.

Boos, W., and H. Shuman. 1998. Maltose/maltodextrin system of *Escherichia coli:* transport, metabolism, and regulation. *Microbiol. Mol. Biol. Rev.* **62:**204–229.

Botstein, D., and I. Herskowitz. 1974. Properties of hybrids between *Salmonella* phage P22 and coliphage lambda. *Nature* **251:**585–589.

Boyd, E. F., and H. Brüssow. 2002. Common themes among bacteriophage-encoded virulence factors and diversity among the bacteriophages involved. *Trends Microbiol.* **10:**521–529.

Breitbart, M., and F. Rohwer. 2005. Here a virus, there a virus, everywhere the same virus? *Trends Microbiol.* **13:**278–284.

Breitbart, M., I. Hewson, B. Felts, J. M. Mahaffy, J. Nulton, P. Salamon, and F. Rohwer. 2003. Metagenomic analyses of an uncultured viral community from human feces. *J. Bacteriol.* **185:**6220–6223.

Breitbart, M., J. H. Miyake, and F. Rohwer. 2004. Global distribution of nearly identical phage-encoded DNA sequences. *FEMS Microbiol. Lett.* **236:**249–256.

Briani, F., G. Dehò, F. Forti, and D. Ghisotti. 2001. The plasmid status of satellite bacteriophage P4. *Plasmid* **45:**1–17.

Bruynoghe, R., and J. Maisin. 1921. Essais de thérapeutique au moyen du bacteriophage du Staphylocoque. *J. Compt. Rend. Soc. Biol.* **85:**120–121.

Calci, K., W. Burkhardt, W. Watkins, and S. Rippey. 1998. Occurrence of male-specific bacteriophage in feral and domestical animal wastes, human feces, and human-associated wastewaters. *Appl. Environ. Microbiol.* **64:**5027–5029.

Campbell, A. M. 1962. Episomes. *Adv. Genet.* **11:**101–145.

Casjens, S. R. 2005. Comparative genomics and evolution of the tailed-bacteriophages. *Curr. Opin. Microbiol.* **8:**451–458.

Chibani-Chennoufi, S., M. L. Dillmann, L. Marvin-Guy, S. Rami-Shojaei, and H. Brüssow. 2004. *Lactobacillus plantarum* bacteriophage LP65: a new member of the SPO1-like genus of the family *Myoviridae*. *J. Bacteriol.* **186:**7069–7083.

Claverie, J. M., H. Ogata, S. Audic, C. Abergel, K. Suhre, and P. E. Fournier. 2006. Mimivirus and the emerging concept of "giant" virus. *Virus Res.* **117:**133–144.

Comeau, A. M., C. Bertrand, A. Letarov, F. Tétart, and H. M. Krisch. 2007. Modular architecture of the T4 phage superfamily: a conserved core genome and a plastic periphery. *Virology* **362:**384–396.

Comeau, A. M., G. F. Hatfull, H. M. Krisch, D. Lindell, N. H. Mann, and D. Prangishvili. 2008. Exploring the prokaryotic virosphere. *Res. Microbiol.* **159:**306–313.

Crawford, J. T., and E. B. Goldberg. 1980. The function of tail fibers in triggering baseplate expansion of bacteriophage T4. *J. Mol. Biol.* **139:**679–690.

Deresinski, S. 2009. Bacteriophage therapy: exploiting smaller fleas. *Clin. Infect. Dis.* **48:**1096–1101.

Donelli, G. 1968. Isolamento di un batteriofago di eccezionali dimensioni attivo su *B. megatherium*. *Atti Accad. Naz. Lincei-Rend Clas. Sci. Fis. Mat. Nat.* **44:**95–97.

Earnshaw, W. C., and S. C. Harrison. 1977. DNA arrangement in isometric phage heads. *Nature* **268:**598–602.

Eaton, M. D., and S. Bayne-Jones. 1934. Bacteriophage therapy. Review of the principles and the results of the use of bacteriophage in the treatment of infections. *JAMA* **23:**1769–1939.

Edwards, R. A., and F. Rohwer. 2005. Viral metagenomics. *Nat. Rev. Microbiol.* **3:**504–510.

Fujisawa, H., and M. Morita. 1997. Phage DNA packaging. *Genes Cells* **2:**537–545.

Gaeng, S., S. Scherer, H. Neve, and M. J. Loessner. 2000. Gene cloning and expression and secretion of *Listeria monocytogenes* bacteriophage-lytic enzymes in *Lactococcus lactis*. *Appl. Environ. Microbiol.* **66:**2951–2958.

Gantzer, C., J. Henny, and L. Schwartzbrod. 2002. *Bacteroides fragilis* and *Escherichia coli* bacteriophages in human faeces. *Int. J. Hyg. Environ. Health* **205:**325–328.

Gautier, M., A. Rouault, P. Sommer, and R. Briandet. 1995. Occurrence of *Propionibacterium freudenreichii* bacteriophages in swiss cheese. *Appl. Environ. Microbiol.* **61:**2572–2576.

Girons, I. S., P. Bourhy, C. Ottone, M. Picardeau, D. Yelton, R. W. Hendrix, P. Glaser, and N. Charon. 2000. The LE1 bacteriophage replicates as a plasmid within *Leptospira biflexa*: construction of an *L. biflexa-Escherichia coli* shuttle vector. *J. Bacteriol.* **182:**5700–5705.

Guttman, B., R. Raya, and E. Kutter. 2005. Basic phage biology, p. 29–66. In E. Kutter and A. Sulakvelidze (ed.), *Bacteriophages: Biology and Applications*. CRC Press, Boca Raton, FL.

Harshey, R. 1988. Phage Mu, p. 193–234. *In* R. Calendar (ed.), *The Bacteriophages*, vol. 1. Plenum Press, New York, NY.

Heller, K. J. 1992. Molecular interaction between bacteriophage and the gram-negative cell envelope. *Arch. Microbiol.* **158:**235–248.

Hendrix, R., and S. Casjens. 2005. Bacteriophage λ and its genetic neighborhood, p. 409–448. *In* R. Calendar (ed.), *The Bacteriophages,* 2nd ed. Oxford University Press, New York, NY.

Hennes, K. P., C. A. Suttle, and A. M. Chan. 1995. Fluorescently labeled virus probes show that natural virus populations can control the structure of marine microbial communities. *Appl. Environ. Microbiol.* **61:**3623–3627.

Henning, U., and S. Hashemolhosseini. 1994. Receptor recognition by T-even-type coliphages, p. 291–298. *In* J. D. Karam et al. (ed.), *Molecular Biology of Bacteriophage T4*. ASM Press, Washington, DC.

Hsu, F. C., Y. S. Shieh, and M. D. Sobsey. 2002. Enteric bacteriophages as potential fecal indicators in ground beef and poultry meat. *J. Food Prot.* **65:**93–99.

Ikeda, H., and J. Tomizowa. 1968. Prophage P1, an extrachromosomal replication unit. *Cold Spring Harbor Symp. Quant. Biol.* **33:**791–798.

Johnson, J. E., and W. Chiu. 2007. DNA packaging and delivery machines in tailed bacteriophages. *Curr. Opin. Struct. Biol.* **17:**237–243.

Karam, J. D., and W. H. Konigsberg. 2000. DNA polymerase of the T4-related bacteriophages. *Prog. Nucleic Acid Res. Mol. Biol.* **64:**65–96.

Kennedy, J. E., Jr., G. Bitton, and J. L. Oblinger. 1984. Recovery of coliphages from chicken, pork sausage, and delicatessen meats. *Appl. Environ. Microbiol.* **51:**956–962.

Krueger, A. P., and E. J. Scribner. 1941. The bacteriophage: its nature and its therapeutic use. *JAMA* **116:**2160–2167.

Kutter, E., R. Raya, and K. Carlson. 2005. Molecular mechanisms of phage infection, p. 166–213. *In* E. Kutter and A. Sulakvelidze (ed.), *Bacteriophages: Biology and Applications*. CRC Press, Boca Raton, FL.

Kwan, T., J. Liu, M. Dubow, P. Gros, and J. Pelletier. 2005. Comparative genomic analysis of 18 *Pseudomonas aeruginosa* bacteriophages. *J. Bacteriol.* **188:**1184–1187.

Lavigne, R., D. Seto, P. Mahadevan, H. W. Ackermann, and A. M. Kropinski. 2008. Unifying classical and molecular taxonomic classification: analysis of the *Podoviridae* using BLASTP-based tools. *Res. Microbiol.* **159:**406–414.

Lawrence, J. G., G. F. Hatfull, and R. W. Hendrix. 2002. Imbroglios of viral taxonomy: genetic exchange and failings of phenetic approaches. *J. Bacteriol.* **184:**4891–4905.

Leiman, P. G., P. R. Chipman, V. A. Kostyuchenko, V. V. Mesyanzhinov, and M. G. Rossmann. 2004. Three-dimensional rearrangement of proteins in the tail of bacteriophage T4 on infection of its host. *Cell* **118:**419–429.

Leiman, P. G., S. Kanamaru, V. V. Mesyanzhinov, F. Arisaka, and M. G. Rossmann. 2003. Structure and morphogenesis of bacteriophage T4. *Cell. Mol. Life Sci.* **60:**2356–2370.

Letellier, L., P. Boulanger, L. Plançon, L. Jacquot, and M. Santamaria. 2004. Main features on tailed phage, host recognition and DNA uptake. *Front. Biosci.* **9:**1228–1339.

Liu, J., M. Dehbi, G. Moeck, F. Arhin, P. Bauda, D. Bergeron, M. Callejo, V. Ferretti, N. Ha, T. Kwan, J. McCarty, R. Srikumar, D. Williams, J. J. Wu, P. Gros, J. Pelletier, and M. DuBow. 2004. Antimicrobial drug discovery through bacteriophage genomics. *Nat. Biotechnol.* **22:**185–191.

Loeffler, J. M., D. Nelson, and V. A. Fischetti. 2001. Rapid killing of *Streptococcus pneumoniae* with a bacteriophage cell wall hydrolase. *Science* **294:** 2170–2172.

Loessner, M. J. 2005. Bacteriophage endolysins—current state of research and applications. *Curr. Opin. Microbiol.* **8:**480–487.

Loessner, M. J., S. K. Maier, H. Daubek-Puza, G. Wendlinger, and S. Scherer. 1997. Three *Bacillus cereus* bacteriophage endolysins are unrelated but reveal high homology to cell wall hydrolases from different bacilli. *J. Bacteriol.* **179:** 2845–2851.

Marston, M. F., and J. L. Sallee. 2003. Genetic diversity and temporal variation in the cyanophage community infecting marine *Synechococcus* species in Rhode Island's coastal waters. *Appl. Environ. Microbiol.* **69:**4639–4647.

Matsuo-Kato, H., H. Fujisawa, and T. Minagawa. 1981. Structure and assembly of bacteriophage T3 tails. *Virology* **109:**157–164.

McPartland, J., and L. B. Rothman-Denes. 2009. The tail sheath of bacteriophage N4 interacts with the *Escherichia coli* receptor. *J. Bacteriol.* **191:** 525–532.

Mesyanzhinov, V. V. 2004. Bacteriophage T4: structure, assembly, and initiation infection studied in three dimensions. *Adv. Virus Res.* **63:**287–352.

Miller, E. S., E. Kutter, G. Mosig, F. Arisaka, T. Kunisawa, and W. Rüger. 2003. Bacteriophage T4 genome. *Microbiol. Mol. Biol. Rev.* **67:** 86–156.

Molineux, I. J. 2001. No syringes please, ejection of phage T7 DNA from the virion is enzyme driven. *Mol. Microbiol.* **40:**1–8.

Molineux, I. J. 2005. The T7 group, p. 302–314. *In* R. Calendar (ed.), *The Bacteriophages,* 2nd ed. Oxford University Press, New York, NY.

Montag, D., I. Riede, M. L. Eschbach, M. Degen, and U. Henning. 1987. Receptor-recognizing proteins of T-even type bacteriophages. Constant and hypervariable regions and an unusual case of evolution. *J. Mol. Biol.* **196:**165–174.

Mosig, G., and D. H. Hall. 1994. Gene expression: a paradigm of integrated circuits, p. 127–131. *In* J. D. Karam et al. (ed.), *Molecular Biology of Bacteriophage T4.* ASM Press, Washington, DC.

Oppenheim, A. B., O. Kobiler, J. Stavans, D. L. Court, and S. Adhya. 2005. Switches in bacteriophage lambda development. *Annu. Rev. Genet.* **39:**409–429.

Paul, J. H., J. B. Rose, S. C. Jiang, C. A. Kellogg, and L. Dickson. 1993. Distribution of viral abundance in the reef environment of Key Largo, Florida. *Appl. Environ. Microbiol.* **59:**718–724.

Paul, J. H., M. B. Sullivan, A. M. Segall, and F. Rohwer. 2002. Marine phage genomics. *Comp. Biochem. Physiol. B Biochem. Mol. Biol.* **133:**463–476.

Rohwer, F. 2003. Global phage diversity. *Cell* **113:** 141.

Rohwer, F., and R. Edwards. 2002. The Phage Proteomic Tree: a genome-based taxonomy for phage. *J. Bacteriol.* **184:**4529–4535.

Russel, M. 1995. Moving through the membrane with filamentous phages. *Trends Microbiol.* **3:**223–238.

Sano, E., S. Carlson, L. Wegley, and F. Rohwer. 2004. Movement of viruses between biomes. *Appl. Environ. Microbiol.* **70:**5842–5846.

São-José, C., R. Parreira, G. Vieira, and M. A. Santos. 2000. The N-terminal region of the *Oenococcus oeni* bacteriophage fOg44 lysin behaves as a bona fide signal peptide in *Escherichia coli* and as a *cis*-inhibitory element, preventing lytic activity on oenococcal cells. *J. Bacteriol.* **182:**5823–5831.

Schellenberger, U., L. L. Livi, and D. V. Santi. 1995. Cloning, expression, purification, and characterization of 2′-deoxyuridylate hydroxymethylase from phage SPO1. *Protein Expr. Purif.* **6:**423–430.

Schuch, R., D. Nelson, and V. A. Fischetti. 2002. A bacteriolytic agent that detects and kills *Bacillus anthracis. Nature* **418:**884–889.

Serwer, P., S. J. Hayes, S. Zaman, K. Lieman, M. Rolando, and S. C. Hardies. 2004. Improved isolation of undersampled bacteriophages: finding of distant terminase genes. *Virology* **329:** 412–424.

Short, C. M., and C. A. Suttle. 2002. Sequence analysis of marine virus communities reveals that groups of related algal viruses are widely distributed in nature. *Appl. Environ. Microbiol.* **68:**1290–1296.

Short, C. M., and C. A. Suttle. 2005. Nearly identical bacteriophage structural gene sequences are widely distributed in both marine and freshwater environments. *Appl. Environ. Microbiol.* **71:**480–486.

Simon, M. N., R. W. Davis, and N. Davidson. 1971. Heteroduplexes of DNA molecules of lambdoid phages: physical mapping of their base sequence relationships by electron microscopy, p. 313–328. *In* A. D. Hershey (ed.), *The Bacteriophage Lambda.* Cold Spring Harbor Laboratory, Cold Spring Harbor, NY.

Smith, D. E., S. J. Tans, S. B. Smith, S. Grimes, D. L. Anderson, and C. Bustamante. 2001. The bacteriophage φ29 portal motor can package DNA against a large internal force. *Nature* **413:** 748–752.

Spears, P. A., M. M. Suyemoto, A. M. Palermo, J. R. Horton, T. S. Hamrick, E. A. Havell, and P. E. Orndorff. 2008. A *Listeria monocytogenes* mutant defective in bacteriophage attachment is attenuated in orally inoculated mice and impaired in enterocyte intracellular growth. *Infect. Immun.* **76:**4046–4054.

Srinivasiah, S., J. Bhavsar, K. Thapar, M. Liles, T. Schoenfeld, and K. E. Wommack. 2008. Phages across the biosphere: contrasts of viruses in soil and aquatic environments. *Res. Microbiol.* **159:** 349–357.

Staley, J. T., and A. Konopka. 1985. Measurement of in situ activities of nonphotosynthetic microorganisms in aquatic and terrestrial habitats. *Annu. Rev. Microbiol.* **39:**321–346.

Sulakvelidze, A., and P. Barrow. 2005. Phage therapy in animals and agribusiness, p. 336–371. *In* E. Kutter and A. Sulakvelidze (ed.), *Bacteriophages: Biology and Applications*. CRC Press, Boca Raton, FL.

Summer, E. J., J. Berry, T. A. Tran, L. Niu, D. K. Struck, and R. Young. 2007. *Rz/Rz1* lysis gene equivalents in phages of Gram-negative hosts. *J. Mol. Biol.* **373:**1098–1112.

Summers, W. C. 1993. Cholera and plague in India: the bacteriophage inquiry of 1927–1936. *J. Hist. Med. Allied Sci.* **48:**275–301.

Summers, W. C. 1999. *Felix d'Herelle and the Origins of Molecular Biology,* p. 47–59. Yale University Press, New Haven, CT.

Suttle, C. A. 2005. Viruses in the sea. *Nature* **437:** 356–361.

Tao, Y., N. H. Olson, W. Xu, D. L. Anderson, M. G. Rossmann, and T. S. Baker. 1998. Assembly of a tailed bacterial virus and its genome release studied in three dimensions. *Cell* **195:**431–437.

Thacker, P. D. 2003. Set a microbe to kill a microbe: drug resistance renews interest in phage therapy. *JAMA* **290:**3183–3185.

Thiel, K. 2004. Old dogma, new tricks—21st century phage therapy. *Nat. Biotechnol.* **22:**31–36.

Valpuesta, J. M., N. Sousa, I. Barthelemy, J. J. Fernández, H. Fujisawa, B. Ibarra, and J. L. Carrascosa. 2000. Structural analysis of the bacteriophage T3 head-to-tail connector. *J. Struct. Biol.* **131:**146–155.

Wang, I. N. 2006. Lysis timing and bacteriophage fitness. *Genetics* **172:**17–26.

Weigel, C., and H. Seitz. 2006. Bacteriophage replication modules. *FEMS Microbiol. Rev.* **30:**321–381.

Williamson, K. E., M. Radosevich, D. W. Smith, and K. E. Wommack. 2007. Incidence of lysogeny within temperate and extreme soil environments. *Environ. Microbiol.* **9:**2563–2574.

Williamson, S. J., D. B. Rusch, S. Yooseph, A. L. Halpern, K. B. Heidelberg, J. L. Glass, C. Andrews-Pfannkoch, D. Fadrosh, C. S. Miller, G. Sutton, M. Frazier, and J. C. Venter. 2008. The Sorcerer II Global Ocean Sampling Expedition: metagenomic characterization of viruses within aquatic microbial samples. *PLoS One* **3:**e1456.

Wommack, K., and R. Colwell. 2000. Virioplankton: viruses in aquatic ecosystems. *Microbiol. Mol. Biol. Rev.* **64:**69–114.

Wommack, K. E., J. Ravel, R. T. Hill, J. Chun, and R. R. Colwell. 1999. Population dynamics of Chesapeake Bay virioplankton: total-community analysis by pulsed-field gel electrophoresis. *Appl. Environ. Microbiol.* **65:**231–240.

Xu, M., A. Arulandu, D. K. Struck, S. Swanson, J. C. Sacchettini, and R. Young. 2005. Disulfide isomerization after membrane release of its SAR domain activates P1 lysozyme. *Science* **307:** 113–117.

Xu, X., O. Kashima, A. Saito, H. Azakami, and A. Kato. 2004. Structural and functional properties of chicken lysozyme fused serine-rich heptapeptides at the C-terminus. *Biosci. Biotechnol. Biochem.* **68:**1273–1278.

Young, I., I. Wang, and W. D. Roof. 2000. Phages will out: strategies of host cell lysis. *Trends Microbiol.* **8:**120–128.

PHAGE-BASED METHODS FOR THE DETECTION OF BACTERIAL PATHOGENS

Mansel W. Griffiths

3

The ability of bacteriophages to infect and propagate only in specific bacteria makes them an ideal tool to use for the detection of pathogens. It allows the development of assays with an inherent specificity toward the target bacterium, and their ability to reproduce within the host cell provides an opportunity to increase the sensitivity of the assay. Bacteriophages can be used in a variety of ways to detect bacteria. They can be used as lytic agents to enable detection by monitoring the release of intracellular materials from the target. Their ability to inhibit metabolic activity and growth of the host bacterium can also be the basis of detection assays. Phages can be used as staining agents or they can be genetically modified to carry reporter genes, the products of which can be easily detected. They can be used as the biorecognition component of biosensors or the phages themselves can be used as the signal in assays through the development of immunoassays with antibodies raised against the phage in question, or by the use of molecular techniques such as PCR. The use of bacteriophages for bacterial detection has been the subject of many reviews (Marks and Sharp, 2000; Goodridge and Griffiths, 2002; Mandeville et al., 2003; Loessner, 2005a; Rees and Loessner, 2005; Rees and Dodd, 2006; Hagens and Loessner, 2007; Petty et al., 2007; Schmelcher and Loessner, 2008).

Mansel W. Griffiths, Department of Food Science and Canadian Research Institute for Food Safety, University of Guelph, 43 McGilvray Street, Guelph, Ontario N1G 2W1, Canada.

DETECTION OF INTRACELLULAR COMPONENTS RELEASED DURING PHAGE LYSIS

The lysis of bacterial cells by phage results in the release of intracellular materials, many of which can be assayed using a variety of techniques. The intracellular compound most widely used is ATP, which can be easily detected using the firefly luciferase/luciferin enzyme system (Griffiths, 1996; Sanders, 2003). The concentration of ATP, and hence the number of bacteria, in the assay is directly proportional to the quantity of light produced by the luciferase/luciferin reaction. However, the detection limit for the assay is of the order of 10^5 CFU/ml. In order to improve sensitivity, Squirrel and Murphy (1994) suggested measuring the release of intracellular adenylate kinase (AK) as a means to monitor phage lysis. In the presence of excess ADP, the equilibrium of the reaction catalyzed by AK is shifted

Bacteriophages in the Control of Food- and Waterborne Pathogens
Edited by Parviz M. Sabour and Mansel W. Griffiths, © 2010 ASM Press, Washington, DC

toward the formation of ATP, which may then be assayed using the luciferase/luciferin assay described above. This method allows the detection of approximately 10^3 cells of *Escherichia coli* or *Salmonella* within 1 or 2 h, respectively (Blasco et al., 1998). Similar results were obtained by Wu et al. (2001). The specificity of the assay could be improved by first removing the target bacterium from suspension using immunomagnetic separation (IMS) and then lysing the captured cells with phage before detection of AK activity (Squirrel et al., 2002). This method can also be used to detect the presence of bacteriophage (Luna et al., 2009).

The method has been commercialized by Alaska Food Diagnostics (Porton Down, United Kingdom) under the name fastrAK™. The fastrAK™ *Salmonella* assay was awarded a Certificate of *Performance Tested*℠ Status by the AOAC Research Institute in 2008.

Other intracellular constituents have also been used to monitor phage-mediated cell lysis, including β-galactosidase (Stanek and Falkinham, 2001). Following filtration and preincubation, a limit of detection of 1 CFU of *E. coli* K-12 per 100 ml was achieved within 6 to 8 h using an amperometric method to detect β-galactosidase activity (Neufeld et al., 2003). A similar electrochemical amperometric method was used to detect the release of α-glucosidase from *Bacillus cereus* and β-glucosidase from *Mycobacterium smegmatis* following phage infection (Yemini et al., 2007). The method enabled the detection of both bacteria at a low concentration of 10 viable cells/ml within 8 h.

Cell lysis can also be achieved using phage endolysins as opposed to the phage itself (Loessner et al., 1995; Nelson et al., 2002; Varkey et al., 2006).

DETECTION BASED ON INHIBITION OF METABOLISM AND GROWTH

As lytic agents, phages cause infections that result in the death of cells. A number of methodologies are available to monitor metabolic activity and microbial growth of microbial populations. Inhibition of microbial growth in the presence of phage can be detected in a variety of ways. When microorganisms grow they convert large molecules into small, more highly charged compounds, and this causes changes in the electrical properties of the growth medium. These changes can be detected as changes in either impedance or conductance. The retardation in the development of these changes in the presence of phage is indicative that the target organism is present. Chang et al. (2002) used an *E. coli* O157:H7-specific phage (AR1) in conjunction with a conductance method for detection of the organism. The multiplication of *E. coli* O157:H7 was inhibited by AR1, resulting in no change in conductance in a MacConkey-sorbitol medium over a 24-h period. Of the 41 strains of *E. coli* O157:H7 tested, all produced positive reactions (i.e., no change in conductance within 24 h). Fourteen of 155 strains of non-O157:H7 *E. coli* also did not produce a change in conductance within 24 h. However, only one of these strains did not utilize sorbitol and, hence, would not grow on the medium even in the absence of phage. The sensitivity and specificity of the method was 100% (41 of 41) and 99.4% (154 of 155), respectively. The technique of combining conductance measurement and phage lysis has also been shown to be an effective way of screening for the presence of *Salmonella* in skimmed milk powder, egg powder, cocoa powder, and various chocolate products (Pugh et al., 1988). McIntyre and Griffiths (1997) showed that a number of bacterial pathogens could be detected in dairy products by monitoring changes in the impedance of selective media in the presence and absence of host-specific bacteriophage. The application of these methods for pathogen detection in foods is limited because they rely on differential or selective media to identify pathogens and they can be affected by the growth of contaminating microorganisms associated with the background microflora of foods (Mandeville et al., 2003).

Again, these methods can be modified to allow detection of the bacteriophages themselves (Carminati and Neviani, 1991; Svensson, 1993, 1994).

Another approach to detect bacteria following lysis by bacteriophage was proposed by Jassim and Griffiths (2007). The method involves exposing the sample suspected to contain target cells to host-specific phage. In this case *Pseudomonas* phage NCIMB 10116 was used to detect *P. aeruginosa*. The bacteriophage was added to samples containing *P. aeruginosa* and after at least one infection cycle (about 2 h), bacteria known to be infected by the phage (helper cells) were added. After a further incubation to allow infection of the helper cells and further propagation of the phage, the bacteria were removed by filtration and the bacteria remaining on the filter were stained using the LIVE/DEAD® *Bac*Light™ Bacterial Viability Kit containing SYTO 9 and propidium iodide. Bacteria with intact cell membranes fluoresce green, whereas bacteria with phage-damaged membranes exhibit red fluorescence. By determining the ratio of red to green fluorescence, it was shown that the dead helper-cell population following phage infection was proportional to the initial number of target cells present in the original sample. Approximately 100 CFU/ml of *P. aeruginosa* could be detected within 4 h without the need for enrichment.

A colorimetric detection method based on the use of the tetrazolium dye 3-(4,5-dimethyl-2-thiazolyl)-2,5-diphenyl-2H-tetrazolium bromide (MTT) that upon reduction changes in color from yellow to purple to detect cell lysis was developed by McNerney et al. (2007). A culture of *M. smegmatis* was grown in the presence of mycobacteriophage D29, and 0.3% sulfuric acid was added to kill extracellular phage. Following neutralization, bacteriophage replication was assayed by addition of a sample to a stationary-phase culture of *M. smegmatis* mc^{2}155. Following overnight incubation at 37°C, MTT was added and the color of each well was recorded after an additional 16 h of incubation. Growth of *M. smegmatis* in the wells resulted in a color change, whereas for samples where bacterial proliferation was prevented by the lytic action of the phage, no such change was observed.

When phage DNA is injected into its host it brings about a massive transitory ion leakage, and the ion fluxes require only that the cells be physiologically viable. The phenomenon occurs within seconds after mixing the cells with sufficient concentrations of phage particles. To detect these fluxes, a micrometer-sized capacitor of titanium electrodes with a gap size of 150 nm was used to measure the electrical field fluctuations in microliter samples containing phage and bacteria (Dobozi-King et al., 2005; Seo et al., 2008). When host bacteria and phage were present together, large stochastic waves with various time and amplitude scales were observed. The authors have termed this SEPTIC (Sensing of Phage-Triggered Ion Cascades) technology and claim that it enables rapid detection and identification of live pathogenic bacteria within minutes, and with a potential sensitivity of 1 bacterium/μl.

Lysis of bacterial cells by bacteriophage can be measured noninvasively by monitoring bacterial growth and phage infection using a one-dimensional porous silicon photonic crystal and a white-light source coupled to a charge-coupled device (CCD) spectrometer (Alvarez et al., 2007). Bacterial growth on porous silicon leads to an increase in scattering efficiency that can be measured as a change in intensity of light reflected from a one-dimensional porous silicon photonic crystal. A linear relationship between bacterium concentration and intensity of light at the photonic resonance is observed, and detection limits are similar to those obtained from optical density measurements. Upon infection with phage, decreases in scattered light intensity are observed that correlate with cell lysis caused by phage replication. The authors claimed that the method was ideal for cell-based biosensing, because bacterial cells can be monitored remotely without the need for sampling.

DETECTION OF PHAGE PARTICLES

Following cell lysis by phage, phage particles are released and their numbers depend on the phage-host combination, with each infected bacterial cell potentially yielding several tens or hundreds of phages. This natural amplification cycle has formed the basis of several methods of bacterial detection in which the phage particles themselves are detected. Stewart et al. (1998) described a "phage amplification assay" for the rapid detection and identification of specific pathogenic bacteria. The technique uses plaque formation by the released bacteriophage as the assay end point, and the assay itself comprises four main steps: (i) bacteriophage infection of target bacterium, (ii) destruction of exogenous bacteriophage with a virucidal agent derived from pomegranate rind, (iii) amplification of bacteriophage within the infected host, and (iv) plaque formation on a lawn of "helper" bacteria. The addition of the virucide ensures that any resulting plaques are derived only from infected target bacteria. The method was applied for the detection of *P. aeruginosa, Salmonella enterica* serovar Typhimurium, and *Staphylococcus aureus*. The assay could be completed within 4 h with detection limits of 40 bacteria/ml for *P. aeruginosa* and 600 bacteria/ml for *Salmonella* serovar Typhimurium. It was subsequently shown that the assay could be applied for the detection of *Salmonella* in poultry meat (de Siqueira et al., 2003) and for the detection of *Listeria* in milk (Almeida et al., 2003). Alternatives to the use of pomegranate rind extract as a virucide have been investigated, and tea infusions were found to be active against certain *Salmonella* phage (de Siqueira et al., 2006).

The assay has been adapted for the detection of *Mycobacterium tuberculosis* and is marketed under the name *FASTPlaque*TB™ by Biotec Laboratories Inc. (Ipswich, United Kingdom). A version of the assay for the detection of rifampicin-resistant strains of *M. tuberculosis* is also available from the company. Singh et al. (2008) compared the *FASTPlaque* assay with acid-fast staining, culture on Lowenstein-Jensen (L-J) slants, a species-specific PCR, and an automated BACTEC™ MGIT™ 960 culture method on sputum samples from 150 patients with clinical diagnosis of pulmonary tuberculosis (TB) and 50 uninfected controls. Of the five in vitro diagnostic tests, the PCR method was found to be the most rapid, sensitive, and specific, detecting all 150 cases of pulmonary TB without any false-positive and -negative results. In comparison with PCR, the sensitivity of MGIT 960 was 90%, followed by the *FASTPlaque*TB assay (76.7%), L-J culture method (73.3%), and acid-fast staining (60%). The *FASTPlaque* failed to detect mycobacteria from the paucibacillary samples. The contamination rates for the MGIT 960, L-J, and *FASTPlaque* assays were 4, 8, and 10%, respectively. In a review of the use of phage assays for the detection of rifampicin resistance in culture isolates of *M. tuberculosis,* Pai et al. (2005) concluded that these assays appear to have high sensitivity but variable and slightly lower specificity than the standard culture-based methods when performed on isolates. In contrast, data on the performance of these assays when they are directly applied to sputum specimens are lacking. A recent study on sputum samples found that the respective sensitivities and specificities for detecting rifampicin resistance were 100% and 100% for direct BACTEC and 94% and 95% for the phage test (Ogwang et al., 2009). However, results could be achieved within 3 days for the phage assay compared with 8 days for BACTEC. Of all the assays investigated, the phage test was the least reproducible, with a failure rate of 27%. Therefore, before phage assays can be successfully used routinely, problems associated with unexplained false positives, potential for contamination, and indeterminate results must be overcome. This problem has been addressed through the development of a more effective antimicrobial supplement to control specimen-related contamination in phage-based diagnostic testing for tuberculosis (Albert et al., 2007). Other studies have shown that the *FASTPlaque*TB assay compared favorably to the BACTEC 460 TB culture method, IS6110 PCR, and other standard methods (Albert et al., 2002; Albay et al., 2003; Marei et al., 2003).

The *FASTPlaque*TB assay has been modified to detect other mycobacteria. Using the assay, viable *Mycobacterium avium* subsp. *paratuberculosis* (MAP) cells were detected as phage plaques in just 24 h (Stanley et al., 2007). The bacteriophage used in the assay is not specific for MAP, so to add specificity to this assay, a PCR-based method was introduced to amplify MAP-specific sequences from the DNA of the mycobacterial cell detected by the phage. Combining the plaque PCR technique with the phage-based detection assay allowed the rapid and specific detection of viable MAP in milk samples in just 48 h.

In a practical application of the *FASTPlaque*TB phage assay, UV light inactivation of MAP in Middlebrook 7H9 broth and whole and semiskim milk was investigated (Altic et al., 2007). The phage assay provided more rapid enumeration of surviving MAP (within 24 h) than culture on Herrold's egg yolk medium (6 to 8 weeks), although counts obtained by the plaque assay were consistently 1 to 2 $\log_{10}$ lower than colony counts. Nevertheless, UV inactivation rates for MAP derived using the phage assay and culture were not significantly different, indicating the quantitative nature of the method. A plaque assay has also been described for the detection of *Mycobacterium bovis* (Regan et al., 2001).

To allow accurate quantification of viable MAP cells, conditions of the *FASTPlaque*TB phage amplification assay were optimized based on the burst time of the D29 mycobacteriophage for different hosts to identify a suitable incubation time before plating. The D29 phage replicates more slowly in MAP than in *M. smegmatis* (used to optimize the commercial test originally) (Foddai et al., 2009). The modified protocol was able to detect 1 to 10 CFU/ml of MAP in spiked milk or broth within 48 h.

The application of bacteriophage-based detection methods for *M. tuberculosis* has been reviewed (Mole and Maskell, 2001; Kalantri et al., 2005).

A simple plaque assay has also been described for *Vibrio cholerae* (Sechter et al., 1975). Studies on artificially contaminated samples showed that counts of $\leq$10 *V. cholerae*/ml of water could be detected. In studies on 660 samples of effluents and 350 samples of vegetables, only 4 false positives were recorded.

Bacterial cells may also be first concentrated by filtration through large-pore-size electropositively charged filters before being eluted and assayed with a bacteriophage plaque assay. Such a protocol was used to detect *Salmonella* in milk and enabled the detection of 5 *Salmonella* cells/ml of milk within 24 h (Hirsh and Martin, 1984).

Favrin et al. (2001) took a different approach to the phage amplification assay. Instead of a virucide, they used IMS to separate infected host cells in an assay consisting of four main stages: (i) infection of the target bacterium with phage, (ii) capture and concentration of target cells using IMS prior to cell lysis, (iii) amplification and recovery of progeny phage, and (iv) assay of progeny phage on the basis of their effect on a healthy population of host cells (signal-amplifying cells). Phage propagation was determined using either fluorescence or optical density measurements. The detection limit of the assay in broth was $<10^4$ CFU/ml, and the assay could be performed in 4 to 5 h. The authors went on to evaluate the assay for the detection of *Salmonella enterica* serovar Enteritidis in chicken rinses, ground beef, and skim milk powder as well as the simultaneous detection of *Salmonella* serovar Enteritidis and *E. coli* O157:H7 in skim milk powder and ground beef (Favrin et al., 2003). The assay enabled the rapid, sensitive, specific, and simple detection of *Salmonella* serovar Enteritidis in all foods tested. The total assay time was approximately 10 h, and the detection limit was approximately 3 CFU/g or ml of food sample. By using another phage, *E. coli* O157:H7 could be detected in beef at the lowest inoculum level tested (2 CFU/g). Although the simultaneous detection of *Salmonella* serovar Enteritidis and *E. coli* in skim milk powder and ground beef was possible, the sensitivity of the assay decreased using this format.

The assay may be further improved by the use of bioluminescent signal-amplifying bacteria so that lysis can be monitored by measuring changes in light output (Sun et al., 2000; Auman et al., 2003; Kim et al., 2009).

An interesting twist to the plaque assay has been proposed by Ulitzur and Ulitzur (2006) to overcome the main challenge in utilizing phage for detecting specific bacteria; that is, the removal of noninfecting phages from the assay mixture. Mutant phages cannot form plaques on their bacterial hosts at concentrations below their reversion rates unless they are repaired by recombination or complementation in the host. Based on this fact, three principles were developed for phage recovery. The first method was based on using a pair of complementing phage mutants. When two different mutants of a given phage coinfect their bacterial host, completion of the infection cycle may be achieved in several ways. (i) A reversion event may result in back mutation of the mutated gene in one of the infecting phages. (ii) If the mutations are not in the same gene locus, a recombination event may occur between the two phage DNAs, resulting in restoration of the wild-type genotype in one phage. (iii) The two phages can complete their infection cycles by complementation, in which each phage "provides" the essential gene product missing in the other. While the first event occurs at very low frequencies ($\sim10^{-7}$), the second event may yield a reversion rate of about 10^{-2} to 10^{-3} under certain conditions (such as *recA* activation). Complementation, on the other hand, is a much more common event and may occur whenever two complementing phages are introduced into the same host cell. The phenotypically rescued single phages will not be able to complete another lytic cycle unless they coinfect the next bacterial host together with another phenotypically rescued phage or if two different phage mutants are recombined to create an infective phage target. On solid medium, where phage and bacterial-cell diffusion is limited, this event occurs repeatedly until a visible plaque is formed. This led to three new methods being proposed for the rapid and simple detection of specific bacteria, based on plaque formation as the end point of the bacteriophage lytic cycle. Different procedures were designed to ensure that the resulting plaques were derived only from infected target bacteria. In the first approach, a pair of amber mutants that could not form plaques at concentrations lower than their reversion rate underwent complementation in the tested bacteria so that the resulting number of plaques formed was proportional to the concentration of the bacteria that were coinfected by these bacteriophage mutants. Using the *Salmonella* phage Felix-O1 to generate mutants, a good correlation was found between the number of bacteria in the tested sample (from 5 cells up to 10^6 cells) and the number of plaques formed. In the second approach, pairs of temperature-sensitive mutants were allowed to coinfect their target bacteria at the permissive temperature, followed by incubation of the plates at the restrictive temperature to avoid bacteriophage infection of the host cells. This method allowed the omission of centrifuging and washing the infected cells. Only bacteriophages that recovered by recombination or complementation were able to form plaques. The detection limit was 1 to 10 viable *Salmonella* or *E. coli* O157 cells after 3 to 5 h. The final approach involved the use of UV-irradiated phages that were recovered by photoreactivation and/or SOS repair mediated by target bacteria and plated on a *recA uvrA* bacterial lawn in the dark to avoid recovery of noninfecting bacteriophages. The last approach was considerably less sensitive than the other two. However, all of the procedures could be used to determine the antibiotic susceptibility of the target bacteria by preincubation of the bacterial cells with antibiotics prior to bacteriophage infection. Bacteria sensitive to the antibiotic lost the ability to form infectious centers.

Several other methods can be used for the detection of amplified phage. Hirsh and Martin (Hirsh and Martin, 1983b; Crane et al., 1984) infected *Salmonella* with Felix-O1 and detected the released phage particles by high-performance liquid chromatography. Although the assay could be performed in 3 h, 3×10^6 *Salmonella* cells/ml were required to produce a detectable response. When an enrichment step was combined with filtration using electropositively charged large-pore filters to remove bacterial cells, it was possible to detect as few as 5 *Salmonella*/ml in milk (Hirsh and Martin, 1983a). Capillary electrophoresis can also be used for the detection of virus particles (Kremser et al., 2009).

Matrix-assisted laser desorption ionization mass spectrometry (MALDI-MS) has been used to rapidly identify and detect bacteria. This technique requires minimal sample preparation and is simple to perform. However, its use is generally limited to pure bacterial cultures at concentrations greater than 1.0×10^6 cells/ml and it requires expensive equipment. A bacterial detection method has been described that integrates IMS with bacteriophage amplification prior to MALDI-MS analysis to produce a molecular weight signal for the virus capsid protein (Madonna et al., 2003). The presence of *E. coli* in broth was determined in less then 2 h total analysis time with a limit of detection of 5.0×10^4 cells/ml. MALDI time-of-flight MS could also be used for the simultaneous analysis of multiple target microorganisms in conjunction with bacteriophage amplification (Rees and Voorhees, 2005). Following infection of target bacteria with specific bacteriophages, proteins contained in the progeny phage are utilized as a secondary biomarker for the target bacterium. *E. coli* when mixed with MS2 and MPSS-1 phages specific for *E. coli* and *Salmonella* spp., respectively, at levels below their corresponding detection limits produced only the protein (13.7 kDa) characteristic of the MS2. Similarly, *Salmonella* spp. when mixed with the two phages produced only a protein (13.5 kDa) characteristic of MPSS-1. When the two bacteria and the two phages were mixed together, proteins characteristic of MS2 and MPSS-1 were produced, indicating that both bacteriophages had been amplified.

A technique similar to the enzyme-linked immunosorbent assay (ELISA) termed the phage-linked immunoabsorbent assay has been developed (Block et al., 1989). Bacteriophage M13, containing the *lacZ* gene, was used as the "reporter" molecule in an immunoassay. Briefly, herpes simplex virus-infected cells were incubated with a mouse monoclonal antibody specific for herpes simplex virus antigen, followed by rabbit anti-mouse serum and mouse anti-M13 serum. Immune complexes were incubated with viable M13 phage. M13 binding was due to the presence of M13 antibodies and the phage was subsequently recovered by elution at pH 11. The recovered phage was quantified by either plaque assay or by an assay for phage-induced β-galactosidase activity in appropriate *E. coli* strains. Therefore, M13 served as a "bioamplifiable tag" to antibody, similar to the way in which enzymes are used in the ELISA. The M13 signal can be amplified by infection of susceptible bacteria, giving the opportunity for a sensitive assay for bacteria.

A lateral flow immunoassay for the detection of *S. aureus*-specific phages has been marketed by MicroPhage Inc. (Longmont, CO). This assay has been tested on 33 wound swabs, which were placed in the test broth containing the bacteriophage (Smith et al., 2009). Phage amplification was detected by plaque assay and by lateral flow immunoassay. Four samples had been confirmed as *S. aureus* positive, and all of these were positive by the MicroPhage test. Testing revealed four additional positive samples, and three of these were detected by the MicroPhage test. The one false-negative sample contained very low bacterial numbers (90 CFU). Overall, seven of eight positives were detected by the MicroPhage test and three false positives were reported, all of which were coagulase-negative *Staphylococcus*. Two of the three false positives showed high levels of phage amplification, and one did not, suggesting cross-reactivity as the cause of the false-positive result.

A phage replication-competitive ELISA (PR-cELISA) was developed for the detection of multiple-antibiotic-resistant *Salmonella* serovar Typhimurium DT104 by Guan et al. (2006). In the PR-cELISA procedure, a phage, BP1, was inoculated into a log-phase bacterial culture at a ratio of 1:100. After a 3-h incubation of the mixture, BP1 replication was measured by cELISA based on the competitive binding between BP1 and biotinylated BP1 to serovar Typhimurium smooth lipopolysaccharide. Among the 84 *Salmonella* strains and 9 non-*Salmonella* strains that were tested by PR-cELISA, BP1 detected 39 of 40 serovar Typhimurium strains, 2 of 10 non-Typhimurium somatic group B strains, and 5 of 18 *Salmonella* somatic group D strains. With the addition of chloramphenicol to the culture medium, the PR-cELISA detected all 27 multiple-antibiotic-resistant serovar Typhimurium DT104 but none of the other strains tested.

Another format for the immunological detection of bacteriophage has been proposed by Bange et al. (2007). They proposed a microbead-based sandwich immunoassay for MS2 bacteriophage using an interdigitated array electrode. After antibody-assisted capture of MS2 bacteriophage using paramagnetic microbeads, a β-galactosidase-labeled secondary antibody was used to convert *p*-aminophenyl-β-D-galactopyranoside into the redox-active *p*-aminophenol. Amperometric detection of *p*-aminophenol with the interdigitated array electrodes enabled detection of MS2 concentrations as low as 10 ng/ml.

Spectroscopic methods have also been employed for the detection of bacteriophage. For example, native fluorescence spectroscopy was used for the detection of bacteriophage targeting pseudomonads (Alimova et al., 2004). Within proteins, the structural environment of the tryptophan residue differs, and the differences are reflected in their spectroscopic signatures. It was observed that the peak of the tryptophan emission from pseudomonad phages was at 330 nm, a significantly shorter wavelength than the residue in the pseudomonad host cells. This allowed the viral attachment process and subsequent lytic release of progeny phage particles to be monitored by measurement of the tryptophan emission spectra during the infection process. This work demonstrates that fluorescence may offer a novel tool to detect phages and hence bacterial cells. Another spectroscopic method, Raman spectroscopy, enabled the detection of as few as 100 bacteriophage particles (Goeller and Riley, 2007).

Ayer et al. (2008) described a method based on pyrolysis gas chromatography differential mobility spectrometry that enabled detection of T4 bacteriophage and suggested that their portable sensor can be used along with the sophisticated data mining method as a fast, reliable, and accurate tool for the recognition of viruses.

Many of the methods that use detection of phage as the end point suffer from a lack of sensitivity and require methods to remove endogenous phage particles to reduce the background signal. The lack of sensitivity can be addressed by a further amplification of the phage signal using PCR targeting sequences unique to the phage. Such an approach has been proposed for the detection of lactococcal phages in milk (Craven et al., 2006). The PCR method initially comprised a phage isolation and concentration step, in which *Lactococcus lactis* strains were added to the milk sample to be tested. Any lactocococcal phages present in the sample would attach to the host cells and could be removed by centrifugation following clarification of the milk. Phage DNA was extracted by boiling and used as a template for PCR. Phages from the three main lactococcal dairy species (936, c2, and P335) were detectable at levels of 10 to 100 PFU/ml. Quantitative PCR has also been used to enumerate phages in lysates with no prior DNA purification, and the method could detect below 1 PFU (Edelman and Barletta, 2003). Good results were obtained when unknown concentrations of a variety of phages were examined.

This technique can be adapted for use as a method to detect the host bacterium. Using genetically modified T4 phage containing genes encoding either a biotin carboxyl carrier protein *(bccp)* or a cellulose-binding module *(cbm)* fused to the *soc* gene encoding an outer capsid protein, one group (Tolba et al., 2010) developed a real-time PCR protocol targeting the unique sequence of the fused gene. Following infection and propagation of the phage in *E. coli* B, they detected the released progeny phage with the PCR assay. They were able to detect as few as 800 *E. coli* cells within 2 h. A similar approach was used to detect *Ralstonia solanacearum* (Kutin et al., 2009). These authors showed that a combination of phage amplification and quantitative PCR consistently detected approximately 3.3 CFU/ml after an hour's incubation with 5.3 $\times$ 10^2 PFU of M_DS1 phage/ml. The problem with amplification of noninfective phage was overcome by using low initial phage numbers. The detection of phage-specific genome sequences following amplification in the host was also used by Reiman et al. (2007). They combined real-time PCR, using primers specifically designed for γ phage, with the highly specific γ phage amplification into one simple assay to indirectly detect *Bacillus anthracis*. This method detected a starting *B. anthracis* concentration of 207 CFU/ml in less than 5 h. No strategy was described to prevent amplification of noninfective phage.

A PCR-based method targeting the antireceptor gene sequence of *Streptococcus thermophilus* bacteriophages was able to detect 10^5 PFU/ml directly in milk (Binetti et al., 2005). This suggests that PCR detection of amplified phages could be used to detect bacterial hosts directly in foods. Similar results were obtained for the real-time PCR detection of *Lactobacillus delbrueckii* bacteriophages (Cruz-Martin et al., 2008).

For multiplex detection, phage DNA can be isolated from infected bacterial cultures and detected by a combination of bead-based sandwich hybridization with enzyme-labeled probes and detection of the enzymatic product using silicon chips (Gabig-Ciminska et al., 2004). The assay resulted in specific signals from all four phages tested without significant background noise. Although high sensitivity was achieved in a 4-h assay time, a useful level of sensitivity (10^7 to 10^8 PFU) is achievable within 25 min. A multiplex DNA chip technique involving a mixture of probes allows for detection of various types of phages in one sample.

PHAGE AS STAINING AGENTS

Fluorescent stains in conjunction with bacteriophage can be used to label nucleic acid within infected cells. Sanders et al. (1991) reported the use of two fluorescent dyes, chromomycin A3 and Hoechst 33258, to bind and label GC- and AT-rich regions, respectively, of phage T4 DNA present inside infected *E. coli* cells, thus presenting a technique for the detection of host cells. In an extension of this principle (Hennes and Suttle, 1995; Hennes et al., 1995), viruses fluorescently stained with YOYO-1 or POPO-1 were used as probes to label, identify, and enumerate specific strains of bacteria and cyanobacteria in mixed microbial communities. Cells to which the stained viruses adsorbed were easily distinguished from non-host cells using epifluorescence microscopy. This technique was applied to the detection of *E. coli* O157:H7 by Goodridge et al. (1999a). They combined IMS and a fluorescently stained bacteriophage for detection of *E. coli* O157:H7 in broth. The assay has two steps: *E. coli* O157:H7 is first removed by IMS and the cells are then labeled with a highly specific, fluorescently stained bacteriophage. This two-step configuration serves to decrease the likelihood of false-positive results. Fluorescence was monitored with either flow cytometry or a modified direct epifluorescent-filter technique. The fluorescent bacteriophage assay (FBA) was capable of detecting 10^4 cells/ml when flow cytometry was used, and a detection limit of between 10^2 and 10^3 cells/ml was obtained with the modified direct epifluorescent-filter technique. However, the latter was found to be an unreliable method for estimating bacterial numbers. These researchers showed that the FBA could be used as a rapid, preliminary screen for the presence of *E. coli* O157:H7 in ground beef mince and raw milk (Goodridge et al., 1999b) after samples of ground beef or raw milk were inoculated with *E. coli* O157:H7 and analyzed using FBA in conjunction with flow cytometry. The limit of detection in the beef was 2.2 CFU *E. coli* per g following a 6-h enrichment period. For raw milk the limit of detection was 10^1 to 10^2 CFU/ml after 10 h of enrichment. Other stains such as SYBR gold have been used to label bacteriophage (Mosier-Boss et al., 2003).

A flow cytometric method for the direct detection of bacteriophage following staining with SYBR green has been described (Brussaard, 2009). Other approaches use morphological changes induced in the host cell following phage infection. For example, infection of *L. lactis* by phage results in a change from cells in chains toward single cells, and late in the infection cycle cells with low-density cell walls appear and can be detected by flow cytometry (Michelsen et al., 2007).

SYBR green staining of the phage was used successfully by Lee et al. (2006) to detect *Microlunatus phosphovorus,* a bacterium with high levels of phosphorus-accumulating activity and phosphate uptake and release activities, in activated sludge.

To improve the intensity of the fluorescent signal, tagging of bacteriophage with quantum dots has been investigated. Very simply, quantum dots are semiconductors that can be excited, for example by UV light, to states that emit light at a different wavelength. Edgar et al. (2006) engineered a phage (T7) to display a small peptide, which can be biotinylated and fused to the major capsid protein. Following in vivo biotinylation of the engineered host-specific bacteriophage, it can be conjugated to streptavidin-coated quantum dots. Biotinylated phage bound to quantum dots were detected and analyzed by both flow cytometry and epifluorescence microscopy. To detect a small number of a given type of bacterium among several different bacterial cells, a culture with mixed bacterial strains on which phage T7 cannot propagate was studied. Mixtures of 2×10^6 cells of each of the strains with different numbers of cells of *E. coli* (from 10 to 10^7 cells per ml) were generated. The signal derived from the non-*E. coli* strains was not significantly different from the background. The signal from as few as 10 T7 host *E. coli* cells was significantly higher than the signal in controls with no *E. coli* or no phage added to the culture. In a practical application of the assay, water samples from the Potomac River were tested. The presence of at least 20 *E. coli* cells in 1 ml of the sample could be detected in approximately 1 h.

Real-time binding of phage to host can be visualized using a CytoViva optical light microscope, and this has been used as a probe for the detection of methicillin-resistant *S. aureus* (MRSA) (Guntupalli et al., 2008). Phage monolayers prepared on glass using the Langmuir-Blodgett technique were exposed to host cells at concentrations ranging from 10^6 to 10^9 CFU/ml, and observed for real-time binding. Results indicate that Langmuir-Blodgett monolayers possessed high levels of elasticity and near-instantaneous MRSA-phage binding produced 33%, 10%, 1.1%, and 0.09% coverage of the substrate at MRSA concentrations of 10^9, 10^8, 10^7, and 10^1 CFU/ml, respectively. No binding of the phage was observed in the presence of *Salmonella* serovar Typhimurium or *Bacillus subtilis*.

Radiolabeled bacteriophages have been proposed as a way to detect and monitor infections in vivo (Rusckowski et al., 2008).

Bacteriophages have also been proposed as a way to visualize eukaryotic cells (Aalto et al., 2001; Jia et al., 2007). Molecular mimicry occurs between the polysialic acid polysaccharide of bacterial pathogens causing sepsis and meningitis, and the carbohydrate units of the neural cell adhesion molecule (Aalto et al., 2001). Mutants of bacteriophage (active against meningitis-causing bacteria) with catalytically disabled endosialidase, which binds to but does not cleave polysialic acid, could recognize and bind to bacterial and eukaryotic polysialic acid. In a nitrocellulose dot blot assay the mutant bacteriophages, but not the wild-type phages, remained specifically bound to polysialic acid-containing bacteria including *E. coli* K1 and K92, group B meningococci, *Mannheimia haemolytica* A2, and *Moraxella nonliquefaciens*. The mutant phages together with secondary, fluorescently labeled antiphage antibodies could be used in conjunction with microscopy as a probe for the eukaryotic polysialic acid and could be used to detect the molecularly mimicked polysialic acid of the neural cell adhesion molecule in affected tissues.

PHAGE AS BIOSENSORS

The use of bacteriophage as biosensing agents has been comprehensively reviewed (Mao et al., 2009) and some applications are described below.

Tamarin and colleagues (Tamarin et al., 2003a, 2003b) constructed a Love wave biosensor, a special type of surface acoustic wave sensor that uses shear horizontal waves guided in a layer on the surface of the sensor to reduce energy dissipation of the acoustic wave to the fluid and to increase the surface sensitivity. They incorporated a monoclonal antibody targeting the M13 bacteriophage on the SiO_2 guiding surface of the device. The M13 bacteriophage could then be detected within 2 h by measuring wave changes when the phage bound to the antibody. However, large numbers of bacteriophage (10^{10} to 10^{12} PFU/ml) are required to obtain a reliable signal. These researchers did not describe the further application of this biosensor for bacterial detection. Other researchers have indicated that Love acoustic wave immunosensors can be used to detect bacteriophage (Moll et al., 2008).

Using a similar approach, Uttenthaler et al. (2001) coated the surface of a high-resonant-frequency quartz crystal microbalance with M13 antibodies and were able to detect phage binding by monitoring changes in mass. Using a 56-MHz high-fundamental-frequency quartz crystal, they were able to detect 5×10^6 M13 PFU/ml. An increased sensitivity is possible by using phage-displayed proteins on its coat to interact with ligands immobilized on the surface of the quartz crystal microbalance. By this technique it was possible to detect the capture of 20 phage particles (Dultsev et al., 2001).

Another biosensor for phage employs a protein analysis technique that combines high-spatial-resolution imaging ellipsometry with multiprotein microarray (Qi et al., 2009). The surface of a silicon wafer was coated with avidin, which then binds a biotinylated anti-M13 antibody. Phage M13KO7 are specifically captured by the ligand when phage M13KO7 solution passes over the surface, resulting in a significant increase of mass surface concentration of the anti-M13 binding phage M13KO7 layer. This change in mass can be detected by imaging ellipsometry with a sensitivity of 10^9 PFU/ml.

Biosensors were prepared for the rapid detection of *Salmonella* serovar Typhimurium based on affinity-selected filamentous phage physically adsorbed to piezoelectric transducers (Olsen et al., 2006). Quantitative deposition studies indicated that approximately 3×10^{10} phage particles/cm^2 could be irreversibly adsorbed on the surface of the transducer in 1 h at room temperature. Specific bacterial binding resulted in rapid ($<$180 s) resonance frequency changes that allowed detection of 10^2 cells/ml.

Bacteriophages were immobilized covalently onto functionalized screen-printed carbon electrode microarrays through electrochemical oxidation in acidic media of 1-ethyl-3-(3-dimethylaminopropyl)carbodiimide (EDC) by applying a potential of +2.2 V to the working electrode (Shabani et al., 2008). Immobilization of T4 bacteriophage onto the screen-printed carbon electrodes was achieved via EDC by formation of amide bonds between the protein coating of the phage and the electrochemically generated carboxylic groups at the carbon surface. The immobilized T4 phages were then used to specifically detect *E. coli* bacteria. Impedance shifts of the order of 10^4 Q were observed due to the binding of *E. coli* to the T4 phages. No significant change in impedance was observed for control experiments using immobilized T4 phage in the presence of a non-host cell. Impedance variations were at a maximum after 20 min of incubation, indicating onset of lysis, and a detection limit of 10^4 CFU/ml was achievable.

In another approach to sensing the presence of bacteriophage, conducting polymer organic electrodes modified with phage host cells (*S. enterica* serovar Newport) were fabricated by chemical deposition of polypyrrole onto the surface of a microporous polycarbonate membrane (Dadarwal et al., 2009). *Salmonella* host cells were absorbed onto the surface of the electrodes and reacted with infecting bacteriophage in LB broth, Miller, at 37°C. Upon bacteriophage-mediated host cell lysis, the impedance of the supporting polypyrrole electrode increased, and from dose-response curves it was found that the sensor could detect 10^3 PFU/ml within 4.5 h, although no linear correlation between phage concentration and sensor response was observed.

A phage-based magnetoelastic sensor for the detection of *Salmonella* serovar Typhimurium was constructed using a filamentous bacteriophage as the biorecognition element (Lakshmanan et al., 2007a). The phage was immobilized onto the surface of the sensors by physical adsorption. The frequency changes caused by binding of serovar Typhimurium to phage immobilized on the sensor surface allowed detection of 10^3 CFU/ml in a sensor with dimensions $1 \times 0.2 \times 0.015$ mm. The sensor had similar detection limits (5×10^3 CFU/ml) when applied to water and skim milk (Lakshmanan et al., 2007b). The sensor also performed well for the detection of serovar Typhimurium in mixed culture, although the sensitivity and the frequency changes were slightly lower for the mixtures (Lakshmanan et al., 2008). A magnetoelastic biosensor has been described based on a genetically modified phage capable of binding *B. anthracis* spores (Wan et al., 2007; Johnson et al., 2008).

Microcantilever-based sensors are the simplest microelectromechanical systems devices. The adsorbed mass of the analytes causes the nanomechanical bending of the microcantilever, and the change in mass on the microcantilever surface due to the binding of the analyte molecules is directly proportional to the deflection of the microcantilever. Thus, qualitative as well as quantitative detection of analytes can be performed. The magnetostrictive microcantilever (MSMC) employs magnetostrictive materials as the actuating layer. Due to the magnetic and magnetostrictive effects, the MSMC can be driven and sensed wirelessly. This is the principal advantage of MSMCs over other microcantilevers. A filamentous phage against serovar Typhimurium was used as the biorecognition element in an MSMC-based biosensor (Fu et al., 2007). The detection of serovar Typhimurium in water demonstrated that the MSMC works in liquid; however, the bacterium was present at levels of 10^8 CFU/ml.

Simply put, phage display is the display of foreign proteins or peptides on the surface of a filamentous phage and has been used for the identification of molecular probes for targeting various biological structures. The probes can be selected from a very large number of clone libraries of recombinant phages expressing on their surface a vast variety of peptides and proteins, including antigen-binding fragments of antibodies. Although phage display traditionally is focused more on the development of medical preparations and studying molecular recognition in biological systems, there are some examples of its successful use for detection (Petrenko and Vodyanoy, 2003; Gulig et al., 2008). The selected phage itself can act as a probe in detector devices, as is the case with the β-galactosidase biosensor described by Nanduri et al. (2007a, 2007b). These researchers produced a phage displaying peptide that binds the enzyme on its surface and used it as the recognition element in a quartz crystal microbalance. The sensor was able to detect the enzyme in the range of 0.003 to 210 nM with a response time of 100 s. The affinity of the phage and enzyme was similar to that for a monoclonal antibody raised against β-galactosidase. This approach could be developed further to detect molecules specific to target bacteria.

Protein microarrays prepared directly from phage displaying antibody clones have been prepared and used to discriminate between recognition profiles of samples from healthy donors and leukemia patients (Bi et al., 2007). It is possible that this approach could be used to develop microarrays for the detection of pathogens.

In "landscape" phage, as in traditional phage-display constructs, foreign peptides or proteins are fused to coat proteins on the surface of a filamentous phage particle. Unlike conventional constructs, however, each virion displays thousands of copies of the peptide in a repeating pattern, occupying a major fraction of the viral surface. These landscape phages have been proposed as biorecognition elements for biosensors, where they may be superior to antibodies, since they are inexpensive, highly specific and strong binders, and resistant to high temperatures and environmental stresses (Petrenko, 2008).

The specific binding between host bacterium and phage can be detected using an electro-optical analyzer (ELUS EO) (Bunina et al., 2004). The analyzer is based on the polarizability of microorganisms, which depends strongly on their composition, morphology, and phenotype. The combination of the EO approach with a phage as the recognition element provided the basis for the mediatorless detection of phage-bacteria complex formation. Using the host-phage combination *E. coli* XL-1 and M13K07, a strong and specific EO signal was produced upon infection of the host. The signal was specific even in the presence of non-host cells. It was concluded that this technique was advantageous because bacteria did not need to be isolated from samples, the phage-host interaction is specific, exogenous substrates and mediators are not required for detection, and it is suitable for use with any bacteria-specific phages.

The detection limit of the majority of the described biosensors suggests that they could not be coupled with a phage amplification protocol to measure low numbers of bacteria.

PHAGE AS TRANSDUCING AGENTS

Arguably, the most researched application of bacteriophages for the detection of bacterial pathogens has involved their use as transducing agents to deliver genes whose products can generate a measurable signal. Several genes have been investigated for reporting, including those encoding bacterial and eukaryotic luciferases, the ice nucleation protein, green fluorescent protein (GFP), and β-galactosidase.

Originally proposed by Ulitzur and Kuhn in 1989 (Ulitzur et al., 1989; Ulitzur and Kuhn, 1989), the technology involves the introduction of DNA carrying a reporter gene into a target bacterium via a phage. Once the reporter gene has been introduced to the bacterium, it is expressed, thereby allowing bacterial cells to be rapidly identified. Since phages need host cells to replicate, the phages will remain "dark" (i.e., the reporter gene will not be expressed until the phage DNA has infected the host). Therefore, expression of the reporter gene is indicative of the presence of the infected organism.

Luciferase Reporter Systems

Most transducing phage detection assays have employed bacterial luciferase *(lux)* or firefly luciferase *(luc)* genes as the reporter system. These genes encode an enzyme system that converts chemical energy to light using a reaction that involves an oxidant (usually O_2), an enzyme (luciferase), a substrate (luciferin for *luc* or a long-chain aldehyde for *lux*), a cofactor (flavin mononucleotide *[lux]* or ATP *[luc]*), a cation, and an ancillary protein (Goodridge and Griffiths, 2002). In the majority of applications, the structural genes (*luxA* and *luxB*) of the luciferase from luminescent bacteria have been used.

Ulitzur and Kuhn (1987) first showed that phages carrying a reporter gene could be used to detect microorganisms. These researchers engineered phage λ Charon 30 to carry the *lux* genes from *Vibrio fischeri*. Using this lux^+ phage, it was possible to detect between 10 and 100 *E. coli* cells/ml of milk or urine within 1 hour. Since then, reporter phage assays have been developed that are capable of detecting several pathogenic food-borne microorganisms.

E. COLI O157:H7

Waddell and Poppe (2000) constructed a mini-transposon (Tn*10luxABcam*/*Ptac*-ATS) in order to develop a luciferase-transducing bacteriophage for detecting *E. coli* O157:H7. The transposon was designed to deliver a 3.6-kb insertion that conferred *n*-decanal-dependent bioluminescence using *luxAB* from *Vibrio harveyi*. A temperate bacteriophage (xV10) that infected common phage types of *E. coli* O157:H7 was mutagenized as a prophage in *E. coli* O157:H7 strain R508. The phage xV10::*luxABcamA1-23* was rescued from the strain by propagating it on a strain lacking the bacteriophage and the vector containing the transposon. Following transduction with the phage, *n*-decanal-dependent bioluminescence was measurable above background approximately 1 h after infection with 10^6 *E. coli* O157:H7 cells/ml.

A *Luciola mingrelica luc* recombinant of bacteriophage broadly specific for *E. coli* O157, designated AR1, has been obtained by simultaneously infecting wild-type *E. coli* O157:H7 with a pBluescript II SK(+/−) phagemid containing the *luc* gene (Promega Corporation, Madison, WI) and AR1. The transducing phage was capable of detecting 10^6 *E. coli* O157:H7 cells/ml (Pagotto et al., 1996).

SALMONELLA SPP.

A lux^+ P22 phage has been engineered and used in experiments to infect *Salmonella* serovar Typhimurium LT2 (Stewart et al., 1989). When light emission following infection was measured with a luminometer, as few as 10^2 serovar *Typhimurium* cells could be detected, even when they were present in mixed culture at a ratio of 1 *Salmonella* cell to 10^6 cells of other bacteria. Serovar Typhimurium could also be detected with the *lux*-modified P22 phage in environmental samples including water, soil, and sewage sludge (Turpin et al., 1993). These researchers adopted a most-probable-number (MPN) technique based on a 15-tube test method, consisting of 5 tubes containing 10 ml, 1 ml, or 0.1 ml of sample in buffered peptone water. After overnight incubation at 37°C, subsamples from each tube were transferred to Luria broth and 6.9×10^9 PFU of the *lux*-modified P22 phage was added. Light output was measured with a luminometer after 90 min of incubation at 30°C. There was an excellent correlation between the MPN obtained by the luminescence method and plate count for all samples, and results were achieved within 24 h. No false-positive or false-negative results were obtained with the lux^+ phage method in any of the samples tested.

Recombinant bacteriophages specific for *Salmonella* spp. and containing bacterial luciferase genes were constructed, which caused the host cells to luminesce when mixed with group B and group D and some group C *Salmonella* spp. (Chen and Griffiths, 1996). The luminescence could be detected using a photon-counting CCD camera, a luminometer, or X-ray film. With 6 h of preincubation, as few as 10 CFU *Salmonella* cells/ml in the original sample could be detected. When *Salmonella* cells were present at levels at or above 10^8 CFU/ml, they could be detected in 1 to 3 h without the need for preincubation. The phage-based assay could also be carried out on Petrifilm *E. coli* count plates and the light emission detected within 24 h. Using CCD imaging systems to visualize bioluminescence, it was possible to detect *Salmonella* contamination directly in whole eggs by direct addition of recombinant bacteriophages into the eggs.

Based on the use of a recombinant P22::*luxAB* phage, a simple method was developed for the fast and inexpensive detection of serovar Typhimurium (Thouand et al., 2008). A simple four-step bioassay was tested and optimized and gave a detection limit of 5×10^2 CFU *Salmonella*/ml for broth cultures, but this increased to 1.5×10^4 when the bacterial cells were in a complex matrix. The bioassay was applied to the detection of serovar Typhimurium LT2 in 14 different feed and environmental samples (including duck feed, litters, and feces). It was possible to detect serovar Typhimurium LT2 in all samples within 16 h.

To improve the production of recombinant phage reagents for the detection of *Salmonella,* a plasmid was constructed to contain a segment of Felix O1 bacteriophage DNA with two adjacent genes and their flanking sequences (Kuhn et al., 2002). These genes were replaced with genes encoding the *luxA* and *luxB* genes as well as a gene encoding a tRNA that recognizes amber codons. Using homologous recombination, a bacteriophage was produced that was capable of propagation only in a host that synthesizes a missing protein required for growth. As a result, loss of foreign DNA from the recombinant bacteriophage during reagent preparation is prevented. The recombinant phage was capable of detecting almost all *Salmonella* serovars (Kuhn, 2007).

LISTERIA MONOCYTOGENES

Listeriaphage A511 is a genus-specific, virulent myovirus that infects 95% of *L. monocytogenes* serovar 1/2 and 4 cells. Loessner et al. (1996) constructed a recombinant phage A511::*luxAB,* which carries the gene for a fused *V. harveyi* LuxAB protein inserted immediately downstream of the major capsid protein gene *(cps)* of the phage. The powerful *cps* promoter ensures efficient transcription of the gene at 15 to 20 min postinfection. The A511::*luxAB* phage was used to directly detect *Listeria* cells, and following infection and a 2-h incubation period, numbers as low as 5×10^2 to 10^3 cells per ml could be detected using a luminometer. The assay sensitivity was further improved by including an enrichment step prior to the *lux* phage assay. In this case, a detection limit of <1 cell of *L. monocytogenes* Scott A/g of artificially contaminated lettuce was achieved. The use of the recombinant phage was evaluated for the detection of viable *Listeria* cells in contaminated foods and environmental samples (Loessner et al., 1997). With a short preenrichment step of 20 h, the assay was capable of detecting ≤1 cell of *L. monocytogenes* Scott A per g in artificially contaminated ricotta cheese, chocolate pudding, and cabbage. The detection limit was slightly higher (≥10 cells per g) in foods having a large and complex microbial background flora, such as minced meat and soft cheese. Of 348 potentially contaminated natural food and environmental samples, 55 were *Listeria* positive by the *lux* phage method, whereas the standard plating procedure detected 57 positive samples. Overall, both methods performed similarly but the minimum time required for detection of *Listeria* with the luciferase phage assay was 24 h, compared with 4 days for the standard plating method. The sensitivity of the bioluminescent phage assay was further improved by the adoption of an MPN technique.

S. AUREUS

A battery of five phages were identified that could infect all strains of *S. aureus* tested, whether they were isolated from mastitic cows or from foods implicated in food-borne illness. A plasmid incorporating the *luxAB* genes fused to a staphylococcal cadmium-resistant gene promoter was introduced into staphylococcal phages by homologous recombination (Pagotto et al., 1996). Using these recombinant phages, it was possible to detect 10^6 *S. aureus* cells/ml in broth cultures. The luminescent strains obtained after transduction could be used to test for antibiotic susceptibility.

B. ANTHRACIS

The reporter phage was engineered by integrating the bacterial *luxA* and *luxB* reporter genes into a nonessential region of the lysogenic Wβ phage genome (Schofield and Westwater, 2009). This resulted in a phage that was capable of specifically infecting and conferring a bioluminescent phenotype to viable cells of *B. anthracis*. When the phage and cells were simply mixed, a bioluminescent signal was evident 16 min after infection of vegetative cells. Although the strength and time required to generate the signal was dependent on the number of cells present, 10^3 CFU/ml of *B. anthracis* was detectable within 60 min. The Wβ::*luxAB* phage was also able to transduce a bioluminescent signal to germinating spores of the organism within 60 min.

INDICATOR ORGANISMS

Indicator microorganisms are nonpathogenic microorganisms present in significant numbers within a food, which can be related through increasing count to the increased probability of pathogen contamination. A recombinant *lux*$^+$ bacteriophage assay could detect enteric indicator bacteria without recovery or enrichment in 50 min, provided that they were present at levels $>10^4$ CFU/g or cm^2 (Kodikara et al., 1991). After a 4-h enrichment, samples having enteric counts of 10 CFU/g or cm^2 were distinguishable from background.

M. TUBERCULOSIS

Reporter phages may help to reduce the time required for establishing antibiotic sensitivity of *M. tuberculosis* from weeks to days, which would accelerate screening for new antituberculosis drugs (Jacobs et al., 1993). In the presence of an antibiotic to which the host cell is sensitive, the infecting phage cannot complete its lytic cycle, and the amount of light emitted is therefore reduced or abolished. The majority of reporter phage testing to determine antibiotic susceptibility has centered on the mycobacteria. Jacobs et al. (1993) proposed the use of a recombinant reporter mycobacteriophage carrying a firefly luciferase reporter gene (*lux*$^+$ phage L5:*FFlux*) as a tool for the rapid determination of *M. tuberculosis* drug susceptibilities. In this research, mutants of *M. bovis* BCG were selected that were resistant to rifampicin, streptomycin, or isoniazid. The results established that L5:*FFlux* is very efficient at revealing the patterns of drug susceptibility or resistance of *M. tuberculosis* strains, and other researchers have shown that L5:*FFlux* can also determine the antibiotic susceptibility profile of other mycobacteria (Sarkis et al., 1995; Carriere et al., 1997; Riska and Jacobs, 1998). The reporter phage can also be used for the detection of *M. tuberculosis,* allowing the detection of as few as 120 BCG cells in as little as 1 day (Carriere et al., 1997). The assay has also been modified to incorporate *p*-nitro-α-acetylamino-β-hydroxy propiophenone, which is inhibitory to members of the *M. tuberculosis* complex, thereby improving selectivity (Riska et al., 1997). The sensitivity of the assay may be improved by the incorporation of high-yielding phages (Chatterjee et al., 2000).

One temperate phage (phGS18) and two lytic phages (phAE40 and phAE85) were modified to carry the *lux* reporter gene and used to detect MAP (Sasahara and Boor, 2001). Phage phAE85 was more sensitive than the other two at detecting small populations of the organism in Middlebrook 7H9 broth, and the detection limit varied between 10^2 CFU/ml and 10 CFU/ml depending on the isolate. Inclusion of a concentration step by filtration for test bacteria, followed by a 14-day enrichment, did not improve the sensitivity of detection using luciferase reporter phages.

The luciferases encoded by different organisms have distinct properties, which can be used to develop systems that can detect and differentiate between more than one bacterium in a single assay (Griffiths, 2000). For example, the bacterial luciferase encoded by *lux* genes has a requirement for a long-chain aldehyde and emits light at a wavelength of about 490 nm, whereas the firefly *luc* gene encodes an enzyme requiring luciferin and produces light at a wavelength near 560 nm. Thus, the light produced from recombinant bacteria containing either the *lux* or *luc* genes can be easily distinguished, either on the basis of substrate specificity or emission wavelength. *lux*-modified phages specific for *Salmonella* species and *S. aureus,* and *luc*$^+$ phages that infect only *E. coli* O157 strains, have been engineered (Chen and Griffiths, 1996; Pagotto et al., 1996). With the aid of a 500-nm cutoff filter, it is easily possible to differentiate between bacteria expressing the *lux* genes and those expressing *luc* by photon imaging.

Phage-triggered luminescent signals can be generated in other ways for the detection of food-borne pathogens. The luminescence genes in *V. fischeri* are controlled by quorum-sensing molecules, which are acyl homoserine lactones (AHLs). These AHL signaling molecules are synthesized by an AHL synthase encoded by *luxI*. This gene has been incorporated into the λ phage genome in order to exploit the autoamplification power of quorum sensing to translate a phage infection event into a chemical signature detectable by a *lux*-based bioluminescent bioreporter (Ripp et al., 2006). The AHL signaling molecules synthesized upon phage infection were detected by an AHL-specific bioluminescent bioreporter based on the *luxCDABE* gene cassette of *V. fischeri*. This resulted in the generation of target-specific visible light signals without the need to add extraneous substrate. This two-component reporter system was able to detect λ phage infection events when the host *E. coli* was present in pure culture at counts ranging from 1 to 1 × 10^8 CFU/ml within 10.3 and 1.5 h, respectively. For artificially contaminated lettuce leaf washings, the presence of *E. coli* at populations from 1 × 10^8 to 130 CFU/ml was achieved within 2.6 and 22.4 h, respectively. In further experiments, the ability of the bioreporter system to detect and quantify the target bacterium in the presence of lactonase-producing non-target organisms that could interrupt AHL signal transduction by enzymatic degradation of AHL and to detect sublethally injured or physiologically stressed target cells was investigated. The bioreporter system was able to autonomously respond to λ phage infection events with a target host *E. coli* at 1 × 10^8 CFU/ml in the presence of a lactonase-producing *Arthrobacter globiformis* at cell densities ranging from 1 to 1 × 10^8 CFU/ml (Birmele et al., 2008). When cells of *E. coli* were stressed by carbon limitation for 2 weeks (i.e., starvation) or exposure to iodine for 1 week at 2 and 20 ppm (i.e., disinfection), a reduced, but detectable, biosensor response was generated. Conversely, short-term iodine exposure produced a significant increase in bioreporter response within the first 24 h. These results showed that both the signal response and the limit of detection for the two-component bioreporter assay were affected by the physiological status of the host and the environment in which they occur. However, the bioreporter maintained target specificity, demonstrating its ruggedness.

The two-component assay could be applied for the detection of other target bacteria by the use of different bacteriophages. A recombinant phage specific for *E. coli* O157:H7 was constructed that, upon infection, catalyzed the synthesis of the AHL *N*-(3-oxohexanoyl)-L-homoserine lactone (OHHL) (Brigati et al., 2007). This phage PP01 derivative carries the *luxI* gene from *V. fischeri* under the control of the phage promoter P_L. As in the system described above, the OHHL produced by infected *E. coli* O157:H7 induces bioluminescence in bioreporter cells carrying the *V. fischeri lux* operon. The phage PP01-*luxI* was able to detect several strains of *E. coli* O157:H7. Reporter phages induced light in the bioreporter cells within 1 h when exposed to 10^4 CFU/ml of *E. coli* O157:H7 and were able to detect 10 CFU/ml in pure culture when a preincubation step was included. This resulted in a total detection time of 4 h. The detection method was also applied to contaminated apple juice and was able to detect 10^4 CFU/ml of *E. coli* O157:H7 in 2 h after a 6-h preincubation. The bacteriophage reporter, derived from phage PP01, was further tested in artificially contaminated apple juice, tap water, ground beef (beef mince), and spinach leaf rinses (Ripp et al., 2008). In apple juice, detection of *E. coli* O157:H7 at initial levels of 1 CFU/ml was possible within about 16 h after a 6-h preincubation, detection of 1 CFU/ml in tap water occurred within approximately 6.5 h after a 6-h preincubation, and detection in spinach leaf rinses using a real-time Xenogen IVIS® imaging system resulted in detection of 1 CFU/ml within about 4 h after a 2-h preincubation. The assay was unable to detect *E. coli* O157:H7 in ground beef, possibly because OHHL is naturally present and resulted in the generation of false-positive bioreporter signals.

Bright et al. (2002) incorporated the bioreporter and phage together on a membrane filter so that the target bacterium could be captured on the membrane. Subsequently the membrane was placed on solid nutrient media and the developing luminescence could be detected using an imaging system. A model system using an M13 phage containing a *lacZ::luxI* fusion and a bioreporter *E. coli* harboring a chromosomally located AHL-inducible *luxRCDABE* gene fusion produced an increase in bioluminescence correlating to a 2-fold increase in bioluminescence per 10-fold increase in phage concentration in the presence of the M13 host.

Ice Nucleation Gene

Wolber and Green (1990) developed a novel reporter phage for *Salmonella* based on the ice nucleation gene *(inaW)* of *Pseudomonas fluorescens*. The assay, termed the bacterial ice nucleation detection (BIND) assay, involves mixing a sample containing bacteria with the genetically engineered transducing *inaW*$^+$ phage and incubation at 37°C to allow phage binding and infection. The sample is then incubated at 23°C to maximize the assembly of ice nuclei from the ice nucleation protein encoded by *inaW*. The sample is cooled to −10°C and assayed for freezing. Freezing will only occur in the presence of transfected cells, and this can be detected using a fluorescent freezing indicator. The BIND assay was tested on samples of culture medium and raw egg, to which *Salmonella* had been added, to give a range of bacterial concentrations. *Salmonella* was detected in all of the samples with high sensitivity (410 bacteria per ml). Additionally, results with culture medium samples showed that phage P22 transduced the *inaW* gene with high efficiency, and the viscosity or complexity of the raw egg samples did not affect the assay. Also, P22 was capable of detecting 10 *S. enterica* serovar Dublin cells/ml, in the presence of 10^7 non-target bacteria/ml. Using a probability-based protocol modification of the assay, quantitative detection with a minimum detectable level (MDL) of 2.0 *Salmonella* serovar Enteritidis cells/ml in buffer within about 3 h was possible (Irwin et al., 2000). The MDL for *S. enterica* serovars Typhimurium DT104 and Abaetetuba were 4.2 and 11.1 cells/ml, respectively. Using salmonella-specific immunomagnetic bead separation technology in conjunction with the modified BIND protocol, an MDL of approximately 4.5 serovar Enteritidis cells/ml with an apparent capture efficiency of 56% was achieved.

In an evaluation of the BIND assay, the performance of three rapid tests—Reveal, an ELISA from Neogen Corp. (Lansing, Mich.); BIND, a bacterial ice nucleation detection method from Idetek Corp. (Westbrook, Maine); and a filter monitor method from Future Medical Technologies, Inc. (Salt Lake City, Utah)—were compared (Peplow et al., 1999). The sensitivities of the different tests were not stable, varying widely between sample times, and were affected by freezing of the samples. All of the rapid tests had low sensitivities, which led to many false-negative results. All tests were able to detect *Salmonella* spp. at a concentration of 10 CFU/ml in at least one of four trials. The BIND and Reveal tests were simple to use with multiple samples and reduced laboratory time by up to 1 day. However, the authors were not able to recommend any of these rapid tests.

β-Galactosidase Gene

Goodridge and Griffiths (2002) created a *lacZ*$^+$ T4 phage that was capable of detecting *E. coli*. The *lacZ*$^+$ assay could detect as few as 10^2

CFU/ml in broth culture when a chemiluminescent substrate (Galacto-Star™) was used. The *lacZ*$^+$ phage was used to develop a *lacZ*$^+$ phage MPN assay. This assay was capable of detecting as few as 10^1 CFU/ml in broth culture within 8 h when the fluorescent substrate 4-methylumbelliferone-β-D-galactoside was used.

This approach was further developed by Willford and Goodridge (2008), who created the Phast Swab. This is a vertically integrated assay for rapid detection of *E. coli* O157:H7. A sampling device, an *E. coli* O157:H7 reporter bacteriophage carrying the *lacZ* gene, bacterial growth medium, *E. coli* O157:H7-specific immunomagnetic beads, and a β-galactosidase substrate (colorimetric or luminescent) are contained within a Snap Valve™ device. The sample to be tested is swabbed, and the swab is returned to the Snap Valve device, in which it is incubated for 8 h. Any *E. coli* O157:H7 cells are isolated and concentrated by IMS, and the reporter bacteriophage is added to the test compartment, where it infects any viable *E. coli* O157:H7 present in the sample, resulting in the production of the enzyme. Following the infection process (approximately 1 h), the cap of the Phast Swab is broken, releasing the substrate, which reacts with the β-galactosidase to produce a colorimetric or a luminescent signal. When the Phast Swab was used to detect *E. coli* O157:H7 on 100-cm^2 portions of beef, it was capable of detecting 10^3 CFU/100 cm^2 of *E. coli* O157:H7 within 12 h when a colorimetric substrate was used, and 10^1 CFU/100 cm^2 within 10 h upon addition of the luminescent substrate.

GFP Gene

GFP was introduced into the C- and N-terminal regions of the small outer capsid (SOC) protein of a T-even-type PP01 by the creation of *gfp::soc* gene fusions (Oda et al., 2004). Fusion of GFP to SOC did not change the host range of PP01. On the contrary, the binding affinity of the recombinant phages to the host cell increased. The adsorption of the GFP-labeled PP01 phages to the cell surface of the *E. coli* O157:H7 host enabled visualization of cells under a fluorescence microscope within 10 min. However, GFP-labeled PP01 phage was adsorbed not only on culturable *E. coli* cells but also on viable but nonculturable (VBNC) or heat-killed cells. The presence of non-host *E. coli* cells did not influence the specificity and affinity of GFP-labeled PP01 for *E. coli* O157:H7. In a slightly different approach, Tanji et al. (2004) fused the *soc* gene of T4 phage with *gfp*. To prevent lysis of the host *E. coli,* they used a T4 mutant that does not contain the lysozyme gene. Using this approach, the target bacterium could be detected by monitoring the intensity of green fluorescence within 1 h. Similar results were reported by Funatsu et al. (2002). This method was subsequently used to monitor *E. coli* in sewage (Miyanaga et al., 2006; Namura et al., 2008).

The PP01e$^-$/GFP phage assay, when combined with nutrient uptake analysis, allowed discriminative detection of culturable, VBNC, and dead cells (Awais et al., 2006). Stress-induced cells, which retained culturability, allowed phage propagation and produced bright green fluorescence. Nonculturable cells (VBNC and dead) allowed only phage adsorption but no proliferation, and fluorescence remained low. The nonculturable cells were further differentiated into VBNC and dead cells via nutrient uptake analysis. Cells which became elongated during prolonged incubation in nutrient medium were defined as being in the VBNC state and could be easily distinguished from dead cells.

Other Reporter Genes

Bacteriophage ϕV10, a lysogenic, nonvirulent phage specific for *E. coli* O157:H7 strains, was modified by replacing a nonessential genetic component of its genome with DNA comprising a kanamycin resistance marker fused to reporter gene *cobA*. The *cobA* gene is found in *Propionibacterium freudenreichii* and when expressed results in the formation of trimethylpyrrocorphin, a fluorophore. When exposed to UV light at 302 nm, strains expressing CobA emit a strong red fluorescence, which can be detected visually (Romero et al., 2008).

PHAGE IMMOBILIZATION AND BACTERIAL CAPTURE

As mentioned above, one of the main hurdles to the use of phage as detection agents is the ability to differentiate between phage resulting from propagation in the host and exogenous phage added as part of the assay protocol. Thus, to facilitate phage-based assays it would be desirable to immobilize phage so that they could be used in much the same way as immunomagnetic separation. Bennett et al. (1997) incorporated the Sapphire lytic phage from Amersham International in a novel immobilized capture method for the separation and concentration of *Salmonella* strains. They passively adsorbed the phage to wells of a polystyrene microtiter plate and/or to plastic dipsticks and showed that *Salmonella* could be specifically captured. However, large numbers of cells ($\geq 10^7$ CFU/ml) were required to be present in the original sample. The poor capture efficiency was probably due to nonoriented binding of the phage particles to the surface. To have the greatest efficiency, the phages need to be immobilized in such a way that their tail receptors are available to bind to the bacterial cells. In order to improve the efficiency of capture, Sun et al. (2000) investigated covalent binding of phage to paramagnetic beads using the biotin-streptavidin reaction. Results showed that the number of cells captured by constructed biosorbent was five times higher than that of the control (i.e., magnetic beads coated with nonbiotinylated phage), indicating that the capture is specific. Up to 20% of target cells could be concentrated on the surface of these streptavidin-coated magnetic beads covered with biotinylated bacteriophage within 30 min. However, the capture efficiency was still lower than reported for immunosorbents.

To overcome problems with ineffective immobilization procedures, Tolba et al. (2008, 2010) have developed methods for oriented immobilization of bacteriophage T4 through introduction of specific binding ligands into the phage head using a phage display technique. Fusion of the *bccp* gene or the *cbm* gene with the *soc* gene of T4 resulted in expression of the respective ligand on the phage head. Both recombinant phages retained their lytic activity and host range. However, phage head modification resulted in decreased burst size and increased latent period. Immobilization of the constructed phages on the surface of streptavidin-coated magnetic beads and microcrystalline cellulose beads resulted in formation of strong and specific bonds between the respective phage and support. The efficiency of bacterial capture by the immobilized phages was significantly different for the constructed BCCP-T4- and CBM-T4-based biosorbents. Biotinylated T4 immobilized onto streptavidin-coated magnetic beads was able to capture 72 to 100% of *E. coli* B cells present at levels between 10 and 10^5 CFU/ml. Capture of bacteria by immobilized CBM-T4, on the other hand, was undetectable by plate counting. This may be explained by the fact that despite the availability of the cellulose affinity tag on the phage head, there still is a possibility for phage tails to interact with sugar moieties on the surface of microcrystalline cellulose in a way similar to its interaction with sugars on the surface of the bacterial cell wall.

The deposition of the CBM-T4 on model cellulose surfaces has been investigated by evanescent wave light scattering (Li et al., in press). Unassembled protein complexes, released with the phages during the lysis of bacteria, limit the deposition of phages to about 1 per 10 to 20 μm^2, as there are far more unassembled protein complexes than phages. The implication is that a few thousand phages can attach to a cellulose fiber, with most of the surface area taken up by protein complexes. The unassembled protein complexes may be beneficial, as they will prevent nonspecific deposition of bacteria. Thus, it may not be necessary to separate the protein complexes from phages in an expensive purification operation.

The BCCP-T4 bacteriophage was immobilized on gold surfaces (Gervals et al., 2007). Such chemical immobilization results in a 15-fold improvement of attachment when compared to the simple physisorption of the wild-type phage onto bare gold. The attachment procedure was then used to investigate the effect of a biotinylated phage-terminated surface on the growth of the host bacteria. This assessment was conducted in an electric cell-substrate impedance-sensing device. The streptavidin-mediated attachment of biotinylated phages significantly delays the growth of the host bacteria by up to 17.2 h. In comparison, nonspecific binding of wild-type phages onto the streptavidin surface is found to cause a lesser growth delay of 13 h. The phage-coated gold surfaces could be used as molecular cantilevers for the detection of *E. coli*. To avoid the use of genetically modified phage, techniques for the chemical attachment of wild-type bacteriophages onto gold surfaces and the subsequent capture of their host bacteria have been developed (Singh et al., 2009). The surfaces were modified with sugars (dextrose and sucrose) as well as amino acids (histidine and cysteine) to facilitate such attachment. Bovine serum albumin was used as a blocking layer to prevent nonspecific attachment. Surfaces modified with cysteine (and cysteamine) followed by activation using 2% glutaraldehyde resulted in an attachment density of 18 phages/μm^2. This represented a 37-fold improvement compared to simple physisorption. Subsequently, the phage-immobilized surfaces were exposed to the host *E. coli* EC12 bacteria and capture was confirmed by fluorescence microscopy. We obtained a bacterial capture density of 11.9 cells/ 100 μm^2, a ninefold improvement when compared to those on physically adsorbed phages. The specificity of recognition was confirmed by exposing similar surfaces to three strains of non-host bacteria. These negative-control experiments do not show any bacterial capture. In addition, no capture of the host was observed in the absence of the phages.

While immobilized phage could be particularly useful to create antimicrobial surfaces and for detection, current immobilization strategies require chemical bioconjugation to surfaces or more difficult processes involving modification of their head proteins to express specific binding moieties, for example, biotin- or cellulose-binding domains, procedures that are both time- and money-intensive. However, morphologically different bacteriophages, active against a variety of food-borne bacteria, will effectively physisorb to silica particles, prepared by silica surface modification with poly(ethylene glycol), carboxylic acid groups, or amines (Cademartiri et al., in press). The phages remain infective to their host bacteria while adsorbed on the surface of the silica particles. The number of active phage on the silica is dependent on the type of surface modification used, with ionic surfaces being more effective for binding.

The cell wall-binding domains (CBDs) of bacteriophage-encoded peptidoglycan hydrolases (endolysins) have also been immobilized and used to capture bacteria (Loessner, 2005b; Kretzer et al., 2007). These polypeptide modules very specifically recognize and bind to ligands on the gram-positive cell wall with high affinity. With paramagnetic beads coated with recombinant listeriaphage endolysin-derived CBD molecules, more than 90% of the viable *L. monocytogenes* cells could be immobilized and recovered from diluted suspensions within 20 to 40 min. Recovery rates were similar for different species and serovars of *Listeria* and were not affected by the presence of other microorganisms. When the CBD-based magnetic separation procedure was evaluated for capture and detection of *L. monocytogenes* from artificially and naturally contaminated food samples, the CBD separation method was superior to the established standard procedures;

it required less time (48 versus 96 h) and was the more sensitive method. *B. cereus, Clostridium perfringens* (Kretzer et al., 2007), and *B. anthracis* (Schuch et al., 2002; Fujinami et al., 2007) could also be captured using specific phage-encoded CBDs specifically recognizing cells of these targets. This technology has been commercialized and is used in the Vidas system manufactured by bioMérieux sa (Marcy l'Etoile, France).

An effective method for the immobilization of phage will lead to one-step assays incorporating cell concentration and detection, and research in the author's laboratory has already shown this is feasible.

OTHER USES OF PHAGE IN DETECTION ASSAYS

Immunochemical-based methods for the detection of *Salmonella* in food can be complicated by the presence of closely related, immuno-cross-reactive non-*Salmonella* species in the sample that may cause false-positive results. To circumvent this problem, specific bacteriophages against immuno-cross-reactive, non-*Salmonella* bacteria were incorporated in the sample enrichment step to suppress their growth and improve the performance of an immunoassay for *Salmonella* (Muldoon et al., 2007). These cross-reactive bacteria were primarily *Citrobacter* spp. and *E. coli* with serology shared with *Salmonella* serogroups B, D, and F. These bacteria were used as hosts for the isolation of specific lytic bacteriophages. When formulated with the primary enrichment, the bacteriophage cocktail significantly reduced false positives in the immunoassay. False positives in naturally contaminated beef samples were reduced from 32 of 115 samples tested to zero. In raw meat and poultry with a high bacterial load (>10^5 CFU/g), the use of the bacteriophage-based enrichment procedure gave improved recovery of *Salmonella* compared with the conventional culture-based reference method. This was observed when the modified enrichment procedure was used in conjunction with either the immunoassay or a standard culture-based assay. This technique is marketed as RapidChek® SELECT™ *Salmonella* by Strategic Diagnostics Inc. (Newark, DE).

Phage-based diagnostic technologies are still in their infancy; although they were first seriously proposed over 30 years ago. The research described in this review suggests that rapid, reliable, and sensitive methods for the phage-based detection of bacterial pathogens are possible, but these technologies have not been widely accepted. One reason for this may be the fact that it is difficult to genetically modify phages. Hopefully, as the genetic sequences of more phages and their bacterial hosts are published, these technologies will become more feasible. Another potential problem is the fact that an individual phage may not possess the host range required to detect all pathogenic isolates of a bacterial species. Therefore, it will be necessary to modify the host ranges of the phages used or employ a cocktail of phages in order to develop an assay that will be able to detect the desired target bacterium. Some technologies that utilize phage are commercially available and the number of these technologies reaching the market should increase over the next decade as phage-based biosensors become more widely accepted.

REFERENCES

Aalto, J., S. Pelkonen, H. Kalimo, and J. Finne. 2001. Mutant bacteriophage with non-catalytic endosialidase binds to both bacterial and eukaryotic polysialic acid and can be used as probe for its detection. *Glycoconj. J.* **18:**751–758.

Albay, A., O. Kisa, O. Baylan, and L. Doganci. 2003. The evaluation of FAST*Plaque*TB™ test for the rapid diagnosis of tuberculosis. *Diagn. Microbiol. Infect. Dis.* **46:**211–215.

Albert, H., A. P. Trollip, K. Linley, C. Abrahams, T. Seaman, and R. J. Mole. 2007. Development of an antimicrobial formulation for control of specimen-related contamination in phage-based diagnostic testing for tuberculosis. *J. Appl. Microbiol.* **103:**892–899.

Albert, H., A. P. Trollip, R. J. Mole, S. J. B. Hatch, and L. Blumberg. 2002. Rapid indication of multidrug-resistant tuberculosis from liquid cultures using *FASTPlaque*TB-*RIF*™, a manual phage-based test. *Int. J. Tuberc. Lung Dis.* **6:**523–528.

Alimova, A., A. Katz, R. Podder, G. Minko, H. Wei, R. R. Alfano, and P. Gottlieb. 2004. Virus particles monitored by fluorescence spectroscopy: a potential detection assay for macromolecular assembly. *Photochem. Photobiol.* **80:**41–46.

Almeida, P. E., R. C. C. Almeida, T. C. F. Barbalho, C. G. Melo, A. O. Almeida, E. R. Magalhaes, I. C. Oliveira, and E. Hofer. 2003. Development of protocols for the bacteriophage amplification assay for rapid, quantitative and sensitive detection of viable *Listeria* cells, abstr. P-052. *Abstr. 103rd Gen. Meet. Am. Soc. Microbiol.* American Society for Microbiology, Washington, DC.

Altic, L. C., M. T. Rowe, and I. R. Grant. 2007. UV light inactivation of *Mycobacterium avium* subsp. *paratuberculosis* in milk as assessed by *FASTPlaque*TB phage assay and culture. *Appl. Environ. Microbiol.* **73:**3728–3733.

Alvarez, S. D., M. P. Schwartz, B. Migliori, C. U. Rang, L. Chao, and M. J. Sailor. 2007. Using a porous silicon photonic crystal for bacterial cell-based biosensing. *Physica Status Solidi A Appl. Mater. Sci.* **204:**1439–1443.

Auman, B. D., N. G. Bright, L. J. Mauer, and B. M. Applegate, Sr. 2003. Development of a bioluminescent assay for the high throughput screening of virucidal activity, abstr. P-103. *Abstr. 103rd Gen. Meet. Am. Soc. Microbiol.* American Society for Microbiology, Washington, DC.

Awais, R., H. Fukudomi, K. Miyanaga, H. Unno, and Y. Tanji. 2006. A recombinant bacteriophage-based assay for the discriminative detection of culturable and viable but nonculturable *Escherichia coli* O157:H7. *Biotechnol. Prog.* **22:**853–859.

Ayer, S., W. X. Zhao, and C. E. Davis. 2008. Differentiation of proteins and viruses using pyrolysis gas chromatography differential mobility spectrometry (PY/GC/DMS) and pattern recognition. *IEEE Sens. J.* **8:**1586–1592.

Bange, A., J. Tu, X. S. Zhu, C. Ahn, H. B. Halsall, and W. R. Heineman. 2007. Electrochemical detection of MS2 phage using a bead-based immunoassay and a NanoIDA. *Electroanalysis* **19:**2202–2207.

Bennett, A. R., F. G. C. Davids, S. Vlahodimou, J. G. Banks, and R. P. Betts. 1997. The use of bacteriophage-based systems for the separation and concentration of *Salmonella. J. Appl. Microbiol.* **83:**259–265.

Bi, Q., X. D. Cen, W. J. Wang, X. S. Zhao, X. Wang, T. Shen, and S. G. Zhu. 2007. A protein microarray prepared with phage-displayed antibody clones. *Biosens. Bioelectron.* **22:**3278–3282.

Binetti, A. G., B. del Rio, M. Cruz-Martin, and M. A. Alvarez. 2005. Detection and characterization of *Streptococcus thermophilus* bacteriophages by use of the antireceptor gene sequence. *Appl. Environ. Microbiol.* **71:**6096–6103.

Birmele, M., S. Ripp, P. Jegier, M. S. Roberts, G. Sayler, and J. Garland. 2008. Characterization and validation of a bioluminescent bioreporter for the direct detection of *Escherichia coli. J. Microbiol. Methods* **75:**354–356.

Blasco, R., M. J. Murphy, M. F. Sanders, and D. J. Squirrell. 1998. Specific assays for bacteria using phage mediated release of adenylate kinase. *J. Appl. Microbiol.* **84:**661–666.

Block, T., R. Miller, R. Korngold, and D. Jungkind. 1989. A phage-linked immunoabsorbent system for the detection of pathologically relevant antigens. *BioTechniques* **7:**756–761.

Brigati, J. R., S. A. Ripp, C. M. Johnson, P. A. Iakova, P. Jegier, and G. S. Sayler. 2007. Bacteriophage-based bioluminescent bioreporter for the detection of *Escherichia coli* O157:H7. *J. Food Prot.* **70:**1386–1392.

Bright, N. G., R. J. Carroll, and B. M. Applegate. 2002. Filter based assay for pathogen detection using a two component bacteriophage/bioluminescent reporter system, abstr. P-304. *Abstr. 102nd Gen. Meet. Am. Soc. Microbiol.* American Society for Microbiology, Washington, DC.

Brussaard, C. P. D. 2009. Enumeration of bacteriophages using flow cytometry. *Methods Mol. Biol.* **501:**97–111.

Bunina, V. D., O. V. Ignatov, O. I. Guliy, I. S. Zaitseva, D. O'Nell, and D. Ivnitski. 2004. Electrooptical analysis of the *Escherichia coli*-phage interaction. *Anal. Biochem.* **328:**181–186.

Cademartiri, R., H. Anany, I. Gros, R. Bhayani, M. Griffiths, and M. A. Brook. 2010. Immobilization of bacteriophages on modified silica particles. *Biomaterials* **31:**1904–1910.

Carminati, D., and E. Neviani. 1991. Application of the conductance measurement technique for detection of *Streptococcus salivarius* ssp. *thermophilus* phages. *J. Dairy Sci.* **74:**1472–1476.

Carriere, C., P. F. Riska, O. Zimhony, J. Kriakov, S. Bardarov, J. Burns, J. Chan, and W. R. Jacobs. 1997. Conditionally replicating luciferase reporter phages: improved sensitivity for rapid detection and assessment of drug susceptibility of *Mycobacterium tuberculosis. J. Clin. Microbiol.* **35:**3232–3239.

Chang, T. C., H. C. Ding, and S. W. Chen. 2002. A conductance method for the identification of *Escherichia coli* O157:H7 using bacteriophage AR1. *J. Food Prot.* **65:**12–17.

Chatterjee, S., M. Mitra, and S. K. Das Gupta. 2000. A high yielding mutant of mycobacteriophage L1 and its application as a diagnostic tool. *FEMS Microbiol. Lett.* **188:**47–53.

Chen, J., and M. W. Griffiths. 1996. *Salmonella* detection in eggs using Lux$^+$ bacteriophages. *J. Food Prot.* **59:**908–914.

Crane, D. D., L. D. Martin, and D. C. Hirsh. 1984. Detection of *Salmonella* in feces by using Felix-01 bacteriophage and high-performance liquid-chromatography. *J. Microbiol. Methods* **2:** 251–256.

Craven, J., L. Goodridge, A. Hill, and M. Griffiths. 2006. PCR detection of lactococcal bacteriophages in milk. *Milchwissenschaft* **61:**382–385.

Cruz-Martin, M., B. del Rio, N. Martinez, A. H. Magadan, and M. A. Alvarez. 2008. Fast real-time polymerase chain reaction for quantitative detection of *Lactobacillus delbrueckii* bacteriophages in milk. *Food Microbiol.* **25:**978–982.

Dadarwal, R., A. Namvar, D. F. Thomas, J. C. Hall, and K. Warriner. 2009. Organic conducting polymer electrode based sensors for detection of *Salmonella* infecting bacteriophages. *Mater. Sci. Eng. C Biomim. Supramol. Syst.* **29:**761–765.

de Siqueira, R. S., C. E. R. Dodd, and C. E. D. Rees. 2003. Phage amplification assay as rapid method for *Salmonella* detection. *Braz. J. Microbiol.* **34:**118–120.

de Siqueira, R. S., C. E. R. Dodd, and C. E. D. Rees. 2006. Evaluation of the natural virucidal activity of teas for use in the phage amplification assay. *Int. J. Food Microbiol.* **111:**259–262.

Dobozi-King, M., S. Seo, J. U. Kim, R. Young, M. Cheng, and L. B. Kish. 2005. Rapid detection and identification of bacteria: SEnsing of Phage-Triggered Ion Cascade (SEPTIC). *J. Biol. Phys. Chem.* **5:**3–7.

Dultsev, F. N., R. E. Speight, M. T. Florini, J. M. Blackburn, C. Abell, V. P. Ostanin, and D. Klenerman. 2001. Direct and quantitative detection of bacteriophage by "hearing" surface detachment using a quartz crystal microbalance. *Anal. Chem.* **73:**3935–3939.

Edelman, D. C., and J. Barletta. 2003. Real-time PCR provides improved detection and titer determination of bacteriophage. *BioTechniques* **35:** 368–375.

Edgar, R., M. McKinstry, J. Hwang, A. B. Oppenheim, R. A. Fekete, G. Giulian, C. Merril, K. Nagashima, and S. Adhya. 2006. High-sensitivity bacterial detection using biotin-tagged phage and quantum-dot nanocomplexes. *Proc. Natl. Acad. Sci. USA* **103:**4841–4845.

Favrin, S. J., S. A. Jassim, and M. W. Griffiths. 2001. Development and optimization of a novel immunomagnetic separation-bacteriophage assay for detection of *Salmonella enterica* serovar Enteritidis in broth. *Appl. Environ. Microbiol.* **67:**217–224.

Favrin, S. J., S. A. Jassim, and M. W. Griffiths. 2003. Application of a novel immunomagnetic separation-bacteriophage assay for the detection of *Salmonella enteritidis* and *Escherichia coli* O157:H7 in food. *Int. J. Food Microbiol.* **85:**63–71.

Foddai, A., C. T. Elliott, and I. R. Grant. 2009. Optimization of a phage amplification assay to permit accurate enumeration of viable *Mycobacterium avium* subsp. *paratuberculosis* cells. *Appl. Environ. Microbiol.* **75:**3896–3902.

Fu, L. L., S. Q. Li, K. W. Zhang, I. H. Chen, V. A. Petrenko, and Z. Y. Cheng. 2007. Magnetostrictive microcantilever as an advanced transducer for biosensors. *Sensors* **7:**2929–2941.

Fujinami, Y., Y. Hirai, I. Sakai, M. Yoshino, and J. Yasuda. 2007. Sensitive detection of *Bacillus anthracis* using a binding protein originating from gamma-phage. *Microbiol. Immunol.* **51:**163–169.

Funatsu, T., T. Taniyama, T. Tajima, H. Tadakuma, and H. Namiki. 2002. Rapid and sensitive detection method of a bacterium by using a GFP reporter phage. *Microbiol. Immunol.* **46:**365–369.

Gabig-Ciminska, M., M. Los, A. Holmgren, J. Albers, A. Czyz, R. Hintsche, G. Wegrzyn, and S. O. Enfors. 2004. Detection of bacteriophage infection and prophage induction in bacterial cultures by means of electric DNA chips. *Anal. Biochem.* **324:**84–91.

Gervals, L., M. Gel, B. Allain, M. Tolba, L. Brovko, M. Zourob, R. Mandeville, M. Griffiths, and S. Evoy. 2007. Immobilization of biotinylated bacteriophages on biosensor surfaces. *Sens. Actuators B Chem.* **125:**615–621.

Goeller, L. J., and M. R. Riley. 2007. Discrimination of bacteria and bacteriophages by Raman spectroscopy and surface-enhanced Raman spectroscopy. *Appl. Spectrosc.* **61:**679–685.

Goodridge, L., J. Chen, and M. Griffiths. 1999a. Development and characterization of a fluorescent-bacteriophage assay for detection of *Escherichia coli* O157:H7. *Appl. Environ. Microbiol.* **65:**1397–1404.

Goodridge, L., J. Chen, and M. Griffiths. 1999b. The use of a fluorescent bacteriophage assay for detection of *Escherichia coli* O157:H7 in inoculated ground beef and raw milk. *Int. J. Food Microbiol.* **47:**43–50.

Goodridge, L., and M. Griffiths. 2002. Reporter bacteriophage assays as a means to detect foodborne pathogenic bacteria. *Food Res. Int.* **35:**863–870.

Griffiths, M. W. 1996. The role of ATP bioluminescence in the food industry: new light on old problems. *Food Technol.* **50:**62–72.

Griffiths, M. W. 2000. How novel methods can help discover more information about foodborne pathogens. *Can. J. Infect. Dis.* **11:**142–153.

Guan, J. W., M. Chan, B. Allain, R. Mandeville, and B. W. Brooks. 2006. Detection of multiple antibiotic-resistant *Salmonella enterica* serovar Typhimurium DT104 by phage replication-competitive enzyme-linked immunosorbent assay. *J. Food Prot.* **69:**739–742.

Gulig, P. A., J. L. Martin, H. G. Messer, B. L. Deffense, and C. J. Harpley. 2008. Phage display methods for detection of bacterial pathogens, p. 755–783. *In* M. Zourob, S. Elwary, and A. Turner (ed.), *Principles of Bacterial Detection: Biosensors, Recognition Receptors and Microsystems.* Springer, New York, NY.

Guntupalli, R., I. Sorokulova, A. Krumnow, O. Pustovyy, E. Olsen, and V. Vodyanoy. 2008. Real-time optical detection of methicillin-resistant *Staphylococcus aureus* using lytic phage probes. *Biosens. Bioelectron.* **24:**151–154.

Hagens, S., and M. J. Loessner. 2007. Application of bacteriophages for detection and control of foodborne pathogens. *Appl. Microbiol. Biotechnol.* **76:**513–519.

Hennes, K. P., and C. A. Suttle. 1995. Direct counts of viruses in natural waters and laboratory cultures by epifluorescence microscopy. *Limnol. Oceanogr.* **40:**1050–1055.

Hennes, K. P., C. A. Suttle, and A. M. Chan. 1995. Fluorescently labeled virus probes show that natural virus populations can control the structure of marine microbial communities. *Appl. Environ. Microbiol.* **61:**3623–3627.

Hirsh, D. C., and L. D. Martin. 1983a. Detection of *Salmonella* spp. in milk by using Felix-O1 bacteriophage and high-pressure liquid chromatography. *Appl. Environ. Microbiol.* **46:**1243–1245.

Hirsh, D. C., and L. D. Martin. 1983b. Rapid detection of *Salmonella* spp. by using Felix-O1 bacteriophage and high-performance liquid chromatography. *Appl. Environ. Microbiol.* **45:**260–264.

Hirsh, D. C., and L. D. Martin. 1984. Rapid detection of *Salmonella* in certified raw milk by using charge-modified filters and Felix-O1 bacteriophage. *J. Food Prot.* **47:**388–390.

Irwin, P., A. Gehring, I. T. Shu, J. Brewster, J. Fanelli, and E. Ehrenfeld. 2000. Minimum detectable level of salmonellae using a binomial-based bacterial ice nucleation detection assay (BIND®). *J. AOAC Int.* **83:**1087–1095.

Jacobs, W. R. J., R. G. Barletta, R. Udani, J. Chan, G. Kalkut, G. Sosne, T. Kieser, G. J. Sarkis, G. F. Hatfull, and B. R. Bloom. 1993. Rapid assessment of drug susceptibilities of *Mycobacterium tuberculosis* by means of luciferase reporter phages. *Science* **260:**819–822.

Jassim, S. A. A., and M. W. Griffiths. 2007. Evaluation of a rapid microbial detection method via phage lytic amplification assay coupled with live/dead fluorochromic stains. *Lett. Appl. Microbiol.* **44:**673–678.

Jia, Y. F., M. Qin, H. K. Zhang, W. C. Niu, X. Li, L. K. Wang, Y. P. Bai, Y. J. Cao, and X. Z. Feng. 2007. Label-free biosensor: a novel phage-modified light addressable potentiometric sensor system for cancer cell monitoring. *Biosens. Bioelectron.* **22:**3261–3266.

Johnson, M. L., J. H. Wan, S. C. Huang, Z. Y. Cheng, V. A. Petrenko, D. J. Kim, I. H. Chen, J. M. Barbaree, J. W. Hong, and B. A. Chin. 2008. A wireless biosensor using microfabricated phage-interfaced magnetoelastic particles. *Sens. Actuators A Phys.* **144:**38–47.

Kalantri, S., M. Pai, L. Pascopella, L. Riley, and A. Reingold. 2005. Bacteriophage-based tests for the detection of *Mycobacterium tuberculosis* in clinical specimens: a systematic review and meta-analysis. *BMC Infect. Dis.* **5:**59.

Kim, S., B. Schuler, A. Terekhov, J. Auer, L. J. Mauer, L. Perry, and B. Applegate. 2009. A bioluminescence-based assay for enumeration of lytic bacteriophage. *J. Microbiol. Methods* **79:**18–22.

Kodikara, C. P., H. H. Crew, and G. S. A. B. Stewart. 1991. Near on-line detection of enteric bacteria using lux recombinant bacteriophage. *FEMS Microbiol. Lett.* **83:**261–265.

Kremser, L., D. Blaas, and E. Kenndler. 2009. Virus analysis using electromigration techniques. *Electrophoresis* **30:**133–140.

Kretzer, J. W., R. Lehmann, M. Schmelcher, M. Banz, K. P. Kim, C. Korn, and M. J. Loessner. 2007. Use of high-affinity cell wall-binding domains of bacteriophage endolysins for immobilization and separation of bacterial cells. *Appl. Environ. Microbiol.* **73:**1992–2000.

Kuhn, J., M. Suissa, J. Wyse, I. Cohen, I. Weiser, S. Reznick, S. Lubinsky-Mink, G. Stewart, and S. Ulitzur. 2002. Detection of bacteria using foreign DNA: the development of a bacteriophage reagent for *Salmonella*. *Int. J. Food Microbiol.* **74:**229–238.

Kuhn, J. C. 2007. Detection of *Salmonella* by bacteriophage Felix 01. *Methods Mol. Biol.* **394:**21–37.

Kutin, R. K., A. Alvarez, and D. M. Jenkins. 2009. Detection of *Ralstonia solanacearum* in natural substrates using phage amplification integrated with real-time PCR assay. *J. Microbiol. Methods* **76:**241–246.

Lakshmanan, R. S., R. Guntupalli, J. W. Hong, D. J. Kim, Z. Y. Cheng, V. A. Petrenko, J. M. Barbaree, and B. A. Chin. 2008. Selective detection of *Salmonella* Typhimurium in the presence of high concentrations of masking bacteria. *Sens. Instrum. Food Qual. Saf.* **2:**234–239.

Lakshmanan, R. S., R. Guntupalli, J. Hu, D. J. Kim, V. A. Petrenko, J. M. Barbaree, and

B. A. Chin. 2007a. Phage immobilized magnetoelastic sensor for the detection of *Salmonella typhimurium*. *J. Microbiol. Methods* **71:**55–60.

Lakshmanan, R. S., R. Guntupalli, J. Hu, V. A. Petrenko, J. M. Barbaree, and B. A. Chin. 2007b. Detection of *Salmonella typhimurium* in fat free milk using a phage immobilized magnetoelastic sensor. *Sens. Actuators B Chem.* **126:**544–550.

Lee, S. H., M. Onuki, H. Satoh, and T. Mino. 2006. Isolation, characterization of bacteriophages specific to *Microlunatus phosphovorus* and their application for rapid host detection. *Lett. Appl. Microbiol.* **42:**259–264.

Li, Z., M. Tolba, M. Griffiths, and T. G. van de Ven. 2010. Effect of unassembled phage protein complexes on the attachment to cellulose of genetically modified bacteriophages containing cellulose binding modules. *Colloids Surf. B Biointerf.* **76:**529–534.

Loessner, M. 2005a. The enemy's enemy is our friend: phageborne tools for detection and control of foodborne pathogens. *Mitt. Lebensmittelunters. Hyg.* **96:**30–38.

Loessner, M. J. 2005b. Bacteriophage endolysins—current state of research and applications. *Curr. Opin. Microbiol.* **8:**480–487.

Loessner, M. J., C. E. D. Rees, G. Stewart, and S. Scherer. 1996. Construction of luciferase reporter bacteriophage A511::*luxAB* for rapid and sensitive detection of viable *Listeria* cells. *Appl. Environ. Microbiol.* **62:**1133–1140.

Loessner, M. J., M. Rudolf, and S. Scherer. 1997. Evaluation of luciferase reporter bacteriophage A511::*luxAB* for detection of *Listeria monocytogenes* in contaminated foods. *Appl. Environ. Microbiol.* **63:**2961–2965.

Loessner, M. J., A. Schneider, and S. Scherer. 1995. A new procedure for efficient recovery of DNA, RNA, and proteins from *Listeria* cells by rapid lysis with a recombinant bacteriophage endolysin. *Appl. Environ. Microbiol.* **61:**1150–1152.

Luna, C. G., A. Costan-Longares, F. Lucena, and J. Jofre. 2009. Detection of somatic coliphages through a bioluminescence assay measuring phage mediated release of adenylate kinase and adenosine 5′-triphosphate. *J. Virol. Methods* **161:** 107–113.

Madonna, A. J., S. Van Cuyk, and K. J. Voorhees. 2003. Detection of *Escherichia coli* using immunomagnetic separation and bacteriophage amplification coupled with matrix-assisted laser desorption/ionization time-of-flight mass spectrometry. *Rapid Commun. Mass Spectrom.* **17:**257–263.

Mandeville, R., M. Griffiths, L. Goodridge, L. McIntyre, and T. T. Ilenchuk. 2003. Diagnostic and therapeutic applications of lytic phages. *Anal. Lett.* **36:**3241–3259.

Mao, C. B., A. H. Liu, and B. R. Cao. 2009. Virus-based chemical and biological sensing. *Angew. Chem. Int. Ed. Eng.* **48:**6790–6810.

Marei, A. M., E. M. El-Behedy, H. A. Mohtady, and A. F. M. Afify. 2003. Evaluation of a rapid bacteriophage-based method for the detection of *Mycobacterium tuberculosis* in clinical samples. *J. Med. Microbiol.* **52:**331–335.

Marks, T., and R. Sharp. 2000. Bacteriophages and biotechnology: a review. *J. Chem. Technol. Biotechnol.* **75:**6–17.

McIntyre, L., and M. W. Griffiths. 1997. A bacteriophage-based impedimetric method for the detection of pathogens in dairy products. *J. Dairy Sci.* **80:**107.

McNerney, R., K. Mallard, H. M. R. Urassa, E. Lemma, and H. D. Donoghue. 2007. Colorimetric phage-based assay for detection of rifampin-resistant *Mycobacterium tuberculosis*. *J. Clin. Microbiol.* **45:**1330–1332.

Michelsen, O., A. Cuesta-Dominguez, B. Albrechtsen, and P. R. Jensen. 2007. Detection of bacteriophage-infected cells of *Lactococcus lactis* by using flow cytometry. *Appl. Environ. Microbiol.* **73:**7575–7581.

Miyanaga, K., T. F. Hijikata, C. Furukawa, H. Unno, and Y. Tanji. 2006. Detection of *Escherichia coli* in the sewage influent by fluorescent labeled T4 phage. *Biochem. Eng. J.* **29:**119–124.

Mole, R. J., and T. W. O. Maskell. 2001. Phage as a diagnostic—the use of phage in TB diagnosis. *J. Chem. Technol. Biotechnol.* **76:**683–688.

Moll, N., E. Pascal, D. H. Dinh, J. L. Lachaud, L. Vellutini, J. P. Pillot, D. Rebiere, D. Moynet, J. Pistre, D. Mossalayi, Y. Mas, B. Bennetau, and C. Dejous. 2008. Multipurpose Love acoustic wave immunosensor for bacteria, virus or proteins detection. *IRBM* **29:**155-161.

Mosier-Boss, P. A., S. H. Lieberman, J. M. Andrews, F. L. Rohwer, L. E. Wegley, and M. Breitbart. 2003. Use of fluorescently labeled phage in the detection and identification of bacterial species. *Appl. Spectrosc.* **57:**1138–1144.

Muldoon, M. T., G. Teaney, L. Jingkun, D. V. Onisk, and J. W. Stave. 2007. Bacteriophage-based enrichment coupled to immunochromatographic strip-based detection for the determination of *Salmonella* in meat and poultry. *J. Food Prot.* **70:** 2235–2242.

Namura, M., T. Hijikata, K. Miyanaga, and Y. Tanji. 2008. Detection of *Escherichia coli* with fluorescent labeled phages that have a broad host range to *E. coli* in sewage water. *Biotechnol. Prog.* **24:**481–486.

Nanduri, V., S. Balasubramanian, S. Sista, V. J. Vodyanoy, and A. L. Simonian. 2007. Highly sensitive phage-based biosensor for the detection of β-galactosidase. *Anal. Chim. Acta* **589:**166–172.

Nanduri, V., I. B. Sorokulova, A. M. Samoylov, A. L. Simonian, V. A. Petrenko, and V. Vodyanoy. 2007. Phage as a molecular recognition element in biosensors immobilized by physical adsorption. *Biosens. Bioelectron.* **22:**986–992.

Nelson, D. C., L. Loomis, and V. A. Fischetti. 2002. Bacteriophage lytic enzymes as tools for rapid and specific detection of bacterial contamination, abstr. P-106. *Abstr. 102nd Gen. Meet. Am. Soc. Microbiol.* American Society for Microbiology, Washington, DC.

Neufeld, T., A. Schwartz-Mittelmann, D. Biran, E. Z. Ron, and J. Rishpon. 2003. Combined phage typing and amperometric detection of released enzymatic activity for the specific identification and quantification of bacteria. *Anal. Chem.* **75:**580–585.

Oda, M., M. Morita, H. Unno, and Y. Tanji. 2004. Rapid detection of *Escherichia coli* O157:H7 by using green fluorescent protein-labeled PP01 bacteriophage. *Appl. Environ. Microbiol.* **70:**527–534.

Ogwang, S., B. B. Asiimwe, H. Traore, F. Mumbowa, A. Okwera, K. D. Eisenach, S. Kayes, E. C. Jones-Lopez, R. McNerney, W. Worodria, I. Ayakaka, R. D. Mugerwa, P. G. Smith, J. Ellner, and M. L. Joloba. 2009. Comparison of rapid tests for detection of rifampicin-resistant *Mycobacterium tuberculosis* in Kampala, Uganda. *BMC Infect. Dis.* **9:**139.

Olsen, E. V., I. B. Sorokulova, V. A. Petrenko, I. H. Chen, J. M. Barbaree, and V. J. Vodyanoy. 2006. Affinity-selected filamentous bacteriophage as a probe for acoustic wave biodetectors of *Salmonella typhimurium*. *Biosens. Bioelectron.* **21:**1434–1442.

Pagotto, F., L. Brovko, and M. W. Griffiths. 1996. Phage-mediated detection of *Staphylococcus aureus* and *E. coli* O157:H7 using bioluminescence, p. 152–156. *In Symposium on Bacteriological Quality of Raw Milk.* International Dairy Federation, Brussels, Belgium.

Pai, M., S. Kalantri, L. Pascopella, L. W. Riley, and A. L. Reingold. 2005. Bacteriophage-based assays for the rapid detection of rifampicin resistance in *Mycobacterium tuberculosis:* a meta-analysis. *J. Infect.* **51:**175–187.

Peplow, M. O., M. Correa-Prisant, M. E. Stebbins, F. Jones, and P. Davies. 1999. Sensitivity, specificity and predictive values of three *Salmonella* rapid detection kits using fresh and frozen poultry environmental samples versus those of standard plating. *Appl. Environ. Microbiol.* **65:**1055–1060.

Petrenko, V. A. 2008. Landscape phage as a molecular recognition interface for detection devices. *Microelectronics J.* **39:**202–207.

Petrenko, V. A., and V. J. Vodyanoy. 2003. Phage display for detection of biological threat agents. *J. Microbiol. Methods* **53:**253–262.

Petty, N. K., T. J. Evans, P. C. Fineran, and G. P. C. Salmond. 2007. Biotechnological exploitation of bacteriophage research. *Trends Biotechnol.* **25:**7–15.

Pugh, S. J., J. L. Griffiths, M. L. Arnott, and C. S. Gutteridge. 1988. A complete protocol using conductance for rapid detection of salmonellas in confectionery materials. *Lett. Appl. Microbiol.* **7:**23–27.

Qi, C., Y. Lin, J. Feng, Z.-H. Wang, C.-F. Zhu, Y.-H. Meng, X.-Y. Yan, L.-J. Wan, and G. Jin. 2009. Phage M13KO7 detection with biosensor based on imaging ellipsometry and AFM microscopic confirmation. *Virus Res.* **140:**79–84.

Rees, C. E. D., and C. E. R. Dodd. 2006. Phage for rapid detection and control of bacterial pathogens in food. *Adv. Appl. Microbiol.* **59:**159–186.

Rees, C. E. D., and M. J. Loessner. 2005. Phage for the detection of pathogenic bacteria, p. 264–281. *In* E. Kutter and A. Sulakvelidze (ed.), *Bacteriophages: Biology and Applications.* CRC Press, Boca Raton, FL.

Rees, J. C., and K. J. Voorhees. 2005. Simultaneous detection of two bacterial pathogens using bacteriophage amplification coupled with matrix-assisted laser desorption/ionization time-of-flight mass spectrometry. *Rapid Commun. Mass Spectrom.* **19:**2757–2761.

Regan, P. M., J. Wilson, T. To, and A. Margolin. 2001. Application of a phage-based assay to determine the viability of *Mycobacterium bovis* following disinfectant treatment, abstr. P-595. *Abstr. 101st Gen. Meet. Am. Soc. Microbiol.* American Society for Microbiology, Washington, DC.

Reiman, R. W., D. H. Atchley, and K. J. Voorhees. 2007. Indirect detection of *Bacillus anthracis* using real-time PCR to detect amplified gamma phage DNA. *J. Microbiol. Methods* **68:**651–653.

Ripp, S., P. Jegier, M. Birmele, C. M. Johnson, K. A. Daumer, J. L. Garland, and G. S. Sayler. 2006. Linking bacteriophage infection to quorum sensing signalling and bioluminescent bioreporter monitoring for direct detection of bacterial agents. *J. Appl. Microbiol.* **100:**488–499.

Ripp, S., P. Jegier, C. M. Johnson, J. R. Brigati, and G. S. Sayler. 2008. Bacteriophage-amplified bioluminescent sensing of *Escherichia coli* O157:H7. *Anal. Bioanal. Chem.* **391:**507–514.

Riska, P. F., and W. R. Jacobs, Jr. 1998. The use of luciferase-reporter phage for antibiotic-susceptibility testing of mycobacteria. *Methods Mol. Biol.* **101:**431–455.

Riska, P. F., W. R. Jacobs, Jr., B. R. Bloom, J. McKitrick, and J. Chan. 1997. Specific identification of *Mycobacterium tuberculosis* with the luciferase reporter mycobacteriophage: use of *p*-nitro-α-acetylamino-β-hydroxy propiophenone. *J. Clin. Microbiol.* **35:**3225–3231.

Romero, P., L. Perry, M. Morgan, and B. Applegate. 2008. A *cobA* based bacteriophage reporter for detection of *Escherichia coli* O157:H7, p. 498–499. *Abstr. 108th Gen. Meet. Am. Soc. Microbiol.* American Society for Microbiology, Washington, DC.

Rusckowski, M., S. Gupta, G. Z. Liu, S. P. Dou, and D. J. Hnatowich. 2008. Investigation of four ^{99m}Tc-labeled bacteriophages for infection-specific imaging. *Nucl. Med. Biol.* **35:**433–440.

Sanders, C. A., D. M. Yajko, P. S. Nassos, W. C. Hyun, M. J. Fulwyler, and W. K. Hadley. 1991. Detection and analysis by dual-laser flow cytometry of bacteriophage T4 DNA inside *Escherichia coli. Cytometry* **12:**167–171.

Sanders, M. F. 9 December 2003. Methods of identifying bacteria of specific bacterial genus, species or serotype. U.S. patent 6,660,470.

Sarkis, G. J., W. R. Jacobs, Jr., and G. F. Hatfull. 1995. L5 luciferase reporter mycobacteriophages: a sensitive tool for the detection and assay of live mycobacteria. *Mol. Micobiol.* **15:**1055–1067.

Sasahara, K. C., and K. Boor. 2001. Detection of viable *Mycobacterium avium* subsp. *paratuberculosis* using luciferase reporter systems, abstr. 576. *Abstr. 101st Gen. Meet. Am. Soc. Microbiol.* American Society for Microbiology, Washington, DC.

Schmelcher, M., and M. J. Loessner. 2008. Bacteriophage: powerful tools for the detection of bacterial pathogens, p. 731–754. *In* M. Zourob, S. Elwary, and A. Turner (ed.), *Principles of Bacterial Detection: Biosensors, Recognition Receptors and Microsystems.* Springer, New York, NY.

Schofield, D. A., and C. Westwater. 2009. Phage-mediated bioluminescent detection of *Bacillus anthracis. J. Appl. Microbiol.* **107:**1468–1478.

Schuch, R., D. Nelson, and V. A. Fischetti. 2002. A bacteriolytic agent that detects and kills *Bacillus anthracis. Nature* **418:**884–889.

Sechter, I., C. B. Gerichter, and D. Cahan. 1975. Method for detecting small numbers of *Vibrio cholerae* in very polluted substrates. *Appl. Microbiol.* **29:** 814–818.

Seo, S., M. Dobozi-King, R. F. Young, L. B. Kish, and M. S. Cheng. 2008. Patterning a nanowell sensor biochip for specific and rapid detection of bacteria. *Microelectronic Eng.* **85:**1484–1489.

Shabani, A., M. Zourob, B. Allain, C. A. Marquette, M. F. Lawrence, and R. Mandeville. 2008. Bacteriophage-modified microarrays for the direct impedimetric detection of bacteria. *Anal. Chem.* **80:**9475–9482.

Singh, A., N. Glass, M. Tolba, L. Brovko, M. Griffiths, and S. Evoy. 2009. Immobilization of bacteriophages on gold surfaces for the specific capture of pathogens. *Biosens. Bioelectron.* **24:**3645–3651.

Singh, S., T. P. Saluja, M. Kaur, and G. C. Khilnani. 2008. Comparative evaluation of FASTPlaque assay with PCR and other conventional in vitro diagnostic methods for the early detection of pulmonary tuberculosis. *J. Clin. Lab. Anal.* **22:**367–374.

Smith, B. C., M. Izzo, and D. Smith. 2009. Detection of *S. aureus* skin and soft tissue infections by use of bacteriophage amplification technology, p. C-161. *Abstr. 109th Gen. Meet. Am. Soc. Microbiol.* American Society for Microbiology, Washington, DC.

Squirrel, D. J., and M. J. Murphy. 1994. Adenylate kinase as a cell marker in bioluminescent assays, p. 486–489. *In* A. K. Campbell, L. J. Kricka, and P. E. Stanley (ed.), *Bioluminescence and Chemiluminescence: Fundamentals and Applied Aspects.* John Wiley & Sons, Chichester, United Kingdom.

Squirrel, D. J., R. L. Price, and M. J. Murphy. 2002. Rapid and specific detection of bacteria using bioluminescence. *Anal. Chim. Acta* **457:**109–114.

Stanek, J. E., and J. O. Falkinham. 2001. Rapid coliphage detection assay. *J. Virol. Methods* **91:**93–98.

Stanley, E. C., R. J. Mole, R. J. Smith, S. M. Glenn, M. R. Barer, M. McGowan, and C. E. D. Rees. 2007. Development of a new, combined rapid method using phage and PCR for detection and identification of viable *Mycobacterium paratuberculosis* bacteria within 48 hours. *Appl. Environ. Microbiol.* **73:**1851–1857.

Stewart, G., S. A. A. Jassim, S. P. Denyer, P. Newby, K. Linley, and V. K. Dhir. 1998. The specific and sensitive detection of bacterial pathogens within 4 h using bacteriophage amplification. *J. Appl. Microbiol.* **84:**777–783.

Stewart, G. S. A. B., T. Smith, and S. P. Denyer. 1989. Genetic engineering for bioluminescent bacteria. *Food Sci. Technol. Today* **3:**19–22.

Sun, W., L. Brovko, and M. Griffiths. 2000. Use of bioluminescent *Salmonella* for assessing the efficiency of constructed phage-based biosorbent. *J. Ind. Microbiol. Biotechnol.* **25:**273–275.

Svensson, U. 1993. Assay of starter culture activity and detecting the presence of bacteriophage by a conductance method. *Int. Dairy J.* **3:**568.

Svensson, U. K. 1994. Conductimetric analyses of bacteriophage infection of two groups of bacteria in DL-lactococcal starter cultures. *J. Dairy Sci.* **77:** 3524–3531.

Tamarin, O., S. Comeau, C. Dejous, D. Moynet, D. Rebiere, J. Bezian, and J. Pistre. 2003a. Real time device for biosensing: design of a bacteriophage model using Love acoustic waves. *Biosens. Bioelectron.* **18:**755–763.

Tamarin, O., C. Dejous, D. Rebiere, J. Pistre, S. Comeau, D. Moynet, and J. Bezian. 2003b. Study of acoustic Love wave devices for real time bacteriophage detection. *Sens. Actuators B Chem.* **91:**275–284.

Tanji, Y., C. Furukawa, S. H. Na, T. Hijikata, K. Miyanaga, and H. Unno. 2004. *Escherichia coli* detection by GFP-labeled lysozyme-inactivated T4 bacteriophage. *J. Biotechnol.* **114:**11–20.

Thouand, G., P. Vachon, S. Liu, M. Dayre, and M. W. Griffiths. 2008. Optimization and validation of a simple method using P22::*luxAB* bacteriophage for rapid detection of *Salmonella enterica* serotypes A, B, and D in poultry samples. *J. Food Prot.* **71:**380–385.

Tolba, M. H., L. Brovko, and M. W. Griffiths. 2008. Bacteriophage-based biosorbent for specific capture, concentration and detection of bacteria, abstr. P-059/0764. *Abstr. 108th Gen. Meet. Am. Soc. Microbiol.* American Society for Microbiology, Washington, DC.

Tolba, M., O. Minikh, L. Y. Brovko, S. Evoy, and M. W. Griffiths. 2010. Oriented immobilization of bacteriophages for biosensor applications. *Appl. Environ. Microbiol.* **76:**528–535.

Turpin, P. E., K. A. Maycroft, J. Bedford, C. L. Rowlands, and E. M. H. Wellington. 1993. A rapid luminescent-phage based MPN method for the enumeration of *Salmonella typhimurium* in environmental samples. *Lett. Appl. Microbiol.* **16:**24–27.

Ulitzur, N., and S. Ulitzur. 2006. New rapid and simple methods for detection of bacteria and determination of their antibiotic susceptibility by using phage mutants. *Appl. Environ. Microbiol.* **72:** 7455–7459.

Ulitzur, S., and J. Kuhn. 1987. Introduction of *lux* genes into bacteria, a new approach for specific determination of bacteria and their antibiotic susceptibility, p. 463–472. *In* J. Scholmerich, R. Andreesen, A. Japp, M. Ernst, and W. G. Woods (ed.), *Bioluminescence and Chemiluminescence: New Perspectives.* John Wiley & Sons Inc., New York, NY.

Ulitzur, S. Y., and J. C. Kuhn. 29 August 1989. Detection and/or identification of microorganisms in a test sample using bioluminescence or other exogenous genetically introduced marker. U.S. patent 4,861,709.

Ulitzur, S., M. Suissa, and J. Kuhn. 1989. A new approach for the specific determination of bacteria and their antimicrobial susceptibilities, p. 235–240. *In* A. Balows, R. C. Tilton, and A. Turano (ed.), *Rapid Methods and Automation in Microbiology and Immunology.* Brixia Academic Press, Brescia, Italy.

Uttenthaler, E., M. Schraml, J. Mandel, and S. Drost. 2001. Ultrasensitive quartz crystal microbalance sensors for detection of M13 phages in liquids. *Biosens. Bioelectron.* **16:**735–743.

Varkey, S., K. Wing, F. Cooling, and D. R. DeMarco. 2006. Sensitivity and inclusivity of a *Listeria* genus PCR detection assay using a novel bacteriophage derived cell binding domain (CBD) and phage endolysin lysis, abstr. no. P-064, p. 453–454. *Abstr. 106th Gen. Meet. Am. Soc. Microbiol.* American Society for Microbiology, Washington, DC.

Waddell, T. E., and C. Poppe. 2000. Construction of mini-Tn*10luxABcam/Ptac*-ATS and its use for developing a bacteriophage that transduces bioluminescence to *Escherichia coli* O157:H7. *FEMS Microbiol. Lett.* **182:**285–289.

Wan, J. H., M. L. Johnson, R. Guntupalli, V. A. Petrenko, and B. A. Chin. 2007. Detection of *Bacillus anthracis* spores in liquid using phage-based magnetoelastic micro-resonators. *Sens. Actuators B Chem.* **127:**559–566.

Willford, J., and L. D. Goodridge. 2008. An integrated assay for rapid detection of *Escherichia coli* O157:H7 on beef samples. *Food Prot. Trends* **28:** 468–472.

Wolber, P. K., and R. L. Green. 1990. New rapid method for the detection of *Salmonella* in foods. *Trends Food Sci. Technol.* **1:**80–82.

Wu, Y., L. Brovko, and M. W. Griffiths. 2001. Influence of phage population on the phage-mediated bioluminescent adenylate kinase (AK) assay for detection of bacteria. *Lett. Appl. Microbiol.* **33:**311–315.

Yemini, M., Y. Levi, E. Yagil, and J. Rishpon. 2007. Specific electrochemical phage sensing for *Bacillus cereus* and *Mycobacterium smegmatis*. *Bioelectrochemistry* **70:**180–184.

APPLICATION OF BACTERIOPHAGES TO CONTROL PATHOGENS IN FOOD ANIMAL PRODUCTION

Lawrence D. Goodridge

4

INTRODUCTION

While many postharvest strategies have proven effective at reducing pathogen populations on carcasses derived from food-producing animals, Jordan et al. (1999) suggest that interventions conducted before animals enter the slaughter facility are even more effective at reducing carcass contamination. As such, preharvest interventions may have the greatest impact on controlling food-borne pathogens in food animals before they enter the processing plant. Furthermore, it has been suggested that the use of agents to suppress the concentration of food-borne pathogens such as *Escherichia coli* O157:H7 in the digestive tract appears necessary for preslaughter control to become effective (Jordan et al., 1999).

Preharvest agents used to control food-borne pathogens include probiotics and vaccines (Callaway et al., 2004). Successful preharvest strategies should reduce or eliminate harmful bacteria without having a negative effect on the animal or environment. The development of a biocontrol strategy, based on the use of bacteriophages, would seem to be ideal, due to the fact that they are natural, are ubiquitous in the environment, and are nontoxic to animals (Goodridge and Abedon, 2003). In addition, they are often highly specific to their host, unlike antibiotics, which are not specific and can lead to the destruction of beneficial bacteria (Jensen et al., 1998). Furthermore, phages are able to infect a single cell and produce multiple copies of themselves, which, when released through lysis of the host cell, allow for infection of other uninfected cells. In this way, phages can be thought of as a natural, self-amplifiable antimicrobial treatment (Levine and Bull, 1996).

Almost immediately following their discovery in 1915 (Twort, 1915), phages were considered to be an important tool in the biocontrol of bacterial pathogens (Goodridge and Abedon, 2003), and the term "bacteriophage (phage) therapy" is used to describe the application of phages to reduce bacterial infections in animals and bacterial contamination in foods. In research studies, phage therapy has shown some promise as an effective preharvest intervention by controlling food-borne pathogens in animals before they enter processing plants (Callaway et al., 2003; Fiorentin et al., 2005; Toro et al., 2005). Phage therapy is also an effective postharvest intervention, reducing pathogen contamination of foods (Brüssow,

Lawrence D. Goodridge, Department of Animal Sciences, Center for Meat Safety and Quality, Colorado State University, 350 West Pitkin Street, Fort Collins, CO 80523.

Bacteriophages in the Control of Food- and Waterborne Pathogens
Edited by Parviz M. Sabour and Mansel W. Griffiths, © 2010 ASM Press, Washington, DC

2005). Studies have demonstrated that phage therapy is effective against a broad range of food-borne pathogens belonging to the genera *Salmonella, Campylobacter, Listeria,* and *Escherichia* (Dykes and Moorhead, 2002; Goode et al., 2003; O'Flynn et al., 2004; Carlton et al., 2005; Higgins et al., 2005). With the current concern over the misuse of antibiotics, and the development of antibiotic resistance in bacteria, phages may play a leading role as a new class of antimicrobials in the agricultural and food industries.

APPLICATION OF PHAGES TO CONTROL PATHOGENS IN RUMINANTS

Salmonella, Campylobacter, and *E. coli* O157:H7 are common contaminants in ruminants (Callaway et al., 2004; Doyle and Erickson, 2006). These pathogens are commonly carried in the gastrointestinal tract asymptomatically; however, the pathogens can be shed through feces and transmitted to other livestock and the food supply. Many disease outbreaks have been linked to contaminated animal food products (Mead et al., 1999; Rangel et al., 2005; Centers for Disease Control and Prevention, 2006). Several studies have investigated the use of phages, either alone or in a cocktail, to control food-borne pathogens in sheep, cattle, and poultry (Bach et al., 2003; Callaway et al., 2003; Fiorentin et al., 2005; Toro et al., 2005). The use of only one phage in some experiments has led to resistance (Bach et al., 2003; Callaway et al., 2003). Other studies have shown that the use of more than one phage has decreased the chance of resistance developing (O'Flynn et al., 2004; Tanji et al., 2004; Toro et al., 2005). A cocktail of phages that use different receptors on the host cell would delay the formation of phage resistance (Tanji et al., 2004; Chase et al., 2005).

E. coli O157:H7 continues to be recognized as an important cause of food-borne illness. This pathogen causes a wide spectrum of disease, including gastrointestinal distress (nonbloody diarrhea and hemorrhagic colitis) and extraintestinal diseases (hemolytic-uremic syndrome and thrombotic thrombocytopenic purpura) (Karmali et al., 1985).

Cattle and other ruminants are considered to be the principal reservoirs of *E. coli* O157:H7 (Zhao et al., 1995). During slaughter, if care is not taken, the contents of the intestines, fecal material, or dust on the hide may contaminate meat, and this is the main transmission route to humans. Transmission to humans also occurs through the use of poor hygienic procedures in the handling of dairy and beef cattle. Several studies have indicated that the concentration at which *E. coli* O157:H7 is shed in the feces of colonized cattle varies from animal to animal, with a range from 10^2 to 10^5 CFU/g (Zhao et al., 1995). High-shedding animals may pose an elevated risk of contaminating the food chain if presented to slaughter. Therefore, a preharvest method to control the presence of *E. coli* O157:H7 in ruminants prior to slaughter would be potentially useful, since such a strategy could be used to reduce the burden of *E. coli* O157:H7 entering the slaughter plant, allowing postharvest interventions to be more effective.

Phages have been successful at reducing *E. coli* O157:H7 during in vitro studies (Bach et al., 2003; Chase et al., 2005). In contrast, the use of phages to reduce *E. coli* O157:H7 shedding in ruminants has not been as successful (Bach et al., 2003; Callaway et al., 2003). Several reasons for this may be development of resistance, low multiplicity of infection (MOI), or inactivation of the phage. However, the use of phage cocktails shows promise in controlling the growth of *E. coli* O157:H7 in ruminants. Callaway et al. (2003) inoculated sheep with phage specific against *E. coli* O157:H7. This treatment caused a decrease in the concentration of *E. coli* O157:H7 throughout the gastrointestinal tract; however, the differences were not statistically significant (Callaway et al., 2003).

A study by Bach et al. (2003) evaluated the ability of a single phage (DC22) to eliminate *E. coli* O157:H7 in an artificial rumen system and in experimentally inoculated sheep. A host-range study was also performed and showed that 23 out of 23 *E. coli* O157:H7 strains were sensitive to DC22, while 12 non-O157:H7 strains of enterohemorrhagic *E. coli* were not sensitive to the phage. Results from this study indicated that phage DC22, with an MOI of 10^5, was successful at eliminating *E. coli* O157:H7 in the rumen. However, after inoculation of the sheep, phage DC22 had no effect on the fecal shedding of *E. coli* O157:H7. This might have been due to an insufficient MOI in the sheep compared to the high MOI in the enclosed fermentation system. Also, some nonspecific binding of the phage to food particles and other debris might have occurred in the rumen. The use of one phage in this study strengthens the idea that a mixture of phages might be more effective at controlling *E. coli* O157:H7 in livestock.

Other research groups have evaluated the use of phage cocktails to decrease various bacterial pathogens in sheep during brief time periods. Callaway and coworkers (2006) anaerobically isolated phages that targeted *E. coli* O157:H7 from fecal samples collected from commercial feedlot cattle in the central United States. The spectrum of activity of the phages was determined, and the phages were combined to form a cocktail of phage for in vivo studies. When a 21-phage cocktail was inoculated into sheep artificially contaminated with *E. coli* O157:H7, intestinal populations of *E. coli* O157:H7 were decreased ($P < 0.05$) in the cecum and rectum. When sheep were dosed with 10^5, 10^6, or 10^7 PFU of the phage cocktail, the addition of 10^5 PFU (MOI of 10) was the most effective at reducing *E. coli* O157:H7 populations throughout the gastrointestinal tract. These results indicate that properly selected phages can be used to reduce *E. coli* O157:H7 in food animals. The authors concluded that phage could be an important part of an integrated food-borne pathogen reduction program. Raya et al. (2006) isolated bacteriophage CEV1 from sheep resistant to *E. coli* O157:H7 colonization and used this phage to reduce populations of *E. coli* O157 in sheep. In vitro, CEV1 efficiently infected *E. coli* O157:H7 grown both aerobically and anaerobically. Four sheep were treated once orally with 10^{11} PFU of phage CEV1, 3 days after challenge with *E. coli* O157:H7. Sheep receiving a single oral dose of CEV1 showed a 2- to 3-log-unit reduction in cecal and rectal *E. coli* O157:H7 levels within 2 days compared to levels in the controls, although rumen concentrations remained unchanged.

These studies show that phage therapy may be useful when used to decrease *E. coli* O157:H7 counts immediately before slaughter, although concerns remain with respect to environmental dissemination, and contamination of hides by *E. coli* O157:H7 during production. The possibility of combining phage therapy with vaccines and probiotics to reduce *E. coli* O157:H7 concentrations in the environment, on hides, and in the gastrointestinal tract of animals is an intriguing one that deserves further study.

To evaluate the idea that a mixture of phages (cocktail) is more effective at controlling *E. coli* O157:H7, Kudva et al. (1999) examined the ability of three phages (KH1, KH4, and KH5) to eliminate concentrations of *E. coli* O157:H7 in broth. Results indicated that *E. coli* O157:H7 cells were sensitive to all three phages, but after 5 days of phage infection the surviving *E. coli* O157:H7 cells were resistant to all three phages. In studies where only one specific phage was used for each broth culture, the surviving *E. coli* O157:H7 cells mutated and became resistant to the phage (Kudva et al., 1999). Resistance can develop in phages, but using a cocktail of phages can lower the possibility of bacteria becoming resistant (Kudva et al., 1999; Tanji et al., 2004). For example, complete bacterial elimination was observed only when cultures were infected with all three phages, indicating the combined effects of the phages (Kudva et al., 1999).

Chase et al. (2005) used a receptor-modeling procedure to produce a phage cocktail to reduce *E. coli* O157:H7 in cattle. This approach allowed the researchers to produce a phage cocktail that contained phages that used multiple different receptors on the cell surface. The rationale behind this approach was based on the results of a study conducted by Tanji et al. (2004), which showed that two phages that used different receptors delayed the formation of phage-resistant cells. In the Chase et al. (2005) study, it was hypothesized that the combination of more than two phages that each used different bacterial receptors would further delay the formation of phage-resistant cells. A total of 56 different phages, with various degrees of specificity for *E. coli* O157:H7, were screened on different *E. coli* strains to determine the receptors used by each phage. A series of *E. coli* K-12 outer membrane protein mutants and a series of *E. coli* lipopolysaccharide (LPS) outer core oligosaccharide mutants were used for the phage receptor studies. Each phage was individually tested against each bacterial isolate. The specificity data were used to generate a database of the receptors that each phage used. Some phages used more than one receptor. The database was used to construct a phage cocktail consisting of 37 different phages, which was screened against 58 *E. coli* O157:H7 isolates. The cocktail produced complete clearing on all of the isolates, and no resistant colonies were observed. In addition, the phage cocktail lysate was dropped on an agar plate containing a lawn of a mixture of all 58 of the *E. coli* O157:H7 isolates. After overnight incubation, no colonies were observed within the zone of clearing, indicating the absence of phage-resistant cells.

The cocktail was examined for its ability to reduce the presence of *E. coli* O157:H7 in bovine fecal slurries and in calves. In an anaerobic in vitro model, the phage cocktail completely eliminated a strain mixture of 10^4 CFU/ml of *E. coli* O157:H7 from bovine fecal slurries within 4 h. To evaluate the phage cocktail in live animals, a total of 14 Black Angus calves ranging from 4 to 6 months of age were orally inoculated with 10^8 CFU *E. coli* O157:H7. The in vivo experiments consisted of two trials. The first trial evaluated ileal samples and the second trial evaluated fecal samples for the presence of *E. coli* O157:H7 and phages. In the first trial, a significant decline in the concentration of *E. coli* O157:H7 in the ileal samples was observed at 8 h (P = 0.05). However, the concentration of *E. coli* O157:H7 increased back to the concentration of the control samples at 16 h. In the second trial, shedding of *E. coli* O157:H7 decreased significantly in the treated group (P = 0.05) at 24 h. As with the ileal samples, an increase in the concentration of *E. coli* O157:H7 was observed at 36 h in the fecal samples. The increases in cell concentration were associated with a decrease in phage concentration. None of the *E. coli* O157:H7 cultured from ileal or fecal samples showed resistance to the phage cocktail.

These results highlight the ability of the 37-phage cocktail to eliminate *E. coli* O157:H7 in fecal slurries and to reduce the concentration of *E. coli* O157:H7 in calves, without the formation of phage-resistant mutants. The receptor-modeling approach appears to be effective in the design of phage cocktails to reduce the presence of *E. coli* O157:H7 in animals. Continuous dosing of the phage cocktail to the calves may have eliminated the increase in concentration of the *E. coli* O157:H7 cells following the initial decrease.

The above-mentioned studies are similar in that all animals were orally inoculated with the phage cocktails. This makes sense from a practical standpoint in that phages could be delivered to animals via drinking water and feed. However, the phages can become diluted in the rumen, which can hold up to 25 gallons of material (according to information on the University of Minnesota Extension website [http://www.extension.umn.edu/distribution/livestocksystems/components/DI0469-02.html]). In addition, phages can become inactivated due to nonspecific binding to ingesta in the rumen. Finally, in order for the phages to be delivered to the lower gastrointestinal tract, they must first traverse the abomasum, which contains hydrochloric acid and digestive enzymes needed for the breakdown of feeds (http://www.extension.umn.edu/

distribution/livestocksystems/components/DI0469-02.html). The viability of orally administered phage may be rapidly reduced under the acidic conditions of the abomasum (Smith et al., 1987) and in the presence of enzymes and other digestive compounds such as bile. Without protection, the phages might not survive gastric passage and would thus not be infective in the intestine. To address this issue, Ma et al. (2008) developed a method to microencapsulate phages as a way to protect them from the low pH of the stomach. In this work, the authors microencapsulated phage Felix-O1 using a chitosan-alginate-$CaCl_2$ system. In vitro studies were used to determine the effects of simulated gastric fluid and bile salts on the viability of free and encapsulated phage. Unencapsulated Felix-O1 was found to be extremely sensitive to acidic environments and was not detectable after a 5-min exposure to pH values below 3.7. In contrast, the number of microencapsulated phage decreased by only 0.67 log units, even at pH 2.4, for the same period of incubation. The viable count of microencapsulated phage decreased by 2.58 log units during a 1-h exposure to simulated gastric fluid with pepsin at pH 2.4. After 3 h of incubation in 1 and 2% bile solutions, the free phage count decreased by 1.29 and 1.67 log units, respectively, while the viability of encapsulated phage was fully maintained. Encapsulated phage was completely released from the microspheres upon exposure to simulated intestinal fluid (pH 6.8) within 6 h. The encapsulated phage retained full viability when stored at 4°C for the duration of the testing period (6 weeks). With the use of trehalose as a stabilizing agent, the microencapsulated phage had a 12.6% survival rate after storage for 6 weeks. This research demonstrates that the use of encapsulation techniques potentially allows phages to remain bioactive, indicating that such methods may facilitate delivery of phage to the gastrointestinal tract.

Other researchers have utilized alternative approaches to solve the issue of phage delivery to the lower gastrointestinal tract. For example, Sheng and colleagues (2006) evaluated the phage-based rectal treatment of cattle based on a previous study (Naylor et al., 2003) which showed the recto-anal junction to be the primary site of *E. coli* O157:H7 colonization in cattle. A previously characterized O157-specific lytic bacteriophage, KH1 (Kudva et al., 1999), and a newly isolated phage designated SH1 were tested, alone or in combination, for efficacy in reducing intestinal *E. coli* O157:H7 in animals. Oral treatment with phage KH1 did not reduce the intestinal *E. coli* O157:H7 concentration in sheep. To optimize bacterial carriage and phage delivery in cattle, *E. coli* O157:H7 was applied rectally to Holstein steers 7 days before the administration of 10^{10} PFU of SH1 and KH1. Phages were applied directly to the recto-anal junction mucosa at an MOI of 10^2. In addition, phages were maintained at a concentration of 10^6 PFU/ml in the drinking water of the treatment group. Results showed that this approach reduced the average number of *E. coli* O157:H7 CFU among phage-treated steers compared to control steers ($P < 0.05$) but did not eliminate the bacteria from the majority of steers.

In a related study, Rozema et al. (2009) compared oral and rectal administration of *E. coli* O157-specific bacteriophages for efficacy in reducing or eliminating the fecal shedding of *E. coli* O157:H7 by experimentally inoculated steers. Fecal shedding of nalidixic acid-resistant (Nal^r) *E. coli* O157:H7 was monitored over 83 days after oral (ORL; 3.3×10^{11} PFU), rectal (REC; 1.5×10^{11} PFU), both oral and rectal (O+R; 4.8×10^{11} PFU), or no (CON; control) treatment with a four-strain O157-specific phage cocktail in multiple doses. ORL steers produced the fewest Nal^r *E. coli* O157:H7 culture-positive samples ($P < 0.06$) compared with REC and O+R steers, but this number was not significant ($P = 0.26$) when compared to CON steers. The overall mean *E. coli* O157:H7 shedding level (log CFU per g of feces) was higher for REC steers ($P < 0.10$) than for steers of the other treatment groups. Despite the shedding of higher mean phage levels (log PFU per g of feces) by ORL and O+R than by CON and REC steers, there was no difference ($P > 0.05$) in the number of *E. coli* O157-positive samples among treatments. Phages were isolated from CON steers, indicating that these steers acquired phages from the environment and shed the phages at a level similar to that of REC steers ($P = 0.39$). The researchers concluded that continuous phage therapy may be an efficacious method for reducing shedding of *E. coli* O157:H7 in cattle, providing that the host bacterium does not develop resistance (Rozema et al., 2009).

With the exception of the Chase et al. (2005) study, the studies described above were conducted on adult animals. Phage therapy may be more efficacious in calves, since adult cattle may have been exposed to *E. coli* O157:H7 to a greater extent than calves (Laegreid et al., 1999; Gannon et al., 2002). One particular study highlights the effectiveness of using phages to reduce *E. coli* O157:H7 concentrations in young cattle. Waddell et al. (2000) orally infected weaned 7- to 8-week-old calves with *E. coli* O157:H7 (10^9 CFU) on day 0, and then treated some of the animals with a cocktail of six phages (10^{11} PFU) on days −7, −6, −1, 0, and 1. The phage cocktail was delivered in a solution containing calcium carbonate, which buffered stomach acid (Smith et al., 1987), and the day 0 treatment was given 4 h before the *E. coli* O157:H7 inoculum. The results indicated that most of the untreated calves shed *E. coli* O157:H7 in their feces for at least 12 to 16 days. In contrast, the treated calves stopped shedding *E. coli* O157:H7 after day 8, which corresponded with a dramatic increase (10^9 to 10^{11} PFU) in the concentration of phages that were shed in the feces of the animals. The increase in the number of phages that were excreted was due to phage replication in the calves, as determined by the fact that such a result was not observed in uninfected control calves that were inoculated only with the phage cocktail. It should be noted that the above study, in which phages were given to the calves prior to *E. coli* O157:H7 inoculation, may have allowed the phages to attack the *E. coli* O157:H7 cells before they became attached to the intestinal epithelium, allowing for a more efficient reduction in bacterial population. Once the bacteria attach to the epithelium and colonize the animal, it is often difficult for the phages to efficiently destroy all cells because they are embedded and therefore hard to reach. Furthermore, the inoculation of the calves with the cocktail prior to *E. coli* O157:H7 infection may be an unrealistic scenario, since calves may be infected by *E. coli* O157:H7 soon after birth and during their first year (Hancock et al., 1998; Laegreid et al., 1999; Gannon et al., 2002), with studies showing that infection often occurs before weaning, prior to entry into feedlots (Laeigreid et al., 1999). Therefore, more studies on calves should be conducted in which *E. coli* O157:H7 is allowed to colonize the gastrointestinal tract prior to phage introduction. Niu and colleagues (2008) conducted an investigation to determine whether several of the phages from the Waddell et al. (2000) study could be used to control *E. coli* O157:H7 in older animals. In this work, the researchers combined four of the original six phages into a cocktail and evaluated their use as a preharvest strategy to control *E. coli* O157:H7 in feedlot steers. The results indicated that the phage therapy approach reduced but did not eliminate *E. coli* O157:H7 shedding, possibly highlighting age-related differences in the architecture of the bovine gastrointestinal tract (Johnson et al., 2008).

Collectively, these studies indicate the possibility of using phages to control *E. coli* O157:H7 colonization and shedding in cattle, although these studies reveal a discipline that is still in its infancy. Many more studies need to be conducted to determine adequate phage dose, duration of phage treatment, and standardized methods to protect the phage from the acidic environment of the stomach.

APPLICATION OF PHAGES TO CONTROL PATHOGENS IN POULTRY

The majority of phage therapy applications to control food-borne pathogens in live animals have been conducted in poultry. *Salmonella* and *Campylobacter* are food-borne pathogens that can be found in contaminated poultry (McCrea et al., 2006) and, as such, have received the most attention. Poultry and egg products are important sources of human pathogens. *Salmonella* spp. are a major cause of food-borne infections following consumption of contaminated poultry products. The problem is associated with the symptomless carriage of salmonellae by a proportion of all live chickens and turkeys and the subsequent contamination of products used as human food (Mead and Barrow, 1990). Thus far, all attempts aimed at eliminating salmonellae from poultry flocks have failed, likely due to the fact that there are numerous routes of *Salmonella* transmission between the birds and their rearing environment. In general, phage-based approaches aimed at controlling *Salmonella* in poultry have reduced but not eliminated the pathogen. For example, newly hatched chicks challenged with *Salmonella enterica* serovar Typhimurium followed by oral introduction (10^4 PFU) or feed-based introduction (10^3 PFU/g) of phage AB2 for 7 days did not show decreased levels of serovar Typhimurium in cecal contents (Berchieri et al., 1991). Phage replication to high concentrations occurred 4 days postinfection, but no phages were detected when the concentration of the challenge strain fell below 10^6 CFU/ml of cecal contents. The researchers reported a high percentage of phage-resistant serovar Typhimurium isolates, which occurred in 34 to 82% of 50 bacterial isolates tested on each of days 2, 4, 7, and 10 postinfection. On the other hand, when another phage (phage 2.2) was employed, significant reductions in mortality were observed over a 21-day period (from 56 to 20%), and phage 2.2 was also efficacious against two other challenge strains. Still, a high concentration of the phage (10^{11} PFU) was required for results to be achieved, suggesting that a passive mechanism such as lysis from without, and not an active mechanism (phage replication), was the main mode of action in reducing the challenge strains. Also, the concentration of the challenge strain was reduced by greater than 2 log units within 3 to 6 h, but these reductions were transient. Challenge strain concentrations were 1 log unit lower in the liver than in control chicks after 24 and 48 h (Berchieri et al., 1991).

Similarly, Fiorentin et al. (2005) isolated phages from free-range chickens and tested them in combination as a therapeutic agent for reducing the concentration of *S. enterica* serovar Enteritidis phage type 4 (PT4) in the ceca of broilers. One-day-old broilers infected with serovar Enteritidis PT4 were orally treated on the seventh day of age with a mixture of 10^{11} PFU of each of three phages. Five days after treatment, the phage-treated group showed a reduction of 3.5 orders of magnitude in CFU of serovar Enteritidis PT4 per g of cecal content. Samples collected at 10, 15, 20, and 25 days after treatment revealed that treated birds still had lower CFU of serovar Enteritidis PT4 per g of cecal content. These data provide evidence that a mixture of phages may be efficacious in reducing serovar Enteritidis PT4 concentration in the ceca of broilers and therefore reducing contamination of poultry products by this food-borne pathogen.

Andreatti Filho et al. (2007) used cocktails of 4 different phages obtained from commercial broiler houses (CB4Ø) and 45 phages from a municipal wastewater treatment plant (WT45Ø) to effect reduction of *Salmonella* serovar Enteritidis in vitro and in experimentally infected chicks. In one experiment, an in vitro crop assay was conducted with selected phage concentrations (10^5 to 10^9 PFU/ml) to determine the ability to reduce serovar Enteritidis in the simulated crop environment. After 2 h at 37°C, CB4Ø or WT45Ø reduced the serovar Enteritidis concentration by 1.5 or 5 logs, respectively, as compared with a control. However, the CB4Ø cocktail did not affect total serovar Enteritidis recovery after 6 h (the concentration returned to pre-phage cocktail treatment levels), whereas WT45Ø resulted in up to a 6-log reduction of serovar Enteritidis. In another experiment, day-of-hatch chicks were challenged orally with 9×10^3 CFU/chick of serovar Enteritidis and treated via oral gavage with 1×10^8 CB4Ø PFU/chick, 1.2×10^8 WT45Ø PFU/chick, or a combination of both, 1 h after challenge with serovar Enteritidis. All treatments significantly reduced serovar Enteritidis recovered from cecal tonsils at 24 h as compared with untreated controls, but no significant differences were observed at 48 h following treatment. The authors concluded that the data suggested that some phages can be efficacious in reducing serovar Enteriditis colonization in poultry during a short period, but with the phages and methods presently tested, persistent reductions were not observed (Andreatti Filho et al., 2007). As described with the Chase et al. (2005) study, continuous dosing of the phages may have led to sustained reduction of the bacterial concentration.

Sklar and Joerger (2001) also evaluated the ability of lytic phages to reduce *Salmonella* serovar Enteritidis. In this work, phages capable of lysing a nalidixic acid-resistant serovar Enteritidis strain were tested for the ability to reduce cecal *Salmonella* counts in young chickens infected with the bacterium. Analysis of cloacal swabs suggested that the phage treatment could reduce shedding of Nal^r serovar Enteritidis, but average counts of between 10^5 and 10^7 CFU of Nal^r serovar Enteritidis per g of cecum were obtained from phage-treated 14-day-old birds, even when more than 10^7 PFU of phage were present per g of cecal content. The average cecal Nal^r serovar Enteritidis counts were generally between 0.3 and 1.3 orders of magnitude lower in phage-treated chickens than in untreated control birds. The difference in counts was statistically insignificant in three animal trials, but significant in two trials when feed particles were used as delivery vehicles for the phage. Although some of the Nal^r serovar Enteritidis in the ceca of phage-treated chickens had developed resistance to some of the phage used, environmental and physical factors may have also contributed to the failure of the phage to substantially reduce the counts. For example, it is possible that the *Salmonella* may have attached to and invaded epithelial cells, which would have made it impossible for phage to attack the cells (Sklar and Joerger, 2001).

Toro et al. (2005) used a combination of three *Salmonella*-specific bacteriophages and competitive exclusion (CE) bacteria to reduce *Salmonella* colonization in experimentally infected chickens. A cocktail of the phages was administered orally to the chickens several days prior to and after *Salmonella* challenge but not simultaneously. The phages that comprised the cocktail were readily isolated from the feces of the phage-treated chickens approximately 48 h after administration. A CE product consisting of a defined culture of seven different microbial species was used either alone or in combination with phage cocktail treatment. The CE product was administered orally at hatch. *Salmonella* counts in the intestine, ceca, and a pool of liver/spleen were evaluated in *Salmonella*-challenged chickens treated with the phage cocktail or with the cocktail and CE. In both trials, a beneficial effect of the phage treatment on weight gain performance was evident. A reduction in *Salmonella* counts was detected in the cecum and ileum of phage-treated, CE-treated, and phage/CE-treated chickens as compared with untreated birds. In trial 1, phage treatment reduced *Salmonella* counts to marginal levels in the ileum and reduced counts sixfold in the ceca. The CE and phage treatments showed differences in the reduction of *Salmonella* counts after challenge between specimens obtained at days 4 and 14 postchallenge in ceca, liver/spleen, and ileum during trial 2. The results of this work indicate that a combination of a phage cocktail and a CE product was capable of reducing *Salmonella* colonization of experimentally infected chickens (Toro et al., 2005).

Atterbury and colleagues (2007) isolated 232 *Salmonella* bacteriophages from poultry farms, abattoirs, and wastewater. Three phages exhibiting the broadest host ranges against *S. enterica* serovars Enteritidis, Hadar, and Typhimurium were characterized further by determining their morphology and lytic activity in vitro. These phages were then administered in antacid suspensions to birds experimentally colonized with specific *Salmonella* host strains. The first phage reduced serovar Enteritidis cecal colonization by more than 4.2 $\log_{10}$ CFU within 24 h compared with controls. Administration of the second phage reduced serovar Typhimurium by more than 2.19 $\log_{10}$ CFU within 24 h. The third phage was ineffective at reducing serovar Hadar colonization. Phage resistance occurred at a frequency commensurate with the titer of phage being administered, with larger phage titers resulting in a greater proportion of resistant *Salmonella* strains. The researchers concluded that the selection of appropriate phages and optimization of both the timing and method of phage delivery are key factors in the successful phage-mediated control of *Salmonella* in broiler chickens.

Recently, Borie et al. (2008) isolated three different lytic phages from the sewage system of commercial chicken flocks and used them to reduce *S. enterica* serovar Enteritidis colonization in experimental chickens. Ten-day-old chickens were challenged with 10^5 CFU/ml of a serovar Enteritidis strain and treated by coarse spray or drinking water with a cocktail of the three phages at an MOI of 10^3 PFU 24 h prior to serovar Enteritidis challenge. Chickens were euthanized at 20 days of age for individual serovar Enteritidis detection, quantitative microbiology, and phage isolation from the intestine and from a pool of organs. The results showed that aerosol-spray delivery of bacteriophages significantly reduced the incidence of serovar Enteritidis infection in the chicken group ($P = 0.0084$) to 72.7% as compared with the control group (100%). In addition, serovar Enteritidis counts showed that phage delivery, both by coarse spray and drinking water, reduced the intestinal serovar Enteritidis colonization ($P < 0.01$ and $P < 0.05$, respectively). Bacteriophages were isolated at 10 days postinfection from the intestine and from pools of organs of bacteriophage-treated chickens. It was concluded that the phage treatment, either by aerosol spray or drinking water, may be a plausible alternative to antibiotics for the reduction of *Salmonella* infection in poultry (Borie et al., 2008).

One possible concern with respect to phage-mediated *Salmonella* control in poultry arises due to the presence of prophages within many wild-type *Salmonella* isolates (Schicklmaier et al., 1998). It is possible that homologous recombinations between prophages and newly introduced phage genomes could take place. In addition, Figueroa-Bossi and Bossi (1999) have shown that certain prophages contribute to virulence in serovar Typhimurium, and it is conceivable that new phage recombinants could enhance virulence even more. Currently, there are no methods available to predict how likely such genetic exchanges would be and whether they would constitute a serious concern for phage therapy (Sklar and Joerger, 2001).

In addition to trying to control the presence of food-borne pathogens at the sites of colonization, an alternative approach is to attempt to reduce food-borne bacterial contamination on the hides and feathers of the food animals. This approach is based on the fact that research studies have demonstrated the need to reduce the presence of *E. coli* O157:H7 and *Salmonella* spp. on the hide, as it has been shown that a high occurrence of these pathogens on the hide greatly increases the risk of contamination of the carcass (Brichta-Harhay et al., 2008). Similarly, reducing the amount of *Salmonella* coming into the poultry processing environment by reducing *Salmonella* contamination on the feathers of live poultry before processing should help existing postharvest interventions operate more effectively. OmniLytics, Inc., a Utah-based phage therapy company, has received "No Objection" status from the U.S. Department of Agriculture's Food Safety Inspection Service (USDA/FSIS) to apply phages to the hides of cattle to reduce levels of *E. coli* O157:H7 and *Salmonella* spp. (OmniLytics, Inc., 2007a, 2007b). More recently, the company received approval from the USDA/FSIS to spray live poultry with phages to reduce *Salmonella* on feathers (OmniLytics, Inc., 2008). Others have investigated the possibility of reducing *Salmonella* spp. on the poultry skin. For example, Goode et al. (2003) applied lytic phages to chicken skins that had been experimentally contaminated with 10^3 to 10^4 CFU/cm^2 of *Salmonella* and *Campylobacter*. The *Salmonella*-

treated samples were swabbed with phage 29C (10^8 PFU/cm^2). The *Campylobacter*-treated samples were swabbed with a *C. jejuni* typing phage 12673 (10^6 PFU/cm^2). After 24 h, the *Salmonella* counts from the phage-treated samples were completely eliminated and the mean plate count from the non-phage-treated samples was 2.03 CFU/cm^2. The *Campylobacter* counts from the phage-treated samples were reduced by 95% and the mean plate count from the non-phage-treated samples was 2.99 CFU/cm^2. A higher MOI in the *Salmonella* study (10^5) compared with the *Campylobacter* study (10^2) resulted in a greater reduction of the pathogen.

In addition to *Salmonella* spp., phage therapy has also been investigated in an attempt to control *Campylobacter* spp. in poultry. Several studies have suggested that the consumption of undercooked poultry and/or the handling of raw poultry are risk factors for human *Campylobacter* infection and illness (Skirrow, 1982; Kapperud et al., 1992; Blaser, 1997; Altekruse et al., 1999). A large number of serotypes of *C. jejuni* isolated from chicken carcasses are frequently linked to human cases of campylobacteriosis, thus confirming that poultry is an important contributor in the epidemiology of human campylobacteriosis (Stern and Kazmi, 1989).

Wagenaar and coworkers (2005) conducted two series of experiments dealing with preventative and therapeutic aspects of phage-based control of *C. jejuni* in broiler chickens. In the preventative study, 10-day-old broiler chickens were challenged with *C. jejuni* strain C356 on day 4 of a 10-day course of oral gavage with 10^{10} PFU of phage 71. During the therapeutic study, 15-day-old broilers were first inoculated with the challenge strain (*C. jejuni* C356) 5 days prior to phage introduction, which was accomplished by daily oral gavage with 10^{10} PFU of phage 71 for a period of 6 days. The concentrations of the phage and challenge strain were monitored in cecal contents until the broilers were 39 days old. Results from the preventative study indicated that the phage delayed but did not prevent colonization of the broilers with *C. jejuni*. For example, the concentration of *C. jejuni* was initially 2 logs lower than in controls, and then stabilized at approximately 1 log unit lower than in controls. Additionally, phage 71 and *C. jejuni* concentrations exhibited fluctuations, which are typically observed in naturally infected flocks due to alternating cycles of host replication and phage amplification (Atterbury et al., 2005). It is possible that the fluctuations in host and phage concentrations were due to a process in which the host strain became resistant to the phage and was therefore able to grow unrestricted until phage mutants appeared that were able to infect the bacteria. Several studies have indicated that during serial passages of phages and their bacterial hosts, the phages are not constrained in their ability to coevolve with bacteria. For example, long-term antagonistic coevolution of *Pseudomonas fluorescens* and a phage were observed over multiple cycles of defense (i.e., development of phage-resistant bacterial mutants) and counterdefense (phages isolated from two subsequent transfers showed consistently greater infectivity to bacteria) (Buckling and Rainey, 2002). Similar data have also been obtained using a T-even phage, PP01, and its bacterial host, *E. coli* O157:H7 (Mizoguchi et al., 2003). A series of bacterial mutants were observed when the phage and bacteria were grown in continuous culture over a period of 200 h. The mutants differed in colony morphology, the nature of the PP01 receptors OmpC and LPS, and phage susceptibility. Phage PP01 responded to the presence of the bacterial mutants by broadening its host range. The system eventually reached a coexistence of PP01 and *E. coli* O157:H7, both at high concentrations, and the system continued to evolve. Collectively, the dynamics of both interacting systems were largely determined by the trade-offs between resistance to phage, which is usually costly from a metabolic standpoint (Brüssow, 2005), and competiveness with the parental strain for limiting resources.

In the therapeutic experiment, counts of *C. jejuni* were immediately reduced by greater than 3 logs for several days, but the counts eventually stabilized at approximately 1 log unit lower than in the control broilers. These results highlighted a potential strategy by which broilers could be given a phage treatment to reduce *C. jejuni* counts shortly before slaughter. To evaluate such an approach, the researchers conducted an experiment in which broilers were challenged 10 days prior to slaughter with the *C. jejuni* challenge strain. The broilers were given a mixture of two phages including the original phage 71 and another phage, designated as 69. Seven days following challenge, the phages were introduced to the broilers daily as described for a period of 4 days. Following an initial drop in concentration of 1.5 logs, the challenge strain concentration stabilized at 1 log unit lower than in the controls, which is consistent with the results obtained in the earlier experiments (Wagenaar et al., 2005). The researchers concluded that the use of phage therapy immediately before slaughter was feasible.

Loc Carrillo and coworkers (2005) developed experimental models of *Campylobacter* colonization of broiler chickens by using low-passage *C. jejuni* isolates HPC5 and GIIC8 from United Kingdom broiler flocks. The models were used to evaluate phage-based therapy to control *C. jejuni* in broilers. The screening of 53 lytic phage isolates against a panel of 50 *Campylobacter* isolates from broiler chickens and 80 strains isolated after human infection identified two phage candidates with broad host lysis. These phages, CP8 and CP34, were orally administered in antacid suspension, at different dosages, to 25-day-old broiler chickens experimentally colonized with the *C. jejuni* broiler isolates. Phage treatment of *C. jejuni*-colonized birds resulted in *Campylobacter* counts decreasing between 0.5 and 5 $\log_{10}$ CFU/g of cecal contents compared to untreated controls over a 5-day period. These reductions were dependent on the phage-*Campylobacter* combination, the dose of phage applied, and the time elapsed after administration. Campylobacters resistant to phage infection were recovered from phage-treated chickens at a frequency of less than 4%. These resistant types were compromised in their ability to colonize experimental chickens and rapidly reverted to a phage-sensitive phenotype in vivo. Scott et al. (2007) further reported that in chickens, phage resistance is infrequent due to the fact that the mutants that escape phage attack are not able to colonize poultry, so readily revert back to colonization-proficient phage-sensitive forms. The researchers observed that phage resistance is generated by reversible genomic scale inversions, leading to the activation of an unrelated phage integrated into the bacterial genome. The researchers concluded that the data suggest that phage therapy of *C. jejuni* is a sustainable measure to reduce poultry contamination (Scott et al., 2007).

Nevertheless, several studies appear to challenge the idea that phage resistance in *C. jejuni* is of little consequence. These studies indicate that circumvention of phage treatments through succession, where a given phage-sensitive population is replaced by strains from the same species or another *Campylobacter* species that are resistant to the phage population, is a likely scenario. For example, when a longitudinal study of phages and their hosts in three flocks was conducted, three genetically distinct *C. jejuni* types insensitive to the resident phages became dominant in two of the flocks, and *C. jejuni* isolates from the third flock were also insensitive to the phages but were genetically distinct from the first two flocks (Connerton et al., 2004). The researchers concluded that the problem was not phage resistance, but rather the arrival of new *C. jejuni* genotypes that were inherently resistant to the resident phages.

El-Shibiny et al. (2005) isolated and enumerated *C. jejuni* and *Campylobacter coli* and *Campylobacter*-specific phages during the rearing cycle of free-range (56 days) and organic chickens (73 days) at 3-day intervals from hatching until slaughter. In both flocks, *C. jejuni* was the initial colonizer but *C. coli* was detected more frequently in animals 5 weeks of age or older. The presence of phages in chickens was observed several days after *Campylobacter* colonization and was more prevalent in chickens from the organic flock than from the free-range flock (El-Shibiny et al., 2005). While phages isolated from chickens toward the end of the rearing period were capable of replication on a broad range of *C. jejuni* strains, the phages could infect only a small percentage of the *C. coli* strains. The appearance of phages in the flock coincided with the appearance of *C. coli* as the dominant species. The replacement of *C. jejuni* with *C. coli* is a potentially negative situation, since *C. coli* is more resistant to a variety of antimicrobials that are useful for human therapy, as compared to *C. jejuni* (Tam et al., 2003). Collectively, these studies indicate that the selection of appropriate phages and their dose optimization are key elements for the success of phage therapy to reduce campylobacters in broiler chickens.

APPLICATION OF PHAGES TO CONTROL PATHOGENS IN SWINE

Most of the phage therapy applications in swine have been directed at controlling the presence of porcine-specific enterotoxigenic *E. coli,* which causes diarrhea in piglets (Nagy and Fekete, 1999). Very little work has been conducted with respect to controlling foodborne zoonotic pathogens in swine. Harris's group has performed the only evaluations of phage therapy as a preharvest intervention to control the presence of *Salmonella* in pork products due to contamination during shipping, lairage, and slaughter. In research posted on the website of the National Pork Board (http://www.pork.org/PorkScience/Documents/REPORT%2004-99-230-Harris-ISU.pdf), Harris developed and evaluated a cocktail of 26 phages in a preliminary study to effect control of *Salmonella* in swine. The researchers observed that the efficacy of phage treatment was related to the route of administration and time elapsed after phage treatment. However, initial studies showed that the phage cocktail did not significantly reduce the concentration of *Salmonella* in pigs. Based on the preliminary data, Lee and Harris (2001) conducted another study in which a single *Salmonella*-specific lytic phage (Felix-O1) was examined as a possible candidate to control *Salmonella.* Three-week-old pigs received 10^{10} PFU of Felix-O1 phage both orally and intramuscularly 3 h following challenge with *Salmonella* serovar Typhimurium at a concentration of 10^8 CFU. At 9 h postchallenge, blood, liver, lung, spleen, ileocecal lymph node, tonsil, and cecum content samples were obtained from sacrificed pigs. The numbers of serovar Typhimurium in the samples were determined by plate count. The phage treatment significantly reduced the amount of serovar Typhimurium in tonsil and cecum contents as compared to the control. It was concluded that the phage treatment could be considered as a short-term intervention strategy to reduce the rapid dissemination of *Salmonella* in swine. A subsequent patent (Harris and Lee, 2003) outlined a protocol in which multiple dose variations administered 24 h prior to shipping and slaughter could reduce the concentrations of serovar Typhimurium in various organs and tissues, although intramuscular introduction of phage may not be practical in conventional swine production (Johnson et al., 2008), and the use of a single phage would probably lead to rapid development of phage-resistant bacterial cells.

THE CHALLENGE OF PHAGE RESISTANCE

It is clear that broad-host-range phages should be selected for phage therapy applications, and these phages should be combined into cocktails to delay or reduce the chance of the development of phage-resistant bacterial mutants. If broad-host-range phages cannot be isolated, an alternative approach is to take advantage of phages that can use more than one receptor on the bacterial cell surface. There are many examples of dual-specificity phages described in the literature. The T-even family of phages is among the best-characterized phages that display dual specificity. Approximately 170 bacteriophages with morphologies similar to T4 have been identified (Ackermann and Krisch, 1997; Ackermann, 1998; Tétart et al., 2001). These T-even phages have been isolated on a wide range of bacterial hosts that grow in diverse environments (Ackermann and Krisch, 1997; Ackermann, 1998; Tétart et al., 2001). Studies have been conducted to characterize the nature of the dual-component receptors in several T-even phages. For example, each of the T-even-type phages (including the type phage, T4) binds to two different receptors (LPS and OmpS) on the bacterial cell surface in a sequential manner. The first, reversible step is mediated by the long tail fibers that bind to LPS or to outer membrane proteins, which determines the host-receptor specificity of the T-even phages (Beumer et al., 1984).

A novel dual-specificity phage, designated K1-5, was described by Scholl et al. (2001). This phage carries two capsule-specific enzymatic tail proteins, an endosialidase and a lyase, allowing it to replicate in both K1 and K5 strains of *E. coli*. One tail protein found on phage K1-5 (the lyase protein) is similar to that of phage K5 (specific for the K5 polysaccharide capsule), and a second tail protein found on this phage (the endosialidase) is similar to a tail protein found on phage K1E (specific for the K1 polysaccharide capsule). In addition, the genomic region encoding these proteins is almost identical to the genomic construct found in the *Salmonella* phage SP6, which codes for a protein that binds to the *Salmonella* O-antigen (Scholl et al., 2002). It was concluded that phages SP6, K1-5, K5, and K1E are very closely related but have different tail fiber proteins, which allows each phage to have different host specificities (Scholl et al., 2002). The observation of a similar tail genome motif in both the *Salmonella* phage SP6 and the coliphages K1E, K5, and K1-5 indicates that this genomic construct might serve in the development of a modular phage platform that could operate over a wide bacterial host range. The collective ability of the phages to adapt to using different receptors could be used to create phage cocktails with expanded host ranges, thereby vastly reducing the possibility of resistant phage isolates while at the same time addressing the problem of bacteria within the same species that express varied cell surface proteins, which determines phage susceptibility.

Furthermore, it is possible that in a live animal the development of phage-resistant isolates would coincide with a loss of fitness and virulence, rendering the contaminating bacteria unable to cause illness. For example, in the case of K1- or LPS-specific phages, the most likely resistant mutants will be deficient in capsule or LPS. Several studies have shown that in both cases the mutant bacterial isolates are less virulent (Smith and Huggins, 1982) or cannot compete with many other strains that are not subject to the pressure of the infecting phage (Brüssow, 2005; Harcombe and Bull, 2005).

REGULATORY AND MANUFACTURING CHALLENGES

Several obstacles remain before implementation of phage therapy as a viable method to control food-borne pathogens in animals can be realized (see chapter 15 in this volume). Such issues are related to the regulatory approval of phage products for use in live animals, types of raw materials used to produce the phages, and the development of universally accepted good manufacturing practices (GMPs) (Goodridge and Abedon, 2008).

Several phage products have been approved for use as pesticides or as antimicrobials on foods (Goodridge and Abedon, 2008). The hide and poultry feather sprays produced by OmniLytics are the only phage products approved for use in the animal industry (OmniLytics, Inc., 2007a, 2007b, 2008), and there are currently no products approved for reduction of bacteria within the living animal. Such approval may become easier to achieve with the recent news that the European Food Safety Authority's Biohazards Panel has endorsed the use of phages as a treatment for foods of animal origin including carcasses, meat, and dairy products. In May 2009, the panel concluded that some phages, under specific conditions, have been demonstrated to be very effective in the targeted elimination of specific pathogens from meat, milk, and products thereof (European Food Safety Authority, 2009).

Raw materials include the bacterial hosts used for phage propagation, and enzymes and growth media. In development of phage products for use in ruminants, a major concern arises regarding the use of growth media and other raw materials derived from ruminants that may be contaminated with transmissible spongiform encephalopathies (Goodridge and Abedon, 2008). Finally, while there are no legal requirements for the development of GMPs for food-grade phage production, the documentary evidence provided by phage-specific GMPs will be needed to confirm the quality assurance of each phage product.

CONCLUSION

The scientific literature demonstrates the possibility of using phage therapy to effectively reduce the presence of food-borne pathogens in food-producing animals. Nevertheless, there are many questions that remain to be answered through the collection of scientific data from rigorously designed experiments. Such studies should be designed to solve issues associated with phage-bacterium interactions and ecology, and phage efficacy under different environmental conditions and physiological conditions of the animals, and continued work should be conducted to more fully address the issue of phage resistance. It is possible that phage therapy will come to be viewed in a way similar to other biocontrol strategies such as CE. If that is the case, then future research should also be accomplished to develop methods to integrate phage therapy approaches with other technologies, including CE and vaccines. In that way, the development of a hurdle approach for food-borne pathogen control in live animals, much like the approach developed for control of pathogens in foods, is a possibility. Finally, the cost of production will become a major consideration. Regardless of the remaining work to be accomplished, as the first century of phage research concludes, their general acceptance and use as antimicrobials appear to be closer to reality than ever before.

REFERENCES

Ackermann, H. W. 1998. Tailed bacteriophages: the order *Caudovirales*. *Adv. Virus Res.* **51:**135–201.

Ackermann, H. W., and H. M. Krisch. 1997. A catalogue of T4-type bacteriophages. *Arch. Virol.* **14:**2329–2345.

Altekruse, S. F., N. J. Stern, P. I. Fields, and D. L. Swerdlow. 1999. *Campylobacter jejuni*—an emerging foodborne pathogen. *Emerg. Infect. Dis.* **5:**28–35.

Andreatti Filho, R. L., J. P. Higgins, S. E. Higgins, G. Gaona, A. D. Wolfenden, G. Tellez, and B. M. Hargis. 2007. Ability of bacteriophages isolated from different sources to reduce *Salmonella enterica* serovar Enteritidis in vitro and in vivo. *Poult. Sci.* **86:**1904–1909.

Atterbury, R. J., E. Dillon, C. Swift, P. L. Connerton, J. A. Frost, C. E. Dodd, C. E. Rees, and I. F. Connerton. 2005. Correlation of *Campylobacter* bacteriophage with reduced presence of hosts in broiler chicken ceca. *Appl. Environ. Microbiol.* **71:**4885–4887.

Atterbury, R. J., M. A. Van Bergen, F. Ortiz, M. A. Lovell, J. A. Harris, A. DeBoer, J. A. Wagenaar, V. M. Allen, and P. A. Barrow. 2007. Bacteriophage therapy to reduce *Salmonella* colonization of broiler chickens. *Appl. Environ. Microbiol.* **73:**4543–4549.

Bach, S. J., T. A. McAllister, D. M. Veira, V. P. J. Gannon, and R. A. Holley. 2003. Effect of bacteriophage DC22 on *Escherichia coli* O157:H7 in an artificial rumensystem (Rusitec) and inoculated sheep. *Anim. Res.* **52:**89–101.

Berchieri, A., Jr., M. A. Lovell, and P. A. Barrow. 1991. The activity in the chicken alimentary tract of bacteriophages lytic for *Salmonella typhimurium*. *Res. Microbiol.* **142:**541–549.

Beumer, J., E. Hannecart-Pokorni, and C. Godard. 1984. Bacteriophage receptors. *Bull. Inst. Pasteur* **82:**173–253.

Blaser, M. 1997. Epidemiologic and clinical features of *Campylobacter jejuni* infection. *J. Infect. Dis.* **176:** S103–S105.

Borie, C., I. Albala, P. Sànchez, M. L. Sánchez, S. Ramírez, C. Navarro, M. A. Morales, J. Retamales, and J. Robeson. 2008. Bacteriophage treatment reduces *Salmonella* colonization of infected shickens. *Avian Dis.* **52:**64–67.

Brichta-Harhay, D. M., M. N. Guerini, T. M. Arthur, J. M. Bosilevac, N. Kalchayanand, S. D. Shackelford, T. L. Wheeler, and M. Koohmaraie. 2008. *Salmonella* and *Escherichia coli* O157:H7 contamination on hides and carcasses of cull cattle presented for slaughter in the United States: an evaluation of prevalence and bacterial loads by immunomagnetic separation and direct plating methods. *Appl. Environ. Microbiol.* **74:** 6289–6297.

Brüssow, H. 2005. Phage therapy: the *Escherichia coli* experience. *Microbiology* **151:**2133–2140.

Buckling, A., and P. B. Rainey. 2002. Antagonistic coevolution between a bacterium and bacteriophage. *Proc. Biol. Soc.* **269:**931–936.

Callaway, T. R., R. C. Anderson, T. S. Edrington, K. J. Genovese, K. M. Bischoff, T. L. Poole, Y. S. Jung, R. B. Harvey, and D. J. Nisbet. 2004. What are we doing about *Escherichia coli* O157:H7 in cattle? *J. Anim. Sci.* **82**(E. Suppl.): E93–E99.

Callaway, T. R., T. S. Edrington, R. C. Anderson, Y. S. Jung, K. J. Genovese, R. O. Elder, and D. J. Nisbet. 2003. Isolation of naturally-occurring bacteriophage from sheep that reduce populations of *E. coli* O157:H7 in vitro and in vivo, p. 25. *Proc. 5th Int. Symp. Shiga Toxin-Producing* Escherichia coli *Infect.,* Edinburgh, United Kingdom, 8 to 11 June, 2003.

Callaway, T. R., T. S. Edrington, A. D. Brabban, E. S. Kutter, R. C. Anderson, and D. J. Nisbet. 2006. Isolation and use of bacteriophage to reduce *E. coli* O157:H7 populations in ruminants. *Proc. Int. Conf. Perspect. Bacteriophage Preparation,* p. 72–83.

Carlton, R. M., W. H. Noordman, B. Biswas, E. D. de Meester, and M. J. Loessner. 2005. Bacteriophage P100 for control of *Listeria monocytogenes* in foods: genome sequence, bioinformatics analyses, oral toxicity study, and application. *Regul. Toxicol. Pharmacol.* **43:**301–312.

Centers for Disease Control and Prevention. 2006. Preliminary FoodNet data on the incidence of infection with pathogens transmitted commonly through food—selected sites, United States, 2005. *MMWR Morb. Mortal. Wkly. Rep.* **55:**392–395.

Chase, J., N. Kalchayanand, and L. D. Goodridge. 2005. Use of bacteriophage therapy to reduce *Escherichia coli* O157:H7 concentrations in an anaerobic digestor that simulates the bovine gastrointestinal tract, abstr. 108-6. *Inst. Food Technologists Annu. Meet. Food Expo.* Institute of Food Technologists, Chicago, IL.

Connerton, P. L., C. M. L. Carrillo, C. Swift, E. Dillon, A. Scott, C. E. D. Rees, C. E. R. Dodd, J. Frost, and I. F. Connerton. 2004. Longitudinal study of *Campylobacter jejuni* bacteriophages and their hosts from broiler chickens. *Appl. Environ. Microbiol.* **70:**3877–3883.

Doyle, M. P., and M. C. Erickson. 2006. Reducing the carriage of foodborne pathogens in livestock and poultry. *Poult. Sci.* **85:**960–973.

Dykes, G. A., and S. M. Moorhead. 2002. Combined antimicrobial effect on nisin and a listeriophage against *Listeria monocytogenes* in broth but not in buffer or on raw beef. *Int. J. Food Microbiol.* **73:** 71–81.

El-Shibiny, A., P. L. Connerton, and I. F. Connerton. 2005. Enumeration and diversity of campylobacters and bacteriophages isolated during the rearing cycles of free-range and organic chickens. *Appl. Environ. Microbiol.* **71:**1259–1266.

European Food Safety Authority. 2009. EFSA evaluates bacteriophages. May 12. European Food Safety Authority, Parma, Italy. http://www.efsa.europa.eu/EFSA/efsa_locale-1178620753812_1211902525764.htm.

Figueroa-Bossi, N., and L. Bossi. 1999. Inducible prophages contribute to *Salmonella* virulence in mice. *Mol. Microbiol.* **33:**167–176.

Fiorentin, L., N. D. Vieira, and W. Barioni, Jr. 2005. Oral treatment with bacteriophages reduces the concentration of *Salmonella* Enteritidis PT4 in caecal contents of broilers. *Avian Pathol.* **34:**258–263.

Gannon, V. P., T. A. Graham, R. King, P. Michel, S. Read, K. Ziebell, and R. P. Johnson. 2002. *Escherichia coli* O157:H7 infection in cows and calves in a beef cattle herd in Alberta, Canada. *Epidemiol. Infect.* **129:**163–172.

Goode, D., V. M. Allen, and P. A. Barrow. 2003. Reduction of experimental *Salmonella* and *Campylobacter* contamination of chicken skin by application of lytic bacteriophages. *Appl. Environ. Microbiol.* **69:**5032–5036.

Goodridge, L. D., and S. T. Abedon. 2003. Bacteriophage biocontrol and bioprocessing: application of phage therapy to industry. *SIM News* **53:** 254–262.

Goodridge, L. D., and S. T. Abedon. 2008. Bacteriophage biocontrol: the technology matures. *Microbiol. Aust.* **March:**48–49.

Hancock, D. D., T. E. Besser, and D. H. Rice. 1998. Ecology of *Escherichia coli* O157:H7 in cattle and impact of management practices, p. 85–91. *In* J. B. Kaper and A. D. O'Brien (ed.), Escherichia coli *O157:H7 and Other Shiga Toxin-Producing* E. coli *Strains.* ASM Press, Washington, DC.

Harcombe, W. R., and J. J. Bull. 2005. Impact of phages on two-species bacterial communities. *Appl. Environ. Microbiol.* **71:**5254–5259.

Harris, D. L., and N. Lee. 2 December 2003. Compositions and methods for reducing the amount of *Salmonella* in livestock. U.S. patent 6,656,463.

Higgins, J. P., S. E. Higgins, K. L. Guenther, W. Huff, A. M. Donoghue, D. J. Donoghue, and B. M. Hargis. 2005. Use of a specific bacteriophage treatment to reduce *Salmonella* in poultry products. *Poult. Sci.* **84:**1140–1145.

Jensen, E. C., H. S. Schrader, B. Rieland, T. L. Thompson, K. W. Lee, K. W. Nickerson, and T. A. Kokjohn. 1998. Prevalence of broad-host-range lytic bacteriophages of *Sphaerotilus natans, Escherichia coli,* and *Pseudomonas aeruginosa. Appl. Environ. Microbiol.* **64:**575–580.

Johnson, R. P., C. L. Gyles, W. E. Huff, S. Ojha, G. R. Huff, N. C. Rath, and A. M. Donoghue. 2008. Bacteriophages for prophylaxis and therapy in cattle, poultry and pigs. *Anim. Health Res. Rev.* **9:**201–215.

Jordan, D., S. A. McEwen, A. M. Lammerding, W. B. McNab, and J. B. Wilson. 1999. Pre-slaughter to control of *Escherichia coli* in beef cattle: a simulation study. *Prev. Vet. Med.* **41:**55–74.

Kapperud, G., E. Skjerve, N. H. Bean, S. M. Ostroff, and J. Lassen. 1992. Risk factors for sporadic *Campylobacter* infections: results of a case-control study in southeastern Norway. *J. Clin. Microbiol.* **30:**3117–3121.

Karmali, M. A., M. Petric, C. Lim, P. C. Fleming, G. S. Arbus, and H. Lior. 1985. The association between idiopathic haemolytic syndrome and infection by verotoxin-producing *Escherichia coli. J. Infect. Dis.* **151:**775–782.

Kudva, I. T., S. Jelacic, P. I. Tarr, P. Youderian, and C. J. Hovde. 1999. Biocontrol of *Escherichia coli* O157 with O157-specific bacteriophages. *Appl. Environ. Microbiol.* **65:**3767–3773.

Laegreid, W. W., R. O. Elder, and J. E. Keen. 1999. Prevalence of *Escherichia coli* O157:H7 in range beef calves at weaning. *Epidemiol. Infect.* **123:** 291–298.

Lee, N., and D. L. Harris. 2001. The effect of bacteriophage treatment as a preharvest intervention strategy to reduce the rapid dissemination of *Salmonella* Typhimurium in pigs, p. 555–557. *Proc. AASV.* American Association of Swine Veterinarians, Perry, IA.

Levine, B. R., and J. J. Bull. 1996. Phage therapy revisited: the population biology of a bacterial infection and its treatment with phage and antibiotics. *Am. Nat.* **147:**881–898.

Loc Carillo, C., R. J. Atterbury, A. el-Shibiny, P. L. Connerton, E. Dillon, A. Scott, and I. F. Connerton. 2005. Bacteriophage therapy to reduce *Campylobacter jejuni* colonization of broiler chickens. *Appl. Environ. Microbiol.* **71:**6554–6563.

Ma, Y., J. C. Pacan, Q. Wang, Y. Xu, X. Huang, A. Korenevsky, and P. M. Sabour. 2008. Microencapsulation of bacteriophage Felix O1 into chitosan-alginate microspheres for oral delivery. *Appl. Environ. Microbiol.* **74:**4799–4805.

McCrea, B. A., K. H. Tonooka, C. VanWorth, C. L. Boggs, E. R. Atwill, and J. S. Schrader. 2006. Prevalence of *Campylobacter* and *Salmonella* species on farm, after transport, and at processing in specialty market poultry. *Poult. Sci.* **85:**136–143.

Mead, G. C., and P. A. Barrow. 1990. *Salmonella* control in poultry by competitive exclusion or immunization. *Lett. Appl. Microbiol.* **10:**221–227.

Mead, P. S., L. Slutsker, V. Dietz, L. F. McCaig, J. S. Bresee, C. Shapiro, P. M. Griffin, and R. V. Tauxe. 1999. Food-related illness and death in the United States. *Emerg. Infect. Dis.* **5:** 607–625.

Mizoguchi, K., M. Morita, C. R. Fischer, M. Yoichi, Y. Tanji, and H. Unno. 2003. Coevolution of bacteriophage PP01 and *Escherichia coli* O157:H7 in continuous culture. *Appl. Environ. Microbiol.* **69:**170–176.

Nagy, B., and P. Z. Fekete. 1999. Enterotoxigenic *Escherichia coli* (ETEC) in farm animals. *Vet. Res.* **30:**259–284.

Naylor, S. W., J. C. Low, T. E. Besser, A. Mahajan, G. J. Gunn, M. C. Pearce, I. J. McKendrick, D. G. Smith, and D. L. Gally. 2003. Lymphoid follicle-dense mucosa at the terminal rectum is the principal site of colonization of enterohemorrhagic *Escherichia coli* O157:H7 in the bovine host. *Infect. Immun.* **71:**1505–1512.

Niu, Y. D., Y. Xu, T. A. McAllister, E. A. Rozema, T. P. Stephens, S. J. Bach, R. P. Johnson, and K. Stanford. 2008. Comparison of fecal versus rectoanal mucosal swab sampling for detecting *Escherichia coli* O157:H7 in experimentally inoculated cattle used in assessing bacteriophage as a mitigation strategy. *J. Food Prot.* **71:**691–698.

O'Flynn, G., R. P. Ross, G. F. Fitzgerald, and A. Coffey. 2004. Evaluation of a cocktail of three bacteriophages for biocontrol of *Escherichia coli* O157:H7. *Appl. Environ. Microbiol.* **70:**3417–3424.

OmniLytics, Inc. 2007a. OmniLytics announces USDA/FSIS allowance for bacteriophage treatment of *Salmonella* on livestock. March 21. OmniLytics, Inc., Salt Lake City, UT. http://www.omnilytics.com/news/news019.html.

OmniLytics, Inc. 2007b. OmniLytics announces USDA/FSIS allowance for bacteriophage treatment of *E. coli* O157:H7 on livestock. January 2. OmniLytics, Inc., Salt Lake City, UT. http://www.omnilytics.com/news/news018.html.

OmniLytics, Inc. 2008. OmniLytics announces USDA/FSIS allowance of bacteriophage treatment of *Salmonella* on poultry. July 29. OmniLytics, Inc., Salt Lake City, UT. http://www.omnilytics.com/news/news021.html.

Rangel, J. M., P. H. Sparling, C. Crowe, P. M. Griffin, and D. L. Swerdlow. 2005. Epidemiology of *Escherichia coli* O157:H7 outbreaks,

United States, 1982–2002. *Emerg. Infect. Dis.* **11:** 603–609.

Raya, R. R., P. Varey, R. A. Oot, M. R. Dyen, T. R. Callaway, T. S. Edrington, E. M. Kutter, and A. D. Brabban. 2006. Isolation and characterization of a new T-even bacteriophage, CEV1, and determination of its potential to reduce *Escherichia coli* O157:H7 levels in sheep. *Appl. Environ. Microbiol.* **72:**6405–6410.

Rozema, E. A., T. P. Stephens, S. J. Bach, E. K. Okine, R. P. Johnson, K. Stanford, and T. A. McAllister. 2009. Oral and rectal administration of bacteriophages for control of *Escherichia coli* O157:H7 in feedlot cattle. *J. Food Prot.* **72:**241–250.

Schicklmaier, P., E. Moser, T. Wieland, W. Rabsch, and H. Schmieger. 1998. A comparative study of the frequency of prophages among natural isolates of *Salmonella* and *Escherichia coli* with emphasis on generalized transducers. *Antonie van Leeuwenhoek* **73:**49–54.

Scholl, D., S. Adhya, and C. R. Merril. 2002. Bacteriophage SP6 is closely related to phages K1-5, K5, and K1E but encodes a tail protein very similar to that of the distantly related P22. *J. Bacteriol.* **184:**2833–2836.

Scholl, D., S. Rogers, S. Adhya, and C. R. Merril. 2001. Bacteriophage K1-5 encodes two different tail fiber proteins, allowing it to infect and replicate on both K1 and K5 strains of *Escherichia coli*. *J. Virol.* **75:**2509–2515.

Scott, A. E., A. R. Timms, P. L. Connerton, C. Loc Carrillo, K. Adzfa Radzum, and I. F. Connerton. 2007. Genome dynamics of *Campylobacter jejuni* in response to bacteriophage predation. *PLoS Pathog.* **3:**e119.

Sheng, H., H. J. Knecht, I. T. Kudva, and C. J. Hovde. 2006. Application of bacteriophages to control intestinal *Escherichia coli* O157:H7 levels in ruminants. *Appl. Environ. Microbiol.* **72:**5359–5366.

Skirrow, M. B. 1982. *Campylobacter* enteritis—the first five years. *J. Hyg.* **89:**175–184.

Sklar, I. B., and R. D. Joerger. 2001. Attempts to utilize bacteriophage to combat *Salmonella enterica* serovar Enteritidis infection in chickens. *J. Food Saf.* **21:**15–29.

Smith, H. W., and M. B. Huggins. 1982. Successful treatment of experimental *Escherichia coli* infections in mice using phage: its general superiority over antibiotics *J. Gen. Microbiol.* **128:**307–318.

Smith, H. W., M. B. Huggins, and K. M. Shaw. 1987. Factors influencing the survival and multiplication of bacteriophages in calves and in their environment. *J. Gen. Microbiol.* **133:**1127–1135.

Stern, N. J., and S. U. Kazmi. 1989. *Campylobacter,* p. 71–110. *In* M. P. Doyle (ed.), *Foodborne Bacterial Pathogens.* Marcel Dekker, New York, NY.

Tam, C. C., S. J. O'Brien, G. K. Adak, S. M. Meakins, and J. A. Frost. 2003. *Campylobacter coli*—an important foodborne pathogen. *J. Infect.* **47:**28–32.

Tanji, Y., T. Shimada, M. Yoichi, K. Miyanaga, K. Hori, and H. Unno. 2004. Toward rational control of *Escherichia coli* O157:H7 by a phage cocktail. *Appl. Microbiol. Biotechnol.* **64:**270–274.

Tétart, F., C. Desplats, M. Kutateladze, C. Monod, H. W. Ackermann, and H. M. Krisch. 2001. Phylogeny of the major head and tail genes of the wide-ranging T4-type bacteriophages. *J. Bacteriol.* **183:**358–366.

Toro, H., S. B. Price, S. McKee, F. J. Hoerr, J. Krehling, M. Perdue, and L. Bauermeister. 2005. Use of bacteriophages in combination with competitive exclusion to reduce *Salmonella* from infected chickens. *Avian Dis.* **49:**118–124.

Twort, F. W. 1915. An investigation on the nature of ultra-microscopic viruses. *Lancet* **186:**1241–1243.

Waddell, T., A. Mazzocco, R. P. Johnson, J. Pacan, S. Campbell, A. Perets, A. MacKinnon, J. Holtslander, B. Pope, and C. Gyles. 2000. Control of *Escherichia coli* O157:H7 infection of calves by bacteriophages. *Proc. 4th Int. Symp. Workshop Shiga Toxin (Verocytotoxin)-Producing* Escherichia coli *Infect.,* Kyoto, Japan, 20 October to 2 November.

Wagenaar, J. A., M. A. P. Van Bergen, M. A. Mueller, T. M. Wassenaar, and R. M. Carlton. 2005. Phage therapy reduces *Campylobacter jejuni* colonization in broilers. *Vet. Microbiol.* **109:** 275–283.

Zhao, T., M. P. Doyle, J. Shere, and L. Garber. 1995. Prevalence of enterohemorrhagic *Escherichia coli* O157:H7 in a survey of dairy herds. *Appl. Environ. Microbiol.* **61:**1290–1293.

BACTERIOPHAGES FOR CONTROL OF PHYTOPATHOGENS IN FOOD PRODUCTION SYSTEMS

Antonet M. Svircev, Alan J. Castle, and Susan M. Lehman

5

INTRODUCTION

In 1917, French-Canadian scientist Félix d'Hérelle observed that an "agent" was associated with bacterial pathogen in recovering dysentery patients (Summers, 1999). The agent was named "bacteriophage," using the Greek derivative *phagein* for "eater" of bacteria (d'Hérelle, 1917). The inventorship dispute between d'Hérelle, Frederick Twort, and their respective supporters consumed a great deal of time and scientific energy. Unfortunately, the vicious debate detracted from the recognition of d'Hérelle's considerable and important contributions on the use of bacteriophages, or "phages" as they are commonly known, as therapeutic agents for control of bacterial pathogens. d'Hérelle's long and productive career included research in phage therapies for control of avian typhosis (d'Hérelle, 1921), bovine hemorrhagic septicemia (d'Hérelle, 1921), bacillary dysentery, and bubonic plague (Summers, 1999, 2001). In Western countries, the large-scale production of broad-spectrum antibiotics by the 1940s resulted in the abandonment of research on bacteriophage-mediated control of bacterial pathogens (Summers, 2001). The historic lack of interest in phage therapies by multinational industries was compounded by high antibiotic efficacies and the absence of patent protection for phage technologies (Clark and March, 2006).

Summers (2001) describes the history of phage-mediated therapies: "phage therapy can be divided into four periods: early enthusiasm, critical skepticism, abandonment, recent interest and reappraisal." The recent, more sustained interest in the therapeutic uses of bacteriophages has been influenced by a number of events in agriculture and human pathogen treatment. In agriculture, interest in phages as biopesticides is being driven by the spread of antibiotic resistance among plant pathogens, the implementation of regulatory restrictions on the use of antibiotics in agriculture (see chapter 1 in this volume), increased consumer demand for organically grown food, and a decreased tolerance of the negative impacts of traditional pesticides on the environment. The consumer pressures will contribute significantly toward adaptation and implementation of softer pesticides, commonly known as bio-

Antonet M. Svircev, Agriculture and Agri-Food Canada, 4902 Victoria Avenue North, P.O. Box 6000, Vineland Station, Ontario L0R 2E0, Canada. *Alan J. Castle,* Department of Biological Sciences, Brock University, 500 Glenridge Avenue, St. Catharines, Ontario L2S 3A1, Canada. *Susan M. Lehman,* Centers for Disease Control and Prevention, 1600 Clifton Road NE, Mail Stop C-16, Atlanta, GA 30333.

Bacteriophages in the Control of Food- and Waterborne Pathogens
Edited by Parviz M. Sabour and Mansel W. Griffiths, © 2010 ASM Press, Washington, DC

logicals, biological control agents, or biopesticides.

Bacteriophages possess many attributes that make them attractive agricultural biopesticides. They are easily isolated, purified, and cultured using well-established protocols in the public domain (Adams, 1959). They are abundant in the environment and host specific at the bacterial genus level. The well-recognized exceptions to host-phage specificity are the phages that infect the *Enterobacteriaceae* (Bradley, 1965). The phages in this group are polyvalent and frequently have the capacity to infect more than one bacterial genus. This particular feature may be an advantage or a disadvantage depending on the type of phage therapy-pathogen system. The mechanisms by which phages infect and kill bacteria (see chapter 2 in this volume) are unrelated to antibiotic modes of action; therefore phages may be used successfully against multidrug-resistant bacteria (Chanishvili et al., 2001). Phages are specific to one or a few related target bacterial species and therefore do not affect eukaryotic cells and have a limited effect on the microbial ecology of the surrounding environment. Phages are self-replicating and self-limiting, surviving only in the presence of their particular host. Finally, their production is inexpensive and does not require complex technologies (Adams, 1959; Clokie and Kropinski, 2009). The production and storage of phages is not cost-prohibitive and is to date relatively unrestrained by patent protections. Unfortunately, commercial companies have recently been allowed to file very broad patents for various methods of phage-mediated elimination of pathogens from food, livestock, and agricultural crops (Gill et al., 2007).

Consequently, the disadvantages of phages as biological control agents have also been well documented in the literature (Okabe and Goto, 1963; Vidaver, 1976; Goodridge, 2004; Jones et al., 2007). The list includes factors such as the potential of phages to mutate from virulent to temperate, the development of phage-resistant hosts, the potential transfer of virulence factors from pathogenic to previously nonpathogenic hosts, host-induced restriction or modification of phages, and the complexity of implementing biocontrol under variable environmental conditions (Vidaver, 1976; Marks and Sharp, 2000). The success of therapy may be limited by abortive infections and lysogenized phages, which may prevent the infection and replication of lytic phages in susceptible hosts (Duckworth, 1987). However, many of these challenges can be tackled by using phage mixtures and by careful screening of candidate phages. In addition, bacteriophage lysins (Gill et al., 2007; Fischetti, 2008; see chapter 8 in this volume) and endolysins (Borysowski et al., 2006), or enzymes produced by phages to degrade bacterial cell walls and the peptidoglycan layer, have been examined as potential replacements for antibiotics. The authors direct readers to excellent reviews on bacteriophage therapy that are available (Summers, 2001; Stone, 2002; Abedon, 2003; Goodridge, 2004; Hudson et al., 2005; Clark and March, 2006; Skurnik and Strauch, 2006; Jones et al., 2007; Gill et al., 2007; Hagens and Loessner, 2007; Thiel, 2004; Mattey and Spencer, 2008; Cairns and Payne, 2008; Obradovic et al., 2008; see also chapters 4, 7, and 13 in this volume). The majority of the literature deals with the application of phage therapy to control human pathogens in food (Abuladze et al., 2008; Bigwood et al., 2009) and clinical treatments (Kutter and Sulakvelidze, 2005).

This chapter will investigate and discuss some of the novel methods and technologies that have been implemented in the search for phage biopesticides in agricultural food production systems. The emphasis is placed on the development of bacteriophages as control agents of bacterial diseases of plants. We expect that the control systems that would be effective against plant-pathogenic bacteria will provide useful information for the control of plant-borne animal pathogens. Such a discussion, however, is premature and is not included here. For a current perspective on the development of bacteriophage control of plant-borne zoonotic pathogens, the reader is directed to Abuladze et al. (2008), Bigwood et al. (2009), and Guenther et al. (2009). The chapter is divided according to agriculturally related application lines. The first section summarizes the historic and current state of phage biopesticides. This is followed by discussions of specific pathogens and application environments in which phages have been used. The third section examines the impact of bacteriophages on soil bacteria located in the rhizosphere and phylloplane, and the potential consequences to plant health and yield. The final section discusses general points to consider for the development of phage biopesticides in agriculture.

CONTROL OF PLANT DISEASE: BACTERIOPHAGES AS BIOPESTICIDES

Research on the use of bacteriophages as biological control agents in plant agricultural systems has been confined mostly to research laboratories dealing with a single pathogen or a group of related host-pathogen systems. In an American Phytopathological Society Internet (APSnet) feature article, Gill and Abedon discuss the role of phage ecology and phage therapy in agriculture (http://www.apsnet.org/online/feature/phages/). Gill et al. (2007) provide an excellent summary of commercial companies involved with phages and phage-derived products, patents, and the associated challenges of phage therapeutics. The bulk of phage therapeutic research and commercial development focuses on the prevention of human pathogens in the food production pipeline. The exception is the recent registration of phages by OmniLytics, Inc. (Salt Lake City, UT) for the control of the plant disease bacterial spot of tomato (Balogh et al., 2008). Agricultural uses of phage-derived products such as lysins and genetically modified phages or bacteria are still confined to the realm of the research laboratory. In the current regulatory environment, these types of phage therapies would have difficulty passing through the regulatory system. Consequently, the phages currently being studied or used as biopesticides are ones that have been collected directly from the environment and optimized through screening, selection, and manner of use, rather than by engineered modifications. The regulatory passage of future phage biopesticides should be considerably eased by OmniLytics' registration of the bacterial spot biopesticide in 2006 and by the FDA ruling that placed a set of *Listeria* phages in the "Generally Recognized as Safe" category of food additives (EBI Food Safety, 2006; see also chapter 15 in this volume). The initial hurdle has been jumped and phages have been approved for use both in plant agriculture and on food products.

Initially, a considerable volume of research centered on the isolation and characterization of phages for control of plant pathogens was conducted (Okabe and Goto, 1963; Ritchie, 1978; Schnabel et al., 1999; Svircev et al., 2002b; Gill et al., 2003). The experimental model that utilized a single phage as a biological control agent for a plant pathogen continued well into the 1990s (Civerolo and Keil, 1969; Civerolo, 1973). The use of mixtures with two (Munsch and Olivier, 1995), four (Flaherty et al., 2000; Flaherty et al., 2001; Balogh et al., 2003; Ravensdale et al., 2007; Balogh et al., 2008), and five or more phages (Harbaugh et al., 1998) has improved efficacy and eliminated the problems that may arise when a single phage is used for the control of a bacterial disease.

In vitro experiments are often used to demonstrate phage-mediated control of plant pathogens. In the soft agar overlay method (Adams, 1959), the pathogen and host are mixed and plated onto a solid agar surface. The resulting clear zones, referred to as "plaques," are an indication that the host was infected by phage and new phages (progeny) were released from the bacterium. Phage growth is prolific, and a single bacterial cell may release up to 50 to 200 new phages (Marks and Sharp, 2000; see chapter 2 in this volume). This type of rudimentary screening for phage activity is employed on a routine basis. Higher-order screening for phage biological activity relies on in vivo or in planta bioassays (Lehman, 2007) or greenhouse and field trials (Balogh et al., 2003; Obradovic et al., 2004; Jones et al., 2007; Balogh et al., 2008). The key to successful biocontrol with phages is extensive end-point trials that will indicate how the biopesticides perform under the specific environmental conditions.

CONTROL OF PATHOGENS BY BACTERIOPHAGES

Phages have been isolated from orchard soils and used for control of *Xanthomonas pruni* (Civerolo and Keil, 1969; Civerolo, 1973) (= *Xanthomonas arboricola* pv. pruni), *Pseudomonas mors-prunorum* (Crosse and Hingorani, 1958), and *Erwinia amylovora* (Erskine, 1973; Billing, 1987; Ritchie and Klos, 1977, 1979; Ritchie, 1978; Schnabel et al., 1999; Schnabel and Jones, 2001; Gill et al., 2003; Lehman et al., 2009; Lehman, 2007; Roach et al., 2008) on both aerial and subsurface plant tissues. The level of interaction among pathogen, host, and the environment influences the amount of disease that develops in a particular crop. Phage biopesticides are used to alter the pathogen pressure under the environmental conditions that lead up to infection of the host plants.

The phyllosphere environment is not conducive to phage survival, resulting in the loss of viability for unprotected phages (Civerolo and Keil, 1969; McNeil et al., 2001; Schnabel and Jones, 2001; Balogh, 2002; Balogh et al., 2003). Viruses (phages) can be inactivated by heat, pH extremes, desiccation, and sunlight, or washed off of surfaces by rain, but the most significant environmental factors mediating viral inactivation are UVA and UVB light (280 to 400 nm) (Ignoffo and Garcia, 1992, 1994; Iriarte et al., 2007). The rhizosphere is also a challenging environment for phage activity. Phages adsorb onto soil particles via pH-dependent electrostatic interactions that appear to be strongest for phages with longer tails (Taylor et al., 1981; Dowd et al., 1998; Ashelford et al., 2003; Williamson et al., 2003). Phage populations do persist in soil, but long-term survival of large populations may be limited unless the host species is abundant (Erskine, 1973; Tan and Reanney, 1976; Ritchie and Klos, 1979; Ashelford et al., 1999a, 2000; Hildebrand et al., 2001; Svircev et al., 2002a; Gill et al., 2003). Orchards are often ideal locations for phage-mediated biological control. Bacterial pathogens tend to occupy narrow niches and primary infection events take place within a restricted window during bloom or fruit development.

Control of *E. amylovora*: Aerial Application of Phages

E. amylovora is an enterobacterial phytopathogen that infects several species of rosaceous plants, resulting in a wilt disease called fire blight. The pathogen is present in most apple- and pear-growing regions of the world and causes serious economic losses to commercial growers (Bonn and van der Zwet, 2000). The disease cycle begins anew each spring when warm weather promotes the growth and dispersal of *E. amylovora*. The pathogen colonizes open blossoms, multiplies, and infects the tree through natural openings in the blossoms (van der Zwet and Beer, 1995). Most control strategies focus on the suppression of *E. amylovora* populations in the flower. Currently, the most effective of these treatments is the timed application of streptomycin to open blossoms. However, specific concerns about streptomycin-resistant *E. amylovora*, along with the general trend to avoid antibiotic use in agriculture, are driving the development of alternative control strategies. More detailed information on fire blight and *E. amylovora* can be found in the excellent text edited by Vanneste (2000).

The possibility of using *Erwinia* phages to manage fire blight by controlling orchard populations of *E. amylovora* has been explored several times (Erskine, 1973; Ritchie, 1978; Schnabel et al., 1999; Schnabel and Jones, 2001; Gill et al., 2003; Lehman, 2007). Phages that infect *E. amylovora* have been isolated from soil since the 1950s (Okabe and Goto, 1963), but detailed characterizations are available only for those isolated since the 1970s (Erskine, 1973; Vlachakis and Verhoyen, 1984; Vandenbergh et al., 1985; Vandenbergh and Cole, 1986; Ritchie and Klos, 1979; Vanneste, 2000; Schnabel and Jones, 2001; Svircev et al., 2002b; Gill et al., 2003; Kim et al., 2004; Lehman et al., 2009). The most diverse set of phages as well as the phages with the broadest host ranges are those described by Gill et al. (2003). Phages belonging to the *Myo-, Sipho-,* and *Podoviridae* (Fig. 1) were isolated using a mixture of multiple host strains. Other studies have used only a single host strain, which tends to bias the enrichment process in favor of those phages that infect that one strain most efficiently (Ritchie and Klos, 1977, 1979; Ritchie, 1978). Also notable among these phages are those that were first isolated on *Pantoea agglomerans* (formerly *Erwinia herbicola*) and then shown to infect *E. amylovora* (Vlachakis and Verhoyen, 1984; Vanneste, 2000; Schnabel and Jones, 2001).

One of the earliest and most complete suggestions for phage-mediated control of *E. amylvora* was articulated by Erskine (1973). Erskine reported the discovery of a lysogenic phage capable of infecting both *E. amylovora* and "a yellow, amylovora-like saprophyte" fitting the description of *P. agglomerans,* a nonpathogenic epiphyte that is abundant in the phyllosphere of fruit orchards. None of the disease symptoms associated with *E. amylovora* inoculation appeared when pear fruit slices were coinoculated with *E. amylovora* and the lysogenized saprophyte, whereas coinoculation with the unlysogenized saprophyte only delayed and reduced those symptoms. Erskine ultimately advocated the use of the "phage-infected saprophyte" as a means of biological control of fire blight, correctly noting that phages applied by themselves would be rapidly inactivated by the ambient conditions, and that the saprophyte would provide some measure of pathogen inhibition on its own. In fact, strains of this saprophyte, *P. agglomerans,* have been developed and registered as biopesticides in their own right (marketed as BlightBan® C9-1, Bloomtime®, Blossom Bless®, and PomaVita®). Erskine did not distinguish between the use of these bacterial lysogens and the use of exclusively lytic phages that happen to infect both bacterial species, such as those later used by Lehman (2007). However, Erskine did articulate one of the most complete early visions of how to put phages to practical use in an aerial orchard environment. Several later efforts to control *E. amylovora* populations in the orchard did not succeed because the phage populations declined in the absence of *E. amylovora* (Ritchie and Klos, 1979; Schnabel et al., 1999; Schnabel and Jones, 2001). Schnabel et al. suggested that phage survival and subsequent disease control might be enhanced by coinoculation with avirulent *E. amylovora* mutants, but this approach would carry the risk of reversion to virulence.

Lehman (2007) describes the development of a biopesticide model that used combinations of *P. agglomerans* and broad-host-range *E. amylovora* phages to establish and maintain a replicating population of phages on the blossom surfaces throughout the risk period for infection. *P. agglomerans* acted as a carrier (Fig. 2), as new generations of the *E. amylovora* phages were produced at the expense of the carrier, increasing the size of the phage population and limiting the amount of time that free phages were exposed to damaging UV light.

One notable aspect of this study was the use of forced pear blossoms as a model system to develop the initial parameters (Fig. 3) for field experiments and to select 10 "best" phage candidates for field-based research (Fig. 4) (Svircev et al., 2006; Lehman, 2007). Other commonly used bioassays for the biological control of *E. amylovora* involve inoculation of immature pear fruit tissue (Gill, 2000) and infection of seedling shoot tips with contaminate scissors. While these are both convenient screening methods and *E. amylovora* does infect both fruit tissue and shoots, neither of these assays reflects the primary mode of host infection or the conditions under which a phage biopesticide is expected to function. In contrast, results obtained using the pear blossom model system demonstrated that the carrier contributes directly to control of *E. amylovora* populations on the blossom and that the addition of certain phages significantly im-

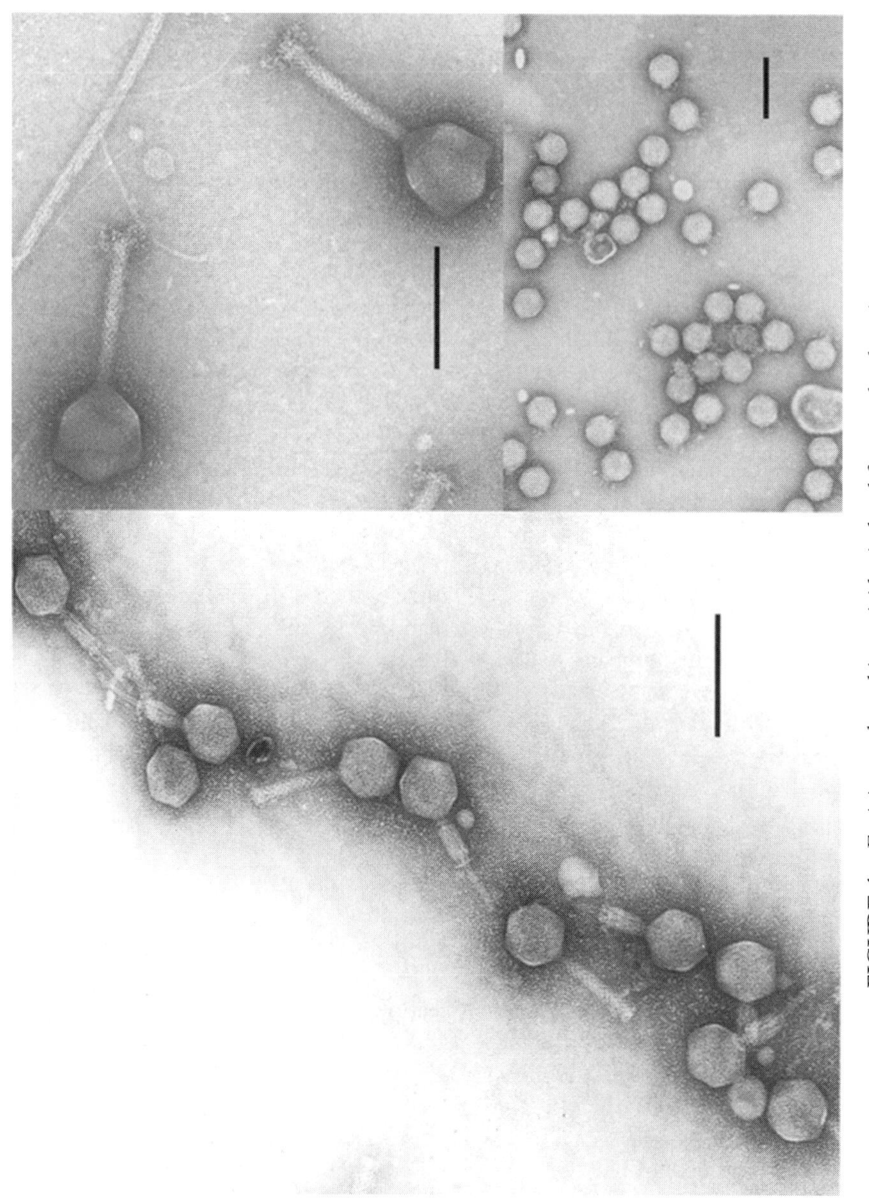

FIGURE 1 *Erwinia* sp. phage biopesticides isolated from orchard environment belong to the *Myo-* (left), *Podo-* (bottom right), and *Siphoviridae* (top right). Marker = 100 nm.

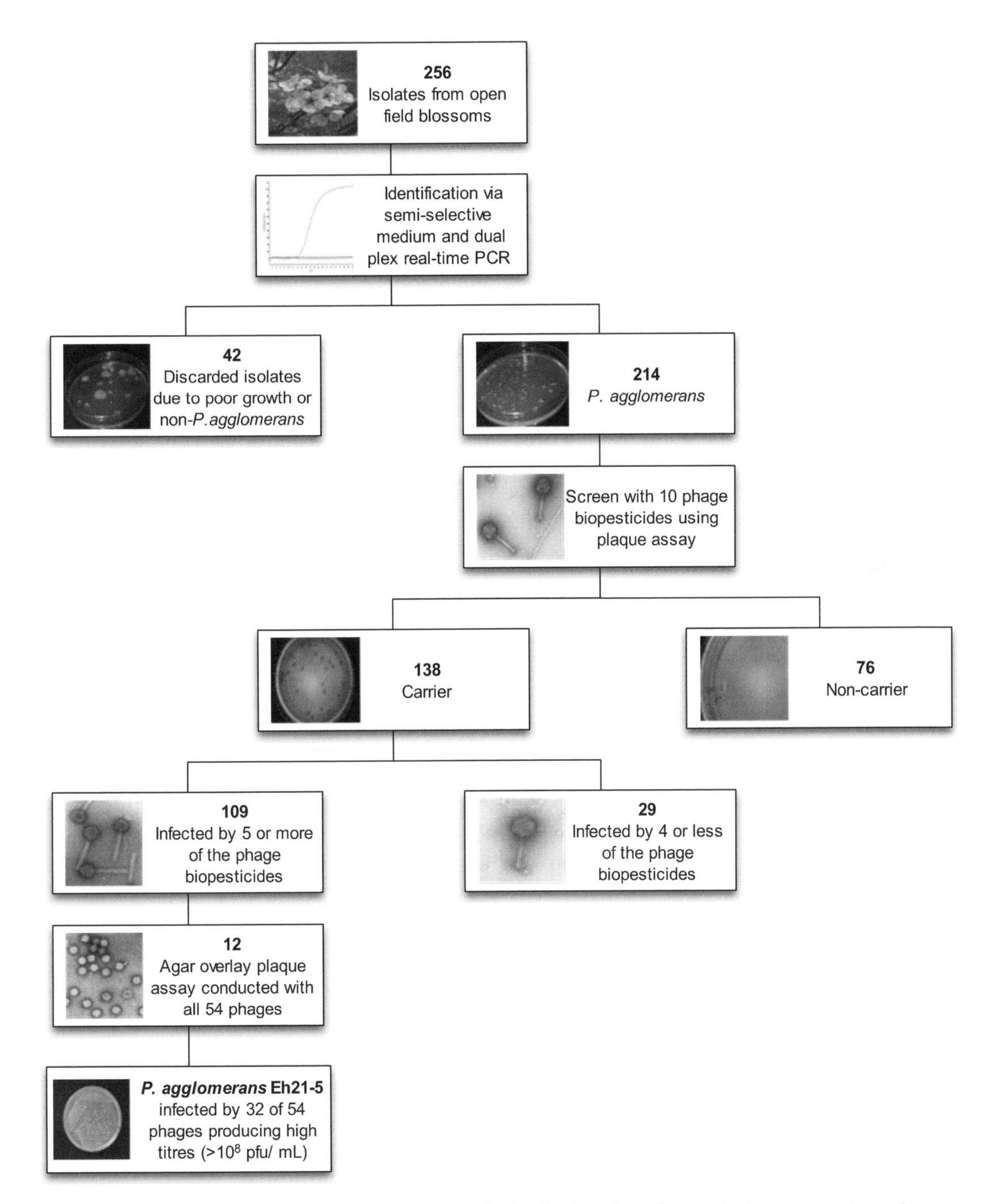

FIGURE 2 Sequential processes and decisions made for the isolation of bacteriophage carrier *P. agglomerans* isolates. Isolates were selected on the basis of culturability and ability to produce high-titer phage solutions with *Erwinia* sp. bacteriophages in the culture collection. (Adapted from Lehman [2007].)

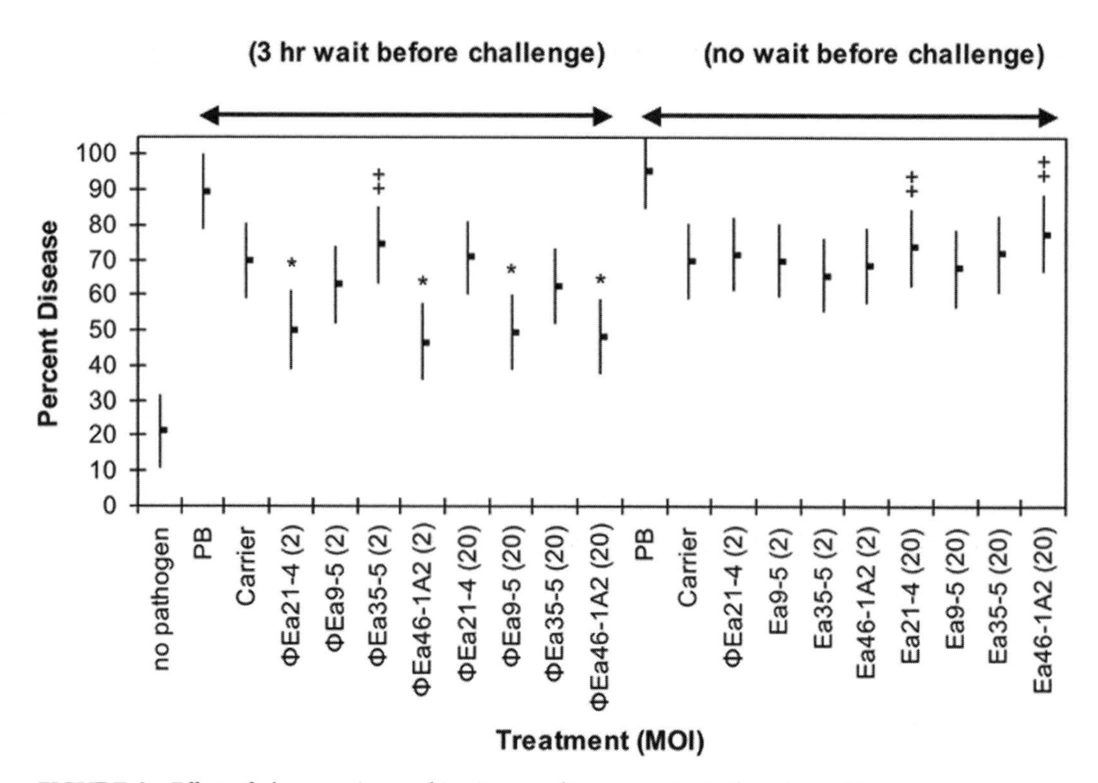

FIGURE 3 Effect of phage-carrier combinations on disease severity in forced pear blossom assays, given as mean ± 95% confidence limits for three replications (10 blossoms per replication). Four phages were applied with the carrier using a multiplicity of infection (MOI) of 2 and then again using an MOI of 20. These eight phage treatments, along with the carrier and phosphate buffer (PB) control, were repeated twice: once where 3 h elapsed between treatment and pathogen application (left-hand side) and once where the pathogen was applied immediately after treatment (right-hand side). All phage and carrier treatments except those marked with ‡ resulted in significantly less severe disease than the PB control. Phage-carrier treatments marked with * were significantly better than the carrier alone.

proves pathogen control (Fig. 3). Treatment timing and phage-carrier ratios were partially optimized prior to field experiments, and the optimum concentration of pathogen used in blossom assays proved to be the same as the concentration needed to detect significant differences in disease control in field trials.

During field trials, multiplex real-time PCR was used to simultaneously monitor the phage, carrier, and pathogen populations (Svircev et al., 2009) over the course of treatments. In all cases the observed population dynamics of the biocontrol agents and the pathogen were consistent with the success or failure of each treatment to control disease incidence. In treatments exhibiting a significantly reduced incidence of fire blight, the average blossom population of *E. amylovora* had been reduced to preexperiment epiphytic levels. In successful treatments the phages grew on the *P. agglomerans* carrier for 2 to 3 days after treatment application. The phages then grew preferentially on the pathogen once it was introduced into this blossom ecosystem. Successful phage-based treatments reduced the incidence of diseased blossom clusters by 50%, an efficacy that was statistically indistinguishable from that of streptomycin, which is the most effective bactericide currently available for fire blight prevention (Fig. 4). Two areas require additional research before this system can be used by commercial growers: testing of

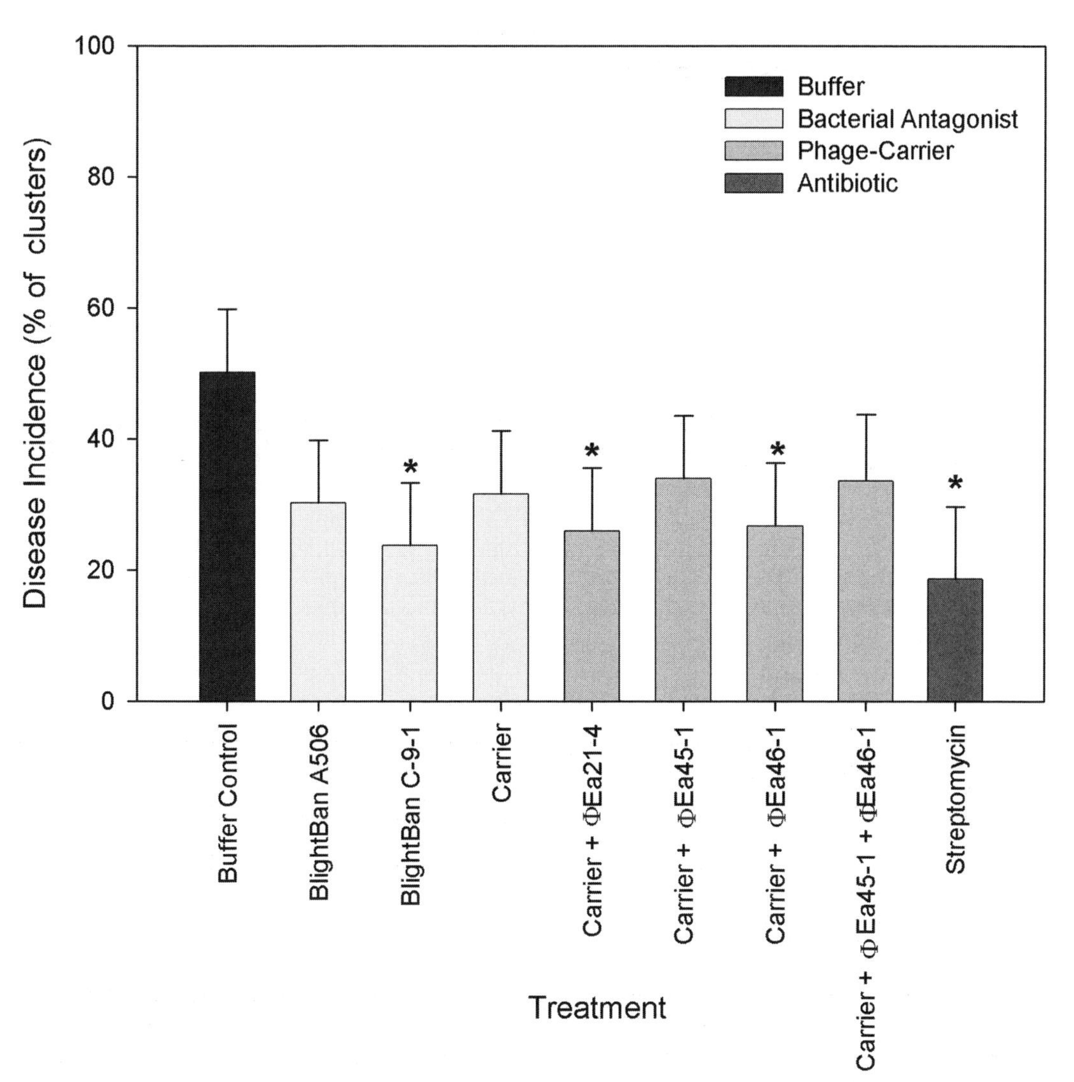

FIGURE 4 Incidence of fire blight in a 4-year-old Gala apple orchard. Posttreatment, trees treated with buffer (controls) and all active treatments were artificially inoculated with 10^6 CFU/ml of *E. amylovora*. BlightBan C9-1 (commercial biological control agent), two carrier-phage treatments, and streptomycin treatments marked with ⋆ caused significant reduction in disease incidence relative to the control ($P < 0.01$). The phage carrier was the orchard-isolated *P. agglomerans* EH21-5. (Reprinted with permission of S. M. Lehman.)

phage mixtures that will control different *E. amylovora* strains from different geographic regions and development of a product formulation that will permit reliable storage. However, the carrier system can accommodate the incorporation of multiple phages, and recent work on phage microencapsulation (Ma et al., 2008) suggests options for conversion to a commercial product. This work demonstrates the feasibility of a phage biopesticide for controlling *E. amylovora* populations and reducing the subsequent disease. It is the first reported use of an exogenously applied carrier bacterium to directly increase the efficacy of a phage-based therapy.

Control of *Xanthomonas* spp.

In the early 1960s, *Xanthomonas* sp. bacteriophages were used for the control of bacterial spot of peach, caused by *Xanthomonas arboricola* pv. *pruni* (Smith 1903) Vauterin et al. 1995 (= *Xanthomonas campestris* pv. *pruni* [Smith 1903] Dye 1978 = *Xanthomonas pruni* [Keil and Wilson, 1963; Civerolo and Keil, 1969]). The greatest efficacies were obtained when peach leaves were preinoculated with high-titer phage solutions (10^9 to 10^{10} PFU/ml) or premixed phage-*X. pruni* suspensions. In greenhouse trials, bacterial spot symptoms were decreased in peach seedlings by 51 to 54% compared to the control. Zaccardelli et al. (1992) studied the survival of the *Xanthomonas* sp. phages under in vitro and in vivo conditions. In addition, the authors examined the effect of common orchard fungicides, insecticides, and acaricides on the survival of the phages (Zaccardelli et al., 1992). Field trials used a single wide-host-range phage, different application times, and three separate orchard locations (Table 1) (Zaccardelli et al., 1992). In the presence of high pathogen pressure and optimal weather conditions for disease development, phages failed to reduce the severity of the disease in the fruit (Saccardi et al., 1993). The successful application strategy requires the spraying of bacteriophages when minimal bacterial populations are present (U. Mazzucchi, personal communication).

Bacteriophages (Fig. 5) were used to control bacterial spot of tomato, caused by *X. campestris* pv. *vesicatoria,* under greenhouse and field conditions (Balogh, 2002; Balogh et al., 2003, 2005; Obradovic et al., 2004, 2005; Jones et al., 2005, 2007; Iriarte et al., 2007). Phages are considered attractive alternatives to conventional treatments due to a lack of commercial cultivars with resistance to the pathogen and the prevalence of antibiotic or copper resistance in the pathogen (Balogh et al., 2003). Control of the pathogen varied between greenhouse and field applications, and between formulated and nonformulated phage mixtures (Fig. 6). Under field conditions, 0.5% pregelatinized corn flour, Casecrete NH-400 with 0.25% pregelatinized corn flour, or 0.75% powdered skim milk with 0.5% sucrose, each containing a mixture of several phages active against the pathogen, provided 12 to 32, 24 to 43, and 9 to 20% disease control, respectively, compared to the untreated control (Balogh et al., 2003).

Host-range mutant phages are included in formulations to prevent the development of bacterial host resistance. The process was ini-

TABLE 1 Effects of phage treatment on bacterial spot in peach orchards[a]

Treatment	Orchard X			Orchard Y			Orchard Z		
	Diseased fruits	Surface per fruit		Diseased fruits	Surface per fruit		Diseased fruits	Surface per fruit	
	%	mm²	%	%	mm²	%	%	mm²	%
Control phage	44.13[a]	3.81[a]	0.024[a]	20.72[a]	1.16[a]	0.008[a]	40.42[a]	5.71[a]	0.051[a]
7-day interval	39.24[a]	2.19[a]	0.015[a]	10.45[a]	1.62[a]	0.011[a]	22.24[b]	1.98[b]	0.018[b]
10-day interval	29.04[a]	1.83[a]	0.012[a]	10.99[a]	1.11[a]	0.008[a]	25.31[b]	3.06[b]	0.028[b]
14-day interval	32.88[a]	1.91[a]	0.012[a]	13.14[a]	1.57[a]	0.010[a]	24.80[b]	2.60[b]	0.023[b]

[a]Mean percentage of diseased fruits and mean area and mean percentage of diseased surface per fruit in orchards X, Y, and Z following treatment with phage F_8 at 7-, 10-, and 14-day intervals. The same phage dose was used for all the treatments and was 10^8 PFU/ml. The control trees were treated weekly with a suspension of heat-killed phages. The number and area of the spots were assessed on the basis of a unit lesion of 0.25 mm^2. Spot counting was performed on the harvested fruits. In orchards X, Y, and Z, 1,591, 2,665 and 6,832 fruits were assessed, respectively. A row of diseased trees as a supplementary source of inoculum was created in the orchards X and Y but not in orchard Z. Means followed by the same letter are not statistically different ($P \leq 0.05$). (Reproduced from Saccardi et al., *Phytopathologia Mediterranea* **32:**206–210, 1993, with permission.)

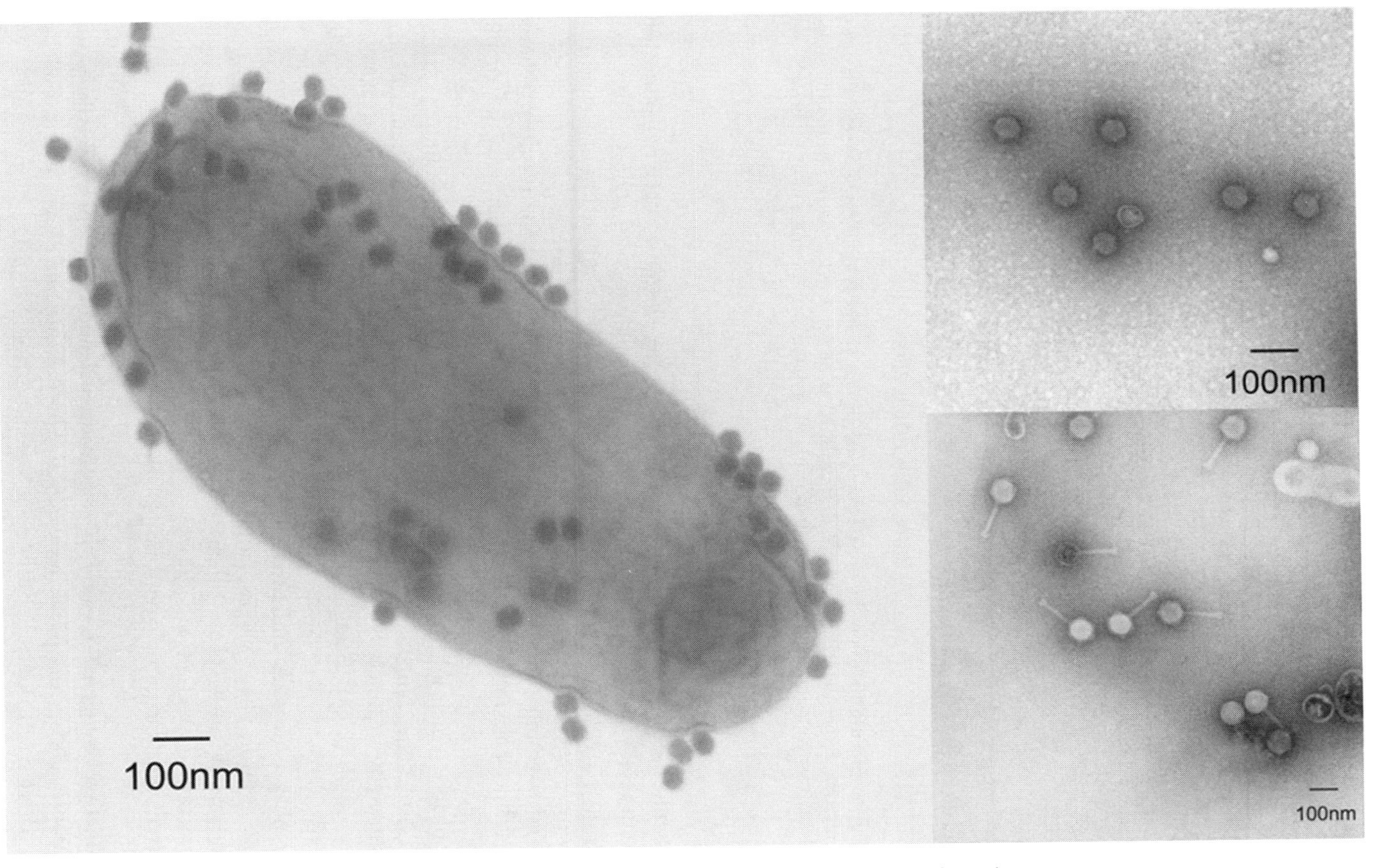

FIGURE 5 Electron micrographs showing (left) the infection of *Xanthomonas* sp. bacterium by phages, and (right) individual phages belonging to the *Podoviridae* (top right) and *Myoviridae* (bottom right). Marker = 100 nm. (Courtesy of B. Balogh and J. Jones.)

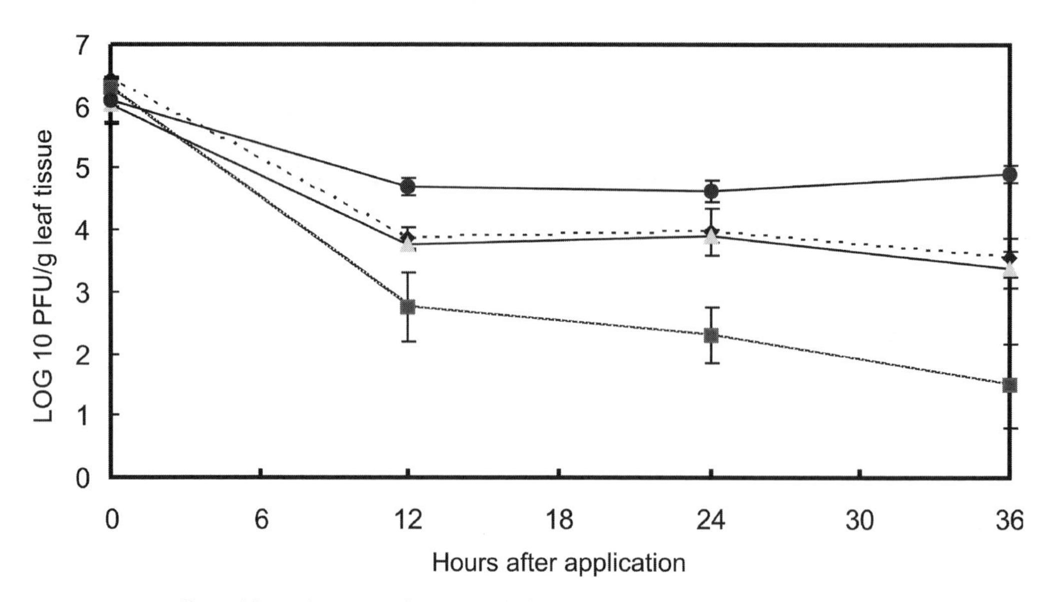

FIGURE 6 Effect of formulations on the survival of *Xanthomonas* sp. bacteriophages in the tomato canopy under field conditions. Experimental groups are distinguished as Silwet (circle), casein (diamond), pregelatinized corn flour (triangle), and nonformulated (square). (Reprinted with permission of B. Balogh and J. Jones.)

tially developed and patented for the control of plant disease caused by *Pseudomonas syringae* (Jackson, 1989). Host-range mutants are produced by growing phages on wild-type bacterial cultures and selecting for host bacteria resistant to the phage. Large numbers of lytic phages are screened on the resistant bacterial host cells. The phages which kill the resistant host are named host-range mutants and have been used subsequently in biocontrol formulations (Flaherty et al., 2000; Balogh et al., 2003; Jones et al., 2007).

To improve phage efficacies and provide consistent disease control, phage biopesticides have been integrated into plant disease management practices (Jones et al., 2007). Systemic acquired resistance or hypersensitive response activators such as Actiguard® (acibenzolar-*S*-methyl; Syngenta Crop Protection, Greensboro, NC) or Messenger® (harpin; Eden Biosciences Corp., Bothell, WA) have been tested along with bacteriophages for the control of leaf blight of onion caused by *Xanthomonas axonopodis* pv. *allii* (Lang et al., 2007) or bacterial blight of tomato (Obradovic et al., 2004; Obradovic et al., 2005). The combination of phages and activator provided disease control as good as (Lang et al., 2007) or better than (Obradovic et al., 2004, 2005) either activator or phages alone. A similar outcome was achieved with the combination of phages and the bacteriocide copper-mancozeb (Obradovic et al., 2004). In contrast, Asiatic citrus canker and citrus bacterial spot, caused respectively by *X. axonopodis* pathovars *citis* and *citrumelo,* were controlled more effectively by copper-mancozeb alone than by either phages or the combination of copper-mancozeb and phages (Balogh, 2006; Balogh et al., 2008). Overall, the efficacy of bacteriophages varies with bacterial host, use of formulations, combination treatments, and treatment protocols.

There is no one single magic bullet recommendation for phage biopesticides. When phages are applied to the same pathogen-host system, inconsistencies are often evident between greenhouse- and field-based results.

Plant bioassays often used to screen for individual phage efficacy involve bacterial populations at pre- and postphage application. Ravensdale et al. (2007) used phages of *Pectobacterium carotovorum* subsp. *carotovorum* (Jones 1901) Hauben et al. 1999 (= *Erwinia carotovora* subsp. *carotovora*) to demonstrate that a calla plug bioassay was not an effective predictor of greenhouse phage efficacy. The phage cocktail that successfully controlled the bacterial soft rot of calla under greenhouse conditions failed to yield similar control on calla plugs. Phages failed to completely eliminate the bacterial population on the calla plugs, and the lower limit of phage biological control activity was approximately $10^{3.3}$ CFU/ml. In addition, under greenhouse conditions, phages were adversely affected by tap water and Fe-EDTA (Ravensdale et al., 2007). In the field, phages may be highly affected by UVB (280 to 320 nm) from sunlight, temperature extremes, desiccation, and other pesticides used on the crop (Iriarte et al., 2007). Microencapsulation technologies have been recently developed to deliver the Felix-O1 phage into the gastrointestinal environment (Ma et al., 2008; see chapter 12 in this volume). The encapsulation process simultaneously protects and releases the contents of the capsules at controlled rates. To date, encapsulation technologies have not been used in the field with agricultural phages.

EFFECT OF THERAPEUTIC PHAGES ON RHIZOSPHERE AND PHYLLOSPHERE MICROBIOTA, AND THE CONSEQUENCES FOR PLANT GROWTH

Agricultural soils harbor an abundant and diverse collection of bacterial genomes and their associated bacteriophages (Williams et al., 1987; Ashelford et al., 2003; Srinivasiah et al., 2008). Direct counts have estimated the number of phages at 1.5×10^8 g^{-1} soil, equivalent to about 4% of the total soil bacterial population (Ashelford et al., 2003). The presence of bacteriophages in the soil is influenced by biotic and abiotic factors. Relevant biotic factors include crop type, presence of phytopathogens, native soil microflora, and the availability of phage hosts (Williams et al., 1987; Day and Miller, 2008) (http://www.apsnet.org/online/feature/phages). Abiotic factors include environmental conditions, soil composition and chemistry, nutrient levels, temperature, pH, and the presence of water. The application of pesticides, biological control agents, and fertilizers modifies some of these conditions and can therefore be expected to change the balance and composition of host-phage communities. Studies on soil bacteria and their associated phages include phages of rhizobacteria (Ashelford et al., 2000), *Bacillus* spp. (Tan and Reanney, 1976), phagelike particles of *Rhizobium leguminosarum* bv. *trifolii* (= *Rhizobium trifolii*) (Evans et al., 1979; Joseph et al., 1985), *R. trifolii* (Kleczkowska, 1950; Barnet, 1972; Evans et al., 1979), *Sinorhizobium meliloti* (Katznelson and Wilson, 1941), *Bradyrhizobium japonicum* (= *Rhizobium japonicum*) (Kowalski et al., 1974), *Serratia liquefaciens* (Ashelford et al., 1999b), *Streptomyces scabiei* (Goyer, 2005), *Pseudomonas* spp. (Stephens et al., 1987), and *Pseudomonas fluorescens* CHA0 (Keel et al., 2002).

The rhizosphere comprises the soil and environment that immediately surrounds plant roots. The rhizobacteria that colonize this zone and plant root tissue have coevolved with plants, and their effects on the plant may be beneficial, deleterious, or neutral (Schroth and Hancock, 1981). Rhizobacteria have been isolated from most plants and generally belong to *Rhizobium* spp. (Katznelson and Wilson, 1941; Kleczkowska, 1950; Barnet and Vincent, 1970; Barnet, 1972; Kowalski et al., 1974; Evans et al., 1979; Joseph et al., 1985; Kumhar and Ramkrishna, 2007), *Pseudomonas* spp. (Keel et al., 2002), and *Azospirillum* spp. (Bashan et al., 2004). These bacteria promote plant growth by colonizing the root zone, and some species induce nodulation. More recently, the term "plant growth-promoting bacteria" (PGPB) has been used to describe the bacteria that contribute to greater plant yields and overall health by nodulation and nitrogen fixation (Munsch and Olivier, 1995) or via production of phytohormones that are associated with increased root surface area and higher absorption capacity (Bashan et al., 2004).

Demolon and Dunez (1936) were the first to suggest that phages may interfere directly with rhizobacteria. Studies that examined the deleterious effect of phages on rhizobacteria initially relied on in vitro tests with single phages and hosts. Lytic phages were isolated from *Rhizobium* spp. (Katznelson and Wilson, 1941; Kleczkowska, 1950; Barnet, 1972; Kowalski et al., 1974). Large populations of phages introduced into the rhizosphere of clover plants reduced the *R. trifolii* population. The phages also impaired overall nodulation function by adding "symbiotically ineffective mutants" to the *R. trifolii* population (Kleczkowska, 1950; Evans et al., 1979). The number of phage-sensitive mutants initially increased and the number of nodules decreased when large phage populations were applied. This shift in phage-sensitive mutant populations and nodulation capacity was lost after 3 to 4 weeks and the populations reverted to normal (Evans et al., 1979). Joseph et al. (1985) observed that *R. trifolii* had an adverse effect on the nodulation capacity of cowpea-associated rhizobia, which led to retarded plant growth. They hypothesized that the population of cowpea-associated rhizobia was reduced by phagelike particles apparently produced by *R. trifolii.* In experiments with sugar beets and a fluorescent pseudomonad, Stephens et al. (1987) demonstrated that soil-resident phages influence which PGPB can be colonized by the rhizosphere phages present in nonsterile soils. This study also demonstrated the critical differences between phage survival and activity in sterile soil versus nonsterile soil.

Temperate phages can modulate the composition and interactions of rhizosphere microflora through mechanisms other than simple lysis. Boyer et al. (2008) examined 24 isolates of *Azospirillum brasilense,* a rhizobacterium that promotes growth of cereals and grasses, resulting in 30% greater crop yields. They showed that *Azospirillum* spp. often contain inducible integrative phages and proposed that the *Azospirillum* community in the soil is influenced by the activity of these prophages. The presence of *Azospirillum* prophages can also influence the interactions between these isolates and indigenous soil bacteriophages, since the cells carrying a prophage may be resistant to superinfection by certain other phages. The transducing ability of bacteriophages in indigenous soil has been examined by studying the interaction between *Serratia* spp. and six phages in a sugar beet phytosphere. Four of the six phages were capable of transduction (Ashelford et al., 1999b). A follow-up study isolated *P. fluorescens, S. liquefaciens,* and their associated phages from the in situ microflora of the sugar beet phytosphere and rhizosphere (Ashelford et al., 1999b). This research demonstrated that phages occupy distinct niches in the soil rhizosphere and phytosphere. This study was unique in that the soil was not preinoculated with bacterial isolates commonly associated with the sugar beet ecosystem. A drawback in the study is that only restriction fragment length polymorphisms and plaque morphologies were used to track the phage groups. This technique underestimates the full diversity of phages present in the plant surface and rhizosphere, since different phages can exist within the same restriction fragment length polymorphism pattern. Nonetheless, experiments of this nature provide an important insight into the complexities of the rhizosphere and the presence of phage niche populations. Ashelford et al. (2000) also studied the population dynamics of *S. liquefaciens* and two of its phages (one virulent and one temperate) over a 6-month period in a natural soil environment. They showed that phage populations will spread between neighboring regions of soil, that the lysogenic form of a bacterial strain can survive in soil as well as the nonlysogenic form, and that phages and bacteria added to a soil environment can alter the local microbial community. Their data also provided experimental support for earlier hypotheses that a phage with a short latent period and small burst size will outcompete a phage with a longer latent period and a larger burst size (Abedon, 1989; Wang et al., 1996). The obligate virulent phages will outcompete temperate phages when the density of susceptible host cells is high (Stewart and Levin, 1984).

The possibility that phages will negatively impact the efficacy of bacterial biopesticides must also be considered. Bacteria may be added to aerial plant tissue or to soils as plant growth promoters and as biopesticides against root pathogens. Phage ΦGP100 significantly reduced survival, root colonization, and biological control activity of *P. fluorescens* CHA0 (Keel et al., 2002). The phage-induced decline in the bacterial population resulted in the infection of the plant roots by a fungal pathogen. The aerial phyllosphere is generally less conducive to phage survival than the soil environment (Schnabel and Jones, 2001; Lehman, 2007; Iriarte et al., 2007). Aerially applied phages can flourish in the underlying soil (K. E. Schneider, S. M. Lehman, and A. M. Svircev, unpublished data), and phyllosphere bacteria can favor the maintenance of phages that are able to infect them (Lehman, 2007). These phages then have the opportunity to impact the efficacy of bacterial biopesticides in both the phyllosphere and the soil beneath.

Novel use of bacteriophages to limit the growth of bacteria introduced into soils was demonstrated by Smit et al. (1996) with Tn*5*-genetically modified *P. fluorescens* ΦR2f and its specific phage ΦR2f in wheat rhizosphere. Bacterial cells were introduced to the rhizosphere on alginate beads. The alginate protected the bacteria from phage attack. In the rhizosphere the bacterial population increased due to added nutrients and the plant-promoting activity continued. To remove the genetically modified bacteria, R2f phages were added and reduced the rhizosphere population by 10^4-fold. The authors propose that in the future phages may be used to remove genetically modified bacteria that are used as biological control agents and for bioremediation of polluted soils.

The terrestrial environment provides an arena for extensive interactions between phages and their related and nonrelated hosts and plants. The theoretical discussions that should be considered include the significance and prevalence of lysogeny in soil bacteria (Williams et al., 2008), the possibility for phage-mediated exchange of genetic information between nonrelated species (Chen and Novick, 2009), the presence of prophage sequences in bacterial genomes (Canchaya et al., 2004), and the contribution of prophages to the genetic individuality of bacterial strains (Canchaya et al., 2004). From a practical perspective, phage biopesticide development clearly needs to incorporate careful characterization of the candidate phages to minimize the risk of unintentional genetic transfer and to consider that the phages might negatively impact PGPB, rhizobacterial symbionts, or bacterial biopesticides.

GENERAL CONSIDERATIONS FOR DEVELOPMENT OF BIOCONTROL PHAGES IN FOOD AGRICULTURE

The key to creating an effective phage-based biopesticide is the philosophy on which the development program is based. The successful development of microbial biopesticides requires that several factors be considered at every stage of research: efficacy, economics, and regulatory constraints. Regulatory requirements include safety/toxicity studies, environmental persistence and impact studies, and the aesthetic acceptability of the basic concept. Economic considerations include the ease and cost of production, storage, and use. Finally, and most importantly, the treatment must work. It must predictably control an appropriately broad spectrum of pathogen strains in the field, either reducing the pathogen population to a level below that which will cause economic damage or removing the pathogen's ability to cause disease.

The criteria by which isolates are selected must approximate the criteria which will determine success or failure in the field. A candidate that is highly effective against a laboratory strain in an in vitro assay may not be effective against a natural infection. Care must be taken to select candidate phages that do not infect beneficial rhizosphere or phyllosphere bacteria. This screening process is more complex than it may seem upon first consideration. A standard panel of bacteria against which candidate phages should be tested might reasonably include species of rhizobacteria, plant growth-promoting pseudomonads, and existing bacterial biopesticides. A secondary, variable panel of screening bacteria should include species that are known to antagonize important pathogens on the crop(s) of interest for that specific biopesticide. Species associated with crops that are grown in similar regions and soil types as the crop of interest are worth considering, since phage populations have at least some ability to migrate to adjacent soils, presumably by spreading through a dispersed host population. Environmental isolates should be used instead of laboratory strains, ideally using isolates from multiple geographic locations and soil types. Host-range testing on target pathogen strains should be conducted at temperatures relevant to pathogen infection, but host-range testing on nontarget beneficial bacteria should be conducted across the entire range of temperatures at which applied phages might interact with environmental microflora throughout the year (likely 4 to 35°C for most situations). Experiments in soil must consider phage-host interactions in nonsterile samples of field soil, since the native microbial ecology will differ from that in commercial or sterilized soil.

None of these requirements should be viewed as impediments to the development of phage biopesticides. Concern about plant-beneficial bacteria may make phage-based biopesticides problematic for some pathogens, such as pseudomonads, but chemical pesticides also affect the survival of nonpathogenic bacteria, and with a much broader spectrum. In practice, it would likely be sufficient to ensure that a given phage biopesticide is intentionally and explicitly not labeled for use on or adjacent to plots where the crops being grown are known to be critically dependent on a bacterial species that the biopesticide phages infect. The transfer of virulence genes or other undesirable traits between bacteria can be minimized by using only phages that are obligately lytic and by screening for generalized transduction of molecular markers by those phages. While it is impossible to prove that transduction will never occur, it is possible to quantify a maximum risk. The use of phage mixtures should also help minimize the likelihood of transduction, since it is unlikely that all phages in a mixture would be capable of transduction following careful screening. If transduction did occur, the transductant should still be lysed by at least one of the other phages in the mixture, preventing its propagation. The use of phage mixtures also substantially reduces the likelihood of bacteria becoming phage resistant in the first place (Tanji et al., 2004, 2005). Phage mixtures are recommended for any phage therapy because the combination of many carefully selected phages drastically increases the likelihood that the pathogen strains endemic to a particular region will be susceptible to the biopesticide. In fact, Stewart (2001) suggests that the tendency toward single-strain biopesticides is one of the most important and most easily rectified barriers standing in the way of consistent performance by biological control agents in general.

The most critical step in biopesticide development is proving that the biopesticide has high efficacy under field conditions. The complex interaction of biotic and abiotic factors in the field can also alter the physiological and biochemical properties of plants in ways that cannot be replicated elsewhere. As a result, greenhouse and growth-chamber plants sometimes support larger epiphytic microbial populations than field plants (O'Brien and Lindow, 1989; Beattie and Lindow, 1994). The pathogen pressure against which biopesticides are tested must also reflect the natural infection process as much as possible. In many cases, applying a pathogen concentration comparable to indigenous populations will not produce enough disease to obtain statistically significant data unless a impractically large number of sample plants is used (Johnson and Stockwell, 2000). If a higher concentration of pathogen is necessary, care should be taken to ensure that the pathogen population must still increase on the plant surface before infection can take place; otherwise there will be no opportunity for the biopesticide to antagonize the pathogen as it would in a natural-infection setting.

Without a doubt, the greatest hindrance to the development of successful phage therapies has been the absence of adequate knowledge of the phage-host ecology in the therapeutic environment (Goodridge, 2004; Summers, 2005; Day and Miller, 2008). The consequence of this has been a historic inability to explain the success or failure of a given treatment and thereby prove that field efficacy is due to the direct action of phages against the pathogen. Addressing this issue absolutely requires that the population dynamics of the biopesticide components and the pathogen be monitored throughout field trials. The effects of environmental stresses on the culturability of bacteria (Wilson and Lindow, 2000) make a culture-independent method preferable.

Finally, commercial production of effective phage-based biopesticides will be dramatically improved by advances in formulation methods. Two possible strategies have begun to be explored: formulation with a protectant material and microencapsulation. Several materials have been investigated as UV protectants for insect viruses and phages, including optical brighteners and other light-interacting dyes (Shapiro, 1992; Dougherty et al., 1996; Ignoffo et al., 1997), heterocyclic amino acids (Ignoffo and Garcia, 1995), lignin (Tamez-Guerra et al., 2000), casein (Behle et al., 1996; Balogh, 2002, 2006; Balogh et al., 2003), Casecrete (Balogh, 2002; Balogh et al., 2003), and pregelatinized and other forms of corn flour (Balogh, 2002; Balogh et al., 2003; Tamez-Guerra et al., 2000). Only formulations incorporating pregelatinized corn flour, Casecrete, and skim milk have been field-tested with phages to a notable extent (Balogh, 2002, 2006; Balogh et al., 2003). The Casecrete formulation showed the most promise: under field conditions, *Xanthomonas* phages formulated with Casecrete persisted longer and improved disease control relative to unformulated phages (Balogh et al., 2003). However, Jones et al. (2007) note that when applied without phages these formulation materials actually facilitated disease development, possibly by providing nutrition and/or a protected surface microenvironment for the pathogen, and note the need for biologically inert formulation materials. The second strategy for commercial phage release is microencapsulation. This technology involves packaging solid, liquid, or gaseous materials in capsules that release their contents at controlled rates under specific conditions (Anal and Singh, 2007). To date, microencapsulation has been studied as a way to enhance the viability of probiotic bacteria (Rao et al., 1989; Lee and Heo, 2000), to facilitate transplantation of mammalian tissue (Uludag et al., 2000), and to improve delivery and immunogenicity of viral vaccines (Offit et al., 1994; Moser et al., 1998; Sturesson and Degling Wikingsson, 2000; Nechaeva, 2002; Lameiro et al., 2006). However, only one study has developed a microencapsulation method that uses materials compatible with bacteriophage survival (Ma et al., 2008). Their system, which uses chitosan-alginate-$CaCl_2$ microspheres, permitted complete or near-complete survival of *Salmonella* phage Felix-O1 under favorable wet conditions or in a simulated acid gut environment and permitted 12.6% survival in dried form. This method may not be directly transferable to phage-based plant biopesticides. Concentrated wet suspensions requiring refrigeration should be acceptable to growers, if not ideal, and the acid tolerance conferred by the method of Ma et al. (2008) would be beneficial for phage biopesticides that are meant to be active in soil or blossom secretions. The main difference is the need for phage-based plant biopesticides to be protected from UV inactivation.

CONCLUDING REMARKS

We have described a number of agricultural applications for phage biopesticides at various stages of development. The development of agricultural biopesticides involves a long and multistep process that is depicted in Fig. 7. Biopesticide development can be delayed and hindered at any point in the chain due to scientific and/or funding problems. In addition, the development chain incorporates stop/go measures that are critical in the commercial development of a biopesticide. Phages as biopesticides have great potential in agriculture; however, the success stories are often limited to specific host-pathogen systems. Research has demonstrated that screening for phages with the highest efficacies is often hampered by inconsistencies between disease control in bioassays and greenhouse and field trials. Phage biopesticides may need to be integrated into current plant disease control practices that employ antibiotics, systemic acquired resistance inducers, and various cultural practices.

Globally an estimated 25% of all food produced is lost to microbial spoilage after harvest/slaughter (Subcommittee on Microbial Criteria, 1985). Phages have been used for the control of human pathogens in food production (EBI Food Safety, 2006; Hagens and Loessner, 2007; Hagens and Offerhaus, 2008; Hudson et al., 2005; Abuladze et al., 2008; Garcia et al., 2008; Guenther et al., 2009; see chapter 7 in this volume). Postharvest reduction of bacterial plant pathogens is generally addressed by physical and chemical means. Ironically, some of the difficulties of managing plant diseases in the field do not exist for postharvest disease control. Environmental conditions are fairly constant and well controlled, which reduces the variety of conditions in which a given treatment must be effective and makes growth-chamber trials highly representative of the therapeutic environment. In many ways, this neglected area of phage biopesticides might be the easiest one for which to develop a commercial product.

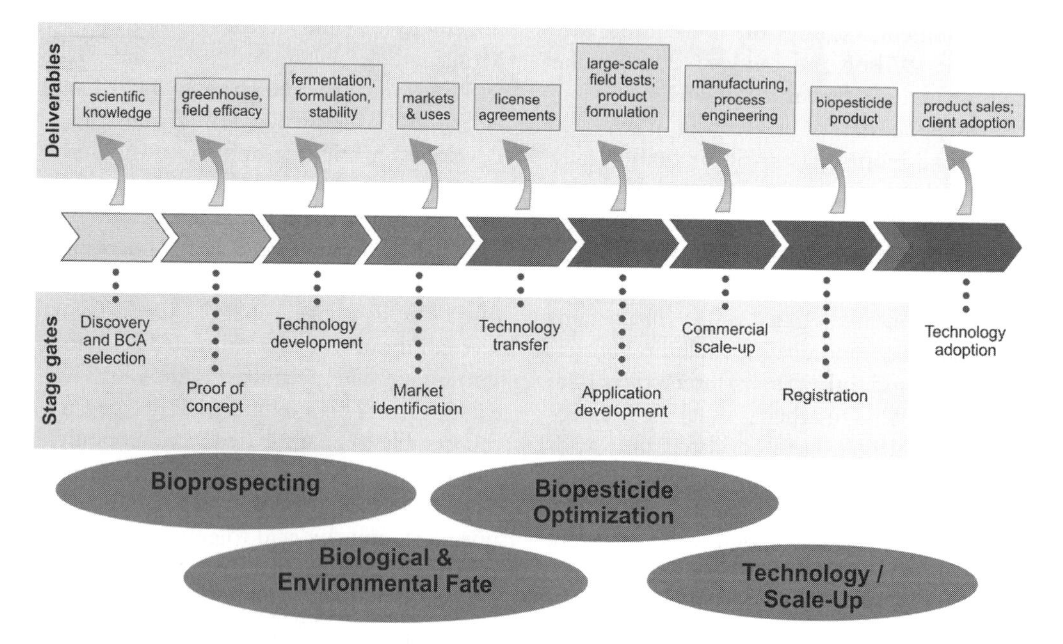

FIGURE 7 The biopesticide value chain was developed at Agriculture and Agri-Food Canada by S. Boyetchko. The chain summarizes the multistep sequential processes and the go/no-go decision processes required in the development of a biopesticide. The main steps in the chain involve bioprospecting, determination of biological and environmental fate, biopesticide optimization, and technology scale-up. (Reprinted from Bailey et al. [2009] with permission of the publisher.)

Phage therapy is not suitable for all bacterial disease, whether in plants, animals, or humans. Certain infection niches may not be accessible to phages, such as with intracellular pathogens, or may not allow phages to persist long enough to be infective. Biofilms also present a particular challenge because of their heterogeneous physical, chemical, and genetic structure (Brüssow and Kutter, 2005). At the same time, not all phages are suitable for phage therapy. The success of a phage therapy against any particular disease also depends on the availability of many phages that, together, are effective against most or all strains of the causative agent.

REFERENCES

Abedon, S. T. 1989. Selection for bacteriophage latent period length by bacterial density: a theoretical examination. *Microb. Ecol.* **18:**79–88.

Abedon, S. T. 2003. Bacteriophages could fight plant pathogens. *Ind. Bioprocessing* **25:**7.

Abuladze, T., M. Li, M. Y. Menetrez, T. Dean, A. Senecal, and A. Sulakvelidze. 2008. Bacteriophages reduce experimental contamination of hard surfaces, tomato, spinach, broccoli, and ground beef by *Escherichia coli* O157:H7. *Appl. Environ. Microbiol.* **74:**6230–6238.

Adams, M. H. 1959. *Bacteriophages,* p. 443–522. Interscience Publishers, New York, NY.

Anal, A. K., and H. Singh. 2007. Recent advances in microencapsulation of probiotics for industrial applications and targeted delivery. *Trends Food Sci.* **18:**240–251.

Ashelford, K. E., M. J. Day, M. J. Bailey, A. K. Lilley, and J. C. Fry. 1999a. *In situ* population dynamics of bacterial viruses in a terrestrial environment. *Appl. Environ. Microbiol.* **65:**169–174.

Ashelford, K. E., M. J. Day, and J. C. Fry. 2003. Elevated abundance of bacteriophage infecting bacteria in soil. *Appl. Environ. Microbiol.* **69:**285–289.

Ashelford, K. E., J. C. Fry, J. M. Bailey, A. R. Jeffries, and M. J. Day. 1999b. Characterization of six bacteriophages of *Serratia liquefaciens* CP6 isolated from sugar beet phytosphere. *Appl. Environ. Microbiol.* **65:**1959–1965.

Ashelford, K. E., S. J. Norris, J. C. Fry, M. J. Bailey, and M. J. Day. 2000. Seasonal population dynamics and interactions of competing bacteriophages and their host in the rhizosphere. *Appl. Environ. Microbiol.* **66:**4193–4199.

Bailey, K. L., S. M. Boyetchko, G. Peng, R. K. Hynes, W. G. Taylor, and W. M. Pitt. 2009. Developing weed control techniques with fungi, p. 1–44. *In* M. Rai (ed.), *Advances in Fungal Biotechnology.* I.K. International Private Ltd., New Delhi, India.

Balogh, B. 2002. Strategies of improving the efficacy of bacteriophages for controlling bacterial spot of tomato. M.S. thesis. University of Florida, Gainesville, FL.

Balogh, B. 2006. Characterization and use of bacteriophages associated with citrus bacterial pathogens for disease control. Ph.D. thesis. University of Florida, Gainesville, FL.

Balogh, B., B. I. Canteros, R. E. Stall, and J. B. Jones. 2008. Control of citrus canker and citrus bacterial spot with bacteriophages. *Plant Dis.* **92:** 1048–1052.

Balogh, B., J. B. Jones, M. T. Momol, and S. M. Olson. 2005. Persistence of bacteriophages as biocontrol agents in the tomato canopy. *Acta Hortic.* **695:**299–302.

Balogh, B., J. B. Jones, M. T. Momol, S. M. Olson, A. Obradovic, P. King, and L. E. Jackson. 2003. Improved efficacy of newly formulated bacteriophages for management of bacterial spot on tomato. *Plant Dis.* **87:**949–954.

Barnet, Y. M. 1972. Bacteriophages of *Rhizobium trifolii. J. Gen. Virol.* **15:**1–15.

Barnet, Y. M., and J. M. Vincent. 1970. Lysogenic conversion of *Rhizobium trifolii. J. Gen. Microbiol.* **61:**319–325.

Bashan, Y., G. Holguin, and L. E. de-Bashan. 2004. *Azospirillum*-plant relationships: physiological, molecular, agricultural, and environmental advances (1997–2003). *Can. J. Microbiol.* **50:**521–577.

Beattie, G. A., and S. E. Lindow. 1994. Comparison of behavior of epiphytic fitness mutants of *Pseudomonas syringae* under controlled and field conditions. *Appl. Environ. Microbiol.* **60:**3799–3808.

Behle, R. W., M. R. McGuire, and B. S. Shasha. 1996. Extending the residual toxicity of *Bacillus thuringiensis* with casein-based formulations. *J. Econ. Entomol.* **89:**1399–1405.

Bigwood, T., J. A. Hudson, and C. Billington. 2009. Influence of host and bacteriophage concentrations on the inactivation of food-borne pathogenic bacteria by two phages. *FEMS Microbiol. Lett.* **291:**59–64.

Billing, E. 1987. Avirulent mutants of *Erwinia amylovora;* relationship between phage sensitivity and biological properties. *Plant Pathog. Bact.* **4:**617–622.

Bonn, W. G., and T. van der Zwet. 2000. Distribution and economic importance of fire blight. p. 37–53. *In* J. L.Vanneste (ed.), *Fire Blight: the Disease and Its Causative Agent,* Erwinia amylovora. CABI Publishing, Wallingford, United Kingdom.

Borysowski, J., B. Weber-Dabrowska, and A. Gorski. 2006. Bacteriophage endolysins as a novel class of antibacterial agents. *Exp. Biol. Med.* (Maywood) **231:**366–377.

Boyer, M., J. Haurat, S. Samain, B. Segurens, F. Gavory, V. Gonzalez, P. Mavingui, R. Rohr, R. Bally, and F. Wisniewski-Dye. 2008. Bacteriophage prevalence in the genus *Azospirillum* and analysis of the first genome sequence of an *Azospirillum brasilense* integrative phage. *Appl. Environ. Microbiol.* **74:**861–874.

Bradley, D. E. 1965. The morphology and physiology of bacteriophages as revealed by the electron microscope. *J. R. Microsc. Soc.* **84:**257–316.

Brüssow, H., and E. Kutter. 2005. Genomics and evolution of tailed phages, p. 91–128. *In* E. Kutter and A. Sulakvelidze (ed.), *Bacteriophages: Biology and Applications.* CRC Press, Boca Raton, FL.

Cairns, B., and R. J. Payne. 2008. Bacteriophage therapy and the mutant selection window. *Antimicrob. Agents Chemother.* **52:**4344–4350.

Canchaya, C., G. Fournous, and H. Brüssow. 2004. The impact of prophages on bacterial chromosomes. *Mol. Microbiol.* **53:**9–18.

Chanishvili, N., T. Chanishvili, M. Tediashvili, and P. A. Barrow. 2001. Phages and their application against drug-resistant bacteria. *J. Chem. Technol. Biotechnol.* **76:**689–699.

Chen, J., and P. Novick. 2009. Phage-mediated intergeneric transfer of toxin genes. *Science* **323:**139–141.

Civerolo, E. L. 1973. Relationship of *Xanthomonas pruni* bacteriophages to bacterial spot disease in *Prunus. Phytopathology* **63:**1279–1284.

Civerolo, E. L., and H. L. Keil. 1969. Inhibition of bacterial spot of peach foliage by *Xanthomonas pruni* bacteriophage. *Phytopathology* **59:**1966–1967.

Clark, J. R., and J. B. March. 2006. Bacteriophages and biotechnology: vaccines, gene therapy and antibacterials. *Trends Biotechnol.* **24:**212–218.

Clokie, M. R. J., and A. M. Kropinski (ed.). 2009. *Bacteriophages: Methods and Protocols,* vol. 1. *Isolation, Characterization, and Interactions.* Vol. 2, *Molecular and Applied Aspects.* Humana Press, Springer, New York, NY.

Crosse, J. E., and M. K. Hingorani. 1958. Method for isolating *Pseudomonas mors-prunorum* phages from the soil. *Nature* **181:**60–61.

Day, M. J., and R. V. Miller. 2008. Phage ecology of terrestrial environments, p. 281–301. *In* S. Abedon (ed.), *Bacteriophage Ecology: Population Growth, Evolution, and Impact of Bacterial Viruses.* Cambridge University Press, Cambridge, United Kingdom.

Demolon, A., and A. Dunez. 1936. Nouvelles observations sur le bactériophage et la fatique des sols cultiveés en Luzerne. *Ann. Agron.* **6:**435–455.

d'Hérelle, F. 1917. Sur un microbe invisible antagonistic des bacilles dysentériques. *C. R. Acad. Sci. Ser. D* **165:**373–375.

d'Hérelle, F. 1921. *Le Bactériophage: Son Rôle dans l'Immunité.* Masson, Paris, France.

Dougherty, E. M., K. P. Guthrie, and M. Shapiro. 1996. Optical brighteners provide baculovirus activity enhancement and UV radiation protection. *Biol. Control* **7:**71–74.

Dowd, S. E., S. D. Pillai, S. Wang, and M. Y. Corapcioglu. 1998. Delineating the specific influence of virus isoelectric point and size on virus adsorption and transport through sandy soils. *Appl. Environ. Microbiol.* **64:**405–410.

Duckworth, D. H. 1987. History and basic properties of bacterial viruses, p. 1–44. *In* S. M. Goyal, C. P. Gerba, and G. Bitton (ed.), *Phage Ecology.* John Wiley & Sons, New York, NY.

EBI Food Safety. 2006. LISTEX™ P100 phages for control of *Listeria* approved by US FDA as GRAS. October 24. EBI Food Safety, Wageningen, The Netherlands. http://www.ebifoodsafety.com/images/LISTEX%20receives%20FDA%20GRAS_Oct%2024%202006.pdf.

Erskine, J. M. 1973. Characteristics of *Erwinia amylovora* bacteriophage and its possible role in the epidemiology of fire blight. *Can. J. Microbiol.* **19:** 837–845.

Evans, J., Y. M. Barnet, and J. M. Vincent. 1979. Effect of bacteriophage on the colonisation and nodulation of clover roots by a strain of *Rhizobium trifolii. Can. J. Microbiol.* **25:**968–973.

Fischetti, V. A. 2008. Bacteriophage lysins as effective antibacterials. *Curr. Opin. Microbiol.* **11:**393–400.

Flaherty, J. E., B. K. Harbaugh, J. B. Jones, G. C. Somodi, and L. E. Jackson. 2001. H-mutant bacteriophages as a potential biocontrol of bacterial blight of geranium. *HortScience* **36:**98–100.

Flaherty, J. E., J. B. Jones, B. K. Harbaugh, G. C. Somodi, and L. E. Jackson. 2000. Control of bacterial spot on tomato in the greenhouse and field with H-mutant bacteriophages. *HortScience* **35:**882–884.

Garcia, P., B. Martinez, J. M. Obeso, and A. Rodriguez. 2008. Bacteriophages and their application in food safety. *Lett. Appl. Microbiol.* **47:** 479–485.

Gill, J. J. 2000. Biological control of *Erwinia amylovora* using bacteriophages. M.S. thesis. Brock University, St. Catharines, Ontario, Canada.

Gill, J. J., T. Hollyer, and P. M. Sabour. 2007. Bacteriophages and phage-derived products as antibacterial therapeutics. *Expert Opin. Ther. Pat.* **17:** 1341–1350.

Gill, J. J., A. M. Svircev, R. Smith, and A. J. Castle. 2003. Bacteriophages of *Erwinia amylovora. Appl. Environ. Microbiol.* **69:**2133–2138.

Goodridge, L. D. 2004. Bacteriophage biocontrol of plant pathogens: fact or fiction? *Trends Biotechnol.* **22:**384–385.

Goyer, C. 2005. Isolation and characterization of phages Stsc1 and Stsc3 infecting *Streptomyces scabiei* and their potential as biocontrol agents. *Can. J. Plant Pathol.* **27:**210–216.

Guenther, S., D. Huwyler, S. Richard, and M. J. Loessner. 2009. Virulent bacteriophage for efficient biocontrol of *Listeria monocytogenes* in ready-to-eat foods. *Appl. Environ. Microbiol.* **75:**93–100.

Hagens, S., and M. J. Loessner. 2007. Application of bacteriophages for detection and control of foodborne pathogens. *Appl. Microbiol. Biotechnol.* **76:**513–519.

Hagens, S., and M. L. Offerhaus. 2008. Bacteriophages—new weapons for food safety. *Food Technol.* **62:**46.

Harbaugh, B. K., J. B. Jones, J. E. Flaherty, and L. E. Jackson. 1998. Beneficial viruses—new allies in the fight against bacterial diseases. *GrowerTalks* **62:**84–88.

Hildebrand, M., C. C. Tebbe, and K. Geider. 2001. Survival studies with the fire blight pathogen *Erwinia amylovora* in soil and in a soil-inhabiting insect. *J. Phytopathol.* **149:**635–639.

Hudson, J. A., C. Billington, G. Carey-Smith, and G. Greening. 2005. Bacteriophages as biocontrol agents in food. *J. Food Prot.* **68:**426–437.

Ignoffo, C. M., and C. Garcia. 1992. Combinations of environmental factors and simulated sunlight affecting activity of inclusion bodies of the *Heliothis* (Lepidoptera: Noctuidae) nucleopolyhedrosis virus. *Environ. Entomol.* **21:**210–213.

Ignoffo, C. M., and C. Garcia. 1994. Antioxidant and oxidative enzyme effects on the inactivation of inclusion bodies of the *Heliothis* baculovirus by simulated sunlight-UV. *Environ. Entomol.* **23:** 1025–1029.

Ignoffo, C. M., and C. Garcia. 1995. Aromatic/heterocyclic amino acids and the stimulated sunlight-UV inactivation of the *Heliothis/Helicoverpa* baculovirus. *Environ. Entomol.* **24:**480–482.

Ignoffo, C. M., C. Garcia, and S. G. Saathoff. 1997. Sunlight stability and rain-fastness of formulations of baculovirus *Heliothis. Environ. Entomol.* **26:**1470–1474.

Iriarte, F. B., B. Balogh, M. T. Momol, L. M. Smith, M. Wilson, and J. B. Jones. 2007. Factors affecting survival of bacteriophage on tomato leaf surfaces. *Appl. Environ. Microbiol.* **73:**1704–1711.

Jackson, L. E. May 1989. Bacteriophage prevention and control of harmful plant bacteria. U.S. patent 4,828,999.

Johnson, K. B., and V. O. Stockwell. 2000. Biological control of fire blight, p. 319–337. *In* J. L. Vanneste (ed.), *Fire Blight: the Disease and Its Causative Agent,* Erwinia amylovora. 2000. CABI Publishing, Wallingford, United Kingdom.

Jones, J. B., L. E. Jackson, B. Balogh, A. Obradovic, F. B. Iriarte, and M. T. Momol. 2007. Bacteriophages for plant disease control. *Annu. Rev. Phytopathol.* **45:**245–262.

Jones, J. B., M. T. Momol, A. Obradovic, B. Balogh, and S. M. Olson. 2005. Bacterial spot management on tomatoes. *Acta Hortic.* **695:**119–124.

Joseph, M. V., J. D. Desai, and A. J. Desai. 1985. Possible involvemnt of phage-like structures in antagonism of cowpea rhizobia by *Rhizobium trifolii. Appl. Environ. Microbiol.* **49:**459–461.

Katznelson, H., and J. K. Wilson. 1941. Occurrence of *Rhizobium meliloti* bacteriophages. *Soil Sci.* **51:**59–63.

Keel, C., Z. Ucurum, P. Michaux, M. Adrian, and D. Haas. 2002. Deleterious impact of a virulent bacteriophage on survival and biocontrol activity of *Pseudomonas fluorescens* strain CHA0 in natural soil. *Mol. Plant Microbe Interact.* **15:**567–576.

Keil, H. L., and R. A. Wilson. 1963. Control of peach bacterial spot by *Xanthomonas pruni* bacteriophage. *Phytopathology* **53:**746–747. (Abstract.)

Kim, W.-S., H. Salm, and K. Geider. 2004. Expression of bacteriophage ϕEa1h lysozyme in *Escherichia coli* and its activity in growth inhibition of *Erwinia amylovora. Microbiology* **150:**2707–2714.

Kleczkowska, J. 1950. Phage resistant mutants of *Rhizobium trifolii. J. Gen. Microbiol.* **4:**298–310.

Kowalski, M., G. E. Ham, L. R. Frederick, and I. C. Anderson. 1974. Relationship between strains of *Rhizobium japonicum* and their bacteriophages from soil and nodules of field grown soybeans. *Soil Sci.* **118:**221–228.

Kumhar, S. R., and K. Ramkrishna. 2007. Symbiotic effectivity of phage resistant mutants of mothbean *Rhizobium* on different host genetic systems. *J. Food Legumes* **20:**83–86.

Kutter, E., and A. Sulakvelidze (ed.). 2005. *Bacteriophages: Biology and Applications.* CRC Press, Boca Raton, FL.

Lameiro, M. H., R. Malpique, A. C. Silva, P. M. Alves, and E. Melo. 2006. Encapsulation of adenoviral vectors into chitosan-bile salt microparticles for mucosal vaccination. *J. Biotechnol.* **126:**152–162.

Lang, J. M., D. H. Gent, and H. F. Schwartz. 2007. Management of *Xanthomonas* leaf blight of onion with bacteriophages and a plant activator. *Plant Dis.* **91:**871–878.

Lee, K. Y., and T. R. Heo. 2000. Survival of *Bifidobacterium longum* immobilized in calcium alginate beads in simulated gastric juices and bile salt solution. *Appl. Environ. Microbiol.* **66:**869–873.

Lehman, S. M. 2007. Development of a bacteriophage-based biopesticide for fire blight. Ph.D. thesis. Brock University, St. Catharines, Ontario, Canada.

Lehman, S. M., A. M. Kropinski, A. J. Castle, and A. M. Svircev. 2009. Complete genome of the broad-host-range *Erwinia amylovora* phage ΦEa21-4 and its relationship to *Salmonella* phage Felix O1. *Appl. Environ. Microbiol.* **75:**2139–2147.

Ma, Y., J. C. Pacan, Q. Wang, Y. Xu, X. Huang, A. Korenevsky, and P. M. Sabour. 2008. Microencapsulation of bacteriophage Felix O1 chitosan-alginate microspheres for oral cavity delivery. *Appl. Environ. Microbiol.* **74:**4799–4805.

Marks, T., and R. Sharp. 2000. Bacteriophages and biotechnology: a review. *J. Clin. Technol. Biotechnol.* **75:**6–7.

Mattey, M., and J. Spencer. 2008. Bacteriophage therapy—cooked goose or phoenix rising? *Curr. Opin. Biotechnol.* **19:**608–612.

McNeil, D. L., S. Romero, J. Kandula, C. Stark, A. Stewart, and S. Larsen. 2001. Bacteriophages: a potential biocontrol agent against walnut blight (*Xanthomonas campestris* pv. *juglandis*). *N. Z. Plant Prot.* **54:**220–224.

Moser, C. A., T. J. Speaker, and P. A. Offit. 1998. Effect of water-based microencapsulation on protection against EDIM rotavirus challenge in mice. *J. Virol.* **72:**3859–3862.

Munsch, P., and J. M. Olivier. 1995. Biocontrol of bacterial blotch of the cultivated mushroom with lytic phages: some practical considerations. *Mushroom Sci.* **14:**595–602.

Nechaeva, E. 2002. Development of oral microencapsulated forms for delivering viral vaccines. *Expert Rev. Vaccines* **1:**385–397.

Obradovic, A., K. Gasic, and M. Stepanovic. 2008. Bakteriofagi u zastiti bilja. (Bacteriophages in plant protection.) *Biljni Lekar* **36:**36–44.

Obradovic, A., J. B. Jones, M. T. Momol, B. Balogh, and S. M. Olson. 2004. Management of tomato bacterial spot in the field by foliar applications of bacteriophages and SAR inducers. *Plant Dis.* **88:**736–740.

Obradovic, A., J. B. Jones, M. T. Momol, S. M. Olson, L. E. Jackson, B. Balogh, K. Guven, and F. B. Iriarte. 2005. Integration of biological control agents and systemic acquired resistance inducers against bacterial spot on tomato. *Plant Dis.* **89:**712–716.

O'Brien, R. D., and S. E. Lindow. 1989. Effect of plant species and environmental conditions on epiphytic population sizes of *Pseudomonas syringae* and other bacteria. *Phytopathology* **79:**619–627.

Offit, P. A., C. A. Khoury, C. A. Moser, H. F. Clark, J. E. Kim, and T. J. Speaker. 1994. Enhancement of rotavirus immunogenicity by microencapsulation. *Virology* **203:**134–143.

Okabe, N., and M. Goto. 1963. Bacteriophages of plant pathogens. *Annu. Rev. Phytopathol.* **1:**397–418.

Rao, A. V., N. Shiwnarain, and J. Maharaj. 1989. Survival of microencapsulated *Bifidobacterium pseudolongum* in simulated gastric and intestinal juices. *Can. Inst. Food Sci. Technol. J.* **22:**345–349.

Ravensdale, M., T. J. Blom, G. Gracia, J. A. Garza, A. M. Svircev, and R. J. Smith. 2007. Bacteriophages and the control of *Erwinia carotovora* subsp. *carotovora*. *Can. J. Plant Pathol.* **29:**121–130.

Ritchie, D. F. 1978. Bacteriophages of *Erwinia amylovora:* their isolation, distribution, characterization, and possible involvement in the etiology and epidemiology of fire blight. Ph.D. thesis. Michigan State University. East Lansing, MI.

Ritchie, D. F., and E. J. Klos. 1977. Isolation of *Erwinia amylovora* bacteriophage from aerial parts of apple trees. *Phytopathology* **67:**101–104.

Ritchie, D. F., and E. J. Klos. 1979. Some properties of *Erwinia amylovora* bacteriophages. *Phytopathology* **69:**1078–1083.

Roach, D. R., A. J. Castle, A. M. Svircev, and F. A. Tumini. 2008. Phage-based biopesticides: characterization of phage resistance and host range for sustainability. *Acta Hortic.* **793:**397–401.

Saccardi, A., E. Gambin, M. Zaccardelli, G. Barone, and U. Mazzucchi. 1993. *Xanthomonas campestris* pv. *pruni* control trials with phage treatments on peach in the orchard. *Phytopathol. Mediterr.* **32:**206–210.

Schnabel, E. L., W. G. D. Fernando, M. P. Meyer, A. L. Jones, and L. E. Jackson. 1999. Bacteriophage of *Erwinia amylovora* and their potential for biocontrol. *Acta Hortic.* **489:**649–653.

Schnabel, E. L., and A. L. Jones. 2001. Isolation and characterization of five *Erwinia amylovora* bacteriophages and assessment of phage resistance in strains of *Erwinia amylovora*. *Appl. Environ. Microbiol.* **67:**59–64.

Schroth, M. N., and J. G. Hancock. 1981. Selected topics in biological control. *Ann. Rev. Microbiol.* **35:**453–476.

Shapiro, M. 1992. Use of optical brighteners as radiation protectants for gypsy moth (Lepidoptera: Lymantriidae) nuclear polyhedrosis virus. *J. Econ. Entomol.* **85:**1682–1686.

Skurnik, M., and E. Strauch. 2006. Phage therapy: facts and fiction. *Int. J. Med. Microbiol.* **296:**5–14.

Smit, E., A. C. Wolters, H. Lee, J. T. Trevors, and J. D. van Elsas. 1996. Interactions between a genetically marked *Pseudomonas fluorescens* strain and bacteriophage ΦR2f in soil: effects of nutrients, alginate encapsulation, and the wheat rhizosphere. *Microbial Ecol.* **31:**125–140.

Srinivasiah, S., J. Bhavsar, K. Thapar, M. Liles, T. Schoenfeld, and K. E. Wommack. 2008. Phages across the biosphere: contrasts of viruses in soil and aquatic environments. *Res. Microbiol.* **159:** 349–357.

Stephens, P. M., M. O'Sullivan, and F. Gara. 1987. Effect of bacteriophages on colonization of sugarbeet roots by fluorescent *Pseudomonas* spp. *Appl. Environ. Microbiol.* **53:**1164–1167.

Stewart, A. 2001. Commercial biocontrol—reality or fantasy? *Australas. Plant Pathol.* **30:**127–131.

Stewart, F. M., and B. R. Levin. 1984. The population biology of bacterial viruses: why be temperate? *Theor. Popul. Biol.* **26:**93–117.

Stone, R. 2002. Bacteriophage therapy. Food and agriculture: testing grounds for phage therapy. *Science* **298:**730.

Sturesson, C., and L. Degling Wikingsson. 2000. Comparison of poly(acryl starch) and poly(lactide-co-glycolide) microspheres as drug delivery system for a rotavirus vaccine. *J. Control. Release* **68:**441–450.

Subcommittee on Microbial Criteria, Committee on Food Protection, Food and Nutrition Board, National Research Council. 1985. *An Evaluation of the Role of Microbiological Criteria for Foods and Food Ingredients.* National Academy Press, Washington, DC.

Summers, W. C. 1999. *Félix d'Hérelle and the Origins of Molecular Biology.* Yale University Press, New Haven, CT.

Summers, W. C. 2001. Bacteriophage therapy. *Annu. Rev. Microbiol.* **55:**437–451.

Summers, W. C. 2005. Bacteriophage research: early history, p. 5–27. *In* E. Kutter and A. Sulakvelidze (ed.), *Bacteriophages: Biology and Applications.* CRC Press, Boca Raton, FL.

Svircev, A. M., J. J. Gill, and P. Sholberg. 2002a. *Erwinia amylovora* (Burrill) Winslow, Broadhurst, Buchanan, Krumwiede, Rogers and Smith, fire blight *(Enterobacteriaceae),* p. 448–451. *In* P. G. Mason and J. T. Huber (ed.), *Biological Control Programmes in Canada, 1981–2000.* CABI Publishing, Wallingford, United Kingdom.

Svircev, A. M., W. S. Kim, S. M. Lehman, and A. J. Castle. 2009. *Erwinia amylovora:* modern methods for detection and differentiation, p. 115–129. *In* R. Burns (ed.), *Methods in Molecular Biology.* Humana Press, Clifton, NJ.

Svircev, A. M., S. M. Lehman, W. S. Kim, E. Barszcz, K. E. Schneider, and A. J. Castle. 2006. Control of the fire blight pathogen with bacteriophages, p. 259–261. *In* W. Zeller and C. Ullrich (ed.), *Proceedings of the 1st International Symposium on Biological Control of Bacterial Plant Diseases.* Arno Brynda, Berlin, Germany.

Svircev, A. M., R. Smith, G. Gracia, J. A. Garza, J. J. Gill, and K. Schneider. 2002b. Biocontrol of *Erwinia* with bacteriophages. *IOBC/WPRS Bull.* **25:**139–142.

Tamez-Guerra, P., M. R. McGuire, R. W. Behle, B. S. Shasha, and L. J. G. Wong. 2000. Assessment of microencapsulated formulations for improved residual activity of *Bacillus thuringiensis. J. Econ. Entomol.* **93:**219–225.

Tan, J. S. H., and D. C. Reanney. 1976. Interactions between bacteriophages and bacteria in soil. *Soil. Biol. Biochem.* **8:**145–150.

Tanji, Y., T. Shimada, H. Fukudomi, K. Miyanaga, Y. Nakai, and H. Unno. 2005. Therapeutic use of phage cocktail for controlling *Escherichia coli* O157:H7 in gastrointestinal tract of mice. *J. Biosci. Bioeng.* **100:**280–287.

Tanji, Y., T. Shimada, M. Yoichi, K. Miyanaga, K. Hori, and H. Unno. 2004. Toward rational control of *Escherichia coli* O157:H7 by a phage cocktail. *Appl. Microbiol. Biotechnol.* **64:**270–274.

Taylor, D. H., R. S. Moore, and L. S. Sturman. 1981. Influence of pH and electrolyte composition on adsorption of poliovirus by soils and minerals. *Appl. Environ. Microbiol.* **42:**976–984.

Thiel, C. 2004. Old dogma, new tricks—21st century phage therapy. *Nat. Biotechnol.* **22:**31–36.

Uludag, H., P. DeVos, and P. A. Tresco. 2000. Technology of mammalian cell encapsulation. *Adv. Drug Deliv. Rev.* **42:**29–64.

Vandenbergh, P. A., and R. L. Cole. 1986. Cloning and expression in *Escherichia coli* of the polysaccharide depolymerase associated with bacteriophage-infected *Erwinia amylovora. Appl. Environ. Microbiol.* **51:**862–864.

Vandenbergh, P. A., A. M. Wright, and A. K. Vidaver. 1985. Partial purification and characterization of a polysaccharide depolymerase associated with phage-infected *Erwinia amylovora. Appl. Environ. Microbiol.* **49:**994–996.

van der Zwet, T., and S. V. Beer. 1995. *Fire Blight—Its Nature, Prevention and Control: A Practical Guide to Integrated Disease Management.* USDA agriculture information bulletin no. 631. U.S. Department of Agriculture, Washington, DC.

Vanneste, J. L. (ed.). 2000. *Fire Blight: the Disease and Its Causative Agent,* Erwinia amylovora. CABI Publishing, Wallingford, United Kingdom.

Vidaver, A. K. 1976. Prospects for control of phytopathogenic bacteria by bacteriophages and bacteriocins. *Ann. Rev. Phytopathol.* **14:**451–465.

Vlachakis, J., and M. Verhoyen. 1984. Isolation of *Erwinia amylovora* (Burrill) Winslow et al., bacteriophages to struggle against pear fireblight. *Int. Symp. Fytofarmacie Fytiatrie* **36:**551–558.

Wang, I.-N., D. E. Dykhuizen, and L. B. Slobodkin. 1996. The evolution of phage lysis timing. *Evol. Ecol.* **10:**545–558.

Williams, K. E., J. B. Schnitker, M. Radosevich, D. W. Smith, and K. E. Wommack. 2008. Cultivation-based assessment of lysogeny among soil bacteria. *Microb. Ecol.* **56:**437–447.

Williams, S. T., A. M. Mortimer, and L. Manchaster. 1987. Ecology of soil bacteriophages, p. 157–179. *In* S. M. Goyal, C. P. Gerba, and G. Bitton (ed.), *Phage Ecology.* John Wiley & Sons, New York, NY.

Williamson, K. E., K. E. Wommack, and M. Radosevich. 2003. Sampling natural viral communities from soil for culture-independent analyses. *Appl. Environ. Microbiol.* **69:**6628–6633.

Wilson, M., and S. E. Lindow. 2000. Implications of the viable but nonculturable state in risk assessment based on field testing of genetically engineered microorganisms, p. 229–241. *In* R. R. Colwell and D. J. Grimes (ed.), *Nonculturable Microorganisms in the Environment.* ASM Press, Washington, DC.

Zaccardelli, M., A. Saccardi, E. Gambin, and U. Mazzucchi. 1992. *Xanthomonas campestris* pv. *pruni* bacteriophages on peach trees and their potential use for biological control. *Phytopathol. Mediterr.* **31:**133–140.

POTENTIAL USE OF BACTERIOPHAGES AS INDICATORS OF WATER QUALITY AND WASTEWATER TREATMENT PROCESSES

Francisco Lucena and Juan Jofre

6

INTRODUCTION

Indicator organisms to assess the microbiological quality of water and the functioning of water treatments have been used for routine control purposes for almost 100 years. These have been used with a great deal of success in view of the present-day low incidence of waterborne infectious diseases in industrialized countries. With the advent of nucleic acid amplification techniques, our ability to detect pathogens in water samples has increased, and suggestions that we move toward the direct detection of pathogens for water quality control are not uncommon. Nevertheless, despite the great sensitivity of these techniques, they still have serious limitations for application to water quality control, including low pathogen concentrations and sample volumes needed to detect them, presence of amplification reaction inhibitors, and the inability to differentiate between live and dead (infectious and noninfectious) microorganisms. Hence, it seems that at least for the moment indicators will be needed to assess water quality and the functioning of wastewater treatment processes.

Yet when the indicator organisms are applied to assess water quality and treatments, some confusion arises in semantics with regard to the traditional term "indicator," which includes both the index and the indicator function. An organism with an index function is related to the risk of occurrence of the surrogated microorganism(s). Alternatively, an organism with an indicator function has behavioral characteristics similar to those of the surrogated microorganisms, and has the same or greater resistance to environmental and/or human-generated stresses. According to some, the term "marker" (Mossel, 1982) or "model" (IAWPRC Study Group, 1991) organism should be used instead of the term "indicator." Ideally, a perfect marker or model organism will fulfill both the index and indicator functions. Still, in most cases, the term "indicator" is used without distinction of the function. The more that is known about fecal pathogens and potential model microorganisms, the less likely it seems that a single fecal microorganism that satisfies both functions in all circumstances will be found. Both the index and indicator functions of bacteriophages will be considered in this review.

Until now the indicator organisms for water quality control had been intestinal bacteria, especially *Escherichia coli* and other coliform

Francisco Lucena and Juan Jofre, Department of Microbiology, Faculty of Biology, University of Barcelona, Avinguda Diagonal 645, 08028 Barcelona, Spain.

Bacteriophages in the Control of Food- and Waterborne Pathogens
Edited by Parviz M. Sabour and Mansel W. Griffiths, © 2010 ASM Press, Washington, DC

bacteria. However, the scientific community involved in water microbiology has raised doubts about the capability of these conventional bacterial indicators to protect against viral diseases because, among other reasons, viruses are more persistent in the water environment (IAWPRC Study Group, 1991). Similar doubts had been raised regarding protozoa (Payment and Franco, 1993). Current research suggests that bacteriophages are well suited as indicators of viruses in water quality assessment and water treatments since monitoring the presence of specific viral pathogens is still impracticable for routine control purposes. Bacteriophages infecting enteric bacteria seem good candidates as indicators of enteric viruses. As early as 1948, coliphages were advocated by Guelin (1948) as potential indicators of enteric microorganisms. He found that phages infecting *E. coli* correlated with coliform bacteria numbers in fresh and marine waters. Since then, the potential value of phages infecting enteric bacteria as quality indicators in water and food has been heavily investigated and reviewed (Armon and Kott, 1996; Grabow, 2001; Havelaar 1987; IAWPRC Study Group, 1991; Jofre, 2002; Pillai, 2007).

GROUPS OF BACTERIOPHAGES PROPOSED AS POTENTIAL INDICATORS OF WATERBORNE ENTERIC VIRUSES

Somatic coliphages (IAWPRC Study Group, 1991; Kott et al., 1974), F-specific RNA bacteriophages (Havelaar and Pot-Hogeboom, 1988; IAWPRC Study Group, 1991), and bacteriophages infecting *Bacteroides fragilis* (IAWPRC Study Group, 1991; Tartera and Jofre, 1987) have been advocated as potential model microorganisms for various aspects of water quality control. The last two groups were proposed and studied mostly because of the view that somatic coliphages, the most abundant and easily detectable group, replicate in the environment. However, present knowledge shows that the numbers of somatic coliphages replicated in the environment within water samples is negligible (Jofre, 2009).

Somatic coliphages attach to *E. coli* cell wall and may lyse the host cell in 20 to 30 min under optimal physiological conditions. They produce plaques of different sizes and morphologies. Standardized methodologies available to detect them are very simple, and results may be obtained as quickly as within 4 h. Bacteriophages frequently used to study somatic coliphages' resistance to environmental stressors, disinfection, or behavior in recovery methods (concentration and extraction from solids) are T4, T7, ϕX174, and PRD1. Somatic coliphages are present in feces from humans and nonhuman homeotherm animals.

F-specific bacteriophages, also called sexual coliphages, are those that infect bacteria through the sex pili, which are coded by the F plasmid first detected in *E. coli* K-12. The F plasmid is transferable to a wide range of gram-negative bacteria, and consequently F-specific phages may have several hosts; yet it is thought that their main host is *E. coli.* The F plasmid is not synthesized below 32°C. F-specific RNA bacteriophages consist of a simple capsid of cubic symmetry of 21 to 30 nm in diameter and contain single-stranded RNA as the genome. They belong to the family *Leviviridae.* Their infectious process is inhibited by the presence of RNase in the assay medium, which can be used to distinguish between the F-specific RNA bacteriophages and the rod-shaped F-specific DNA bacteriophages of the family *Inoviridae,* which also infect the host cell through the sex pili. Standardized methods for the detection and enumeration of both F-specific and F-specific RNA bacteriophages are available, and results can be obtained in 12 h. MS2, and to a lesser extent Qβ and f2 bacteriophages, are frequently used to study F-specific RNA bacteriophages' resistance to environmental stressors, disinfection, or behavior in recovery methods. F-specific RNA bacteriophages are present in feces from humans and nonhuman homeotherms.

Addition of somatic and F-specific bacteriophage counts gives the numbers of bacteriophages that are called total coliphages or just coliphages. There are some host strains that can enumerate both, with counts that approach the numbers obtained counting them separately and then adding them to obtain the total values (Guzmán et al., 2008; Sobsey et al., 2004).

The phages infecting *B. fragilis* and other *Bacteroides* species described so far attach to the cell wall of the host bacteria and under optimal, always anaerobic, conditions may lyse it in 30 to 40 min. Most *Bacteroides* phages have a narrow host range. Some strains, namely *B. fragilis* HSP40, *Bacteroides thetaiotaomicron* GA14, and *Bacteroides ovatus* GB124, have the capability to discriminate human from animal fecal contaminants (Payán et al., 2005; Tartera et al., 1989), whereas others, such as *B. fragilis* RYC2056, do not have this ability (Puig et al., 1999). This gives them value as fecal source tracers, yet those with discriminating capacity have limited geographical application (Blanch et al., 2006). Yet host strains suitable for a given geographic area are easily isolated (Payán et al., 2005). A standardized method for the detection and enumeration of *B. fragilis* phages is now available, and results can be achieved in 18 h. The phages used mostly to study phage resistance to environmental stressors, disinfection, or behavior in recovery methods are B40-8 and B56-3.

The most important factor in defining a method for the detection of a given group of bacteriophages is the choice of the host strain. Standardized plaque assays and presence/absence methods have been identified for somatic coliphages (Clesceri et al., 1998; International Organization for Standardization, 2000; U.S. Environmental Protection Agency, 2000a, 2000b), F-specific RNA phages (International Organization for Standardization, 1995; U.S. Environmental Protection Agency, 2000a, 2000b), and bacteriophages infecting *B. fragilis* (International Organization for Standardization, 2001). All these standardized methods are easily implemented in routine laboratories as confirmed by multilaboratory studies (Mooijman et al., 2005). The cost, considering material and labor, for analyzing somatic coliphages is about the same as or even less than that for evaluating a conventional bacterial indicator. Testing for F-specific RNA bacteriophages and *Bacteroides* phages costs roughly twice as much.

OCCURRENCE AND DENSITIES OF INDICATOR BACTERIOPHAGES IN FECAL CONTAMINANTS

Somatic coliphages and F-specific RNA bacteriophages have been detected in feces of humans and all homeotherm animals studied, yet the percentages of feces specimens with presence of coliphages and phage concentrations reported are variable (Calci et al., 1998; Grabow et al., 1995; Havelaar et al., 1986). Regarding phages infecting *Bacteroides,* those infecting strain *B. fragilis* RYC2056 have been recovered from 28% of human stool samples and also from animal feces (Puig et al., 1999), with the maximum incidence of 30% occurring in pigs. In contrast, *B. fragilis* HSP40 phages have been isolated from 10 to 13% of human stool samples, but never from nonhuman feces (Grabow et al., 1995; Tartera and Jofre, 1987).

Numbers of the three above groups of bacteriophages, when a geographically suitable host strain of *Bacteroides* is used, are fairly constant in raw municipal and hospital sewage, as well as wastewater from abattoirs in different continents. These are consistent with the numbers of the various bacterial indicators. Somatic coliphages are the most abundant indicator phages, with the great majority of values ranging from 10^6 to 10^7 PFU per 100 ml, with values around half a $\log_{10}$ unit lower than those of *E. coli* (Blanch et al., 2006; Chung et al., 1998; Grabow et al., 1993; Lodder and de Roda Husman, 2005; Lucena et al., 2003; Nieuwstad et al., 1988; Uemura et al., 2002). F-specific RNA bacteriophages rank second, with values ranging from 5×10^5 to 5×10^6 PFU per 100 ml (Blanch et al., 2006; Chung et al., 1998; Grabow et al., 1993; Lodder and de Roda Husman, 2005; Lucena et al., 2003; Luther and Fujioka, 2004; Nieuwstad et al., 1988; Uemura et al., 2002). When a suitable strain of *Bacteroides* is used as a host, values of phages infecting *Bacteroides* ranging from 10^4 to 10^5 PFU per 100 ml have been reported (Blanch et al., 2006; Chung et al., 1998; Lucena et al., 2003; Payán et al., 2005). Numbers of human-pathogenic viruses in raw sewage in industrialized countries are, with the exception of a very few atypical samples, lower than those of the less abundant bacteriophages. Thus, in industrialized countries of different continents, results most frequently reported for culturable virus consign values per liter of wastewater ranging from 100 to 10,000 PFU for enteroviruses (Bosch et al., 1986; Costán-Longares et al., 2008a; Irving and Smith, 1981; Lodder and de Roda Husman, 2005; Sedmak et al., 2005), from 10 to 10,000 PFU for adenovirus (He and Jiang, 2005; Irving and Smith, 1981), from 10 to 1,000 cell culture reverse transcriptase PCR units for astrovirus (El-Senousy et al., 2007), and from 100 to 1,000 fluorescent foci for rotavirus (Bosch et al., 1986; Green and Lewis, 1999). Reported values of genome copies detected by quantitative PCR for the same viruses are 2 or 3 orders of magnitude higher than those mentioned above (Bofill-Mas et al., 2006; El-Senousy et al., 2007; He and Jiang, 2005; Meleg et al., 2008; Schvoerer et al., 2001), yet as indicated previously, this technique does not allow distinction between infectious and noninfectious viruses.

Somatic coliphages have also been reported in all samples of septage tested (Deborde et al., 1998; Lucena et al., 2003). F-specific bacteriophages and phages infecting *Bacteroides* have been reported in smaller percentages of septage samples (Calci et al., 1998; Lucena et al., 2003).

The occurrence and densities of the three groups of bacteriophages in raw sludge vary widely in the reports available. This is probably due to the lack of standardized methods for phage extraction from sludges. However, in the available reports in which the three groups of phages have been analyzed (Guzmán et al., 2007; Mignotte-Cadiergues et al., 1999), the highest counts again correspond to somatic coliphages, followed by F-specific RNA bacteriophages and bacteriophages infecting *Bacteroides*. Moreover, they have similar accumulation rates, which are also comparable to those of bacterial indicators (Guzmán et al., 2007; Williams and Hurst, 1988).

Numbers of bacteriophages infecting enteric bacteria found in animal slurries are in between those recovered in animal feces and in abattoir wastewater (Blanch et al., 2006; Cole et al., 2003; Hill and Sobsey, 1998; Schaper et al., 2002).

BACTERIOPHAGES AS INDICATORS

The indicator function requires the removal and inactivation of the surrogate indicator to be similar or less than that of the surrogated pathogens both in nature and in water treatment facilities. Viruses, and consequently bacteriophages, are removed from water by adsorption to particles, sedimentation, and filtration through porous matrixes, either soil or man-made filters. These processes do not inactivate viruses but transfer them from one compartment to another. Viruses and phages in water and the solid compartments (sediments and biosolids) undergo inactivation by both natural and human-introduced physical, chemical, and biological stressors. Natural factors, such as temperature, sunlight, pH, and ionic environment, play a major role in virus and phage inactivation. Human-introduced physical and chemical treatments such as irradiation (UV, gamma, and electron beam radiation), thermal treatment, and chemical disinfection (lime, halogens, ozone, and others) play a major role in the inactivation of viruses in water and sludge treatments. In the evaluation of phages as viral indicators, the question is whether their outcome in all these natural and man-driven processes resembles the viral one and whether after all these processes they still persist in numbers higher than those of viruses in recreational waters, drinking water sources, drinking water, reclaimed water, water used for aquaculture and shellfish culture, shellfish, etc. But the effect of each of these natural and man-driven processes on the removal and inactivation of viruses and bacteriophages from water is difficult to determine. Comparing numbers of viruses and phages before and after the wastewater processes and determining the fate of seeded laboratory-cultured phages or viruses in laboratory or pilot experiments are the standard procedures to evaluate and compare their removal and inactivation. However, both procedures have important shortcomings, and comparisons are difficult to interpret because the detection methods and the sample volumes analyzed for phages and viruses are very different. Moreover, reported differences in the removal and inactivation of seeded laboratory-grown and naturally occurring viruses and phages are significant. This difficulty can be partially overcome with phages since naturally occurring phages can be partially purified from wastewater at densities high enough to give some laboratory or pilot inactivation studies a better approximation of what happens in the real world. There is vast and ever-growing literature and reviews (Armon and Kott, 1996; Grabow, 2001; Havelaar, 1987; IWAPRC, 1991; Jofre, 2002; Leclerc et al., 2000) on removal and inactivation of viruses and bacteriophages in the environment and in water treatments, where, with very few exceptions, bacteriophages are always more abundant than viruses. Based on available information, it is suggested that the fate of bacteriophages in natural-water environments and their outcome in water and sludge treatments resemble those of human-pathogenic viruses.

BACTERIOPHAGES AS INDICATORS IN WASTEWATER TREATMENTS

Primary and Secondary Treatments of Raw Sewage

In primary sedimentation, flocculation-aided sedimentation, activated sludge digestion, activated sludge digestion plus precipitation, and upflow anaerobic sludge blanket processes, the die-off of the microorganisms seems to play a minor role in the reduction in counts. In well-operated plants, bacterial indicators, bacteriophages, and infectious enteroviruses undergo similar reductions in numbers. Roughly, reductions of naturally occurring pathogens and indicators range from less than 1.0 $\log_{10}$ unit for the primary sedimentation (Lucena et al., 2004) and upflow anaerobic sludge blanket processes (Uemura et al., 2002) to between 1.5 and 2.0 $\log_{10}$ units for well-operated primary plus activated sludge treatments (Fleisher et al., 2000; Grabow et al., 1984; Lodder and de Roda Husman, 2005; Lucena et al., 2004; Nieuwstad et al., 1988; Uemura et al., 2002). Reductions reach up to 3 $\log_{10}$ units for activated sludge plus precipitation/coagulation (Fleisher et al., 2000; Lucena et al., 2004; Niewstad et al., 1988). Yet different reports describe different removals for various viruses and bacteriophages grown in the laboratory when applied to wastewater treatment (Arraj et al., 2005; Benyahya et al., 1998).

In contrast, in other treatments dealing with raw wastewater such as oxidation ponds, constructed wetlands, or slow-rate trickling filters, the microbial removal characteristics vary and depend on factors such as inactivation by natural stressors, filtration, adsorption, and desorption, which in turn depend on design specifics such as filter medium characteristics, retention time, and local climatic conditions. Consequently, reported removals are extremely varied and even contradictory. In oxidation ponds the different groups of bacteriophages differ in removal, as they differ from bacterial indicators. Generally speaking, greater removals are obtained at higher temperatures and the differences in inactivation among different microorganisms are also greater at higher temperatures (Alcalde et al., 2004; Campos et al., 2002, Hill and Sobsey, 1998, Lucena et al., 2004). Reported results for trickling filters as well as for natural and constructed wetlands are very diverse. In general, the removal of bacteriophages resembles the removal of viruses and is greater than the removal of bacterial indicators. It seems that adsorption plays a major role in phage and virus removal in these depuration processes (Gersberg et al., 1987; Karim et al., 2004;

Scheuerman et al., 1987; Vega et al., 2003; Vidales et al., 2003). Studies of desorption in rain episodes are needed to better evaluate these treatments regarding virus and bacteriophage removal.

Membrane bioreactors constitute a new generation of wastewater treatment plants that combine membranes with biological processes for wastewater treatment. Reported information indicates that both naturally occurring and inoculated bacteriophages (Coté et al., 1997; Shang et al., 2005; Ueda and Horan, 2000) as well as human viruses (Ottoson et al., 2006) are removed less efficiently than bacteria. Data on human viruses available refer only to viral genomes, and consequently removal is potentially underestimated. Their removal resembles bacteriophage elimination, so it is also less than bacterium removal. Data on removal of viruses, bacteriophages, and bacterial indicators are highly variable, and available information on correlations between effluent flux/transmembrane pressure and virus/phage removal give evidence that phage removal in the membrane bioreactor is most likely susceptible to both biological and physical factors, including the quantity and property of the biomass, biofilm, and membrane pore-size reduction.

Tertiary Treatments

Tertiary treatments are applied to secondary effluents and operated in water reclamation facilities. They include either some type of physical or chemical disinfection or removal by filtration and result in substantial differences between the elimination of conventional bacterial indicators and viruses and bacteriophages. Data on the efficiency of these processes in the elimination of bacterial indicators, bacteriophages, and viruses are available at the bench scale, generally with microbes grown in the laboratory; at the pilot scale with both naturally occurring and laboratory-grown microorganisms; and in wastewater reclamation facilities. Data reported on the comparative persistence of bacterial indicators, viruses, and phages do not always match. This lack of resemblance might be due to the fact that many studies were done using laboratory-grown microorganisms that differ from naturally occurring microorganisms in the level of resistance to natural stressors, as well as in the level of autoaggregation, adsorption, or inclusion in suspended solids. Differences in the elimination of different bacteriophages or groups of bacteriophages, and between different human viruses, are observed. Yet in all cases, in spite of these differences, the elimination of phages resembles the elimination of viruses more than that of bacteria. Most of the studies regarding the effects of disinfectants on bacteriophages and viruses refer to laboratory-grown viruses or bacteriophages (Abad et al., 1994a; Duran et al., 2003; Harakeh, 2006; IAWPRC, 1991; Meng and Gerba, 1996; Shin and Sobsey, 2003; Sobsey, 1989). But because of the mounting interest in reclaimed water, an increasing amount of information regarding the effect of these processes on naturally occurring bacteriophages and culturable human viruses is accumulating.

UV, gamma, and electron beam radiation have been reported to be useful for water and sludge treatment, yet in practice gamma and electron beam radiation are not used. In contrast, UV radiation is extensively used for disinfecting drinking water, secondary effluents, and water used for shellfish depuration. UV irradiation results in comparable reductions of densities of naturally occurring phages and viruses (Costán-Longares et al., 2008b; Gehr et al., 2003; Jancangelo et al., 2003; Montemayor et al., 2008).

Chemical disinfectants used to treat bacteriophages in water treatments include chlorine, chloramines, chlorine dioxide, lime, ozone, peracetic acid, copper and silver ions, and TiO_2 photocatalytic disinfection. Yet the majority of chemical disinfectant data were obtained in bench experiments with laboratory microbial cultures. However, information obtained in either pilot plants or water reclamation facilities with naturally occurring phages and viruses is increasing steadily. The effects on naturally occurring phages and human viruses of most of these disinfection processes are comparable, though with different degrees of similarity depending on the viruses and the phage groups studied (Costán-Longares et al., 2008b; Gehr et al., 2003; Grabow et al., 1978; Harwood et al., 2005; Omura et al., 1989; Tyrrell et al., 1995).

Conventional filtration procedures have been applied to secondary effluents, and naturally occurring coliphages seem to be removed similarly to viruses (Levine et al., 2008). But more and more frequently, membrane filtration (microfiltration, ultrafiltration, and reverse osmosis) is used in water reclamation processes. In membranes with small pore size, size exclusion plays the major role in determining which microorganisms are retained, and consequently viruses and bacteriophages respond differently to bacteria. But in microfiltration, as well as in conventional filtration procedures, adsorption also plays a very significant role, which depends on many factors, including the charge of the virus or bacteriophage, the characteristics of the membrane, the clogging of the membrane pores, and the characteristics of the clogging substance. Even though the available data vary, probably because of different characteristics of the membranes used and differences in operating characteristics, the removal of phages and viruses by membrane filtration is comparable (Farahbakhsh and Smith, 2004; Herath et al., 1998).

BACTERIOPHAGES AS INDICATORS IN DRINKING WATER TREATMENTS

Results for drinking water treatments are similar to those for tertiary treatments, though in this case the information on the elimination of naturally occurring phages and viruses is even scarcer because usually the amounts of bacterial indicators, viruses, and bacteriophages in the source water are low and seldom detected after treatment. However, a few reports on the reduction of numbers of naturally occurring phages and enteroviruses indicate similar outcomes for different treatments, as they suffer reductions in numbers that are less than the reduction in numbers of conventional indicator bacteria, with the exception of spores of sulfite-reducing clostridia (Abbaszadegan et al., 2008; Jofre et al., 1995; Payment et al., 1985a; Stetler, 1984; Wentsel et al., 1982).

BACTERIOPHAGES AS INDICATORS IN SLUDGE TREATMENTS

Wastewater treatment facilities generate enormous amounts of sludges where pathogens and fecal indicators tend to accumulate (Berg and Berman, 1980; Guzmán et al., 2007). In industrialized countries, prior to their release in the environment, sludges should undergo treatment according to environmental rules in order to be disinfected. Though disinfection can be attained via thermal treatment, chemical disinfectants, and high-energy radiation, in practice disinfection of sludges is achieved mainly by thermal treatment (pasteurization, thermophilic digestion, and composting) and to a much lesser extent by liming. Available results on effects of these processes on persistence of pathogens and indicators of fecal origin are still partial and, in our opinion, heavily influenced by the lack of standardized methods to extract the pathogens and indicators from the solids. Yet, as in the case of water disinfection, there are differences between the extents of removal of different microorganisms. However, the great majority of reported results indicate that phages respond to thermal treatments (Guzmán et al., 2007; Mignotte-Cadiergues et al., 2002; Mocé-Llivina et al., 2003; Spillman et al., 1987) and liming (Mignotte-Cadiergues et al., 2002) more similarly to viruses than conventional bacterial indicators do.

FATE AND OCCURRENCE OF INDICATOR BACTERIOPHAGES IN WATER ENVIRONMENTS

Either directly or after treatments of fecal residues, fecal pathogens and indicators are released into water bodies or the soil. The extent of dilution, removal by filtration and sedimentation, inactivation by natural stressors, and transport determines their occurrence and densities in surface and underground water bodies.

Like viruses, bacteriophages are adsorbed into solids (Payment et al., 1988). This feature has a major contribution to their removal from waters. Indeed, adsorption of phages into suspended solids facilitates their sedimentation, as indicated by increased phage densities in sediments compared to the overlaying water column (Araujo et al., 1997; Paul et al., 1993). Though most of the studies done on adsorption of phages in soil columns or in pilot fields have been done with seeded laboratory-cultured phages or viruses, results suggest that phage behavior resembles that of viruses (Schijven et al., 2000; Sobsey et al., 1995; Yates et al., 1985).

Viruses and bacteriophages persist longer than conventional bacterial indicators in water environments, and their persistence resembles that of viruses, as shown by model experiments (Duran et al., 2002; Gironés et al., 1989; Mocé-Llivina et al., 2005; Sinton et al., 1999) and by the decreases in ratios between the numbers of conventional bacterial indicators and of viruses and phages in water environments with aged pollutants, either surface water (Contreras-Coll et al., 2002; Duran et al., 2002; Lodder and de Roda Husman, 2005; Mocé-Llivina et al., 2005; Skraber et al., 2004) or groundwater (Abbaszadegan et al., 2003; Borchardt et al., 2003; Lucena et al., 2006). As well, bacteriophages infecting enteric bacteria have been reported in drinking waters that fulfill the quality criteria based on bacterial indicators (Dutka et al., 1990; El-Abagy et al., 1988; Mendez et al., 2004).

In conclusion, removal and inactivation of phages and viruses in water environments and water treatments follow similar behaviors, which differ from those of conventional bacterial indicators. Generally speaking, the persistence of bacteriophages and viruses in water is clearly superior to that of bacterial indicators, with the exception of spores of sulfite-reducing clostridia.

BACTERIOPHAGES IN FOOD

Indicator bacteriophages can be found when food is polluted with fecal contaminants as a consequence of watering green vegetables and fruits with fecally contaminated water, using contaminated water in the processing of ground beef or poultry meat, or growing shellfish in contaminated waters (Pillai, 2007).

Recently there has been a surge of interest in the potential use of bacteriophages as quality indicators of shellfish, and a great amount of scientific literature regarding accumulation and persistence of bacteriophages and viruses in shellfish is available (Pillai, 2007). As a consequence of the feeding process, shellfish efficiently concentrate viruses and phages, causing their decreased elimination by shellfish depuration compared with the elimination of conventional bacterial indicators. The extent of concentration and elimination depends on factors such as water characteristics, the physiological status of the animal, temperature, and the identity of the animal and the microorganism. This multifactorial process gives different results, so contradictory reports abound. Yet in the end, concentration and depuration allow for viruses and phages to accumulate in shellfish to a comparable extent, which is significantly greater that that of conventional bacterial indicators (Burkhardt et al., 1992; Burkhardt and Calci, 2000; Doré and Lees, 1995; Lucena et al., 1994; Muniain-Mujika et al., 2002).

BACTERIOPHAGES AS INDEXES OF HUMAN VIRUSES

Seldom is a correlation between numbers of viruses and phages detected in many water samples found when traditional statistics is applied (Abbaszadegan et al., 2003; Borchardt et al., 2004; Formiga-Cruz et al., 2003; Gerba, 1979; Griffin et al., 1999; Jiang et al., 2001). However, positive degrees of correlation have been reported in a few studies between enteroviruses and somatic coliphages in water at different stages of a water treatment plant (Stetler, 1984); F-specific RNA bacteriophages and enteroviruses in freshwater (Havelaar et al., 1993); F-specific RNA bacteriophages and enteroviruses in shellfish (Chung et al., 1998); F-specific RNA bacteriophages and genomes of norovirus in shellfish (Doré et al., 1998); enteroviruses and rotaviruses and phages infecting *B. fragilis* HSP40 in sea sediments (Jofre et al., 1989); somatic coliphages and enteroviruses in marine coastal water (Mocé-Llivina et al., 2005); different phages and the genomes of adenovirus, hepatitis A, and norovirus in shellfish (Brion et al., 2005); somatic coliphages or *B. fragilis* phages and the presence of infectious enteroviruses or the presence of the enterovirus genome in different types of depurated wastewater (Gantzer et al., 1998); and enteroviruses and phages in secondary effluents and bacteriophages infecting *B. thetaiotaomicron* GA17 in reclaimed waters (Costán-Longares et al., 2008b).

Yet, as stated above, phages do not often fulfill the index function, and the absence of phages might not guarantee the absence of viruses in a given water sample. A great amount of uncertainty exists for simple linear regressions obtained from the analysis of single indicators and pathogens, making them unconvincing statistical analyses to establish relationships between these kinds of parameters. Most likely, the application of advanced mathematical models such as artificial neural networks or support vector machines will reduce the uncertainty and give much better information about relationships between indicator bacteriophages and human viruses. Up to now, these types of analyses have permitted the establishment of relationships not detected by linear regression techniques and, for example, have illuminated the importance of bacteriophages as potential viral indicators in shellfish (Brion et al., 2005) or phages as tracers of the source of fecal contamination (Blanch et al., 2006). In our opinion, efforts should be focused on building useful databases for application of advanced modeling techniques.

In conclusion, the absence of phages infecting enteric bacteria might not guarantee the absence of human viruses. However, bacteriophages infecting enteric bacteria are always present in raw municipal sewage, and at least in industrialized countries, sewage constitutes the main input of fecal pollution to receiving waters. On the other hand, phages behave similarly to human viruses. Therefore, the presence of numbers of indicator phages higher than a certain threshold in water samples may indicate the presence of viruses. Thus, Skraber et al. (2004) have shown that in river water the number of samples positive for viral genomes was low when the concentrations of somatic coliphages remained under the value of 100 PFU per 100 ml and suggest that this threshold might be very useful for predicting virological contamination in river water.

WHICH TYPE OF BACTERIOPHAGES ARE BETTER SURROGATES OF HUMAN VIRUSES?

When it comes to deciding which of the three groups of bacteriophages better fulfills the index and indicator functions, each of the groups of phages advocated as model organisms has strong and weak points.

Somatic coliphages are the most abundant, and the methods for their detection and enumeration are the most simple, fast, and cost-effective, with results available in the working day. The most important weakness attributed to this group is their potential replication outside the gut, but in the view of available knowledge, it seems that in the great majority of situations the contribution of replication outside the gut to the numbers of somatic coliphages found in water samples is negligible (Jofre, 2009). The somatic coliphages are a heterogeneous group, with members showing different sensitivities to disinfectants and natural stressors (Dee and Fogelman, 1992; Duran et al., 2003; Mocé-Llivina et al., 2003), although their resistance to treatments and persistence in the environment is high compared to that of conventional bacterial indicators. However, human viruses also show great differences in resistance to disinfectants and natural stressors (Abad et al., 1994a, 1994b; Chung and Sobsey, 1993; Engelbrecht et al., 1980; Koch and Strauch, 1981; Meng and Gerba, 1996). These differences are observed even among isolates of the same viral species and serotype (Enriquez et al., 1995; Nwachuku et al., 2005; Payment et al., 1985b; Shaffer et al., 1980). Hence, in the end, this heterogeneity can be contemplated as a positive value.

F-specific RNA bacteriophages rank second in abundance. The method for detecting them is simple and fast, but to a lesser extent than that for somatic coliphages. They offer good perspectives as index organisms for viruses in groundwater and for monitoring some water treatments, for example, UV disinfection or removal by membrane filters. In contrast, their persistence in surface waters, mainly in warm climates, is low (Chung and Sobsey, 1993; Mocé-Llivina et al., 2005), as is their resistance to some inactivating treatments, such as those based on heating or liming (Feng et al., 2003; Mocé-Llivina et al., 2003).

For some purposes, such as determination of viral contamination in minimally contaminated waters like groundwater and drinking water (Federal Register, 2006), or reclamation treatments including two disinfecting treatments (for example, UV irradiation and chlorination) (Costán-Longares et al., 2008b), total coliphages might be a fine alternative.

Bacteriophages infecting *Bacteroides* rank third in abundance. Some bacterial strains detect phages with an unequivocal origin in the human gut. They are more resistant than the other groups to most natural stressors and water treatments. The method for detection requires cultivation of the host in anaerobiosis. Their low numbers in water and the need for different hosts for different geographic areas are at present the main weak points. In contrast, they seem very well suited as source tracers of fecal contaminants (Blanch et al., 2006; Ebdon et al., 2007).

Results reviewed here and in other publications have prompted some regulatory agencies to include bacteriophages in their detection policies. Thus, the U.S. Environmental Protection Agency has included coliphages in the Ground Water Rule (Federal Register, 2006); the Canadian Province of Québec has included coliphages in the regulation about drinking water quality (Gouvernement du Québec, 2001); and a few U.S. states have regulations regarding the quality required to reclaim water for certain uses (U.S. Environmental Protection Agency, 2004). Also, the U.S. Environmental Protection Agency is using bacteriophages in the validation and evaluation of membrane integrity (U.S. Environmental Protection Agency, 2006) and for validation of UV lamps to be used for water disinfection (U.S. Environmental Protection Agency, 2006).

In conclusion, it can be stated that, though not perfect, bacteriophages infecting enteric bacteria are useful additional indicators for water quality control because of their numbers in fecally contaminated waters and their outcome in water environments and water treatment plants. Moreover, they offer appealing additional advantages. (i) There is the limited significance, if any, of the occurrence of "stress," "injury," or "reactivation" that often causes misinterpretation of environmental data on bacterial indicators and pathogens. (ii) Analysis after collection of samples can be delayed for longer periods of time than samples for bacteriological analysis. (iii) Present methods for somatic coliphages provide results in 4 h (International Organization for Standardization, 2000). Having results of infectious bacteriophages in 4 h, in our opinion, gives coliphages extra value as indicators in water treatment plants operated according to the hazard analysis and critical control point principles.

Now that standardized and feasible methods for enumerating indicator bacteriophages are available, it will be suitable to construct databases with sound data on bacteriophages and some of the human viruses in water environments and water treatments. In our opinion, this will allow the application of advanced models, such as artificial neural networks or support vector machines, which will likely clarify the worth of bacteriophages as indicators and will guarantee the best choice for the various potential applications of bacteriophages as model organisms.

REFERENCES

Abad, F. X., R. M. Pinto, J. M. Diez, and A. Bosch. 1994a. Disinfection of human enteric viruses in water by copper and silver in combination with low levels of chlorine. *Appl. Environ. Microbiol.* **60:**2377–2383.

Abad, F. X., R. M. Pinto, and A. Bosch. 1994b. Survival of enteric viruses on environmental fomites. *Appl. Environ. Microbiol.* **60:**3704–3710.

Abbaszadegan, M., M. LeChevalier, and C. P. Gerba. 2003. Occurrence of viruses in US groundwaters. *J. Am. Water Works Assoc.* **95:**107–120.

Abbaszadegan, M., P. Monteiro, N. Nwachuku, A. Alum, and H. Ryu. 2008. Removal of adenovirus, calicivirus, and bacteriophages by conventional drinking water treatment. *J. Environ. Sci. Health A Tox. Hazard Subst. Environ. Eng.* **43:**171–177.

Alcalde, L., G. Oron, and M. Salgot. 2004. Evaluation of an integrated waste stabilization pond system for microbial removal, p. 103–110. *In* P. Lens and R. Stuetz (ed.), *Young Researchers 2004.* IWA Publishing, London, United Kingdom.

Araujo, R., A. Puig, J. Lasobras, F. Lucena, and J. Jofre. 1997. Phages of enteric bacteria in fresh water with different levels of faecal pollution. *J. Appl. Microbiol.* **82:**281–286.

Armon, R., and Y. Kott. 1996. Bacteriophages as indicators of pollution. *Crit. Rev. Environ. Sci. Technol.* **26:**299–335.

Arraj, A., J. Bohatier, H. Laveran, and O. Traore. 2005. Comparison of bacteriophages and enteric virus removal in pilot scale activated sludge plants. *J. Appl. Microbiol.* **98:**516–524.

Benyahya, M., H. Laveran, J. Bohatier, J. Senaud, and M. Ettayebi. 1998. Comparison of elimination of bacteriophages MS2 and ϕX174 during sewage treatment by natural lagooning or activated sludges. A study on laboratory scale pilot plants. *Environ. Technol.* **19:**513–519.

Berg, G., and D. Berman. 1980. Destruction by anaerobic mesophilic and thermophilic digestion of viruses and indicator bacteria indigenous to domestic sludges. *Appl. Environ. Microbiol.* **39:**361–368.

Blanch, A. R., L. Belanche-Muñoz, X. Bonjoch, J. Ebdon, C. Gantzer, F. Lucena, J. Ottoson, C. Kourtis, A. Iversen, I. Kühn, L. Mocé-Llivina, M. Muniesa, J. Schwartzbrod, S. Skraber, G. T. Papageorgiou, H. Taylor, J. Wallis, and J. Jofre. 2006. Integrated analysis of established and novel microbial and chemical methods for microbial source tracking. *Appl. Environ. Microbiol.* **72:**5915–5926.

Bofill-Mas, S., N. Albiñana-Gimenez, P. Clemente-Casares, A. Hundesa, J. Rodríguez-Manzano, A. Allard, M. Calvo, and R. Girones. 2006. Quantification and stability of human adenovirus and polyomavirus JCPyV in wastewater matrices. *Appl. Environ. Microbiol.* **72:**7894–7896.

Borchardt, M. A., P. D. Bertz, S. K. Spencer, and D. A. Battigelli. 2003. Incidence of enteric viruses in groundwater from household wells in Wisconsin. *Appl. Environ. Microbiol.* **69:**1172–1180.

Borchardt, M. A., N. L. Haas, and R. J. Hunt. 2004. Vulnerability of drinking-water wells in La Crosse, Wisconsin, to enteric-virus contamination from surface water contributions. *Appl. Environ. Microbiol.* **70:**5937–5946.

Bosch, A., F. Lucena, and J. Jofre. 1986. Fate of human enteric viruses (rotaviruses and enteroviruses) after primary sedimentation. *Water Sci. Technol.* **18:**47–52.

Brion, G., C. Viswanathan, T. R. Neelakantan, S. Lingireddy, R. Girones, D. Lees, A. Allard, and A. Vantarakis. 2005. Artificial neural network prediction of viruses in shellfish. *Appl. Environ. Microbiol.* **71:**5244–5253.

Burkhardt, W., III, W. D. Watkins, and S. R. Rippey. 1992. Seasonal effects on accumulation of microbial indicator organisms by *Mercenaria mercenaria. Appl. Environ. Microbiol.* **58:**826–831.

Burkhardt, W., III, and K. R. Calci. 2000. Selective accumulation may account for shellfish-associated viral illness. *Appl. Environ. Microbiol.* **66:**1375–1378.

Calci, K. R., W. Burkhardt III, W. D. Watkins, and S. R. Rippey. 1998. Occurrence of male-specific bacteriophages in feral and domestic animal wastes, human feces and human associated wastewaters. *Appl. Environ. Microbiol.* **64:**5027–5029.

Campos, C., A. Guerrero, and M. Cárdenas. 2002. Removal of bacterial and viral fecal indicator organisms in a waste stabilization pond system in

Choconta, Cundinamarca (Colombia). *Water Sci. Technol.* **45:**61–66.

Chung, H., and M. D. Sobsey. 1993. Comparative survival of indicator viruses and enteric viruses in seawater and sediment. *Water Sci. Technol.* **27:**425–429.

Chung, H., L.-A. Jaykus, G. Lovelance, and M. D. Sobsey. 1998. Bacteriophages and bacteria as indicators of enteric viruses in oysters and their harvest waters. *Water Sci. Technol.* **38:**37–44.

Clesceri, L. S., A. E. Greenberg, and A. D. Eaton (ed.). 1998. *Standard Methods for the Examination of Water and Wastewater,* 20th ed. American Public Health Association. Washington, DC.

Cole, D., S. Long, and M. D. Sobsey. 2003. Evaluation of F+ RNA and DNA coliphages as source-specific indicators of faecal contamination in surface waters. *Appl. Environ. Microbiol.* **69:** 6507–6514.

Contreras-Coll, N., F. Lucena, K. Mooijman, A. Havelaar, V. Pierzo, M. Boqué, A. Gawler, C. Höller, M. Lambiri, G. Mirolo, B. Moreno, M. Niemi, R. Sommer, B. Valentin, A. Wiedenmann, V. Young, and J. Jofre. 2002. Occurrence and levels of indicator bacteriophages in bathing waters through Europe. *Water Res.* **36:** 4963–4974.

Costán-Longares, A., L. Mocé-Llivina, A. Avellón, J. Jofre, and F. Lucena. 2008a. Occurrence and distribution of culturable enteroviruses in wastewater and surface waters of north-eastern Spain. *J. Appl. Microbiol.* **105:**1945–1955.

Costán-Longares, A., M. Montemayor, A. Payán, J. Mendez, J. Jofre, R. Mujeriego, and F. Lucena. 2008b. Microbial indicators and pathogens: removal, relationships and predictive capabilities in water reclamation facilities. *Water Res.* **42:**4439–4448.

Coté, P., H. Buisson, C. Pound, and G. Arakaki. 1997. Immersed membrane activated sludge for the reuse of municipal wastewater. *Desalination* **113:**189–196.

Deborde, D. C., W. W. Woessner, B. Lauerman, and P. Ball. 1998. Coliphage prevalence in high school septic effluent and associated ground water. *Water Res.* **32:**3781–3785.

Dee, S. W., and J. C. Fogelman. 1992. Rates of inactivation of waterborne coliphages by monochloramine. *Appl. Environ. Microbiol.* **58:**3136–3141.

Doré, W. J., and D. N. Lees. 1995. Behavior of *Escherichia coli* and male specific bacteriophages in environmentally contaminated bivalve. *Appl. Environ. Microbiol.* **61:**2830–2834.

Doré, W. J., K. Hensilwood, and D. N. Lees. 1998. The development of management strategies for control of virological quality in oysters. *Water Sci. Technol.* **38:**29–35.

Duran, A. E., M. Muniesa, X. Méndez, F. Valero, F. Lucena, and J. Jofre. 2002. Removal and inactivation of indicator bacteriophages in fresh waters. *J. Appl. Microbiol.* **92:**338–347.

Duran. A. E., M. Muniesa, L. Mocé-Llivina, C. Campos, J. Jofre, and F. Lucena. 2003. Usefulness of different groups of bacteriophages as model micro-organisms for evaluating chlorination. *J. Appl. Microbiol.* **95:**29–37.

Dutka, B. J., G. A. Palmenteer, S. M. Meissner, E. M. Janzen, and M. Sakellaris. 1990. Coliphages and bacteriophages in Canadian drinking waters. *Water. Int.* **15:**157–160.

Ebdon, J., M. Muniesa, and H. Taylor. 2007. The application of a recently isolated strain of *Bacteroides* (GB-124) to identify human sources of faecal pollution in a temperate river catchment. *Water Res.* **41:**3683–3690.

El-Abagy, M., B. J. Dutka, and M. Kamel. 1988. Incidence of coliphage in potable water supplies. *Appl. Environ. Microbiol.* **56:**1632–1633.

El-Senousy, W. M., S. Guix, I. Abid, R. M. Pintó, and A. Bosch. 2007. Removal of astrovirus from water and sewage treatment plants, evaluated by a competitive reverse transcription PCR. *Appl. Environ. Microbiol.* **73:**164–167.

Engelbrecht, R. S., M. J. Weber, B. L. Salter, and C. A. Schmidt. 1980. Comparative inactivation by chlorine. *Appl. Environ. Microbiol.* **40:** 249–256.

Enriquez, C. E., C. J. Hurst, and C. P. Gerba. 1995. Survival of the enteric adenoviruses 40 and 41 in tap, sea, and waste water. *Water Res.* **29:** 2548–2553.

Farahbakhsh, K., and D. W. Smith. 2004. Removal of coliphages in secondary effluent by microfiltration—mechanisms of removal and impact of operating parameters. *Wat. Res.* **38:**585–592.

Federal Register. 2006. National primary drinking water regulations: Ground Water Rule, final rule. 40 CFR parts 9, 141, and 142. *Fed. Regist.* **71:** 65574–65660.

Feng, Y. Y., S. L. Ong, J. Y. Hu, X. L. Tan, and W. J. Ng. 2003. Effects of pH and temperature on the survival of coliphages MS2 and Qβ. *J. Ind. Microbiol. Biotechnol.* **30:**549–552.

Fleisher, J., K. Schlafmann, R. Otchwemah, and K. Botzenhart. 2000. Elimination of enteroviruses, F-specific coliphages, somatic coliphages and *E. coli* in four sewage treatment plants of southern Germany. *J. Water Suppl. Res. Technol. AQUA* **49:** 127–138.

Formiga-Cruz, M., A. K. Allard, A.-C. Condin-Hansson, K. Henshilwood, B. E. Hernroth, J. Jofre, D. N. Lees, F. Lucena, M. Papapetropoulou, R. E. Rangdale, A. Tsibouxi, A. Vantarakis, and R. Girones. 2003. Evaluation of potential indicators of viral contamination in shellfish with applicability to diverse geographical areas. *Appl. Environ. Microbiol.* **69:**1556–1563.

Gantzer, C., A. Maul, J. M. Audic, and L. Schwartzbrod. 1998. Detection of infectious enteroviruses, enterovirus genomes, somatic coliphages, and *Bacteroides fragilis* phages in treated wastewater. *Appl. Environ. Microbiol.* **64:**4307–4312.

Gehr, R., M. Wagner, P. Veerasubramanian, and P. Payment. 2003. Disinfection efficiency of peracetic acid, UV and ozone after enhanced primary treatment of municipal wastewater. *Water Res.* **37:**4573–4586.

Gerba, C. P. 1979. Failure of indicator bacteria to reflect the occurrence of enteroviruses in marine waters. *Am. J. Public Health* **69:**1116–1119.

Gersberg, R. M., S. R. Lyon, R. Brenner, and B. V. Elkins. 1987. Fate of viruses in artificial wetlands. *Appl. Environ. Microbiol.* **54:**731–736.

Gironés, R., J. Jofre, and A. Bosch. 1989. Natural inactivation of enteric viruses in seawater. *J. Environ. Qual.* **18:**34–39.

Gouvernement du Québec. 2001. Loi sur la qualité de l'environnement: règlement sur la qualité de l'eau potable, c. Q-2, r. 18.1.1. *Gazette Officielle du Québec* **24:**3561.

Grabow, W. O. K., I. G. Middendorff, and N. C. Basson. 1978. Role of lime treatment in the removal of bacteria, enteric viruses, and coliphages in a wastewater reclamation plant. *Appl. Environ. Microbiol.* **35:**663–669.

Grabow, W. O. K., P. Coubrough, E. M. Nupen, and B. W. Bateman. 1984. Evaluation of coliphages as indicators of the virological quality of sewage-polluted water. *Water S. Afr.* **10:**7–14.

Grabow, W. O. K., C. S. Holtzhausen, and C. J. De Villiers. 1993. *Research on Bacteriophages as Indicators of Water Quality 1990–1992.* Water Research Commission report no. 321/1/93. Water Research Commission. Pretoria, South Africa.

Grabow, W. O. K., T. E. Neubrech, C. S. Holtzhausen, and J. Jofre. 1995. *Bacteroides fragilis* and *Escherichia coli* bacteriophages. Excretion by humans and animals. *Water Sci. Technol.* **31:** 223–230.

Grabow, W. O. K. 2001. Bacteriophages: update on application as models for viruses in water. *Water S. Afr.* **27:**251–268.

Green, D. H., and G. D. Lewis. 1999. Comparative detection of enteric viruses in wastewaters, sediments and oysters by reverse transcription-PCR and cell culture. *Water Res.* **33:**1195–1999.

Griffin, D. W., C. J. Gibson III, E. K. Lipp, K. Riley, J. H. Paul III, and J. B. Rose. 1999. Detection of viral pathogens by reverse transcriptase PCR and of microbial indicators by standard methods in the canals of the Florida Keys. *Appl. Environ. Microbiol.* **65:**4118–4125.

Guelin, A. 1948. Etude quantitative de bacteriophages de la mer. *Ann. Inst. Pasteur* (Paris) **74:**104–112.

Guzmán, C., J. Jofre, M. Montemayor, and F. Lucena. 2007. Occurrence and levels of indicators and selected pathogens in different sludges and biosolids. *J. Appl. Microbiol.* **103:**2420–2429.

Guzmán, C., L. Mocé-Llivina, F. Lucena, and J. Jofre. 2008. Evaluation of the *Escherichia coli* host strain (CB390) for simultaneous detection of somatic and F-specific coliphages. *Appl. Environ. Microbiol.* **74:**531–534.

Harakeh, S. 2006. The behaviour of viruses on disinfection by chlorine dioxide and other disinfectants in effluent. *FEMS Microbiol. Lett.* **44:**335–341.

Harwood, V. J., A. D. Levine, T. M. Scott, V. Chivukula, J. Lukasik, S. R. Farrah, and J. B. Rose. 2005. Validity of the indicator organism paradigm for pathogen reduction in reclaimed water and public health protection. *Appl. Environ. Microbiol.* **71:**3163–3170.

Havelaar, A. H., K. Furuse, and W. M. Hogeboom. 1986. Bacteriophages and indicator bacteria in human and animal faeces. *J. Appl. Bacteriol.* **60:**255–262.

Havelaar, A. H. 1987. Bacteriophages as model organisms in water treatment. *Microbiol. Sci.* **4:**362–364.

Havelaar, A. H., and W. M. Pot-Hogeboom. 1988. F-specific RNA bacteriophages as model viruses in water hygiene: ecological aspects. *Water Sci. Technol.* **20:**399–407.

Havelaar, A. H., M. van Olphen, and Y. Drost. 1993. F-specific RNA bacteriophages are adequate model organisms for enteric viruses in fresh water. *Appl. Environ. Microbiol.* **59:**2956–2962.

He, J.-W., and S. Jiang. 2005. Quantification of enterococci and human adenoviruses in environmental samples by real-time PCR. *Appl. Environ. Microbiol.* **71:**2250–2255.

Herath, G., K. Yamamoto, and T. Urase. 1998. Mechanism of bacterial and viral transport through microfiltration membranes. *Water Sci. Technol.* **38:** 489–496.

Hill, V. R., and M. D. Sobsey. 1998. Microbial indicator reductions in alternative treatment systems for swine wastewater. *Water Sci. Technol.* **38:** 119–122.

IAWPRC Study Group on Health Related Water Microbiology. 1991. Bacteriophages as model viruses in water quality control. *Water Res.* **25:**529–545.

International Organization for Standardization. 1995. *ISO 10705-1: Water Quality. Detection and*

Enumeration of Bacteriophages, Part 1. Enumeration of F-specific RNA Bacteriophages. International Organization for Standardization, Geneva, Switzerland.

International Organization for Standardization. 2000. *ISO 10705-2: Water Quality. Detection and Enumeration of Bacteriophages, Part 2. Enumeration of Somatic Coliphages*. International Organization for Standardization. Geneva, Switzerland.

International Organization for Standardization. 2001. *ISO 10705-4: Water Quality. Detection and Enumeration of Bacteriophages, Part 4. Enumeration of Bacteriophages Infecting* Bacteroides fragilis. International Organization for Standardization, Geneva, Switzerland.

Irving, L. G., and F. A. Smith. 1981. One-year survey of enteroviruses, adenoviruses, and reoviruses isolated from effluent at an activated-sludge purification plant. *Appl. Environ. Microbiol.* **41:**51–59.

Jancangelo, J. G., P. Loughram, B. Petric, D. Simpson, and C. McIlroy. 2003. Removal of enteric viruses and selected microbial indicators by UV irradiation of secondary effluent. *Water Sci. Technol.* **47:**193–198.

Jiang, S., R. Noble, and W. Chu. 2001. Human adenoviruses and coliphages in urban runoff-impacted coastal waters of Southern California. *Appl. Environ. Microbiol.* **67:**179–184.

Jofre, J., M. Blasi, A. Bosch, and F. Lucena. 1989. Occurrence of bacteriophages infecting *Bacteroides fragilis* and other viruses in polluted marine sediments. *Water Sci. Technol.* **21:**15–19.

Jofre, J., E. Ollé, F. Ribas, A. Vidal, and F. Lucena. 1995. Potential usefulness of bacteriophages that infect *Bacteroides fragilis* as model organisms for monitoring virus removal in drinking water treatment plants. *Appl. Environ. Microbiol.* **61:**3227–3231.

Jofre, J. 2002. Bacteriophages as indicators, p. 354–363. *In* G. Bitton (ed.), *Encyclopedia of Environmental Microbiology,* vol. 1. John Wiley & Sons, Inc., New York, NY.

Jofre, J. 2009. Is the replication of somatic coliphages in water environments significant? *J. Appl. Microbiol.* **106:**1059–1069.

Karim, M. R., F. D. Manshadi, M. M. Karpisck, and C. P. Gerba. 2004. The persistence and removal of enteric pathogens in constructed wetlands. *Water Res.* **38:**1831–1837.

Koch, K., and D. Strauch. 1981. Removal of polio and parvovirus in sewage sludge by lime-treatment. *Zentralbl. Bakteriol. Mikrobiol. Hyg. B* **174:**335–347.

Kott, Y., N. Roze, S. Sperber, and N. Betzer. 1974. Bacteriophages as viral pollution indicators. *Water Res.* **8:**165–171.

Leclerc, H., S. Edberg, V. Pierzo, and J. M. Delattre. 2000. Bacteriophages as indicators of enteric viruses and public health risk in groundwaters. *J. Appl. Microbiol.* **88:**5–21.

Levine, A. D., V. J. Harwood, S. R. Farrah, T. M. Scott, and J. B. Rose. 2008. Pathogen and indicator organism reduction through secondary effluent filtration: implications for reclaimed water production. *Water Environ. Res.* **80:**596–608.

Lodder, W. J., and A. M. de Roda Husman. 2005. Presence of noroviruses and other enteric viruses in sewage and surface water in The Netherlands. *Appl. Environ. Microbiol.* **71:**1453–1461.

Lucena, F., J. Lasobras, D. McIntosh, M. Forcadell, and J. Jofre. 1994. Effect of distance from the polluting focus on relative concentrations of *Bacteroides fragilis* phages and coliphages in mussels. *Appl. Environ. Microbiol.* **60:**2272–2277.

Lucena, F., X. Mendez, A. Morón, E. Calderón, C. Campos, A. Guerrero, M. Cárdenas, C. Gantzer, L. Schwartzbrod, S. Skraber, and J. Jofre. 2003. Occurrence and densities of bacteriophages proposed as indicators and bacterial indicators in river waters from Europe and South America. *J. Appl. Microbiol.* **94:**808–815.

Lucena, F., A. E. Duran, A. Morón, E. Calderón, C. Campos, C. Gantzer, S. Skraber, and J. Jofre. 2004. Reduction of bacterial indicators and bacteriophages infecting faecal bacteria in primary and secondary wastewater treatments. *J. Appl. Microbiol.* **97:**1069–1076.

Lucena, F., F. Ribas, A. E. Duran, S. Skraber, C. Gantzer, C. Campos, A. Morón, E. Calderón, and J. Jofre. 2006. Occurrence of bacterial indicators and bacteriophages infecting enteric bacteria in groundwater in different geographical areas. *J. Appl. Microbiol.* **101:**96–102.

Luther, K., and R. Fujioka. 2004. Usefulness of monitoring tropical streams for male-specific RNA coliphages. *J. Water Health* **2:**171–181.

Meleg, E., K. Bányai, V. Martella, B. Jiang, B. Kocsis, P. Kisfali, B. Melegh, and G. Szücs. 2008. Detection and quantification of group C rotaviruses in communal sewage. *Appl. Environ. Microbiol.* **74:**3394–3399.

Mendez, J., A. Audicana, M. Cancer, A. Isern, J. Llaneza, J. Moreno, M. Navarro, M. L. Tarancon, F. Valero, F. Ribas, J. Jofre, and F. Lucena. 2004. Assessment of drinking water quality using indicator bacteria and bacteriophages. *J. Water Health* **2:**201–214.

Meng, Q. S., and C. P. Gerba. 1996. Comparative inactivation of enteric adenoviruses, polioviruses and coliphages by ultraviolet irradiation. *Water Res.* **30:**2665–2668.

Mignotte-Cadiergues, B., A. Maul, and L. Schwartzbrod. 1999. Comparative study of techniques used to recover viruses from residual urban sludge. *J. Virol. Methods* **78:**71–80.

Mignotte-Cadiergues, B., C. Gantzer, and L. Schwartzbrod. 2002. Evaluation of bacteriophages during the treatment of sludge. *Water Sci. Technol.* **46:**189–194.

Mocé-Llivina, L., M. Muniesa, H. Pimenta-Vale, F. Lucena, and J. Jofre. 2003. Survival of bacterial indicator species and bacteriophages after thermal treatment of sludge and sewage. *Appl. Environ. Microbiol.* **69:**1452–1456.

Mocé-Llivina, L., F. Lucena, and J. Jofre. 2005. Enteroviruses and bacteriophages in bathing waters. *Appl. Environ. Microbiol.* **71:**6838–6844.

Montemayor, M., A. Costán, F. Lucena, J. Jofre, J. Muñoz, E. Dalmau, R. Mujeriego, and L. Sala. 2008. The combined performance of UV light and chlorine during reclaimed water disinfection. *Water Sci. Technol.* **57:**935–940.

Mooijman, K. A., Z. Ghameshlou, M. Bahar, J. Jofre, and A. Havelaar. 2005. Enumeration of bacteriophages in water by different laboratories of the European Union in two interlaboratory comparison studies. *J. Virol. Methods* **127:**60–68.

Mossel, D. A. A. 1982. Marker (index and indicator organisms) in food and drinking water. Semantics, ecology, taxonomy and enumeration. *Antoine van Leeuwenhoek* **48:**609–611.

Muniain-Mujika, I., R. Girones, G. Tofiño-Quesada, M. Calvo, and F. Lucena. 2002. Depuration dynamics of viruses in shellfish. *Int. J. Food Microbiol.* **77:**125–133.

Nieuwstad, T. J., E. P. Mulder, A. H. Havelaar, and M. van Olphen. 1988. Elimination of microorganisms from wastewater by tertiary precipitation followed by filtration. *Water Res.* **22:**1389–1397.

Nwachuku, N., C. P. Gerba, A. Oswald, and F. D. Mashadi. 2005. Comparative inactivation of adenovirus serotypes by UV light disinfection. *Appl. Environ. Microbiol.* **71:**5633–5636.

Omura, T., M. Onuma, J. Aizawa, T. Umita, and T. Yagi. 1989. Removal efficiency of indicator micro-organisms in sewage treatment plants. *Water Sci. Technol.* **21:**119–124.

Ottoson, A., B. Hansen, B. Björlenius, H. Norder, and T. A. Stentröm. 2006. Removal of viruses, parasitic protozoa and microbial indicators in conventional and membrane processes in a wastewater pilot plant. *Water Res.* **40:**1449–1457.

Paul, J. H., J. B. Rose, S. C. Jiang, C. A. Kellogg, and L. Dickson. 1993. Distribution of viral abundance in the reef environment of Key Largo, Florida. *Appl. Environ. Microbiol.* **59:**718–724.

Payán, A., J. Ebdon, H. Taylor, C. Gantzer, J. Ottoson, G. T. Papageorgiou, A. R. Blanch, F. Lucena, J. Jofre, and M. Muniesa. 2005. Method for isolation of *Bacteroides* bacteriophage host strains suitable for tracking sources of fecal pollution in water. *Appl. Environ. Microbiol.* **71:**5659–5662.

Payment, P., M. Trudel, and R. Plante. 1985a. Elimination of viruses and indicator bacteria at each step of treatment during preparation of drinking water at seven water treatment plants. *Appl. Environ. Microbiol.* **49:**1418–1482.

Payment, P., M. Tremblay, and M. Trudel. 1985b. Relative resistance to chlorine of poliovirus and coxsackievirus isolates from environmental sources and drinking water. *Appl. Environ. Microbiol.* **49:**981–983.

Payment, P., E. Morin, and M. Trudel. 1988. Coliphages and enteric viruses in the particulate phase of river water. *Can. J. Microbiol.* **34:**907–910.

Payment, P., and E. Franco. 1993. *Clostridium perfringens* and somatic coliphages as indicators of the efficiency of drinking water treatment for viruses and protozoan cysts. *Appl. Environ. Microbiol.* **59:**2418–2424.

Pillai, S. D. 2007. Bacteriophages as fecal indicators, p. 205–222. *In* S. M. Goyal (ed.), *Viruses in Foods.* Springer, New York, NY.

Puig, A., N. Queralt, J. Jofre, and R. Araujo. 1999. Diversity of *Bacteroides fragilis* strains in their capacity to recover phages from human and animal wastes and from fecally polluted wastewater. *Appl. Environ. Microbiol.* **65:**1772–1776.

Schaper, M., J. Jofre, M. Uys, and W. Grabow. 2002. Distribution of genotypes of F-specific RNA bacteriophages in human and non-human sources of faecal pollution in South Africa and Spain. *J. Appl. Microbiol.* **92:**657–667.

Scheuerman, P. R., S. R. Farrah, and G. Bitton. 1987. Reduction of microbial indicators and viruses in a cypress strand. *Water Sci. Technol.* **19:**539–546.

Schijven, J. F., G. Medema, A. J. Vogelaar, and S. M. Hassanizadeh. 2000. Removal of microorganisms by deep well injection. *J. Contam. Hydrol.* **44:**301–327.

Schvoerer, E., M. Ventura, O. Dubos, G. Cazaux, R. Serceau, N. Gournier, V. Dubois, P. Caminade, H. J. Fleury, and M. E. Lafon. 2001. Qualitative and quantitative molecular detection of enteroviruses in water from bathing areas and from sewage treatment plant. *Res. Microbiol.* **152:**179–186.

Sedmak, G., D. Bina, J. MacDonald, and L. Couillard. 2005. Nine-year study of the occurrence of culturable viruses in source water for two drinking water treatment plants and the influent and effluent of a wastewater treatment plant in Milwaukee, Wisconsin (August 1994 through July 2003). *Appl. Environ. Microbiol.* **71:**1041–1050.

Shaffer, P. T., T. G. Metcalf, and O. Sproul. 1980. Chlorine resistance of poliovirus isolates recovered from drinking water. *Appl. Environ. Microbiol.* **40:**1115–1121.

Shang, C., H. M. Wong, and G. Chen. 2005. Bacteriophage MS2 removal by submerged membrane bioreactor. *Water Res.* **39:**4211–4219.

Shin, G. A., and M. D. Sobsey. 2003. Reduction of Norwalk virus, poliovirus 1, and bacteriophage MS2 by ozone disinfection of water. *Appl. Environ. Microbiol.* **69:**3975–3978.

Sinton, L. W., R. K. Finley, and P. A. Lynch. 1999. Sunlight inactivation of fecal bacteriophages and bacteria in sewage-polluted seawater. *Appl. Environ. Microbiol.* **65:**3605–3613.

Skraber, S., B. Gassilloud, and C. Gantzer. 2004. Comparison of coliforms and coliphages as tools for assessment of viral contamination in river water. *Appl. Environ. Microbiol.* **70:**3644–3649.

Sobsey, M. D. 1989. Inactivation of health-related microorganisms in water by disinfection processes. *Water Sci. Technol.* **21:**179–195.

Sobsey, M. D., R. M. Hall, and R. L. Hazard. 1995. Comparative reductions of hepatitis A virus, enteroviruses and coliphage MS2 in miniature soil columns. *Water Sci. Technol.* **31:**203–209.

Sobsey, M. D., M. V. Yates, F.-C. Hsu, G. Lovelace, D. Battigelli, A. Margolin, S. D. Pillai, and N. Nwachuku. 2004. Development and evaluation of methods to detect coliphages in large volumes of water. *Water Sci. Technol.* **50:** 211–217.

Spillman, S. K., F. Traub, M. Schwyzer, and R. Wyler. 1987. Inactivation of animal viruses during sewage sludge treatment. *Appl. Environ. Microbiol.* **53:**2077–2081.

Stetler, R. 1984. Coliphages as indicators of enteroviruses. *Appl. Environ. Microbiol.* **48:**668–670.

Tartera, C., and J. Jofre. 1987. Bacteriophages active against *Bacteroides fragilis* in sewage-polluted waters. *Appl. Environ. Microbiol.* **53:**1632–1637.

Tartera, C., F. Lucena, and J. Jofre. 1989. Human origin of *Bacteroides fragilis* bacteriophages present in the environment. *Appl. Environ. Microbiol.* **55:** 2696–2701.

Tyrrell, S. A., S. R. Rippey, and W. D. Watkins. 1995. Inactivation of bacterial and viral indicators in secondary sewage effluents, using chlorine and ozone. *Water Res.* **29:**2483–2490.

Ueda, T., and N. J. Horan. 2000. Fate of indigenous bacteriophage in a membrane bioreactor. *Water Res.* **34:**2151–2159.

Uemura, S., K. Takahashi, A. Takaishi, I. Machdar, A. Ohashi, and H. Harada. 2002. Removal of indigenous coliphages and fecal coliforms by a novel sewage treatment system consisting of UASB and DHS units. *Water Sci. Technol.* **46:**303–309.

U.S. Environmental Protection Agency. 2000a. *Method 1601. Male-Specific (F+) and Somatic Coliphage in Water by Two-Step Enrichment Procedure.* EPA report no. 821-R-00-009. U.S. Environmental Protection Agency, Washington, DC.

U.S. Environmental Protection Agency. 2000b. *Method 1602. Male-Specific (F+) and Somatic Coliphage in Water by Single Agar Layer Procedure.* EPA report no. 821-R-00-010. U.S. Environmental Protection Agency, Washington, DC.

U.S. Environmental Protection Agency. 2004. *Guidelines for Water Reuse.* EPA report no. 625-R-04-108. U.S. Environmental Protection Agency, Washington, DC.

U.S. Environmental Protection Agency. 2006. *Environmental Technology Verification Report, Removal of Microbial Contaminants From Drinking Water, Koch Membrane Systems, Inc., HF-82-35-PMPW Ultrafiltration Membrane (PDF).* EPA report no. 600-R-06-100. U.S. Environmental Protection Agency, Washington, DC.

U.S. Environmental Protection Agency. 2006. *Ultraviolet Disinfection Guidance Manual for the Final Long Term 2 Enhanced Surface Water Treatment Rule.* EPA report no. 815-R-06-007. U.S. Environmental Protection Agency, Washington, DC.

Vega, E., B. Lesikar, and S. D. Pillai. 2003. Transport and survival of bacterial and viral tracers through submerged-flow constructed wetland and sand-filter system. *Bioresourc. Technol.* **89:**49–56.

Vidales, J. A., C. P. Gerba, and M. M. Karpiscak. 2003. Virus removal from wastewater in a multispecies subsurface-flow constructed wetland. *Water Environ. Res.* **75:**238–245.

Wentsel, R. S., P. E. O. Neill, and J. F. Kitchens. 1982. Evaluation of coliphage detection as a rapid indicator of water quality. *Appl. Environ. Microbiol.* **43:**430–434.

Williams, F. P., and C. J. Hurst. 1988. Detection of environmental viruses in sludge. Enhancement of enterovirus plaque assay titer with 5-iodo 2′-deoxiuridine and comparison to adenoviruses and coliphage titers. *Water Res.* **22:**847–851.

Yates, M. V., C. P. Gerba, and L. M. Kelly. 1985. Virus persistence in groundwater. *Appl. Environ. Microbiol.* **49:**778–781.

APPLICATION OF BACTERIOPHAGES TO CONTROL PATHOGENIC AND SPOILAGE BACTERIA IN FOOD PROCESSING AND DISTRIBUTION

J. Andrew Hudson, Lynn McIntyre, and Craig Billington

7

INTRODUCTION

The use of bacteriophages (phages) to control pathogens in foods is relatively recent compared to their use in therapeutic and diagnostic applications (Mandeville et al., 2003; Rees and Dodd, 2006). Phage biocontrol in foods has been investigated primarily in fresh produce, dairy products, and meat products, while most work with spoilage bacteria has focused on meat. The work published to date has focused on application during processing, as this is the point at which the food processor is able to use the approach. It seems that the effect of added phages is relatively rapid, and while phages may survive on foods for some time after application and possibly during distribution, there is little biocontrol being exerted during distribution and control does not rely on phage replication to be effective (Guenther et al., 2009). Thus, added phages act more as a critical control point resulting in a short-term inactivation than as a biopreservative.

J. Andrew Hudson, Lynn McIntyre, and Craig Billington, Food Safety Programme, Institute of Environmental Science and Research Ltd., Christchurch 8540, New Zealand.

Food-borne pathogen control using phages has moved beyond the laboratory, and regulatory approvals have been granted for commercial phage products which are currently available to the food industry. A *Listeria monocytogenes* control preparation, LMP-102™, which consists of six phages, has been approved by the U.S. Food and Drug Administration for use in ready-to-eat meat and poultry products (Federal Register, 2006). Approval of the antilisterial product LISTEX™, originally for use in cheese, has since been widened to include all foods subject to contamination by the organism (Tarantino, 2007). The same product has now been granted "organic" status and can be used in the European Union for pathogen control in both conventional and organic foods (EBI Food Safety, 2007). Phage P100, on which this product is based, has recently been the subject of a successful patent application (Loessner and Carlton, 2008).

Beyond their use to control bacteria in foods directly, phages might also be used as sanitizers in food production facilities, and their activities against biofilms have received some attention. One such product has recently been granted approval by the U.S. Environmental Protection Agency for use

Bacteriophages in the Control of Food- and Waterborne Pathogens
Edited by Parviz M. Sabour and Mansel W. Griffiths, © 2010 ASM Press, Washington, DC

in the control of *L. monocytogenes* in the food production environment (Intralytix, 2008).

CONTROL OF PATHOGENIC AND SPOILAGE BACTERIA ON FOODS

Dairy Products

The first description of phage biocontrol of pathogens in dairy products was by Modi et al. (2001), who investigated the control of *Salmonella enterica* serovar Enteritidis GCDE (a bioluminescent construct) during manufacture and ripening of cheddar cheese. Raw and pasteurized milk were inoculated with approximately 10^4 CFU ml^{-1} *Salmonella*. After addition of a mixed mesophilic starter culture, the phage isolate was added to achieve 10^8 PFU ml^{-1} in the milk. Cheese was manufactured using a standard commercial procedure, vacuum-packed, and stored at 8°C for 99 days. Serovar Enteritidis survived cheese manufacture, increasing in numbers in both raw and pasteurized milk cheeses in the absence of phages (although this apparent growth may have been due to entrapment of cells in the curd). In the presence of phages a 1- to 2-$\log_{10}$ CFU g^{-1} reduction in pathogen counts after 24 h in raw and pasteurized milk cheeses, respectively, was recorded when compared to counts for milk shortly after inoculation. After 90 days of storage, pasteurized milk cheese inoculated with phages contained <50 CFU g^{-1} salmonellae, while the pathogen was present at 50 CFU g^{-1} in one of the raw milk phage-containing cheese samples after 99 days of ripening and storage. In control cheeses the counts of *Salmonella* were consistently 2 to 3 $\log_{10}$ CFU g^{-1} higher than those in the phage-treated cheeses. The concentration of phages increased between days 1 and 90 of ripening.

Similarly, control of *Staphylococcus aureus* in ultra-high-temperature-treated whole milk and acid curd manufacture has been achieved (García et al., 2007). Two virulent phages infecting the pathogen were produced by inducing random deletions in the DNA of temperate phages. Used at concentrations of approximately 10^8 PFU ml^{-1} in milk, the mutant phages reduced a 6-$\log_{10}$ CFU ml^{-1} inoculum of *S. aureus* Sa9 to beneath the level of detection within 2 h at 37°C. Although data were not provided, the phage titer was stated to increase over the 8 h of sampling. These phages were investigated further in both acid and enzymatic curd manufacture. Phages were added to pasteurized milk inoculated with 6 $\log_{10}$ CFU ml^{-1} of *S. aureus* at ratios of 250:1 and 350:1. *S. aureus* increased in numbers by 2 $\log_{10}$ CFU ml^{-1} in control milk, but growth was controlled to some extent by the addition of a starter culture in production of curd by both methods. When phages were added, pathogen numbers rapidly reduced to undetectable levels within 1 and 4 h of incubation in enzymatic and acid curd manufacture, respectively.

There are some reports describing the use of phages and nisin in combination. One of these focused on their combined use for the control of *S. aureus* (Martínez et al., 2008). Incubation at 37°C in pasteurized milk showed that a combination of the two, with phages added at 10^3 PFU ml^{-1}, gave a greater reduction (around 5 to 6 $\log_{10}$ CFU ml^{-1} after a 6-h incubation) compared to either treatment on its own (both giving a 3- to 4-$\log_{10}$ CFU ml^{-1} reduction over the same time period). The paper reported the production of a nisin-adapted isolate that was also resistant to phage infection (although mutants resistant to phage infection were isolated that were not nisin resistant). It was proposed that this resistance was due to changes on the bacterial cell surface, so affecting phage adsorption. The nisin-adapted isolate was obtained by exposing *S. aureus* cells to increasing concentrations of nisin, but the frequency of these cells arising during application of both phage and bacteriocin in food was not assessed. Interestingly, it was reported that a ratio of phage to host of 10^{-4}:1 needed to be used to prevent "rapid killing," so presumably ratios greater than this would produce results better than those reported.

The same group has examined the use of a cloned staphylococcal phage endolysin (see chapter 8) for the inactivation of staphylococci in pasteurized milk (Obeso et al., 2008). The cloned endolysin was capable of killing isolates of *S. aureus* and *Staphylococcus epidermidis* but not other bacteria tested (lactic acid bacteria, *Streptococcus, Listeria,* and *Enterococcus* spp.). Bovine isolates were more susceptible to the enzyme than those from humans. When incubated at 37°C in milk the effect of the enzyme was dose dependent, with a smaller quantity required to inactivate staphylococci present at lower concentrations. Inactivation to undetectable levels was demonstrated within 4 h when sufficient quantities of the enzyme were added. The effects of temperature and pH on the enzyme (pH optimum of 6 to 7, active at 30°C) are consistent with its possible use in the coagulation step of cheese production where cheese milk is kept warm to allow the activity of a starter culture, although activity was lost at pH 4.0.

In work focused on developing treatments for mastitis, phage K present at 10^8 PFU ml^{-1} was shown to eliminate *S. aureus* at an initial concentration of 10^6 CFU ml^{-1} in bacteriological broth and heat-treated milk but had no effect in raw milk. This resistance to phage infection in raw milk was attributed to the binding of heat-labile immunoglobulin to host cells (O'Flaherty et al., 2005). The phage was also able to eliminate the pathogen in whey and heat-treated whey, although the effect was delayed in the former. Inhibition of phage K binding to *S. aureus* by whey proteins has also been described by Gill et al. (2006), with the degree of inhibition varying significantly when the inhibitory activities in raw whey from 23 different cows were compared. If these observations apply to other phage/host combinations, then inactivation of bacteria in raw milk may be problematic.

Control of *L. monocytogenes* on soft cheeses has been reported using the unusually broad-host-range phage P100 (Carlton et al., 2005). The cheeses investigated were Muenster-style cheeses ripened by the addition of *Brevibacterium linens* and *Debaryomyces hansenii* surface smears. Cheeses made using a commercial process were smeared with brine containing the ripening microflora and washed with a suspension of *L. monocytogenes* to achieve an inoculum of 2×10^1 CFU cm^{-2}. The phage were applied to cheese surfaces either repeatedly at either a high dose, 6×10^7 PFU cm^{-2}, or lower dose, 2×10^6 PFU cm^{-2}, or as a single treatment of 6×10^7 PFU cm^{-2} just after the addition of the pathogen. The cheese was ripened at 14°C for 16 days, and during this period smearing was performed a further six times. Following this, samples were packaged and stored for five more days at 6°C. The number of pathogens reached 10^5 CFU cm^{-2} in phage-free cheeses on day 10 and increased to more than 10^7 CFU cm^{-2} by day 21. Application of the repeated high dose and the single dose of the phage eliminated *Listeria* entirely, as judged by the inability of selective enrichments to detect it, and no further growth occurred during storage. The pathogen did grow on cheeses repeatedly treated with the lower phage dose, although it was slower than the phage-free control, taking 20 days to attain 10^4 CFU cm^{-2}. Importantly, no adverse impact of phage addition on the ripening of cheeses occurred. Similar results using the same phage, but over an extended storage period (37 days), have also been reported (Schellekens et al., 2007).

Phage P100 and the closely related phage A511 have also been shown to produce significant reductions in the concentration of *L. monocytogenes* in chocolate milk and mozzarella cheese brine (Guenther et al., 2009) when incubated for a variety of time/temperature combinations.

A description of the ability of phages to control the opportunistic pathogen *Enterobacter sakazakii* (now *Cronobacter sakazakii* [Iversen et al., 2007]) in reconstituted infant formula has been published (Kim et al., 2007). Two phages, isolated from sewage, were examined at three temperatures (12, 24, and 37°C) in broth and reconstituted formula inoculated with *C. sakazakii*. In both cases, phages were added at levels of 10^7, 10^8, and 10^9 PFU ml^{-1}, and biocontrol was evaluated at intervals by either optical density (OD) measurement in broth or dilution plating of reconstituted formula. Although bacterial growth was in log phase by the time phages were added, control of growth was noted at the different temperatures used. One phage was able to reduce the OD at 24°C by almost 1.0 unit over 8 h and prevent further growth, while the other reduced the OD to a lesser extent over 4 h at 37°C. The OD ultimately increased to levels slightly greater than 50% of the control. Higher phage concentrations demonstrated better control of the pathogen, but no treatment was able to eliminate *C. sakazakii* completely. However, it needs to be noted that spectrophotometry can detect only relatively high numbers of bacteria, and a reduction of 1 unit is unlikely to translate to a very large reduction in bacterial numbers.

In reconstituted infant formula inoculated with 2 $\log_{10}$ CFU ml^{-1} *C. sakazakii,* counts reached 7 to 8 $\log_{10}$ CFU ml^{-1} at the end of incubation at 12 and 24°C, and 9 $\log_{10}$ CFU ml^{-1} at 37°C in phage-free controls. One phage added at 10^9 PFU ml^{-1} reduced counts compared to the phage-free control by 2 to 4 $\log_{10}$ CFU ml^{-1} at 37°C and to undetectable levels at 24°C within 2 h but had little effect with incubation at 12°C. In contrast, the other phage reduced host concentrations to beneath the detection limit of the method used at all temperatures within 2 to 12 h when added at the same concentration, and also at 24°C when added at 10^8 PFU ml^{-1}. The addition of fewer phages generally resulted in less control of the pathogen.

More recently, the effect of different concentrations of phages on the control of *C. sakazakii* in broth culture has been examined (Zuber et al., 2008). When 10^8 PFU ml^{-1} of two different phages were incubated separately with 10^6 CFU ml^{-1} of the host, the OD increased in parallel with the phage-free control before reducing after a 3-h incubation. Outgrowth was noted upon extended incubation of one of the cultures. Similar results were found when the host cell concentration was reduced to 10^2 CFU ml^{-1} and phages were added at 10^8 PFU ml^{-1}, but an increase in pathogen numbers was measured when the phage concentration was reduced to 10^4 PFU ml^{-1}. In this last case, cell numbers began to decline only once they had grown to around 10^5 CFU ml^{-1}. As noted by the authors, these results are not surprising when taking into account the need to consider the actual multiplicity of infection (MOI_{actual}) rather than MOI_{input} at low cell concentrations (Kasman et al., 2002), as the MOI_{actual} would have been much lower than the MOI_{input} of 100 (10^4 PFU ml^{-1} phages and 10^2 CFU ml^{-1} host cells).

With respect to dairy product spoilage, the effect of phage and host concentrations on the control of *Pseudomonas fragi* has been studied in sterile milk incubated at 7°C for 72 h (Ellis et al., 1973). When the host was inoculated at 10^5 CFU ml^{-1} and phages added at levels ranging from 0.1 to 10^3 PFU ml^{-1}, phages increased in number but the effect on host concentrations was modest. For example, pseudomonads in the phage-free control reached 1.5×10^7 CFU ml^{-1}, while in the presence of phages at the highest concentration host cells grew to approximately 5×10^6 CFU ml^{-1}. With the host inoculum reduced to 10^3 CFU ml^{-1} no inhibition of pseudomonad growth resulted from the addition of phages, although phage titers increased when added at 10 or 10^3 PFU ml^{-1}. Again, because of the low concentration of the host cells used, the majority of them would not have been infected and little effect from phage addition would be expected.

Meat Products

In a beef-based trial using phages to control *Escherichia coli* O157:H7, significant decreases in the concentrations of pathogen were demonstrated (O'Flynn et al., 2004). Beefsteak inoculated with approximately 10^2 CFU of *E. coli* O157:H7 was treated with 2×10^8 PFU ml^{-1} of a mixture of three phages. Following incubation for 1 h at 37°C, samples were briefly enriched prior to enumeration. All phage-free control samples were contaminated with *E. coli* O157:H7 at levels of around 10^5 CFU ml^{-1} of enrichment, but the pathogen was undetectable in seven out of nine phage-treated samples. When *E. coli* O157:H7 was present, the numbers were <10 CFU ml^{-1}.

In other experiments, beef surfaces were inoculated with host and phage, incubated, and then ground prior to enumeration (Abuladze et al., 2008). From the data presented, the beef was inoculated with 7.3×10^3 CFU of *E. coli* O157:H7 cm^{-2}, while phages were added at approximately 4×10^6 PFU cm^{-2}. Incubation was for 24 h at 10°C, and a statistically significant reduction of 94.5% was measured with respect to the phage-free control.

The application of phages to inactivate *Salmonella* and *Campylobacter* pathogens on raw and cooked beef surfaces has also been reported (Bigwood et al., 2008). Pathogens were applied at both low and high host concentrations (approximately 10^2 and 10^4 CFU cm^{-2}, respectively) and phages at low and high phage-to-pathogen ratios (10^2 and 10^4, respectively). Samples were incubated at 5 or 24°C to simulate chilled and room-temperature storage. The duration of the experiments was initially 24 h, but they were then repeated for up to 8 days at 5°C to simulate typical extended meat storage conditions. Over 24 h, various reductions of both pathogens were produced in the presence of phages under all conditions, with the greatest reductions occurring at 24°C when both the host concentration and phage-to-pathogen ratio were high. Over an 8-day period at 5°C a slow decline in *Salmonella* numbers was observed in the absence of phages, while *Campylobacter jejuni* populations decreased by 0.3 to 1 $\log_{10}$ CFU cm^{-2}. Both phages reduced pathogen populations, with reductions of the pathogens greatest on raw meat with the hosts at the highest concentration used. These reductions were not, however, as large as those observed in the initial short-term experiments. The explanation of these results is unclear, but it was suggested that during extended storage the moisture content of the meats may have been reduced, hence negatively affecting the phages' ability to diffuse to host cells.

Dykes and Moorhead (2002) reported findings regarding the combined use of nisin and a listeriophage to control *L. monocytogenes* on chilled, vacuum-packaged raw beef. The phage was active against two host isolates, but additional host-range data were not provided. Meat cubes were immersed in a suspension containing the two listerial isolates to achieve an inoculum of 6 $\log_{10}$ CFU cm^{-2} and then immersed in one of the following: nisin (5×10^3 IU ml^{-1}), phage (3×10^3 PFU ml^{-1}), nisin and phage combined, or a buffer control. Samples were dried, vacuum-packed, and incubated for 4 weeks at 4°C. *L. monocytogenes* numbers in control samples did not change over the 28-day duration of storage. Similarly, phage treatment alone resulted in no effect on numbers. Nisin treatment alone initially reduced *Listeria* populations by 1 $\log_{10}$ CFU cm^{-2}, but numbers were static during further incubation. The presence of phage and nisin gave the greatest reduction in pathogen numbers, although by only around 0.6 $\log_{10}$ CFU cm^{-2} more than the nisin-treated meat. The absence of inactivation by phage alone might be explained by the low concentration of phage applied. While a value was not given per unit area of the meat, the concentration in which the cubes were immersed was low and the *L. monocytogenes* did not grown on the meat. The dose-dependent nature of biocontrol exerted by phages on surfaces has been discussed (Hagens and Offerhaus, 2008).

More-encouraging results have been reported with phages infecting *L. monocytogenes* on hot dogs and sliced turkey breast meat (Guenther et al., 2009). In this report the phages were applied at 3×10^8 PFU g^{-1}, which is substantially higher than the concentration used in the work described above. Phage control worked best on hot dogs, where a minimum 2-$\log_{10}$ CFU g^{-1} reduction was recorded, and the reduction in pathogen numbers on turkey breast meat was of the order of 1 $\log_{10}$ CFU g^{-1}. A clear dose-dependent response was demonstrated for phage A511 on hot dogs, where application of 3×10^6 PFU g^{-1} produced a reduction that was less rapid than when 3×10^7 PFU g^{-1} was used, and application of 3×10^8 PFU g^{-1} produced a reduction to beneath the level of detection after 2 days of incubation at 6°C, which persisted for another 4 days.

The first reported applications of phages to poultry skin for the control of *Salmonella* and/or *Campylobacter* were published in 2003. In one (Goode et al., 2003), the effectiveness of phages to reduce populations of *Salmonella* serovar Enteritidis and *C. jejuni* applied separately to chicken skin was evaluated. Skin was inoculated at levels of $\sim 10^3$ and 10^4 CFU cm^{-2} and phages applied at 10^3 and 10^6 PFU cm^{-2}, respectively, and the samples were stored at 4°C for up to 48 h. The concentration of serovar Enteritidis was quite stable on chicken skin over 48 h and phage treatment resulted in reductions of 0.6 and 1 $\log_{10}$ CFU cm^{-2} after 24 and 48 h, respectively. *C. jejuni* survived less well on chicken skin, with a 1-$\log_{10}$ CFU cm^{-2} reduction in concentration being measured. In the presence of phages, host counts were reduced by over 2 $\log_{10}$ CFU cm^{-2} (99.5%) after 24 h of incubation. Further experiments employing a spray inoculation and phage application approach using different *Salmonella* phages with higher plaquing efficiencies resulted in a reduction of serovar Enteritidis to undetectable levels when high phage and low pathogen concentrations were used, although the actual concentrations applied are not clearly defined.

A similar approach was adopted when investigating the reduction of *C. jejuni* on chicken skin using a different phage (Atterbury et al., 2003). Skin sections (2 cm^2) were inoculated with host cells and phages applied at a number of concentrations to create a range of phage-to-pathogen ratios. Treated and untreated skin samples were incubated at either 4 or −20°C for up to 5 days, and counts were made on the entire skin sample. In phage-free controls inoculated at the highest level of *C. jejuni,* counts on skin incubated at 4°C were stable for the duration of storage, but at the lowest inoculum level counts were reduced to beneath the level of enumeration after 3 days. Reductions in the number of *C. jejuni* in the presence of phage were measurable when phages were present at a concentration of 10^7 PFU and the host was present at 10^6 and 10^4 CFU, with reductions of 1.1 to 1.2 $\log_{10}$ CFU occurring after 1 day. Further reductions did not occur when incubation was extended to 5 days. Significantly larger reductions of 2.3 and 2.5 $\log_{10}$ CFU were reported for phage-treated (10^7 PFU) skin samples inoculated with 10^6 and 10^4 CFU *C. jejuni* and frozen at −20°C. It was suggested that the phage activity observed did not occur in situ, but rather occurred when cell growth resumed (i.e., during enumeration).

The control of *Salmonella* serovar Enteritidis on poultry carcasses during processing has been reported by Higgins et al. (2005). A poultry-derived *Salmonella* isolate was used to contaminate processed broilers with 31 and 20 CFU of *Salmonella* per carcass in two separate experiments. Carcasses were then sprayed with 5.5 ml of phage suspensions ranging from 10^4 to 10^{10} PFU ml^{-1}. In the first experiment, salmonellae could be detected in 18 of 19 control carcasses and only 2 of 21 phage-treated carcasses at the highest phage titer applied. In the second experiment, all control carcasses harbored *Salmonella* but only 1 out of 15 yielded the host in the presence of 10^8 and 10^{10} PFU of phage ml^{-1}. Lower phage titers failed to reduce *Salmonella* numbers. In further experiments, phage suspensions containing 72 phage isolates were propagated using different serovar Enteritidis isolates and then applied at 10^7 PFU ml^{-1} rinse to naturally contaminated processed turkeys. In a preliminary experiment, 24 out of 30 control carcasses were found to be naturally contaminated, while only 7 or 8 (out of 28 to 32) phage-rinsed carcasses remained positive for the pathogen. A similar outcome occurred in the second experiment (which differed from the first by the rinse volume used and details of phage and host preparation), with only 2 to 5 positive carcasses detected out of 30 tested when phages were applied. Statistically significant reductions of 50 to 60% carcass prevalence were obtained versus controls.

Further work has examined the effect of phages on the control of *Salmonella* in a processed poultry product, using both the broad-host-range Felix-O1 wild-type phage and a large-plaque-size mutant (Whichard et al., 2003). Chicken frankfurters (10 g) were surface inoculated with 300 CFU of *Salmonella enterica* serovar Typhimurium DT104, and phages applied at a level of 5.25×10^6 PFU per sample. Incubation was for 24 h at 22°C. The pathogen grew to a mean count of 6.8 $\log_{10}$ CFU g^{-1} in control samples, while in the presence of the wild-type and mutant phages, statistically significant reductions resulted in mean populations of 5.0 and 4.7 $\log_{10}$ CFU g^{-1}, respectively. Further experiments conducted at more-realistic storage temperatures using typical pathogen levels need to be done to explore further the potential of phage biocontrol under these circumstances.

Research into phage control of spoilage bacteria has tended to focus on *Pseudomonas,* which is significant in the aerobic spoilage of protein-containing foods. One Canadian group has produced a number of papers on the control of beef spoilage by pseudomonads using phages. Experiments using surface-sanitized rib-eye (longissimus dorsi muscle) steaks (Greer, 1986) found that *Pseudomonas* inoculated alone at around 10^5 CFU cm^{-2} grew exponentially when incubated at 7°C following a 1-day lag period. However, in the presence of phages at a concentration of approximately 7×10^5 PFU cm^{-2}, there was an initial reduction in bacterial numbers and growth did not commence until 2 days had elapsed, after which the growth rate was similar to that of the control. The final bacterial count after 3 days of incubation was significantly less in the phage-treated samples, and the concentration of phages increased during storage. Importantly, when phages were added, both color and acceptability scores declined at a decreased rate, resulting in a 2.9-day shelf life compared to 1.6 days for the phage-free control.

Subsequently the effect of both phage and host numbers on the prevention of spoilage was examined (Greer, 1988); this information is important, as most other studies just use single concentrations of both phages and their hosts. Rib-eye steaks were inoculated with 0 to 9.7×10^5 PFU cm^{-2} phages and 0 to 4.6×10^5 CFU cm^{-2} pseudomonads and stored at temperatures ranging between 1 and 10°C. Assessment of the acceptability of the steaks over time was assessed by a sensory panel. Full sensory data are given for one set of conditions only and show that the addition of phages alone did not reduce the shelf life, as might be expected. At a very low concentration of host (0.46 CFU cm^{-2}), addition of phages again did not affect the shelf life, but when host was inoculated at 4.6×10^1 CFU cm^{-2}, phages added in excess of 9.7×10^5 PFU cm^{-2} significantly extended the shelf life of the steaks. With higher host concentrations, fewer phages than the number cited above resulted in increased shelf life. It was concluded that the use of phages to control beef spoilage was questionable because of the need for relatively high numbers of the host to be present for an effect to be observed. However, with pseudomonads present at only 46 CFU cm^{-2} and phages added at $>9.7 \times 10^5$ CFU cm^{-2}, the shelf life was equivalent to that of the control, and so an effect was achieved at the lowest concentration of pseudomonads causing spoilage. Full control was therefore achieved under conditions most likely to be encountered in "real life" but not when pseudomonads were initially present at high numbers, so leading to the increased likelihood of phage-resistant mutants emerging.

A further paper illustrates the need for phages with a suitably broad host range to be used. Here, a mixture of seven phages added at 10^6 PFU cm^{-2} did not increase the shelf life of naturally contaminated rib-eye steaks (Greer and Dilts, 1990). Given that the phage mixture infected only 25 to 72% of pseudomonad isolates from various meat sources, it is likely that phage-resistant isolates grew and caused spoilage. To be effective, any phage preparation will need to be able to infect a large majority of the extant strains of the bacterial species concerned.

Brochothrix thermosphacta can be a significant spoilage organism of aerobically stored chilled meat (Russo et al., 2006), and phages that infect this bacterium have been isolated from beef (Greer, 1983). Experiments have been performed using phages to increase the shelf life of chilled pork adipose tissue through the control *B. thermosphacta* (Greer and Dilts, 2002). Experiments were performed at 2 and 6°C in the presence or absence of two phages. With host and phages added at equal concentrations (10^5 CFU and PFU cm^{-2}, respectively), an initial decrease in host-cell numbers in phage-treated samples was measured at both temperatures when compared to untreated controls. Regrowth followed at a rate similar to that of the untreated host at 6°C. Bacteria isolated from phage-treated samples incubated for 7 days at 2°C were both phage resistant and phage sensitive, and so the regrowth was most likely due to the presence of phage-resistant mutants. A significant effect on the organoleptic quality of the adipose samples resulted from phage treatment; phage-free control samples were unacceptable after 4 days' incubation at 2°C, while phage-treated samples were acceptable after 8 days' incubation at the same temperature.

Leuconostoc gelidum is a lactic acid bacterium tentatively identified as comprising a major component of the lactic acid bacteria present on chill-stored vacuum-packed beef. While it may not be a significant spoilage organism in this food (Yost and Nattress, 2002), it has been implicated in the spoilage of acetic acid-preserved herrings (Lyhs et al., 2004) and value-added beef products stored under a high-oxygen (70%), balance-carbon dioxide, modified atmosphere (Vihavainen and Björkroth, 2007). Experiments have been conducted to assess the ability of a phage isolated from pork loin to control the growth of this bacterium on pork adipose tissue (Greer et al., 2007). Little control was exerted at 4°C under aerobic or vacuum-packed conditions when phages were added at approximately 10^4 PFU cm^{-2} to 10^6 CFU cm^{-2}, although phages replicated. When the host was inoculated onto aerobically stored and vacuum-packed samples at 10^2 and 10^3 CFU cm^{-2}, respectively, and treated with phages at 10^5 PFU cm^{-2}, there was an initial slight reduction in bacterial concentration followed by growth parallel to that of the host achieved in the absence of phages. Spoilage organisms reached the same maximum concentration after 12 days' incubation in the presence or absence of phages. Phage-resistant host cells were not recovered.

Produce

To date, published reports on attempts to control pathogens in produce have focused largely on *Salmonella* and *L. monocytogenes* growing on fruit surfaces. In particular, Leverentz and colleagues have published a series of papers concerning the use of phages infecting these pathogens inoculated onto cut melon and apple. The first of these publications (Leverentz et al., 2001) assessed the efficacy of applying a proprietary mixture of four phages, SCPLX-1 (Intralytix, Inc., Baltimore, MD), to artificially wounded apple and melon slices inoculated with approximately 2.5 × 10^4 CFU of *Salmonella* serovar Enteritidis. The phage mixture was applied directly to contaminated wound sites at a stated level of ~5 × 10^6 PFU (although 10^8 in the graphs), and slices were incubated at 5, 10, and 20°C for up to 7 days. The growth of serovar Enteritidis in phage-free controls was similar for both apple and melon slices, with no growth at 5°C, some growth at 10°C, and significant growth at 20°C, although growth on melon reached a higher concentration than that on apple.

In the presence of phages, a rapid reduction in the concentration of *Salmonella* on the order of 2 $\log_{10}$ CFU per wound occurred, followed by growth parallel to that of the phage-free control. The most effective application of phages was to wounded melon stored at 5°C, with the concentration of *Salmonella* reducing by 3.5 $\log_{10}$ CFU per wound. However, at incubation temperatures of 10 and 20°C, respectively, the pathogen reached concentrations of ca. 4 and 7 $\log_{10}$ CFU per wound after 7 days of storage. There did not appear to be any phage replication over the 168-h incubation period, and in fact, phage titers declined.

Phage treatment had no effect on *Salmonella* concentrations when compared to the phage-free controls in apple tissue wounds. Phage titers declined rapidly at all incubation temperatures, suggesting that the phages may have been inactivated at the low pH of apple (pH 4.2, compared to pH 5.8 for melon).

Subsequent research examined the use of the bacteriocin nisin and two proprietary phage mixtures, both alone and in combination, to control *L. monocytogenes* on cut melon and apple slices stored at 10°C for up to 7 days (Leverentz et al., 2003). Fruit tissue plugs (10-mm diameter) and squares (30 mm^2) were surface contaminated with *L. monocytogenes* to 1.25×10^4 CFU per sample. The phage preparations, which included LMP-102™ (Intralytix), were applied either by pipetting 25 μl of a 5×10^7 PFU ml^{-1} suspension onto fruit plugs or spraying <250 μl of the same phage suspension onto fruit pieces. Nisin was added to a subset of samples at concentrations of 1, 200, and 400 IU per sample. Limited growth of *L. monocytogenes* (0.6 $\log_{10}$ CFU per sample) was observed in phage-free controls on apple (pH 4.4) stored for 7 days at 10°C, while the concentration increased by 4.5 $\log_{10}$ CFU on melon (pH 5.8).

Both phage and nisin treatments alone were initially able to reduce and then control pathogen concentrations on melon, but between 2 and 5 days of incubation, pathogen growth resumed. Phage treatment alone maintained lower pathogen concentrations on melon than treatment with nisin alone over the last 5 days of storage. The most effective treatment, as assessed by the exertion of the best control of pathogen growth over 7 days, was the combination of phages and 400 IU nisin. These results indicate that while nisin produced an immediate reduction in pathogen concentration (also demonstrated on beef by Dykes and Moorhead [2002]), it is best used in combination with another biocontrol approach to maintain the reduction.

On apple slices, phage titers were reduced to beneath the limit of detection within 30 min of application. However, small, phage-attributed reductions in *L. monocytogenes* concentration were reported in both the presence and absence of nisin. The effect of pH on phage titer was confirmed in broth culture, where it was shown that the phage was most active between pH 6 and 8 and practically inactive at pH 4.5. The combined results from this and the previous work with *Salmonella* suggest that low pH-tolerant phages will be needed to improve efficacy of phage biocontrol on more-acidic foods. Although the effects of nisin addition are usually best at lower pH values, here nisin exerted a lesser impact on pathogen control on apple than on melon, suggesting that the application of nisin may be limited for pathogen control on fruits. Phage-mediated reductions following 5 and 7 days of storage were larger when the initial pathogen inoculum concentration was 10^5 CFU ml^{-1}, compared to 10^6 CFU ml^{-1} in the applied cell suspension, although the phage concentration applied is not clear, and it may be that more phages were added to the samples inoculated with the highest concentration of host cells. Spraying phages onto melon was found to be more effective than pipetting, possibly because of better surface coverage, but the effect was not statistically significant.

Further information on optimization of phage biocontrol on fruits was subsequently published (Leverentz et al., 2004), again using melons and apples. The variables examined included phage concentration (versus pathogen levels), timing of phage application, and the addition of $MnCl_2$. Fruit surfaces (30 mm^2) were treated with *L. monocytogenes* at a level of 1.25×10^4 CFU and then sprayed with 250 μl of phage suspensions ranging from 10^4 to 10^8 PFU ml^{-1} (equivalent to approximately

2.5×10^3 to 2.5×10^7 PFU per sample). Samples were subsequently incubated for up to 7 days at 10°C. To determine whether the timing of phage application is important, phage suspension was pipetted onto fruit surfaces 0, 0.5, and 1 h prior to contamination, and 0.5, 1, 2, and 4 h after. Phage suspensions were also supplemented with $MnCl_2$ before being pipetted onto 10-mm-diameter apple tissue plugs.

Results from preliminary experiments showed a general trend of improved pathogen control with increasing phage concentration, but only when phages were applied at $\geq 10^6$ PFU ml^{-1} ($>8.3 \times 10^4$ PFU cm^{-2}). Only application of phages at the highest titer (10^8 PFU ml^{-1}) resulted in a reduction of the pathogen to a concentration below detectable levels at days 0 and 2 and maintained the concentration at less than 2 log_{10} CFU per sample after 7 days of storage. Subsequent work looking at the timing of phage application used the highest phage titer. Application of phages concurrent with, or up to 1 h before, addition of the pathogen was most successful in controlling growth, with less than 1 log_{10} CFU per sample detectable after 5 days' incubation. Over 7 days of incubation, *L. monocytogenes* was undetectable on fruit inoculated simultaneously with both phage and pathogen, while phage addition 1 h prior to contamination by *L. monocytogenes* eventually resulted in 2.3 log_{10} CFU per sample at the end of storage. Application of phages 0.5 to 4 h after addition of the pathogen successfully reduced *L. monocytogenes* concentrations relative to the controls and maintained concentrations below 2 log_{10} CFU per sample over a 2-day period, but counts increased over the remainder of the incubation period to reach levels of 3.1 to 5.6 log_{10} CFU per sample. Addition of $MnCl_2$ did not improve the effectiveness of phages on apple at 10 and 25°C.

The same group published an abstract describing the use of *Gluconobacter asaii* and a mixture of bacteriophages alone and in combination to control *L. monocytogenes* on honeydew melon (Hong et al., 2006). Treatment with either of these alone was described as effective, but combined treatment reduced *L. monocytogenes* by 6 log_{10} CFU at day 7.

The reduction in concentration exerted by phage A511 on *L. monocytogenes* on cabbage was on the order of 3 to 4 log_{10} CFU g^{-1} (Guenther et al., 2009), but the effect on iceberg lettuce was closer to 2 log_{10} CFU g^{-1} when the phage was added at 3×10^8 PFU g^{-1}. A similar effect on cabbage was demonstrated using phage P100.

An attempt to control salmonellae in quite different produce was reported by Pao et al. (2004). The ability of two phages isolated from sewage to control *S. enterica* serovars Typhimurium, Enteritidis, and Montevideo was evaluated on artificially contaminated, soaked broccoli and mustard seeds at 25°C. Seeds were inoculated by soaking in suspensions containing 10^3 to 10^4 CFU ml^{-1} of pathogen for 2 min, with the resulting seeds containing 10^2 to 10^3 CFU g^{-1} *Salmonella*. For experiments, dried inoculated seeds were soaked for 24 h in water at 25°C containing 1 ml ($\sim 10^8$ PFU) of viable or killed (control) phages. Pathogen reductions of 1.37 and 1.5 log_{10} CFU g^{-1} were achieved using one phage on mustard seeds and both phages on broccoli seeds, respectively. While these reductions are slightly lower than those typically attained by the use of hypochlorite, they appear to be similar to those observed by Leverentz et al. (2001).

Other work with seed sprouts has described an attempt to control *S. enterica* serovar Oranienburg using two phage isolates from sewage (Kocharunchitt et al., 2009). However, the effect measured was modest (around 1 log_{10} CFU g^{-1} compared to the phage-free control) and the pathogen was able to grow to a high concentration in phage-treated samples. The data presented indicated that no significant phage replication occurred during sprouting; in fact, a slight decline in phage concentration occurred, although an initial increase over the first 3 h of seed soaking when compared to phages in host-free control seeds was reported. The marginal control displayed is surprising given that the concentration of phages in the water used to soak the seeds was high (around 10^8 ml^{-1}). However, results from in vitro studies also showed a modest effect on host-cell concentration, with a 2.3-log_{10} CFU reduction occurring in the first 2 h of incubation, followed by parallel growth of *Salmonella* in phage-containing and phage-free cultures. The results for this phage were therefore consistent. The authors present some possible explanations for the lack of inactivation, such as the formation of biofilms protecting the pathogen from phage attack.

The pathogen *E. coli* O157:H7 has received comparatively little attention in terms of control by phages on produce. A conference paper described phage biocontrol on lettuce and melon (Sharma et al., 2008). This work used a commercially available mixture of phages (ECP-100; Intralytix) with incubation at 4°C. Compared to phage-free controls, significant reductions of the pathogen were evident when the phage mixture was used. The effect on lettuce is reported to have been rapid, with an effect noted on day 0, but this was not the case for melon. This might be considered surprising as a larger number of phages was added to the treated melon, although the method of application differed. Given the temperature of incubation, the effect noted cannot have been due to phage replication.

Other work on the control of this organism has focused on broccoli, tomato, and spinach (Abuladze et al., 2008). In each case the concentration of host and phages (a three-isolate cocktail was used) per unit area varied, as they were applied on a per-gram basis, but the reductions in pathogen numbers after incubation at 10°C were very similar, being in the range of 94 to 100% (although the method used would have affected the limit of detection, so the 100% figure presumably would be better represented by "greater than" some percentage as determined by the method). There was some regrowth over the 168-h incubation period, but it was minimal.

The control of *Bacillus cereus* in mashed potato held at 37°C has been described in a recent poster paper (Lee et al., 2008). Phages added at around 10^9 PFU g^{-1} brought about a rapid decline in pathogen numbers, to become undetectable after around 5 h of incubation. When a lower titer was used (approximately 10^6 PFU g^{-1}), there was only a slight reduction of around 1 log_{10} CFU g^{-1} compared to the phage-free control.

Little work has been published on the control of spoilage organisms in produce, and it is not the intention here to cover control of plant pathogens causing disease during growth, for example, fire blight (see chapter 5). Some work has focused on the control of bacterial blotch in mushrooms, which is caused by *Pseudomonas tolaasii*. Infection with this organism can result in losses of 10 to 15% during production and storage (Munsch and Olivier, 1993). Control could be demonstrated under growing conditions, and in an additional study (Munsch and Olivier, 1995) a 30 to 80% reduction in blotch compared to untreated controls was reported. The time of phage administration was determined to be important for spoilage control.

Seafood

Vibrio spp. are common in aquatic habitats and can be detected in many seafoods (see also chapter 13 of this volume). *Vibrio vulnificus* can cause severe and sometimes fatal disease in at-risk individuals when raw or undercooked seafood, especially contaminated oysters (*Crassostrea* spp.), are consumed (Strom and Paranjpye, 2000). Phages infecting *V. vulnificus* can be isolated from estuarine waters and oysters growing in them and have been enumerated at 10^1 to 10^5 PFU g^{-1} in oyster tissue (DePaola et al., 1998). Phage biocontrol of *V. vulnificus* in live shucked oysters using a nine-phage mixture achieved significant reductions of the pathogen in pooled tissue (Luftig and Pelon, 1997). Subsequently, the same mixture was evaluated in vitro for the control of *V. vulnificus,* either alone or with oyster extract (Pelon et al., 2005). Over 18 h, the phage mixture reduced the concentration of *V. vulnificus* from 10^6 to 3 CFU ml^{-1} when incubated at 4°C. The addition of oyster extract had a very small additive effect. Phages could, therefore, potentially be used as a biodepuration treatment for *V. vulnificus* in oysters and other shellfish.

The addition of phages A511 and P100 to mixed seafood at a concentration of 3×10^8 PFU g^{-1} resulted in a significant reduction in two different isolates of *L. monocytogenes* (Guenther et al., 2009), but the effect was not as marked when used on smoked salmon under the same conditions (6 days at 6°C).

PHAGES AND BIOFILMS

The formation of biofilms can pose significant problems to the food industry (Poulsen, 1999). Of special interest has been biofilms produced by *L. monocytogenes* because of the organism's ability to grow at low temperatures and so potentially colonize food manufacturing plant and equipment (Borucki et al., 2003). Once established in biofilms, bacteria are more refractory to the effects of disinfectants. A property of some phages that makes them candidates for biofilm disinfection is their possession of extracellular polymeric substances (EPS)-degrading enzymes capable of breaking down the structure of the biofilm (Hughes et al., 1998a, 1998b). The use of this enzymatic activity in combination with detergents has been shown to remove biofilms effectively (Tait et al., 2002).

Phages infecting *L. monocytogenes* have been assessed alone and in combination with a quaternary ammonium compound (QAC) for their efficacy in inactivating *L. monocytogenes* in biofilms (Roy et al., 1993). Three phages applied to biofilms on polypropylene and stainless steel surfaces individually produced reductions of approximately 3.4 $\log_{10}$, while when the phages were used as a mixture the reduction was improved. The QAC when used alone also resulted in removal of the pathogen, but when used in combination with the three phages the same result could be obtained using a lower concentration of the QAC, and when used in combination with phages reductions approaching 5 $\log_{10}$ were recorded. The phages themselves were stable in the QAC prepared at the concentration used in these experiments. Phages have been bred to inactivate L-forms of *L. monocytogenes* colonizing stainless steel surfaces (Hibma et al., 1997), producing a 3-$\log_{10}$ reduction.

A spoilage organism of significance to the food processing industry, *Pseudomonas fluorescens,* frequently forms biofilms and has been the subject of phage biocontrol studies (Sillankorva et al., 2008). Phage ϕIBB-PF7 was able to reduce biofilm biomass on stainless steel slides by 63 to 91% within 4 h of application. Phage treatment was equally effective against established and new biofilms under both static and dynamic (medium replenishment and shaking) growth conditions for the host. Under static conditions significantly more phage remained in the biofilm and cell reduction was slightly (ca. 1 $\log_{10}$ CFU ml^{-1}) higher than in dynamic conditions. Phages also appeared to be adhering to the stainless steel under static, but not dynamic, conditions. A limitation of the experiments is that by stopping measurements at 4 h the potential for formation of phage-resistant colonies was not explored.

In contrast, the application of a phage to a biofilm of *E. coli* O157:H7 proved to be of marginal efficacy (Sharma et al., 2005). However, the possession or otherwise of an EPS-degrading activity was not determined, and in other work with *E. coli,* phage T4 was shown to infect and replicate in biofilms formed on polyvinyl chloride coupons (Doolittle et al., 1995).

The desirability of the expression of an EPS-degrading enzyme has been shown in work comparing the biofilm-degrading activity of phages that did and did not express such an activity (Lu and Collins, 2007). Phage T7 was engineered to contain the gene *dspB,* encoding dispersin B, which was expressed during infection, and the activity of the engineered phage was compared with that of controls. The dispersin B-expressing phage was able to reduce the number of cells in biofilms to a greater degree than a control that did not express the gene. The rate of reduction of *E. coli* in the biofilm was also greater with the phage expressing the polymer-degrading gene.

The possibilities of using phages to inactivate bacteria in biofilms have not yet been explored in depth. A question to be addressed is whether biofilms formed from numerous bacterial species, therefore possibly containing many types of EPS, will be amenable to phage attack if the target species is physically protected (Tait et al., 2002).

CONCLUSIONS

For phages to be used efficiently for biocontrol or as biopreservatives, it is necessary to understand how the effect is being mediated. If control is dependent on phage multiplication in situ, then the bacterial host may need to be growing. While spoilage organisms are likely to be growing on stored foods and evidence suggesting virulent infection of spoilage bacteria is available (Greer and Dilts, 2002), most food-borne pathogens do not grow at temperatures lower than 7°C (with the notable exceptions of *L. monocytogenes* and *Yersinia enterocolitica*). The replication of phages on most pathogens cannot, therefore, be assumed under most conditions typical for storage of ready-to-eat foods.

Given that inactivation of pathogens has been frequently reported following the addition of phages, then some other mechanism(s) must be operating. If sufficient phages are applied, then lysis from without could occur. This involves the adsorption of a large number of phages to the bacterial cell, resulting in its rapid lysis without the need for phage replication. Indeed, many reports describing successful biocontrol on foods use very large numbers of phages. For example, control of *E. coli* O157:H7 on beef was demonstrated from the application of 2×10^8 PFU phages per piece of meat inoculated with 2×10^2 CFU of the host (O'Flynn et al., 2004). Additionally, reductions often seem to occur rapidly, for example, those recorded on melon (Leverentz et al., 2001) and chicken skin (Atterbury et al., 2003). Both of these observations are consistent with lysis from without being the mechanism behind pathogen inactivation. However, lysis from without did not occur when coliphage T4 was incubated with *E. coli* at temperatures lower than 14°C (Tarahovsky et al., 1994), and other work has shown that a phage infecting *Enterobacter* did not produce lysis of the host when added in high numbers at 4°C; instead, lysis occurred during subsequent culture and enumeration (Verthé and Verstraete, 2006).

This last observation suggests that as long as a cell is infected by a phage then lysis of the cell is imminent on the resumption of growth. Therefore, if an infected pathogen resumes growth following consumption, it will be killed before colonization of the human host can occur. The high numbers of phages that are required to control pathogens in foods are therefore probably needed simply to ensure that a sufficient proportion of host cells become infected by the applied phages. In liquid foods, and at realistic (i.e., low) pathogen concentrations, high phage concentrations (or long incubation periods) are required to ensure that a sufficient MOI_{actual} is achieved (Kasman et al., 2002). This view has received support in a recent paper where the proportion of cells infected under conditions where growth did not occur and subsequent inactivation on plating were in accord with consideration of MOI_{actual} (Bigwood et al., 2009). This paper also showed that the inactivation measured was independent of the concentration of host cells (at concentrations of $<10^4$ CFU ml^{-1} at least) but correlated with the concentration of phages used.

The need for high concentrations of phages to mediate significant biocontrol on food surfaces has also been the subject of discussion. Data have been presented illustrating a dose-dependent response for the inactivation of *L. monocytogenes* on salmon (Hagens and Offerhaus, 2008) and other foods (Guenther et al., 2009). The authors suggest that, for food surfaces, consideration needs to be given to the probability of infecting a cell in a given unit area. In addition to consideration of co-location of phage and host, phages may also diffuse and the degree of diffusion is likely to depend on the presence of water. Other physiochemical conditions of the food may also influence the ability of phages to infect their hosts. The logical conclusion would seem to be that moist foods will need fewer phages to be applied to produce a given kill compared to drier food.

For control of organisms that are growing, an additional consideration will be the effect of phages produced from virulent infections, although under normal conditions with few pathogens present the contribution made by lysis may be minimal. Any positive effect of phage replication may be negated by the possibility that phage-resistant cells could grow.

It is necessary that a broad range of subtypes of the host organism be infected by the phage preparation used. It seems likely that phage mixtures will need to be used to meet this requirement, and most of the published work takes this approach. The development of phage mixtures will need to consider both the host ranges of the phages included in it and the diversity of bacterial cell surface molecules to which they adsorb in order to negate resistant cell outgrowth (Yoichi et al., 2004).

The combined use of different biological controls with varying modes of action is intuitively attractive, as the likelihood of resistant cells emerging should be reduced, although some caution has been expressed because of the potential for cross-resistance to occur (Martínez et al., 2008). Phages could, therefore, be used in tandem with other biologically derived controls such as bacteriocins. Relatively little work has been reported on this approach, with two papers assessing the combination of phages and nisin for control of *L. monocytogenes* (Dykes and Moorhead, 2002; Leverentz et al., 2003) and one for control of *S. aureus* (Martínez et al., 2008). In the first two of these papers the combined effect was marginal, but the potential for synergistic activity was shown with phage and bacteriocin in milk (Martínez et al., 2008). A similar effect was shown on *L. monocytogenes* when phages were used in combination with the bacterial antagonist *G. asaii* (Hong et al., 2006), as described previously.

In summary, the data indicate that phages show promise as biocontrol agents for the control of food-borne pathogens and spoilage bacteria, especially since their mode of action is becoming better understood. That the concept has already moved from the laboratory to commercial products with appropriate regulatory approvals suggests that, in the future, the application of phages may become a routine control option for at least some sectors of the food industry.

REFERENCES

Abuladze, T., M. Li, M. Y. Menetrez, T. Dean, A. Senecal, and A. Sulakvelidze. 2008. Bacteriophages reduce experimental contamination of hard surfaces, tomato, spinach, broccoli and ground beef by *Escherichia coli* O157:H7. *Appl. Environ. Microbiol.* **74:**6230–6238.

Atterbury, R. J., P. L. Connerton, C. E. R. Dodd, C. E. D. Rees, and I. F. Connerton. 2003. Application of host-specific bacteriophages to the surface of chicken skin leads to a reduction in recovery of *Campylobacter jejuni*. *Appl. Environ. Microbiol.* **69:**6302–6306.

Bigwood, T., J. A. Hudson, and C. Billington. 2009. Influence of host and phage concentration on the inactivation of foodborne pathogenic bacteria by two bacteriophages. *FEMS Microbiol. Lett.* **291:**59–64.

Bigwood, T., J. A. Hudson, C. Billington, G. V. Carey-Smith, and J. A. Heinemann. 2008. Phage inactivation of foodborne pathogens on cooked and raw meat. *Food Microbiol.* **25:**400–406.

Borucki, M. K., J. D. Peppin, D. White, F. Loge, and D. R. Call. 2003. Variation in biofilm formation among strains of *Listeria monocytogenes*. *Appl. Environ. Microbiol.* **69:**7336–7342.

Carlton, R. M., W. H. Noordman, B. Biswas, E. D. de Meester, and M. J. Loessner. 2005. Bacteriophage P100 for control of *Listeria monocytogenes* in foods: genome sequence, bioinformatic analysis, oral toxicity study, and application. *Regul. Toxicol. Pharmacol.* **43:**301–312.

DePaola, A., M. L. Motes, A. M. Chan, and C. A. Suttle. 1998. Phages infecting *Vibrio vulnificus* are abundant and diverse in oysters *(Cras-*

sostrea virginica) collected from the Gulf of Mexico. *Appl. Environ. Microbiol.* **64:**346–351.

Doolittle, M. M., J. J. Cooney, and D. E. Caldwell. 1995. Lytic infection of *Escherichia coli* biofilms by bacteriophage T4. *Can. J. Microbiol.* **41:** 12–18.

Dykes, G. A., and S. M. Moorhead. 2002. Combined antimicrobial affect of nisin and listeriophage against *Listeria monocytogenes* in broth but not in buffer or on raw beef. *Int. J. Food Microbiol.* **73:** 71–81.

EBI Food Safety. 2007. FDA extends GRAS approval LISTEX™ to all food products. July 3. EBI Food Safety, Wageningen, The Netherlands. http://www.ebifoodsafety.com/images/FDA%20and%20USDA%20GRAS%20Approval%20for%20LISTEX%20on%20all%20Foods_July%203%202007.pdf.

Ellis, D. E., P. A. Whitman, and R. T. Marshall. 1973. Effect of homologous bacteriophage on growth of *Pseudomonas fragi* WY in milk. *Appl. Microbiol.* **25:**24–25.

Federal Register. 2006. Food additives permitted for direct addition to human consumption; bacteriophage preparation, final rule. 21 CFR part 172. *Fed. Regist.* **71:**47729–47732.

García, P., C. Madera, B. Martínez, and A. Rodríguez. 2007. Biocontrol of *Staphylococcus aureus* in curd manufacturing processes using bacteriophages. *Int. Dairy J.* **17:**1232–1239.

Gill, J. J., P. M. Sabour, K. E. Leslie, and M. W. Griffiths. 2006. Bovine whey proteins inhibit the interaction of *Staphylococcus aureus* and bacteriophage K. *J. Appl. Microbiol.* **101:**377–386.

Goode, D., V. M. Allen, and P. A. Barrow. 2003. Reduction of experimental *Salmonella* and *Campylobacter* contamination of chicken skin by application of lytic bacteriophage. *Appl. Environ. Microbiol.* **69:**5032–5036.

Greer, G. G. 1988. Effects of phage concentration, bacterial density, and temperature on phage control of beef spoilage. *J. Food Sci.* **53:**1226–1227.

Greer, G. G. 1986. Homologous bacteriophage control of *Pseudomonas* growth and beef spoilage. *J. Food Prot.* **49:**104–109.

Greer, G. G. 1983. Psychrotrophic *Brochothrix thermosphacta* bacteriophages isolated from beef. *Appl. Environ. Microbiol.* **46:**245–251.

Greer, G., and B. D. Dilts. 2002. Control of *Brochothrix thermosphacta* spoilage of pork adipose tissue using bacteriophages. *J. Food Prot.* **65:**861–863.

Greer, G. G., and B. D. Dilts. 1990. Inability of a bacteriophage pool to control beef spoilage. *Int. J. Food Microbiol.* **10:**331–342.

Greer, G. G., B. D. Dilts, and H.-W. Ackermann. 2007. Characterization of a *Leuconostoc gelidum* bacteriophage from pork. *Int. J. Food Microbiol.* **114:**370–375.

Guenther, S., D. Huwyler, S. Richard, and M. J. Loessner. 2009. Virulent bacteriophage for efficient biocontrol of *Listeria monocytogenes* in ready-to-eat foods. *Appl. Environ. Microbiol.* **75:**93–100.

Hagens, S., and M. L. Offerhaus. 2008. Bacteriophages—new weapons for food safety. *Food Technol.* **April:**46–54.

Hibma, A. M., S. A. A. Jassim, and M. W. Griffiths. 1997. Infection and removal of L-forms of *Listeria monocytogenes* with bred bacteriophage. *Int. J. Food Microbiol.* **34:**197–207.

Higgins, J. P., S. E. Higgins, K. L. Guenther, W. Huff, A. M. Donoghue, D. J. Donoghue, and B. M. Hargis. 2005. Use of a specific bacteriophage treatment to reduce *Salmonella* in poultry. *Poult. Sci.* **84:**1141–1145.

Hong, Y., B. Leverentz, W. S. Conway, W. J. Janisiewicz, M. Abadias, and M. Camp. 2006. Biocontrol of *Listeria monocytogenes* on fresh-cut honeydew melon using a bacterial antagonist and a bacteriophage. *Phytopathology* **96:**S51.

Hughes, K. A., I. W. Sutherland, J. Clark, and M. V. Jones. 1998a. Bacteriophage and associated polysaccharide depolymerases—novel tools for study of bacterial biofilms. *J. Appl. Microbiol.* **85:** 583–590.

Hughes, K. A., I. W. Sutherland, and M. V. Jones. 1998b. Biofilm susceptibility to bacteriophage attack: the role of phage-borne polysaccharide depolymerase. *Microbiology* **144:**3039–3047.

Iversen, C., A. Lehner, N. Mullane, J. Marugg, S. Fanning, R. Stephen, and H. Joosten. 2007. Identification of "*Cronobacter*" spp. *(Enterococcus sakazakii). J. Clin. Microbiol.* **45:**3814–3816.

Kasman, L. M., A. Kasman, C. Westwater, J. Dolan, M. G. Schmidt, and J. S. Norris. 2002. Overcoming the phage replication threshold: a mathematical model with implications for phage therapy. *J. Virol.* **76:**5557–5564.

Kim, K.-P., J. Klumpp, and M. J. Loessner. 2007. *Enterobacter sakazakii* bacteriophages can prevent bacterial growth in reconstituted infant formula. *Int. J. Food Microbiol.* **115:**195–203.

Kocharunchitt, C., T. Ross, and D. L. McNeill. 2009. Use of bacteriophages as biocontrol agents to control *Salmonella* associated with seed sprouts. *Int. J. Food Microbiol.* **128:**453–459.

Lee, W. J., J. A. Hudson, J. A. Heinemann, C. Billington, and L. McIntyre. 2008. Isolation of bacteriophages infecting Gram-positive foodborne pathogens, abstr. P4-69. *Abstr. 35th Annu. Meet. Int. Assoc. Food Prot.* International Association for Food Protection, Columbus, OH.

Leverentz, B., W. S. Conway, Z. Alavidze, W. J. Janisiewicz, Y. Fuchs, M. J. Camp, E. Chighladze, and A. Sulakvelidze. 2001. Examination of bacteriophage as a biocontrol method for *Sal-*

monella on fresh-cut fruit: a model study. *J. Food Prot.* **64:**1116–1121.

Leverentz, B., W. S. Conway, M. J. Camp, W. J. Janisiewicz, T. Abuladze, M. Yang, R. Saftner, and A. Sulakvelidze. 2003. Biocontrol of *Listeria monocytogenes* on fresh-cut produce by treatment with lytic bacteriophages and a bacteriocin. *Appl. Environ. Microbiol.* **69:**4519–4526.

Leverentz, B., W. S. Conway, W. J. Janisiewicz, and M. J. Camp. 2004. Optimizing concentration and timing of a phage spray application to reduce *Listeria monocytogenes* on honeydew melon tissue. *J. Food Prot.* **67:**1682–1686.

Loessner, M. J., and R. M. Carlton. October 2008. Virulent phages to control *Listeria monocytogenes* in foodstuffs and in food processing plants. U.S. patent 7,438,901.

Lu, T. K., and J. J. Collins. 2007. Dispersing biofilms with engineered enzymatic bacteriophage. *Proc. Natl. Acad. Sci. USA* **104:**11197–11202.

Luftig, R. B., and W. Pelon. 1997. Bacteriophage biodepuration of *Vibrio vulnificus*-containing shellfish. *J. Shellfish Res.* **16:**271.

Lyhs, U., J. M. K. Koort, H.-S. Lundström, and K. J. Björkroth. 2004. *Leuconostoc gelidum* and *Leuconostoc gasicomitatum* strains dominated the lactic acid bacterium population associated with strong slime formation in an acetic-acid herring preserve. *Int. J. Food Microbiol.* **90:**207–218.

Mandeville, R., M. Griffiths, L. Goodridge, L. McIntyre, and T. T. Ilenchuk. 2003. Diagnostic and therapeutic applications of lytic phages. *Anal. Lett.* **26:**3241–3259.

Martínez, B., J. M. Obeso, A. Rodríguez, and P. García. 2008. Nisin-bacteriophage corresistance in *Staphylococcus aureus*. *Int. J. Food Microbiol.* **122:**253–258.

Modi, R., Y. Hirvi, A. Hill, and M. W. Griffiths. 2001. Effect of phage on survival of *Salmonella* Enteritidis during manufacture and storage of cheddar cheese made from raw and pasteurised milk. *J. Food Prot.* **64:**927–933.

Munsch, P., and J. M. Olivier. 1995. Biocontrol of bacterial blotch of the cultivated mushroom with lytic phages: some practical considerations. *Mushroom Sci.* **14:**595–602.

Munsch, P., and J. M. Olivier. 1993. Biological control of mushroom bacterial blotch with bacteriophages, p. 469. *In* B. Fritig and M. Legrand (ed.), *Mechanisms of Plant Defence Responses*. Kluwer Academic Publishers, Dordrecht, The Netherlands.

Obeso, J. M., B. Martínez, A. Rodríguez, and P. García. 2008. Lytic acitvity of the recombinant staphylococcal bacteriophage ΦH5 endolysin active against *Staphylococcus aureus* in milk. *Int. J. Food Microbiol.* **128:**212–218.

O'Flaherty, S., A. Coffey, W. J. Meaney, G. F. Fitzgerald, and R. P. Ross. 2005. Inhibition of bacteriophage K proliferation on *Staphylococcus aureus* in raw bovine milk. *Lett. Appl. Microbiol.* **41:** 274–279.

O'Flynn, G., R. P. Ross, G. F. Fitzgerald, and A. Coffey. 2004. Evaluation of a cocktail of three bacteriophages for biocontrol of *Escherichia coli* O157:H7. *Appl. Environ. Microbiol.* **70:**3417–3424.

Pao, S., S. P. Randolph, E. W. Westbrook, and H. Shen. 2004. Use of bacteriophages to control *Salmonella* in experimentally contaminated sprout seeds. *J. Food Sci.* **69:**M127–M130.

Pelon, W., R. B. Luftig, and K. H. Johnston. 2005. *Vibrio vulnificus* load reduction in oysters after combined exposure to *Vibrio vulnificus*-specific bacteriophage and to an oyster extract component. *J. Food Prot.* **68:**1188–1191.

Poulsen, L. V. 1999. Microbial biofilm in food processing. *Food Sci. Technol.* **32:**321–326.

Rees, C. E. D., and C. E. R. Dodd. 2006. Phage for rapid detection and control of bacterial pathogens. *Adv. Appl. Microbiol.* **59:**159–186.

Roy, B., H.-W. Ackerman, S. Pandian, G. Picard, and J. Goulet. 1993. Biological inactivation of adhering *Listeria monocytogenes* by listeriaphages and quaternary ammonium compound. *Appl. Environ. Microbiol.* **59:**2914–2917.

Russo, F., D. Ercolini, G. Mauriello, and F. Villani. 2006. Behaviour of *Brochothrix thermosphacta* in presence of other meat spoilage microbial groups. *Int. J. Food Microbiol.* **23:**797–802.

Schellekens, M. M., J. Wouters, S. Hagens, and J. Hugenholtz. 2007. Bacteriophage P100 application to control *Listeria monocytogenes* on smeared cheese. *Milchwissenschaft* **62:**284–287.

Sharma, M., J. Patel, W. S. Conway, S. Ferguson, and A. Sulakvelidze. 2008. Biocontrol of *Escherichia coli* O157:H7 on fresh-cut lettuce and cantaloupe by treatment with bacteriophage, abstr. P5-25. *Abstr. 35th Annu. Meet. Int. Assoc. Food Prot.* International Association for Food Protection, Columbus, OH.

Sharma, M., J.-H. Ryu, and L. R. Beuchat. 2005. Inactivation of *Escherichia coli* O157:H7 in biofilm on stainless steel by treatment with and alkaline cleaner and bacteriophage. *J. Appl. Microbiol.* **99:**449–459.

Sillankorva, S., P. Neubauer, and J. Azeredo. 2008. *Pseudomonas fluorescens* biofilms subjected to phage phiIBB-PF7A. *BMC Biotechnol.* **8:**79.

Strom, M. S., and R. N. Paranjpye. 2000. Epidemiology and pathogenesis of *Vibrio vulnificus*. *Microbes Infect.* **2:**177–188.

Tait, K., L. C. Skillman, and I. W. Sutherland. 2002. The efficacy of bacteriophage as a method of biofilm eradication. *Biofouling* **18:**305–311.

Tarahovsky, Y. S., G. R. Ivanitsky, and A. A. Khusainov. 1994. Lysis of *Escherichia coli* cells induced by bacteriophage T4. *FEMS Microbiol. Lett.* **122:**195–200.

Tarantino, L. M. 2007. Agency response letter GRAS Notice No. GRN 000218. www.fda.gov/Food/FoodIngredientsPackaging/Generally RecognizedasSafeGRAS/GRASListings/ucm153865.htm.

Verthé, K., and W. Verstraete. 2006. Use of flow cytometry for analysis of phage-mediated killing of *Enterobacter aerogenes*. *Res. Microbiol.* **157:**613–618.

Vihavainen, E. J., and K. J. Björkroth. 2007. Spoilage of value-added, high-oxygen modified-atmosphere packaged raw beef steaks by *Leuconostoc gasicomitatum* and *Leuconostoc gelidum*. *Int. J. Food Microbiol.* **119:**340–345.

Whichard, J. M., N. Sriranganathan, and F. W. Pierson. 2003. Suppression of *Salmonella* growth by wild-type and large-plaque variants of bacteriophage Felix O1 in liquid culture and on chicken frankfurters. *J. Food Prot.* **66:**220–225.

Yoichi, M., M. Morita, K. Mizoguchi, C. R. Fischer, H. Unno, and Y. Tanji. 2004. The criterion for selecting effective phage for *Escherichia coli* O157:H7 control. *Biochem. Eng. J.* **19:**221–227.

Yost, C. K., and F. M. Nattress. 2002. Molecular typing techniques to characterize the development of lactic acid bacteria community on vacuum-packaged beef. *Int. J. Food Microbiol.* **72:**97–105.

Zuber, S., C. Boisson-Delaporte, L. Michot, C. Iversen, B. Diep, H. Brüssow, and P. Breeuwer. 2008. Decreasing *Enterobacter sakazakii* (*Cronobacter* spp.) food contamination level with bacteriophages: prospects and problems. *Microb. Biotechnol.* **1:**532–543.

BACTERIOPHAGE LYTIC ENZYMES AS ANTIMICROBIALS

Caren J. Stark, Richard P. Bonocora, James T. Hoopes, and Daniel C. Nelson

8

INTRODUCTION

"Cell wall hydrolase" is a generic term used to describe a wide range of lytic enzymes that act upon the bacterial peptidoglycan, which along with the surface carbohydrate is a component of the cell wall. Many of these enzymes, called autolysins, are encoded by bacteria and are used for growth, division, maintenance, and repair of the peptidoglycan. Bacteriophages, or phages, also encode their own cell wall hydrolases called endolysins or simply lysins, which along with holins comprise the bacteriophage "lytic system" (Young, 1992). During the latter stages of a phage infection cycle within a host organism, holins are produced to perforate the bacterial membrane, allowing the accumulating lysins access to the highly cross-linked peptidoglycan, which functions to contain the elevated cytoplasmic pressure (~3 to 5 atm) found in most bacterial cells. The lysins then cleave the exposed peptidoglycan, resulting in osmotic lysis of the bacterial cell and liberation of progeny phage.

Bacteriophages represent a powerful tool for the control and treatment of food-borne and other pathogens, a topic discussed in several other chapters of this book. However, phage-encoded lysins themselves can also be exploited for their bacteriolytic activity and should be considered as an alternative to whole-phage-based applications (Fig. 1). Appreciably, exogenous addition of purified lysins to susceptible gram-positive bacteria produces "lysis from without," a term used to describe bacterial lysis in the absence of bacteriophage. This phenomenon is the basis for the use of lysins as antimicrobial agents for gram-positive pathogens (for reviews, see Fischetti, 2005, 2008; Fischetti et al., 2006; and Loessner, 2005). Due to the presence of the outer membrane in gram-negative bacteria, an exogenously added lysin will not gain access to the cell wall without surfactant or some other mechanism to translocate the membrane. As such, lysins generally do not produce lysis from without on gram-negative organisms, although we discuss potential strategies to achieve this at the end of this chapter.

Caren J. Stark, James T. Hoopes, and Daniel C. Nelson, Center for Advanced Research in Biotechnology, University of Maryland Biotechnology Institute, Rockville, MD 20850. *Richard P. Bonocora,* Laboratory of Molecular and Cellular Biology, National Institute of Diabetes and Digestive and Kidney Diseases, National Institutes of Health, Bethesda, MD 20892.

Bacteriophages in the Control of Food- and Waterborne Pathogens
Edited by Parviz M. Sabour and Mansel W. Griffiths, © 2010 ASM Press, Washington, DC

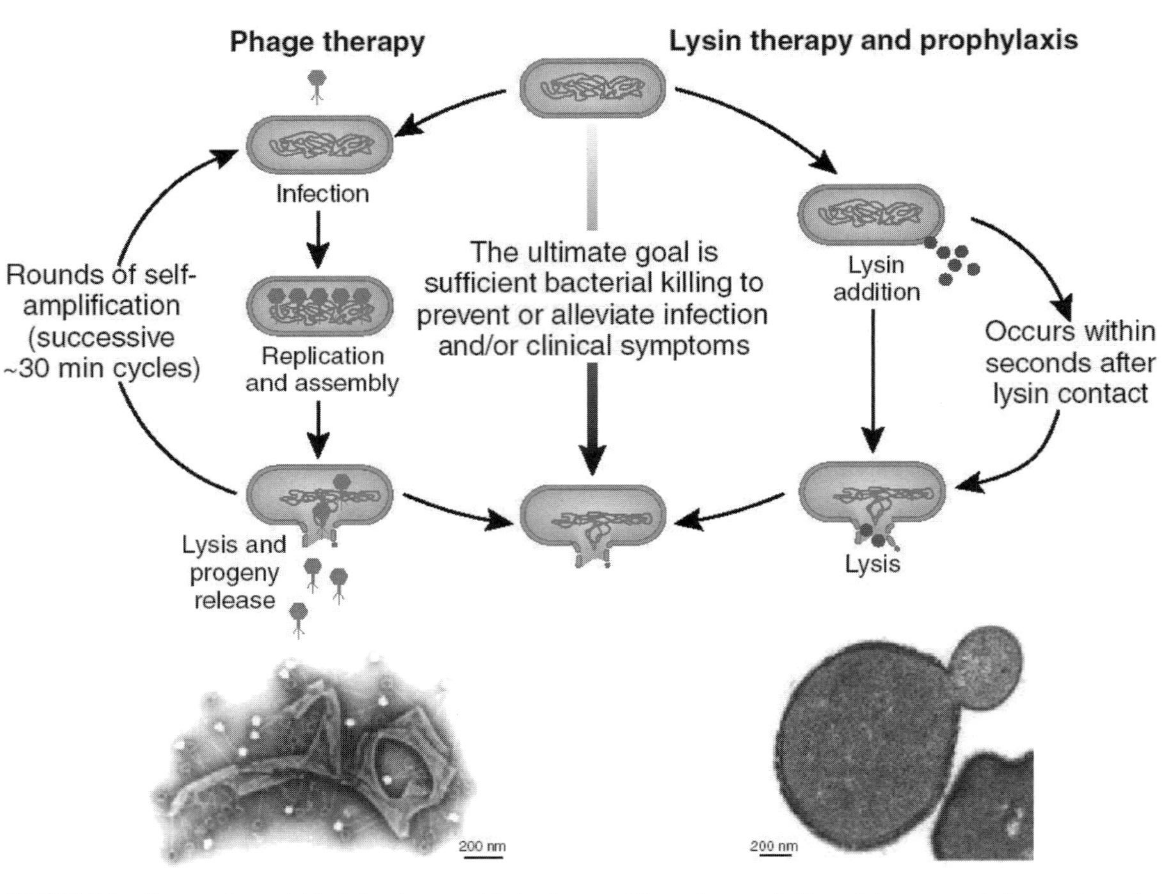

FIGURE 1 Steps to bacterial lysis in phage and lysin therapy. Phage therapy (left) exploits a natural phage lytic cycle, which occurs over 30 min and is divided into three major steps, including the release of new virions into the environment. Subsequent infection of new hosts illustrates the process of self-amplification. The electron micrograph depicts phage particles adhering to the debris of a lysed streptococcal cell. In comparison, lysin therapy and prophylaxis (right) are defined by only two steps, in which purified lysin binds to and rapidly kills, through osmotic lysis, the target pathogen. The electron micrograph depicts a cross section of *B. anthracis* treated with the purified PlyG lysin showing an externalized cytoplasmic membrane just before lysis. (Reprinted from Fischetti et al. [2006] with permission of the publisher.)

Peptidoglycan Architecture and Lysin Catalytic Domains

The bacterial peptidoglycan is a polymer composed of alternating *N*-acetylmuramic acid (MurNAc) and *N*-acetylglucosamine (GlcNAc) residues coupled by β(1→4) linkages, which are in turn covalently linked to a short stem peptide through an amide bond between MurNAc and L-alanine (Fig. 2). The glycan polymer displays little variation between bacterial species. The stem peptides are composed of alternating L- and D-form amino acids that are fairly well conserved in gram-negative organisms, but fluctuate in length and composition for gram-positive organisms. A *meso*-diaminopimelic acid (mDAP) residue is present at position number 3 of the stem peptide in gram-negative species and some species of gram-positive bacteria (i.e., *Bacillus* and *Listeria* spp.). In these organisms, mDAP directly cross-links to the terminal D-Ala of the opposite peptide chain. For most other gram-positive bacteria, the third residue is L-Lys, which links to the opposite peptide chain

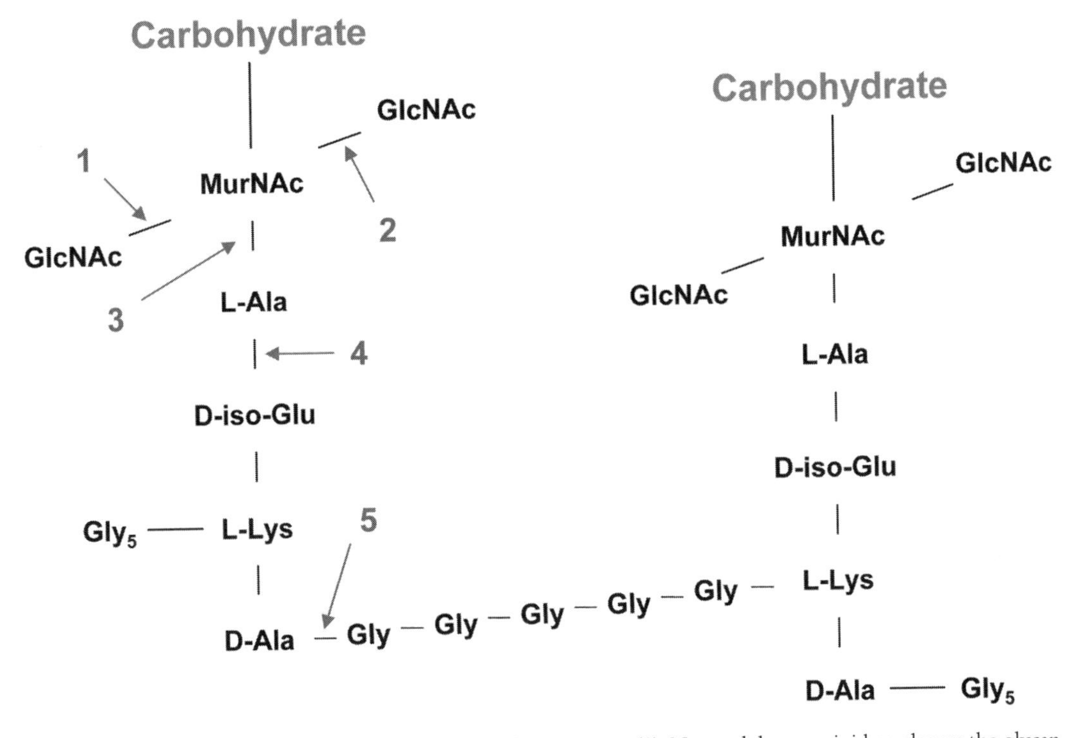

FIGURE 2 Bacterial peptidoglycan structure and lysin targets. (1) *N*-acetylglucosaminidase cleaves the glycan component of the peptidoglycan on the reducing side of GlcNAc. (2) *N*-acetylmuramidase likewise cleaves the glycan component of the peptidoglycan, but on the reducing side of MurNAc. This activity is commonly referred to as a muramidase or lysozyme. (3) An *N*-acetylmuramoyl-L-alanine amidase cleaves a critical amide bond between the glycan moiety (MurNAc) and the peptide moiety (L-alanine) of the cell wall. This activity is sometimes referred to generically as an amidase. However, a true endopeptidase, or protease, will also cleave an amide bond, but only if it is between two amino acids. This type of activity may occur in the stem peptide (4) of the cell wall or in an interpeptide bridge (5) connecting two cell wall fragments. CBDs typically bind the peptidoglycan-associated carbohydrate or an epitope directly related to the peptidoglycan structure. Note, the structure of the *S. aureus* cell wall, which is distinguished by a pentaglycine interpeptide bridge, is shown for illustration purposes. Other bacterial species have interpeptide bridges composed of different amino acids or may lack an interpeptide bridge all together. In these organisms, an mDAP replaces L-Lys and directly cross-links to the terminal D-Ala of the opposite peptide chain.

through an interpeptide bridge, the composition of which varies between species. For example, the interpeptide bridge of *Staphylococcus aureus* is composed of pentaglycine (depicted in Fig. 2), whereas the interpeptide bridge of *Streptococcus pyogenes* is dialanine.

Due to the moderately conserved overall structure of the peptidoglycan, lysins and other cell wall lytic enzymes are limited in the nature and location of covalent bonds that are available for hydrolysis (Fig. 2). There are three basic hydrolytic activities associated with lysins: glycosidic, proteolytic, and that of a specific amidohydrolase. One type of glycosidic activity, known as an *N*-acetylglucosaminidase, cleaves the glycan component of the peptidoglycan on the reducing side of GlcNAc (Fig. 2, target 1). This type of activity is frequently found in autolysins, such as AltA from *Enterococcus faecalis* (Mesnage et al., 2008) or AcmA, AcmB, AcmC, and AcmD from *Lactococcus lactis* (Steen et al., 2007). However, it is not commonly associated with phage lysins, although at least one, the streptococcal LambdaSa2 lysin, has been reported in the literature (Pritchard et al., 2007). A second type of glycosidic activity is an *N*-acetylmuramidase, which cleaves the glycan component of the peptidoglycan on the reducing side of MurNAc (Fig. 2, target 2). This activity is commonly referred to as a muramidase or lysozyme and is frequently found in both autolysins and phage lysins, including the pneumococcal Cpl-1 lysin (Garcia et al., 1987) and the streptococcal B30 lysin (Pritchard et al., 2004). The amidohydrolase activity is that of an *N*-acetylmuramoyl-L-alanine amidase, which cleaves a critical amide bond between the glycan moiety (MurNAc) and the peptide moiety (L-alanine) of the peptidoglycan (Fig. 2, target 3). This activity is more often associated with bacteriophage lysins than autolysins. The reasons for this are not clear; however, because hydrolysis of this bond separates the glycan polymer from the stem peptide, this activity is speculated to be more destabilizing to the peptidoglycan than hydrolysis of other bonds and, therefore, is evolutionarily favored by bacteriophages. Finally, peptidoglycan hydrolases can also be true endopeptidases, cleaving peptide bonds between two amino acids. This cleavage may occur in the stem peptide, such as the *Listeria* Ply500 and Ply118 L-alanyl-D-glutamate endopeptidases (Loessner et al., 1995) (Fig. 2, target 4), or in the interpeptide bridge, such as the staphylococcal ϕ11 D-alanyl-glycyl endopeptidase (Navarre et al., 1999) (Fig. 2, target 5).

All lysins contain at least one catalytic domain that displays one of the aforementioned activities. Some streptococcal and staphylococcal lysins have been reported to have multiple catalytic domains and can display multiple types of activity. For example, the group B streptococcal lysin B30 (also known as PlyGBS) has an *N*-acetylmuramidase and a D-alanyl-L-alanyl endopeptidase catalytic domain (Pritchard et al., 2004), while the staphylococcal ϕ11 lysin has both *N*-acetylmuramoyl-L-alanine amidase and D-alanyl-glycyl endopeptidase catalytic domains (Navarre et al., 1999). However, the presence of two catalytic domains does not necessarily indicate that both are equally active. In the streptococcal LambdaSa2 lysin, the D-glutaminyl-L-lysine endopeptidase domain was found to be responsible for almost all of the hydrolytic activity of this enzyme, whereas its *N*-acetylglucosaminidase domain was found to be almost devoid of activity (Donovan and Foster-Frey, 2008).

Issues with Lysin Nomenclature

The assignment of putative lysin catalytic activities has been less than ideal. Lysins that were isolated decades ago were simply called lysozymes as a generic term for cell wall hydrolases. Unfortunately, this older nomenclature persists to this day. The lysin of the T7 bacteriophage continues to be called the T7 lysozyme in the literature despite experimental evidence dating back to 1973 showing that it is actually an *N*-acetylmuramoyl-L-alanine amidase rather than an *N*-acetylmuramidase (i.e, lysozyme) (Inouye et al., 1973). Another challenge is the generic classification of many lysins simply as amidases, which is ubiquitously used to describe both *N*-acetylmuramoyl-L-alanine amidases and endopeptidases, the latter being exclusive to hydrolysis of an amide bond between two amino acids. To further confound this issue, a recently described protein family called CHAP (cysteine, histidine-dependent amidohydrolase/peptidase) has emerged as a common domain architecture found in bacteriophage lysins. Experimental evidence shows that the CHAP domain of the group B streptococcal B30 lysin is a D-alanyl-L-alanyl endopeptidase (Pritchard et al., 2004), whereas the CHAP domain of the group A streptococcal PlyC lysin is an *N*-acetylmuramoyl-L-alanine amidase (Fischetti et al., 1972; Nelson et al., 2006). Finally, many lysin catalytic domains are alleged to possess a particular activity based solely on loose homology to another lysin domain with a putative function. When actual experiments are conducted to determine cleavage specificities, the results are often contrary to the predetermined in silico activity. For example, bioinformatic analysis suggested that the streptococcal lysins LambdaSa1 and LambdaSa2 contained *N*-acetylmuramoyl-L-alanine amidase activities. Using electrospray ionization mass spectrometry, Pritchard et. al. (2007) showed not only an absence of *N*-acetylmuramoyl-L-alanine amidase activity, but that these enzymes functioned as D-glutaminyl-L-lysine endopeptidases. Clearly, more-rigorous biochemical characterization of bacteriophage lysins will help define boundaries and elucidate interrelationships between the catalytic domains of these enzymes.

Lysin Architecture

The gram-negative peptidoglycan, which lies subjacent to the outer membrane, is relatively thin and undecorated by surface proteins or carbohydrates. Consequently, most lysins from phages that infect gram-negative hosts are single-domain globular proteins that typically comprise only a single catalytic domain and have a mass of 15 to 20 kDa. Conversely, gram-positive organisms contain no protective outer membrane and the substantial peptidoglycan layer is highly cross-linked with surface carbohydrates and proteins. Accordingly, lysins from gram-positive infecting bacteriophages utilize a modular design, having one or more catalytic domains and a cell wall binding domain (CBD) that recognizes epitopes on the surface of susceptible organisms, giving rise to the strain- or species-specific binding. A potent lytic activity, combined with an often extreme binding specificity, distinguishes the lysins of phages that infect gram-positive organisms, which range from ~30 to >70 kDa for multicatalytic enzymes.

With one notable exception, all lysins are the product of single genes, though group I introns are often found within these genes (Foley et al., 2000). The mutlidomain lysins from phages that infect gram-positive organisms characteristically have an N-terminal catalytic domain or domains, an interdomain linker sequence, and a C-terminal CBD. This modular nature and the presence of a linker sequence impart an inherent flexibility to these proteins, making crystallography of full-length lysins challenging. Many attempts have yielded only the structures of individual catalytic domains or isolated CBDs (Low et al., 2005; Porter et al., 2007). It has only been in very recent years that full-length structures have become available, such as for PlyPSA, a listerial *N*-acetylmuramoyl-L-alanine amidase (Korndorfer et al., 2006), and Cpl-1, a pneumococcal *N*-acetylmuramidase (Hermoso et al., 2003). Remarkably, both structures reveal extreme compartmentalization displayed by the individual domains (Color Plate 1). Structures such as these demonstrate real potential for engineered lysins that contain multiple catalytic domains, multiple CBDs, or exchanged catalytic domains and/or CBDs. Indeed, proof-of-principle experiments along this line of reasoning have already been performed. Fusion of clostridial or lactococcal *N*-acetylmuramidase catalytic domains to choline binding domains from pneumococcal lysin CBDs resulted in choline dependence of the chimeric enzyme (Croux et al., 1993; Lopez et al., 1997).

The exception to the single-gene nature of these enzymes is PlyC, a lysin from the streptococcal C_1 bacteriophage. This enzyme is composed of a gene product, PlyCA, which contains the catalytic domain and eight identical copies of a second gene product, PlyCB, which comprises the CBD (Nelson et al., 2006). To date, no other multimeric lysin has been identified and the implications for a multigene heterononamer are not abundantly clear. Nonetheless, nanogram quantities of PlyC can achieve ~7-log killing of streptococcal cells within seconds, making PlyC the most active lysin, by several orders of magnitude, ever described (Nelson et al., 2001).

CBDs and Their Role in Lysin Specificity

The CBD epitopes are usually carbohydrates or teichoic acids that are unique to a species, much like a bacterial fingerprint (Fig. 2). For example, choline moieties distinctive of the pneumococcal cell wall serve as the binding receptor for many pneumococcal lysin CBDs (Garcia et al., 1983, 1988). Alternatively, the binding epitopes may be part of the structural peptidoglycan itself. The CBD of the pneumococcal Cpl-7 lysin binds the pneumococcal peptidoglycan in a choline-independent manner (Diaz et al., 1991; Garcia et al., 1990), and similar Cpl-7 CBDs have been found in the LambdaSa2 lysin from a group B streptococcal phage (Donovan and Foster-Frey, 2008). Additionally, bacterial Src homology 3 (SH3b) domains, which are found on many phage lysins, appear to bind directly to the peptidoglycan in an as yet unknown species-specific manner.

The CBD recognition of an epitope is analogous to recognition of a receptor by a tail fiber. In fact, there is some evidence that these two disparate types of proteins have evolved to target identical epitopes. For example, the γ phage of *Bacillus anthracis* forms plaques on all tested *B. anthracis* strains as well as *Bacillus cereus* 4342, which is considered a *B. anthracis* transition-state strain, but not other *B. cereus* strains (Schuch et al., 2002). Significantly, the lytic range of γ phage lysin, PlyG, mirrors the host range of the phage. In a similar fashion to pneumococcal phage tail fibers (Lopez et al., 1982), pneumococcal lysin CBDs are known to bind choline in the pneumococcal cell wall (Hermoso et al., 2003; Lopez et al., 1982, 1997). However, these are exceptions rather than the rule. In most cases, the specificity of the phage is more restrictive than its encoded lysin. The C_1 bacteriophage forms plaques only on group C streptococci, yet its lysin, PlyC, efficiently lyses groups A, C, and E streptococci (Krause, 1957), as well as *Streptococcus uberis* (our unpublished observations). An extreme example would be PlyV12, a lysin derived from the enterococcal phage ϕ1. This enzyme not only lyses *E. faecalis* and *Enterococcus faecium* but also lyses almost all *Streptococcus* strains (groups A, B, C, E, F, G, L, and N streptococci, *S. uberis, S. gordonii, S. intermedius,* and *S. parasanguis*) as well as *Staphylococcus* strains *(S. aureus* and *S. epidermidis)* (Yoong et al., 2004).

In Vivo Use of Lysins in Models of Human Disease

Although purification and biochemical characterization of many phage lysins has been ongoing for decades, it was Fischetti and coworkers who coined the term "enzybiotics" to describe the therapeutic potential of these enzymes and first began to test lysins in vivo to demonstrate the efficacy of prophylactic lysin treatment in infected mice (Nelson et al., 2001). In this study, nine mice were given *S. pyogenes* and found to be heavily colonized for four consecutive days by oral swabs. On the fourth day mice were given 500 U of lysin orally, and 2 h later the nasopharyngeal cavity was found to be sterilized. Two days later, on day 6, only one mouse was found to be colonized, presumably from recolonization of streptococci that were intracellular during the initial lysin treatment.

Phage lysins have also been used against *Bacillus* species. PlyG, isolated from the *B. an-*

thracis γ phage, was shown to kill both vegetative cells and germinating spores (Schuch et al., 2002). When an intraperitoneal mouse model of infection was employed, 50 U of PlyG rescued 13 out of 19 mice (68.4%) and extended the life of the remaining mice severalfold over controls. A second *Bacillus* lysin, PlyPH, is characterized by activity over a broad pH range (4.0 to 10.5). This enzyme also saved 40% of mice in a *Bacillus* infection model, compared to a 100% death rate in control mice (Yoong et al., 2006). Additional in vivo lysin experiments were done with PlyGBS, which showed efficacy against a mouse vaginal model of *Streptococcus agalactiae* (i.e., group B streptococcus) (Cheng et al., 2005), and MV-L, which prevented nasal colonization of methicillin-resistant *S. aureus* (MRSA) and protected mice from intraperitoneal challenge with MRSA (Rashel et al., 2007).

The most extensively studied lysins in animal models are Cpl-1, an *N*-acetylmuramidase, and PAL, an *N*-acetylmuramoyl-L-alanine amidase, both of which are from phages that infect *Streptococcus pneumoniae*. PAL, at 100 U/ml, was shown to cause an ~4-log drop in viability of 15 different pneumococcal serotypes in 30 s, including strains that displayed multidrug resistance and those that contained a heavy polysaccharide capsule (Loeffler et al., 2001). In the same study, 1,400 U of PAL was shown to eliminate pneumococci from a nasopharyngeal carriage model and 700 U was shown to significantly reduce bacterial counts, suggesting a dose response. In another paper, Cpl-1 was shown to be effective in a mucosal colonization model and in blood using a pneumococcal bacteremia model (Loeffler and Fischetti, 2003). Significantly, because the catalytic domains of PAL and Cpl-1 hydrolyze different bonds in the pneumococcal peptidoglycan, they show a synergistic effect in vitro when used in combination (Loeffler et al., 2003), which was later confirmed in vivo in a murine intraperitoneal model (Jado et al., 2003). Cpl-1 was also shown to work on pneumococcal biofilms in a rat endocarditis model (Entenza et al., 2005). Infusion of 250 mg/kg was able to sterilize 10^5 CFU/ml pneumococci in blood within 30 min and reduce bacterial titers on heart valve vegetations by >4 log CFU/g in 2 h. In an infant rat model of pneumococcal meningitis, a single intracisternal injection (20 mg/kg) of Cpl-1 resulted in a 3-log decrease of pneumococci in the cerebrospinal fluid, and an intraperitoneal injection (200 mg/kg) led to a decrease of 2 orders of magnitude in the cerebrospinal fluid (Entenza et al., 2005). Finally, Cpl-1 treatment of mice colonized with *S. pneumoniae* significantly reduced cocolonization by challenge with influenza virus, which subsequently decreased symptoms of acute otitis media (McCullers et al., 2007).

LYSINS AND POTENTIAL USE AGAINST FOOD-BORNE PATHOGENS

The Food and Drug Administration's Center for Food Safety and Applied Nutrition (www.foodsafety.gov) has identified several food-borne pathogenic microorganisms of relevance in the United States. On this list are a number of gram-positive bacteria including *Listeria monocytogenes, Clostridium perfringens, Clostridium botulinum, B. cereus,* and *S. aureus.* Significantly, bacteriophages active against all of these organisms have been identified and in many cases their lysins have been cloned and characterized at various biochemical levels. Next we discuss the use of lysins in the detection and elimination of these food-borne pathogens as well as potential applications in food production. Many important food-borne illnesses are caused by gram-negative organisms, and speculation about future use of lysins with these bacteria will also be discussed below.

LYSINS FOR DETECTION OF FOOD-BORNE PATHOGENS

Food-borne illnesses caused by pathogenic bacteria often have a rapid onset and, in many cases, can be treated only by supportive care.

Therefore, prevention of infection is crucial. Detection of food-borne bacterial contamination currently relies on multiday culture enrichment before testing via immunoassay or DNA-based methods. Although some of these assays can require up to 10^7 cells, the infectious doses found in food can be orders of magnitude less. More sensitive and rapid methods of detection are clearly needed to effectively minimize food-borne illnesses.

Lysin CBDs convey species specificity to the enzyme and are capable of binding bacterial cell wall components in the absence of catalytic domains. The association constants of CBDs for their substrates are in the nanomolar range, comparable to an antibody-antigen interaction (Loessner et al., 2002). CBDs can be chemically cross-linked to an organic dye for fluorescent detection of bacterial pathogens by spectrophotometric or microscopic analysis. Color Plate 2 shows precise labeling of *S. pyogenes* cells by an AlexaFluor conjugate of the streptococcal PlyC CBD in a mixture also containing *S. aureus* and *B. cereus* cells. Further, lysins are amenable to genetic manipulation, allowing fusion of various tags for recombinant expression and purification, and their modular nature facilitates construction of chimeras with multiple activities and specificities, making lysins and their CBDs ideal candidates for host-cell detection.

Listeriosis

Listeriosis is caused by infection with *L. monocytogenes*. The disease presents initially with flulike symptoms but can progress to septicemia, encephalitis, meningitis, and eventually death. While healthy adults and children typically have mild symptoms, pregnant women, newborns, elderly individuals, and people who are immunocompromised are at risk for more severe cases. The fatality rate for listeriosis varies by risk group but can be as high as 50% for newborns and up to 20% for the elderly (Bortolussi, 2008). The Centers for Disease Control and Prevention's Foodborne Disease Outbreak Surveillance System determined *L. monocytogenes* to be the cause of the greatest number of food-borne fatalities in the United States for 1998 to 2002 (43%), resulting in more deaths than any other bacterium, virus, parasite, toxin, or cause of unknown etiology (Lynch et al., 2006).

Listeria contamination is typically found in ready-to-eat (RTE) foods such as lunch meats and hot dogs, soft cheeses, unpasteurized milk products, and uncooked meats and vegetables. Pasteurization kills *Listeria,* although contamination can occur after processing and before packaging. A particular issue with this microorganism is its ability to grow at 4°C, a temperature commonly used to hinder growth of bacteria. Long-term refrigeration can allow *Listeria* to reach its infectious dose, which is thought to be less than 1,000 CFU in high-risk individuals (Lyytikainen et al., 2000; Maijala et al., 2001).

The CBDs of lysins Ply500 and Ply118 bind *Listeria* cells in a serovar-specific manner and can do so in a mixed culture also containing staphylococci, enterococci, bacilli, and lactococci (Loessner et al., 2002). In foods with mixed bacterial loads, distinguishing specific species that represent a health threat is critical. Additionally, for detection in cases where nonpathogenic bacteria serve a productive role, as is the case in the fermentation process, it is crucial to differentiate between species. False positives due to signal interference from another gram-positive species could result in great financial loss for a manufacturer.

The Ply500 and Ply118 CBDs have also been bound to magnetic beads for concentration and separation of pathogens from food sources. Kretzer et al. (2007) used this approach with fluorescently labeled CBDs from *L. monocytogenes, B. cereus,* and *C. perfringens* lysins. Magnetic beads were coated with green fluorescent protein (GFP)-tagged Ply500 and Ply118 CBDs and were shown to bind and recover *Listeria* cells with >90% efficiency alone and among mixed cell populations. This magnetic separation technique was compared to the standard plating method currently used by food manufacturers. In *L. monocytogenes*-spiked food samples, magnetic separation detected *Listeria* more rapidly and with greater sensitivity than the standard plating method in several food types. The authors reported detection of as little as 100 CFU/g of food after 6 h of enrichment, compared with 2 to 4 days of enrichment typically required for standard testing methods. Significantly, these numbers are below the infectious dose of *Listeria* for high-risk individuals. The magnetic separation technique was also successful in detecting natural *Listeria* contamination of store-bought foods. CBD-coated beads represent a viable means of sensitive and rapid detection of *Listeria*-contaminated foods. This technology could certainly be extended to other lysin CBDs to detect a variety of gram-positive pathogenic bacteria in food and environmental samples.

Bacillus Food Poisoning

An important biotechnology consideration of lysins and their associated CBDs is the ability to exploit activity as well as specific binding for diagnostic purposes. Whole lysins can be used to selectively lyse target organisms in a mixed culture, which then release cytoplasmic ATP that can be detected in the presence of luciferin/luciferase with a luminometer. For example, PlyPH was shown to selectively lyse *B. cereus* RSVF1 in a bacterial mixture using this luminescent assay (Yoong et al., 2006).

B. cereus contamination occurs with vegetative cells, spores, and biofilms and can be found on a wide variety of foods and preparation surfaces. *B. cereus* causes two forms of food-borne illness, a rapid-onset emetic form and a longer-onset diarrheal illness. The vomiting illness is associated with *B. cereus* found on starchy foods such as rice, potatoes, and pasta, while the diarrheal form is linked to bacteria found in meats, milk and milk products, soups, and sauces. The infectious dose is relatively high, at an estimated 10^5 to 10^7 total cells or spores causing the diarrheal disease and 10^5 to 10^8 cells/g of food for the emetic form (Granum and Lund, 1997). Both illnesses are caused by toxins—the vomiting illness is caused by an emetic toxin, cereulide, and the diarrheal type is caused by an enterotoxin: nonhemolytic enterotoxin, hemolysin BL, or both (reviewed in Stenfors Arnesen et al., 2008).

Additionally, the spores are heat stable, rendering the cooking process ineffective at killing them, and they readily adhere to food processing equipment, making decontamination difficult (Faille et al., 2002). *B. cereus* biofilms can also contaminate food production facilities, and this matrix of cells is resistant to disinfectants (Ryu and Beuchat, 2005).

PlyG demonstrates lytic activity on a variety of *B. anthracis* strains as well as one *B. cereus* strain (Schuch et al., 2002). Importantly, the enzyme was shown to be effective in killing and detecting germinating spores in addition to vegetative cells. The spore coat that normally forms an impenetrable surface for lytic enzymes undergoes an increase in porosity following germination (Santo and Doi, 1974), allowing lysins access to the peptidoglycan. The authors showed that as few as 100 *B. cereus* germinating spores were detected by measuring ATP release after treatment with PlyG.

Although this represents a promising method of detection for *Bacillus* contamination, one caveat of this work is that the lysin recognized only one of the tested *B. cereus* strains, suggesting that other *B. cereus* phage lysins may prove more effective in recognizing a wider variety of serovars. Alternatively, chimeric forms of PlyG with additional CBDs might broaden its utility by allowing wider strain recognition. Further examination of PlyG for activity on biofilms would extend its use in detection of *B. cereus* in food processing facilities.

PlyL, a lysin from the λ Ba02 prophage of *B. anthracis,* shares 90% identity over the catalytic domain with PlyG and is also active on *B. cereus* (Low et al., 2005). A deletion mutant containing only the catalytic domain of PlyL was reported to have higher activity than the full-length protein for a variety of bacilli, not including *B. cereus*. This activity was not reported for PlyG, underscoring the importance of the few differing amino acids between the two enzymes. Due to the broader host range of activity for this lysin, further characterization of lytic ability on spores or biofilms is warranted.

PlyB and PlyPH are also active against *B. cereus* (Porter et al., 2007; Yoong et al., 2006). While the catalytic domain of PlyB on its own has little or no activity, the full-length enzyme completely lyses a culture of *B. cereus* in seconds (Porter et al., 2007). PlyPH maintains activity over a broad range of pH (4 to 10.5) and may function in conditions that inactivate other lytic enzymes (Yoong et al., 2006). Exploration of the host range of these lysins along with possible chimeric constructs with the other *Bacillus* lysin domains are further potential avenues of research into *Bacillus* detection in food.

Botulism and *C. perfringens* Food Poisoning

C. perfringens and *C. botulinum* are spore-forming, toxin-producing, anaerobic bacteria. *C. perfringens* infection causes a gastrointestinal illness, and contamination is found in meats, poultry, and gravy. Quite a large number of cells, 10^7, are required to cause illness, and most people recover within 24 h (Brynestad and Granum, 2002). *C. botulinum* produces a more severe illness. Only a few nanograms of the botulinum toxin can cause weakness, blurred vision, and vertigo, with progressive difficulty in speaking, swallowing, and breathing, and ultimately death (Arnon et al., 2001; Schantz and Johnson, 1992). Treatment involves intense supportive care and administration of an antitoxin if the intoxication is diagnosed early after ingestion. As it is an anaerobic microorganism, *C. botulinum* contamination most often occurs in improperly canned foods and is typically limited to home-canned foods, although occurrences of commercial contamination are still reported (Aureli et al., 2000; Sheth et al., 2008; Townes et al., 1996).

Very little work has been done with *Clostridium* lysins. A single amidase, Ply3626, has been identified from the ϕ3626 phage of *C. perfringens* and initial characterization showed specific lytic activity for its host strain (Zimmer et al., 2002). All 48 *C. perfringens* strains tested were sensitive to extracts of *Escherichia coli* expressing the lysin, and a single non-*perfringens* species, *Clostridium fallax,* was also sensitive. Phages for both *C. perfringens* and *C. botulinum* have been identified and their genomes sequenced (http://www.ncbi.nlm.nih.gov/genomes/genlist.cgi?taxid=10239&type=6&name=Phages). Sequence analysis has identified putative lytic enzymes for some of these phages. Cocktails of whole lysins or their CBDs recognizing a broad range of pathogenic species may prove useful in developing novel commercial diagnostics for *Clostridium* contamination of foods.

LYSINS FOR USE AS ANTIMICROBIALS AGAINST FOOD-BORNE PATHOGENS

Dairy Fermentation

S. aureus is a fairly ubiquitous organism, found on the skin and hair and in nasal passages of 30 to 50% of healthy individuals (Lowy, 1998; von Eiff et al., 2001). Food-borne illness, called staphyloenterotoxicosis, is caused by enterotoxin produced from certain staphylococcal strains. Less than 1 μg of toxin causes illness, with gastrointestinal symptoms (nausea, vomiting, diarrhea, and abdominal cramps) occurring rapidly, within 1 to 6 h after toxin exposure (Evenson et al., 1988). *S. aureus* is found in unrefrigerated or improperly refrigerated meats, salads (egg, tuna, potato), cream pastries, and dairy products and, due to its ubiquitous nature, is often associated with food handlers as well as preparation and fermentation equipment and surfaces.

In dairy starter cultures for cheese making, lactic acid bacteria are used for fermentation. During the maturation process these bacteria are lysed, releasing peptidases that break down milk proteins (Law and Wigmore, 1983). Timing and rate of lysis contribute to flavor development for different cheeses (Chapot-Chartier et al., 1994). Contamination of entire batches of cheese starter by pathogens such as

S. aureus and *L. monocytogenes* is common and can result in great economic loss for manufacturers. The presence of contaminating bacteria can interfere with growth of the natural microbial communities, destroying the flavor of the cheese, and pathogenic microorganisms present a health risk if no further pasteurization occurs. Additionally, broad-spectrum bactericidal agents cannot be used due to the nonspecific killing of lactic acid bacteria. Use of highly species- or strain-specific lysins would be beneficial for this process, either added exogenously or expressed from genetically engineered starter strains.

LysK, a multidomain lysin from *S. aureus* phage K, has an N-terminal CHAP domain, a central amidase domain, and a C-terminal SH3b cell binding domain (O'Flaherty et al., 2005), although the N-terminal CHAP domain alone is sufficient for lytic activity (Horgan et al., 2009). The full-length enzyme is active against *S. aureus,* including MRSA, vancomycin-resistant *S. aureus,* and teicoplanin-resistant strains, and other staphylococci. Expression of LysK in *L. lactis* under a nisin-inducible promoter killed a variety of *S. aureus* strains, including bovine strains and methicillin- and vancomycin-resistant strains (O'Flaherty et al., 2005).

Similarly, Gaeng et al. (2000) expressed plasmid-borne *Listeria* phage lysin in *L. lactis.* Ply511 was fused to the *Lactobacillus brevis* S-layer protein signal peptide and expressed under the control of a lactococcal promoter. Colonies of *L. lactis* formed clearing zones on a lawn of *L. monocytogenes* (Gaeng et al., 2000). A further set of experiments examined activity of Ply511 expressed in *L. lactis* and several lactobacilli (Turner et al., 2007). While the expressed lysin was able to produce clearing zones on killed *Listeria* cells, no growth inhibition was observed when the strain expressing the lysin was cocultured with live *L. monocytogenes.* Additionally, optimum pH for Ply511 was determined to be 8.0, with only 5% of activity remaining at pH of 5.5. For Ply511 to be a potential antimicrobial in dairy fermentation, the pH requirements would need to be addressed, as the pH of milk is between 6.5 and 6.8.

The additional staphylococcal lysins MV-L (Rashel et al., 2007), ϕ11 (Donovan et al., 2006), and ΦH5 (Obeso et al., 2008) demonstrated lytic activity on live staphylococcal cells and would be good candidates for fermentation antimicrobials. MV-L, of phage ϕMR11, lyses *S. aureus,* including antibiotic-resistant strains, and *Staphylococcus simulans* (Rashel et al., 2007). Lysis occurred rapidly, starting within 10 s of enzyme addition and reaching completion within 1 min, as shown by electron microscopy.

The *S. aureus* prophage lysin, ϕ11, is a multidomain lysin, with CHAP and amidase activity domains and an SH3b CBD (Navarre et al., 1999). Deletion analysis showed the protein can be truncated to the CHAP domain and still retain activity in a turbidity assay with *S. aureus,* although activity was reduced relative to the CHAP-amidase construct or full-length protein (Donovan et al., 2006). The authors showed peak enzyme activity at 2 to 3 mM $CaCl_2$ and pH of 6 to 7, consistent with the physiological conditions of milk.

Finally, the LysH5 lysin, isolated from the RF122 prophage of *S. aureus,* kills a range of *S. aureus* strains of both human and bovine origin. It displayed higher activity on *S. aureus* of bovine origin and functions in commercial, pasteurized whole milk (Obeso et al., 2008). It would be interesting to examine activity in raw milk, as precedent exists for inactivation of lysins in this slurry of proteins and fats (Donovan et al., 2006).

Use of exogenously added, inducible or constitutively expressed lysins as biocides against pathogenic bacteria in the dairy fermentation process could reduce the dependency on time- and labor-intensive detection methods. Since lysins are proteins and would be degraded upon entry into the digestive system, adverse effects of their addition on the consumer should be minimal. Although these in vitro results show promise for a future dairy starter culture biocide, the efficacy of any lysin must be tested under conditions that replicate the environment in which it needs to function (i.e., milk).

Listeria and RTE Foods

As mentioned previously, *L. monocytogenes* is particularly problematic because of its ability to replicate at 4°C. RTE foods contaminated before packaging and stored at refrigeration temperatures can carry infectious titers of *Listeria,* and these foods are often not cooked to temperatures high enough to kill the bacteria. An antilisterial agent applied just prior to packaging would be an effective measure to prevent cell growth.

Two purified phage-based products for decontamination of RTE meats and poultry have been approved for use in the United States and Europe. Intralytix, Inc. (Baltimore, MD) manufactures LMP-102™, a mixture of six *Listeria* phages that received FDA approval for use as a spray on RTE foods and U.S. Environmental Protection Agency approval as an environmental decontaminant. LISTEX™ P100 is a preparation of the P100 phage made by EBI Food Safety (Wageningen, The Netherlands) for use on meats, cheese, produce, and other foods and approved for use in the European Union.

Listeria lysins could be used in this type of application and may be an improvement to the current phage products. Unlike with phages, bacterial cells have not been shown to become resistant to lysins (Loeffler et al., 2001, Schuch et al., 2002). Further, introduction of extra genetic material would be minimized, reducing the possibility of allergic responses in consumers. As protein products, lysins would be easier to manipulate genetically to carry out desirable activities, whereas engineering phages, particularly those of gram-positive bacteria, can be technically challenging.

Long-term stability studies and examination of activity at 4°C are necessary to find an enzyme(s) with ideal properties for prepackaged foods. Given the FDA approval for a whole-phage product, a phage-derived protein should meet with minimal, if any, challenges by regulatory agencies.

Lysin Disinfectants

Bacterially contaminated food preparation surfaces and equipment contribute to transmission of food-borne illnesses and are a particular problem for spore- and biofilm-forming bacteria. Chemical disinfectants are typically used for decontamination purposes in food production facilities, but their efficacy is dependent upon cleanliness of surfaces and can be diminished by the presence of organic matter. Additionally, insufficient rinsing of surfaces can promote passage of chemical residues into foodstuffs and, for some classes of disinfectants, corrosion of equipment. There are a variety of classes of compounds used for disinfection in food production and preparation facilities, including chlorine-releasing compounds, quaternary ammonium compounds, iodophors, and amphoteric compounds (Forsythe and Hayes, 1998). These vary in spectrum of bactericidal activity (effectiveness on spores, gram-positive bacteria, and gram-negative bacteria), level of corrosiveness, sensitivity to organic materials, and cost.

PlyC was recently shown to be effective as an antistreptococcal enzyme disinfectant for agricultural applications (Hoopes et al., 2009). Unlike its chemical counterparts, PlyC was shown to be effective in the presence of an organic load, and as it is protein based it should not pose a health risk to consumers. Furthermore, the high-affinity interaction between the lysin CBD and its cellular substrate aids in rapid binding and killing of target cells. Since this type of disinfectant is useful against a single or narrow range of microorganism(s), it might be best applied in conjunction with other disinfectants and surfactants for full biocidal activity. Food production facilities with specific problematic contaminants would be ideal candidates for use of lysin disinfectants, particularly those with biofilm- or spore-forming pathogens.

S. aureus biofilm formation is a problem in food processing; however, little is known about biofilms in food-related isolates. One study examined several food and clinical isolates of *S. aureus* under a variety of environmental conditions relevant to food processing and found biofilm-forming strains among both types (Rode et al., 2007), suggesting that conditions in these processing plants may be facilitating biofilm growth.

To date, only one lysin has been shown to function on *S. aureus* biofilms. In addition to lysing staphylococcal cell walls and killed whole cells (*S. aureus, S. epidermidis,* and *S. simulans*), ϕ11 also lyses *S. aureus* biofilms (Sass and Bierbaum, 2007). This is a particularly important result since biofilms are not completely killed by disinfectants or antibiotics (Kumar and Anand, 1998).

Bacillus contamination in food production facilities is common due to the ubiquitous nature of this organism in the environment and leads to contamination of food. Biofilms of *B. cereus* can be found on surfaces in these facilities including stainless steel (Peng et al., 2001, Ryu and Beuchat, 2005), and one study found that *B. cereus* represents more than 12% of constitutive microflora in biofilms in a dairy plant (Sharma and Anand, 2002). The spore form is particularly problematic as it adheres to stainless steel surfaces even after cleaning and is heat resistant even at pasteurization temperatures (Novak et al., 2005; Peng et al., 2001).

As mentioned previously, there are several *Bacillus* lytic enzymes including one, PlyG, that recognizes and lyses *B. cereus* spores. Lysins active on the spore and/or biofilm forms of bacilli used in conjunction with chemical disinfectants may aid in the elimination of food contamination due to the presence of these challenging types of microorganisms.

GRAM-NEGATIVE FOOD-BORNE PATHOGENS

In addition to the gram-positive pathogens mentioned above, the Center for Food Safety and Applied Nutrition lists several gram-negative bacteria among current concerns as food-borne pathogens, including species of *Salmonella, Yersinia, Vibrio, Aeromonas, Plesiomonas, Shigella; Campylobacter jejuni;* and several pathogenic *E. coli* and other enteric bacteria. Of particular interest are species of the genus *Salmonella,* which were responsible for over 40% (7,444 out of 18,499) of cases of laboratory-confirmed infections in the United States in 2008 (Centers for Disease Control and Prevention, 2009a). *Salmonella* contamination has been found in a wide variety of foods including raw meats, poultry, eggs, and milk and dairy products, among others, and 2008 saw two high-profile, multistate outbreaks. Produce contaminated with *Salmonella enterica* serovar Saintpaul (Centers for Disease Control and Prevention, 2008) and *S. enterica* serovar Typhimurium (Centers for Disease Control and Prevention, 2009b) contamination of peanut butter and peanut butter-containing products affected >2,000 individuals in 43 states, with the stories receiving significant attention in the popular press.

The use of phage lytic enzymes to control gram-negative bacterial contaminants has been very limited. Essentially, their effectiveness at this task when added exogenously is hampered by the presence of an outer membrane not found on gram-positive cells. The peptidoglycan layer resides between the inner and outer membranes and as such is not directly exposed to the environment. Furthermore, the outer membrane is an effective barrier to most macromolecules, allowing only small hydrophobic substances to diffuse across and thereby restricting lysins' access to the peptidoglycan (Vaara, 1992). Therefore, an effective, nontoxic strategy to allow lysins to bypass the outer membrane is paramount to their use against gram-negative food-borne pathogens.

Extensive reviews have been published on the use of cationic peptides, detergents, and chelators to permeabilize the outer membrane (Vaara, 1992) and on the modification of hen egg white lysozyme (HEWL) by denaturation, conjugation of various chemical moieties, or genetic fusion to hydrophobic peptides to alter its activity or membrane solubility (Masschalck and Michiels, 2003). Each strategy can be applied to phage-derived enzymes, but each strategy also poses questions regarding efficacy, practicality, and toxicity that must be determined empirically. However, the use of high hydrostatic pressure (HHP) to increase access of phage endolysins to peptidoglycan during food processing is particularly promising. HHP has several advantages: it is inherently bactericidal (Briers et al., 2008; Hauben et al., 1996; Masschalck et al., 2000, 2001; Nakimbugwe et al., 2006a, 2006b), it does not use heat so will not compromise the quality of foodstuffs, and most importantly, it is not an additive. However, generating the high pressures (200 to 500 MPa) required to inactivate bacterial contaminants is costly. HHP has been used with a variety of antibacterials including EDTA, several lysozymes, nisin, lactoferrin, and lactoferricin (Briers et al., 2008; Hauben et al., 1996; Masschalck et al., 2000, 2001; Nakimbugwe et al., 2006a, 2006b).

Nakimbugwe et al. (2006b) tested HHP in conjunction with six individual lysozymes (including phage lysozymes from λ and T4) on ten different bacterial strains (five each of gram-negative and -positive bacteria). Both of these phage enzymes were active on four out of five of the gram-negative bacteria and *Bacillus subtilis,* though the λ enzyme showed greater activity on most of the strains. *Salmonella* serovar Typhimurium, a representative of an important genus of pathogenic bacteria, was the sole gram-negative bacterium examined that was resistant to all lysozymes under the test conditions. As discussed below, this observed resistance is conditional (Nakimbugwe et al., 2006b).

In a separate study, the efficacy of HEWL and λ lysozyme in conjunction with HHP was tested in skim milk (pH 6.8) and banana juice (pH 3.8) on four gram-negative bacteria: *E. coli* O157:H7, *Shigella flexneri, Yersinia enterocolitica,* and *Salmonella* serovar Typhimurium (Nakimbugwe et al., 2006a). λ lysozyme outperformed HEWL in a bacterial inactivation assay by almost 2 and 5 log units in skim milk and banana juice, respectively. Unlike the previous study, λ lysozyme was effective on *Salmonella* serovar Typhimurium, showing an increased activity over HHP alone in milk and banana juice. λ lysozyme was also more active in banana juice for all strains tested, whereas the nature of the food environment had little effect on HEWL activity. This underscores the need to test any lysin in the environment where it is to be employed. Interestingly, neither lysozyme is catalytically active at a pH of 3.8, suggesting that they work by a nonenzymatic microbicidal activity (During et al., 1999, Nakimbugwe et al., 2006a).

Precedent exists for lysozyme bactericidal activity against gram-negative bacteria when added exogenously (During et al., 1999). This seems counterintuitive since the outer membrane blocks access to the peptidoglycan layer. However, there is evidence that this microbicidal effect is not due to the muramidase activity of the lysozyme, but is instead a result of an interaction with the outer membrane (During et al., 1999). During et al. demonstrated that heat-denatured T4 lysozyme retains 50% of the microbicidal activity but lacks muramidase activity. The authors also identified three positively charged, amphipathic helices and showed that one, A4, which corresponds to residues 143 to 155, exhibits 2.5 times more killing of *E. coli* than intact T4 lysozyme. A4 is proposed to act by membrane disruption. This action may be similar to that of other positively charged amphipathic helices collectively referred to as host-defense peptides (Sahl and Bierbaum, 2008).

The Lysins of *Pseudomonas aeruginosa* Phages ϕKZ and EL

As stated above, lysins from phages of gram-negative bacteria typically consist of a sole catalytic domain without a CBD. The lysins from the large, lytic *P. aeruginosa* phages ϕKZ and EL (KZ144 and EL183, respectively) are notable exceptions. These lysins have a modular structure similar to those found in phages of gram-positive bacteria (Fokine et al., 2008). However, unlike gram-positive lysins, the catalytic domain in both KZ144 and EL183 is located at the C terminus (Briers et al., 2007). Both lysins are able to lyse five different genera of outer membrane-compromised, gram-negative bacteria including *Salmonella* serovar

Typhimurium. Gram-positive bacterial cell walls are generally not substrates for these lysins, though KZ144 has slight activity on *B. subtilis,* which, interestingly, has a peptidoglycan structure similar to that of gram-negative bacteria. Also, like other modular lysins, the CBD can function in the absence of the catalytic domain. GFP-CBD fusions allowed the visualization of gram-negative bacteria when the outer membrane was compromised by addition of chloroform, but not visualization of gram-positive bacteria.

Briers and colleagues demonstrated that the lysins KZ144 and EL183 reduce viable *P. aeruginosa* by 2.4 to 4.2 orders of magnitude in an HHP-dependent manner (Briers et al., 2008). Furthermore, this activity is most likely due to the hydrolase activity of the proteins and not an analogous nonenzymatic bactericidal activity previously observed for T4 lysozymes (During et al, 1999), as the cell wall binding activity was demonstrated independently with CBD-GFP fusions (Briers et al., 2008).

Though *P. aeruginosa* is not a common food-borne pathogen, this work establishes important proof-of-principle experiments for the use of modular lysins in combination with HHP to detect and eliminate gram-negative bacterial contaminations. Also, since both KZ144 and EL183 are effective in both detection and elimination of a variety of gram-negative pathogens, these lysins are promising candidates for use with all gram-negative contaminants (Briers et al., 2007). A single lysin should be effective on the highly similar peptidoglycan of all gram-negative bacteria. However, one should be cautious before extolling the virtues of these proteins until demonstration of their activities in real food environments occurs.

Modular lysins specific against gram-negative bacteria are potentially a very powerful tool to identify and combat contaminants in food preparation and storage. While HHP shows great promise for food applications, development of additional methodologies is certainly warranted.

CONCLUDING REMARKS

Both bacteriophage treatment, described in other chapters of this book, and lysin treatment, described here, offer great potential against food-borne pathogens (Table 1). Each treatment modality offers advantages over traditional antimicrobial or disinfectant therapy. These include the ability to target specific

TABLE 1 Comparative advantages and disadvantages of bacteriophage and lysin therapy

Therapy	Advantages	Disadvantages
Bacteriophage	Self-replicating	Resistance easily evolved
	Works on gram-negative and gram-positive organisms	Potential transfer of toxin genes if lytic phages are not used
	Almost 100 years of historical use	Limited host range
	Some have attained regulatory approval	High bar set for regulatory approval due to foreign DNA
	Many uses (humans, animals, food, environment)	Consumer acceptance—phages can have negative connotations
	Specific targets, no harm to nontarget bacteria	
Lysin	Protein therapeutic (no DNA, toxins, etc.)	Unlike phage, not self-replicating
	Enzymatic activity, high turnover rate	Protein stability (i.e., denaturation/inactivation)
	Resistance not yet reported	Limited host range
	Ability to engineer or use directed evolution	Does not currently work on gram-negative organisms
	Many uses (humans, animals, food, environment)	
	Specific targets, no harm to nontarget bacteria	

pathogenic bacteria with little to no effect on indigenous microflora and the diversity of uses (i.e., humans, animals, foodstuffs, environmental applications, etc.). Phage therapy has additional advantages of being self-replicating, has over 100 years of historical use, has obtained some regulatory approval, and can target either gram-positive or gram-negative organisms. Lysin therapy, in contrast, is not self-replicating and at the moment requires additional techniques to show efficacy on gram-negative bacteria. Nonetheless, lysins offer distinct advantages over phages by not having any known resistance; do not possess toxins, DNA, or other factors that would give pause to regulatory agencies; and can be easily modified or evolved through standard molecular biology techniques. Clearly, both phage therapy and lysin therapy represent reasonable alternatives for management of food-borne pathogens.

REFERENCES

Arnon, S. S., R. Schechter, T. V. Inglesby, D. A. Henderson, J. G. Bartlett, M. S. Ascher, E. Eitzen, A. D. Fine, J. Hauer, M. Layton, S. Lillibridge, M. T. Osterholm, T. O'Toole, G. Parker, T. M. Perl, P. K. Russell, D. L. Swerdlow, and K. Tonat. 2001. Botulinum toxin as a biological weapon: medical and public health management. *JAMA* **285:**1059–1070.

Aureli, P., M. Di Cunto, A. Maffei, G. De Chiara, G. Franciosa, L. Accorinti, A. M. Gambardella, and D. Greco. 2000. An outbreak in Italy of botulism associated with a dessert made with mascarpone cream cheese. *Eur. J. Epidemiol.* **16:**913–918.

Bortolussi, R. 2008. Listeriosis: a primer. *CMAJ* **179:**795–797.

Briers, Y., A. Cornelissen, A. Aertsen, K. Hertveldt, C. W. Michiels, G. Volckaert, and R. Lavigne. 2008. Analysis of outer membrane permeability of *Pseudomonas aeruginosa* and bactericidal activity of endolysins KZ144 and EL188 under high hydrostatic pressure. *FEMS Microbiol. Lett.* **280:**113–119.

Briers, Y., G. Volckaert, A. Cornelissen, S. Lagaert, C. W. Michiels, K. Hertveldt, and R. Lavigne. 2007. Muralytic activity and modular structure of the endolysins of *Pseudomonas aeruginosa* bacteriophages ϕKZ and EL. *Mol. Microbiol.* **65:**1334–1344.

Brynestad, S., and P. E. Granum. 2002. *Clostridium perfringens* and foodborne infections. *Int. J. Food Microbiol.* **74:**195–202.

Centers for Disease Control and Prevention. 2008. Outbreak of *Salmonella* serotype Saintpaul infections associated with multiple raw produce items—United States, 2008. *MMWR Morb. Mortal. Wkly. Rep.* **57:**929–934.

Centers for Disease Control and Prevention. 2009a. Preliminary FoodNet Data on the incidence of infection with pathogens transmitted commonly through food—10 states, 2008. *MMWR Morb. Mortal. Wkly. Rep.* **58:**333–337.

Centers for Disease Control and Prevention. 2009b. Multistate outbreak of *Salmonella* infections associated with peanut butter and peanut butter-containing products—United States, 2008–2009. *MMWR Morb. Mortal. Wkly. Rep.* **58:**85–90.

Chapot-Chartier, M. P., C. Deniel, M. Rousseau, L. Vassal, and J. C. Gripon. 1994. Autolysis of two strains of *Lactococcus lactis* during cheese ripening. *Int. Dairy J.* **4:**251–259.

Cheng, Q., D. Nelson, S. Zhu, and V. A. Fischetti. 2005. Removal of group B streptococci colonizing the vagina and oropharynx of mice with a bacteriophage lytic enzyme. *Antimicrob. Agents Chemother.* **49:**111–117.

Croux, C., C. Ronda, R. Lopez, and J. L. Garcia. 1993. Interchange of functional domains switches enzyme specificity: construction of a chimeric pneumococcal-clostridial cell wall lytic enzyme. *Mol. Microbiol.* **9:**1019–1025.

Diaz, E., R. Lopez, and J. L. Garcia. 1991. Chimeric pneumococcal cell wall lytic enzymes reveal important physiological and evolutionary traits. *J. Biol. Chem.* **266:**5464–5471.

Donovan, D. M., and J. Foster-Frey. 2008. LambdaSa2 prophage endolysin requires Cpl-7-binding domains and amidase-5 domain for antimicrobial lysis of streptococci. *FEMS Microbiol. Lett.* **287:**22–33.

Donovan, D. M., M. Lardeo, and J. Foster-Frey. 2006. Lysis of staphylococcal mastitis pathogens by bacteriophage ϕ11 endolysin. *FEMS Microbiol. Lett.* **265:**133–139.

During, K., P. Porsch, A. Mahn, O. Brinkmann, and W. Gieffers. 1999. The non-enzymatic microbicidal activity of lysozymes. *FEBS Lett.* **449:** 93–100.

Entenza, J. M., J. M. Loeffler, D. Grandgirard, V. A. Fischetti, and P. Moreillon. 2005. Therapeutic effects of bacteriophage Cpl-1 lysin against

Streptococcus pneumoniae endocarditis in rats. *Antimicrob. Agents Chemother.* **49:**4789–4792.

Evenson, M. L., M. W. Hinds, R. S. Bernstein, and M. S. Bergdoll. 1988. Estimation of human dose of staphylococcal enterotoxin A from a large outbreak of staphylococcal food poisoning involving chocolate milk. *Int. J. Food Microbiol.* **7:**311–316.

Faille, C., C. Jullien, F. Fontaine, M. N. Bellon-Fontaine, C. Slomianny, and T. Benezech. 2002. Adhesion of *Bacillus* spores and *Escherichia coli* cells to inert surfaces: role of surface hydrophobicity. *Can. J. Microbiol.* **48:**728–738.

Fischetti, V. A. 2005. Bacteriophage lytic enzymes: novel anti-infectives. *Trends Microbiol.* **13:**491–496.

Fischetti, V. A. 2008. Bacteriophage lysins as effective antibacterials. *Curr. Opin. Microbiol.* **11:**393–400.

Fischetti, V. A., D. Nelson, and R. Schuch. 2006. Reinventing phage therapy: are the parts greater than the sum? *Nat. Biotechnol.* **24:**1508–1511.

Fischetti, V. A., J. B. Zabriskie, and E. C. Gotschlich. 1972. Physical, chemical and biological properties of Type 6 M-protein extracted with purified streptococcal phage-associated lysin, p. 26–36. *In* M. J. Haverkorn (ed.), *Fifth International Symposium on* Streptococcus pyogenes. Excerpta Medica, Amsterdam, The Netherlands.

Fokine, A., K. A. Miroshnikov, M. M. Shneider, V. V. Mesyanzhinov, and M. G. Rossmann. 2008. Structure of the bacteriophage ϕKZ lytic transglycosylase gp144. *J. Biol. Chem.* **283:**7242–7250.

Foley, S., A. Bruttin, and H. Brüssow. 2000. Widespread distribution of a group I intron and its three deletion derivatives in the lysin gene of *Streptococcus thermophilus* bacteriophages. *J. Virol.* **74:** 611–618.

Forsythe, S. J., and P. R. Hayes. 1998. *Food Hygiene, Microbiology and HACCP,* 3rd ed., p. 340–347. Aspen Publishers, Inc., Gaithersburg, MD.

Gaeng, S., S. Scherer, H. Neve, and M. J. Loessner. 2000. Gene cloning and expression and secretion of *Listeria monocytogenes* bacteriophage-lytic enzymes in *Lactococcus lactis*. *Appl. Environ. Microbiol.* **66:**2951–2958.

Garcia, E., J. L. Garcia, P. Garcia, A. Arraras, J. M. Sanchez-Puelles, and R. Lopez. 1988. Molecular evolution of lytic enzymes of *Streptococcus pneumoniae* and its bacteriophages. *Proc. Natl. Acad. Sci. USA* **85:**914–918.

Garcia, J. L., E. Garcia, A. Arraras, P. Garcia, C. Ronda, and R. Lopez. 1987. Cloning, purification, and biochemical characterization of the pneumococcal bacteriophage Cp-1 lysin. *J. Virol.* **61:**2573–2580.

Garcia, P., E. Garcia, C. Ronda, R. Lopez, and A. Tomasz. 1983. A phage-associated murein hydrolase in *Streptococcus pneumoniae* infected with bacteriophage Dp-1. *J. Gen. Microbiol.* **129:**489–497.

Garcia, P., J. L. Garcia, E. Garcia, J. M. Sanchez-Puelles, and R. Lopez. 1990. Modular organization of the lytic enzymes of *Streptococcus pneumoniae* and its bacteriophages. *Gene* **86:**81–88.

Granum, P. E., and T. Lund. 1997. *Bacillus cereus* and its food poisoning toxins. *FEMS Microbiol. Lett.* **157:**223–228.

Hauben, K. J. L., E. Y. Wuytack, C. C. F. Soontjens, and C. W. Michiels. 1996. High-pressure transient sensitization of *Escherichia coli* to lysozyme and nisin by disruption of outer-membrane permeability. *J. Food Prot.* **59:**350–355.

Hermoso, J. A., B. Monterroso, A. Albert, B. Galan, O. Ahrazem, P. Garcia, M. Martinez-Ripoli, J. L. Garcia, and M. Menendez. 2003. Structural basis for selective recognition of pneumococcal cell wall by modular endolysin from phage Cp-1. *Structure* **11:**1239–1249.

Hoopes, J. T., C. J. Stark, H. A. Kim, D. J. Sussman, D. M. Donovan, and D. C. Nelson. 2009. Use of a bacteriophage lysin, PlyC, as an enzyme disinfectant against *Streptococcus equi*. *Appl. Environ. Microbiol.* **75:**1388–1394.

Horgan, M., G. O'Flynn, J. Garry, J. Cooney, A. Coffey, G. F. Fitzgerald, R. P. Ross, and O. McAuliffe. 2009. Phage lysin LysK can be truncated to its CHAP domain and retain lytic activity against live antibiotic-resistant staphylococci. *Appl. Environ. Microbiol.* **75:**872–874.

Inouye, M., N. Arnheim, and R. Sternglanz. 1973. Bacteriophage T7 lysozyme is an *N*-acetylmuramyl-L-alanine amidase. *J. Biol. Chem.* **248:**7247–7252.

Jado, I., R. Lopez, E. Garcia, A. Fenoll, J. Casal, and P. Garcia. 2003. Phage lytic enzymes as therapy for antibiotic-resistant *Streptococcus pneumoniae* infection in a murine sepsis model. *J. Antimicrob. Chemother.* **52:**967–973.

Korndorfer, I. P., J. Danzer, M. Schmelcher, M. Zimmer, A. Skerra, and M. J. Loessner. 2006. The crystal structure of the bacteriophage PSA endolysin reveals a unique fold responsible for specific recognition of *Listeria* cell walls. *J. Mol. Biol.* **364:**678–689.

Krause, R. M. 1957. Studies on bacteriophages of hemolytic streptococci. I. Factors influencing the interaction of phage and susceptible host cells. *J. Exp. Med.* **106:**365–384.

**Kretzer, J. W., R. Lehmann, M. Schmelcher, M. Banz, K. P. Kim, C. Korn, and M. J. Loes-

sner. 2007. Use of high-affinity cell wall-binding domains of bacteriophage endolysins for immobilization and separation of bacterial cells. *Appl. Environ. Microbiol.* **73:**1992–2000.

Kumar, C. G., and S. K. Anand. 1998. Significance of microbial biofilms in food industry: a review. *Int. J. Food Microbiol.* **42:**9–27.

Law, B. A., and A. S. Wigmore. 1983. Accelerated ripening of Cheddar cheese with a commercial proteinase and intracellular enzymes from starter streptococci. *J. Dairy Res.* **50:**519–525.

Loeffler, J. M., S. Djurkovic, and V. A. Fischetti. 2003. Phage lytic enzyme Cpl-1 as a novel antimicrobial for pneumococcal bacteremia. *Infect. Immun.* **71:**6199–6204.

Loeffler, J. M., and V. A. Fischetti. 2003. Synergistic lethal effect of a combination of phage lytic enzymes with different activities on penicillin-sensitive and -resistant *Streptococcus pneumoniae* strains. *Antimicrob. Agents Chemother.* **47:**375–377.

Loeffler, J. M., D. Nelson, and V. A. Fischetti. 2001. Rapid killing of *Streptococcus pneumoniae* with a bacteriophage cell wall hydrolase. *Science* **294:**2170–2172.

Loessner, M. J. 2005. Bacteriophage endolysins—current state of research and applications. *Curr. Opin. Microbiol.* **8:**480–487.

Loessner, M. J., K. Kramer, F. Ebel, and S. Scherer. 2002. C-terminal domains of *Listeria monocytogenes* bacteriophage murein hydrolases determine specific recognition and high-affinity binding to bacterial cell wall carbohydrates. *Mol. Microbiol.* **44:**335–349.

Loessner, M. J., G. Wendlinger, and S. Scherer. 1995. Heterogeneous endolysins in *Listeria monocytogenes* bacteriophages: a new class of enzymes and evidence for conserved holin genes within the siphoviral lysis cassettes. *Mol. Microbiol.* **16:**1231–1241.

Lopez, R., E. Garcia, P. Garcia, and J. L. Garcia. 1997. The pneumococcal cell wall degrading enzymes: a modular design to create new lysins? *Microb. Drug Resist.* **3:**199–211.

Lopez, R., E. Garcia, P. Garcia, C. Ronda, and A. Tomasz. 1982. Choline-containing bacteriophage receptors in *Streptococcus pneumoniae. J. Bacteriol.* **151:**1581–1590.

Low, L. Y., C. Yang, M. Perego, A. Osterman, and R. C. Liddington. 2005. Structure and lytic activity of a *Bacillus anthracis* prophage endolysin. *J. Biol. Chem.* **280:**35433–35439.

Lowy, F. D. 1998. *Staphylococcus aureus* infections. *N. Engl. J. Med.* **339:**520–532.

Lynch, M., J. Painter, R. Woodruff, C. Braden, and Centers for Disease Control and Prevention. 2006. Surveillance for foodborne-disease outbreaks—United States, 1998–2002. *MMWR Surveill. Summ.* **55:**1–42.

Lyytikainen, O., T. Autio, R. Maijala, P. Ruutu, T. Honkanen-Buzalski, M. Miettinen, M. Hatakka, J. Mikkola, V. J. Anttila, T. Johansson, L. Rantala, T. Aalto, H. Korkeala, and A. Siitonen. 2000. An outbreak of *Listeria monocytogenes* serotype 3a infections from butter in Finland. *J. Infect. Dis.* **181:**1838–1841.

Maijala, R., O. Lyytikainen, T. Autio, T. Aalto, L. Haavisto, and T. Honkanen-Buzalski. 2001. Exposure of *Listeria monocytogenes* within an epidemic caused by butter in Finland. *Int. J. Food Microbiol.* **70:**97–109.

Masschalck, B., C. Garcia-Graells, E. G. Van Haver, and C. W. Michiels. 2000. Inactivation of high pressure resistant *Escherichia coli* by nisin and lysozyme under high pressure. *Innov. Food Sci. Emerg. Technol.* **1:**39–47.

Masschalck, B., and C. W. Michiels. 2003. Antimicrobial properties of lysozyme in relation to foodborne vegetative bacteria. *Crit. Rev. Microbiol.* **29:**191–214.

Masschalck, B., R. Van Houdt, and C. W. Michiels. 2001. High pressure increases bactericidal activity and spectrum of lactoferrin, lactoferricin and nisin. *Int. J. Food Microbiol.* **64:**325–332.

McCullers, J. A., A. Karlstrom, A. R. Iverson, J. M. Loeffler, and V. A. Fischetti. 2007. Novel strategy to prevent otitis media caused by colonizing *Streptococcus pneumoniae. PLoS Pathog.* **3:**e28.

Mesnage, S., F. Chau, L. Dubost, and M. Arthur. 2008. Role of *N*-acetylglucosaminidase and *N*-acetylmuramidase activities in *Enterococcus faecalis* peptidoglycan metabolism. *J. Biol. Chem.* **283:**19845–19853.

Nakimbugwe, D., B. Masschalck, G. Anim, and C. W. Michiels. 2006a. Inactivation of gram-negative bacteria in milk and banana juice by hen egg white and lambda lysozyme under high hydrostatic pressure. *Int. J. Food Microbiol.* **112:**19–25.

Nakimbugwe, D., B. Masschalck, M. Atanassova, A. Zewdie-Bosuner, and C. W. Michiels. 2006b. Comparison of bactericidal activity of six lysozymes at atmospheric pressure and under high hydrostatic pressure. *Int. J. Food Microbiol.* **108:**355–363.

Navarre, W. W., H. Ton-That, K. F. Faull, and O. Schneewind. 1999. Multiple enzymatic activities of the murein hydrolase from staphylococcal phage ϕ11. Identification of a D-alanyl-glycine endopeptidase activity. *J. Biol. Chem.* **274:**15847–15856.

Nelson, D., L. Loomis, and V. A. Fischetti. 2001. Prevention and elimination of upper respiratory colonization of mice by group A streptococci using a bacteriophage lytic enzyme. *Proc. Natl. Acad. Sci. USA* **98:**4107–4112.

Nelson, D., R. Schuch, P. Chahales, S. Zhu, and V. A. Fischetti. 2006. PlyC: a multimeric bacteriophage lysin. *Proc. Natl. Acad. Sci. USA* **103:**10765–10770.

Novak, J. S., J. Call, P. Tomasula, and J. B. Luchansky. 2005. An assessment of pasteurization treatment of water, media, and milk with respect to *Bacillus* spores. *J. Food Prot.* **68:**751–757.

Obeso, J. M., B. Martinez, A. Rodriguez, and P. Garcia. 2008. Lytic activity of the recombinant staphylococcal bacteriophage ΦH5 endolysin active against *Staphylococcus aureus* in milk. *Int. J. Food Microbiol.* **128:**212–218.

O'Flaherty, S., A. Coffey, W. Meaney, G. F. Fitzgerald, and R. P. Ross. 2005. The recombinant phage lysin LysK has a broad spectrum of lytic activity against clinically relevant staphylococci, including methicillin-resistant *Staphylococcus aureus*. *J. Bacteriol.* **187:**7161–7164.

Peng, J. S., W. C. Tsai, and C. C. Chou. 2001. Surface characteristics of *Bacillus cereus* and its adhesion to stainless steel. *Int. J. Food Microbiol.* **65:** 105–111.

Porter, C. J., R. Schuch, A. J. Pelzek, A. M. Buckle, S. McGowan, M. C. Wilce, J. Rossjohn, R. Russell, D. Nelson, V. A. Fischetti, and J. C. Whisstock. 2007. The 1.6 Å crystal structure of the catalytic domain of PlyB, a bacteriophage lysin active against *Bacillus anthracis*. *J. Mol. Biol.* **366:**540–550.

Pritchard, D. G., S. Dong, J. R. Baker, and J. A. Engler. 2004. The bifunctional peptidoglycan lysin of *Streptococcus agalactiae* bacteriophage B30. *Microbiology* **150:**2079–2087.

Pritchard, D. G., S. Dong, M. C. Kirk, R. T. Cartee, and J. R. Baker. 2007. LambdaSa1 and LambdaSa2 prophage lysins of *Streptococcus agalactiae*. *Appl. Environ. Microbiol.* **73:**7150–7154.

Rashel, M., J. Uchiyama, T. Ujihara, Y. Uehara, S. Kuramoto, S. Sugihara, K. Yagyu, A. Muraoka, M. Sugai, K. Hiramatsu, K. Honke, and S. Matsuzaki. 2007. Efficient elimination of multidrug-resistant *Staphylococcus aureus* by cloned lysin derived from bacteriophage ϕMR11. *J. Infect. Dis.* **196:**1237–1247.

Rode, T. M., S. Langsrud, A. Holck, and T. Moretro. 2007. Different patterns of biofilm formation in *Staphylococcus aureus* under food-related stress conditions. *Int. J. Food Microbiol.* **116:**372–383.

Ryu, J. H., and L. R. Beuchat. 2005. Biofilm formation and sporulation by *Bacillus cereus* on a stainless steel surface and subsequent resistance of vegetative cells and spores to chlorine, chlorine dioxide, and a peroxyacetic acid-based sanitizer. *J. Food Prot.* **68:**2614–2622.

Sahl, H.-G., and G. Bierbaum. 2008. Multiple activities in natural antimicrobials. *Microbe* **3:**467–473.

Santo, L. Y., and R. H. Doi. 1974. Ultrastructural analysis during germination and outgrowth of *Bacillus subtilis* spores. *J. Bacteriol.* **120:**475–481.

Sass, P., and G. Bierbaum. 2007. Lytic activity of recombinant bacteriophage ϕ11 and ϕ12 endolysins on whole cells and biofilms of *Staphylococcus aureus*. *Appl. Environ. Microbiol.* **73:**347–352.

Schantz, E. J., and E. A. Johnson. 1992. Properties and use of botulinum toxin and other microbial neurotoxins in medicine. *Microbiol. Rev.* **56:** 80–99.

Schuch, R., D. Nelson, and V. A. Fischetti. 2002. A bacteriolytic agent that detects and kills *Bacillus anthracis*. *Nature* **418:**884–889.

Sharma, M., and S. K. Anand. 2002. Biofilms evaluation as an essential component of HACCP for food/dairy processing industry—a case. *Food Control* **13:**469–477.

Sheth, A. N., P. Wiersma, D. Atrubin, V. Dubey, D. Zink, G. Skinner, F. Doerr, P. Juliao, G. Gonzalez, C. Burnett, C. Drenzek, C. Shuler, J. Austin, A. Ellis, S. Maslanka, and J. Sobel. 2008. International outbreak of severe botulism with prolonged toxemia caused by commercial carrot juice. *Clin. Infect. Dis.* **47:**1245–1251.

Steen, A., S. van Schalkwijk, G. Buist, M. Twigt, M. Szeliga, W. Meijer, O. P. Kuipers, J. Kok, and J. Hugenholtz. 2007. LytR, a phage-derived amidase is most effective in induced lysis of *Lactococcus lactis* compared with other lactococcal amidases and glucosaminidases. *Int. Dairy J.* **17:**926–936.

Stenfors Arnesen, L. P., A. Fagerlund, and P. E. Granum. 2008. From soil to gut: *Bacillus cereus* and its food poisoning toxins. *FEMS Microbiol. Rev.* **32:**579–606.

Townes, J. M., P. R. Cieslak, C. L. Hatheway, H. M. Solomon, J. T. Holloway, M. P. Baker, C. F. Keller, L. M. McCroskey, and P. M. Griffin. 1996. An outbreak of type A botulism associated with a commercial cheese sauce. *Ann. Intern. Med.* **125:**558–563.

Turner, M. S., F. Waldherr, M. J. Loessner, and P. M. Giffard. 2007. Antimicrobial activity of lysostaphin and a *Listeria monocytogenes* bacteriophage endolysin produced and secreted by lactic acid bacteria. *Syst. Appl. Microbiol.* **30:**58–67.

Vaara, M. 1992. Agents that increase the permeability of the outer membrane. *Microbiol. Rev.* **56:**395–411.

von Eiff, C., K. Becker, K. Machka, H. Stammer, and G. Peters. 2001. Nasal carriage as a source of *Staphylococcus aureus* bacteremia. *N. Engl. J. Med.* **344:**11–16.

Yoong, P., R. Schuch, D. Nelson, and V. A. Fischetti. 2004. Identification of a broadly active phage lytic enzyme with lethal activity against antibiotic-resistant *Enterococcus faecalis* and *Enterococcus faecium*. *J. Bacteriol.* **186:**4808–4812.

Yoong, P., R. Schuch, D. Nelson, and V. A. Fischetti. 2006. PlyPH, a bacteriolytic enzyme with a broad pH range of activity and lytic action against *Bacillus anthracis*. *J. Bacteriol.* **188:**2711–2714.

Young, R. 1992. Bacteriophage lysis: mechanism and regulation. *Microbiol. Rev.* **56:**430–481.

Zimmer, M., S. Scherer, and M. J. Loessner. 2002. Genomic analysis of *Clostridium perfringens* bacteriophage φ3626, which integrates into *guaA* and possibly affects sporulation. *J. Bacteriol.* **184:** 4359–4368.

LYSOGENIC CONVERSION IN BACTERIA OF IMPORTANCE TO THE FOOD INDUSTRY

Marcin Łoś, John Kuzio, Michael R. McConnell, Andrew M. Kropinski, Grzegorz Węgrzyn, and Gail E. Christie

9

INTRODUCTION TO LYSOGENIC CONVERSION

The classification of phages as virulent or temperate on the basis of their replicative cycle is an established practice of modern virology (see chapter 1). In his epic review on lysogeny, Lwoff (1953) described the possible relationships between phage and host and explained the rules which would govern the lysogenic state. Many of his postulates were subsequently proved to be correct, and in particular, evidence began to accumulate that prophages such as λ could integrate into the host genome. Concurrent with this renewed interest in the lysogenic state came the discovery of conversion to toxin production by prophage β (originally called B) (Freeman, 1951) in *Corynebacterium diphtheriae,* suggesting that a prophage could alter its host's phenotype. The phenomenon of lysogenic or phage conversion has been discussed in the scientific literature but is often not defined (see, e.g., Saunders et al., 2001). A simple definition is that it is a phage-associated heritable change in the host cell's genotype and phenotype that is independent of the effects expected from repression and integration. In other words, phenomena which are directly associated with lysogenization such as immunity to superinfection (a function of the prophage repressor) and loss of function resulting from insertion into a host gene (a function of the integrase) are expected consequences of lysogenization and should not be defined as lysogenic conversion.

Phage-associated conversion genes and genome modules which are nonessential, have little to do with the phage life cycle, or contradict the expected gene synteny have been variably labeled "morons" (Juhala et al., 2000) or "cargo genes" (Brüssow et al., 2004). An example of this can be seen by comparing the sequences of *Salmonella* phages g341 and P22 (Fig. 1), where in the case of P22 the *sieA-orf59a-mnt-arc-ant* gene cluster, with its associated promoters and terminators, is inserted between virion morphogenesis genes *16* and *9*.

Marcin Łoś and Grzegorz Węgrzyn, Department of Molecular Biology, University of Gdansk, Kladki 24, 80-822 Gdansk, Poland. *John Kuzio,* Department of Microbiology and Immunology, Queen's University, Kingston, Ontario K7L 3N6, Canada. *Michael R. McConnell,* Department of Biology, Point Loma Nazarene University, 3900 Lomaland Drive, San Diego, CA 92106. *Andrew M. Kropinski,* Laboratory for Foodborne Zoonoses, Public Health Agency of Canada, 110 Stone Road West, Guelph, Ontario N1G 3W4, Canada, and Department of Molecular and Cellular Biology, University of Guelph, Guelph, Ontario N1G 2W1, Canada. *Gail E. Christie,* Department of Microbiology & Immunology, Virginia Commonwealth University, P.O. Box 980678, 1101 East Marshall Street, 6-034 Sanger Hall, Richmond, VA 23298-0678.

Bacteriophages in the Control of Food- and Waterborne Pathogens
Edited by Parviz M. Sabour and Mansel W. Griffiths, © 2010 ASM Press, Washington, DC

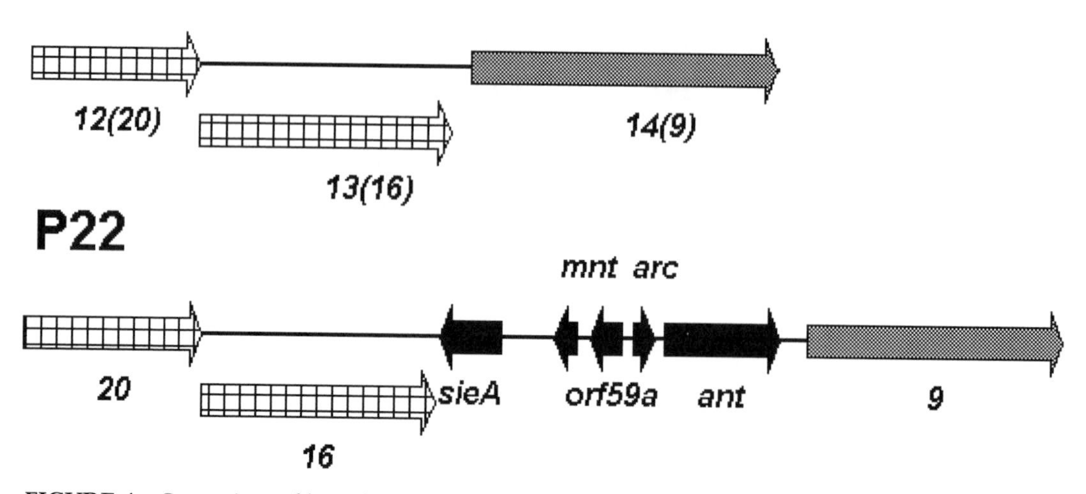

FIGURE 1 Comparison of homologous late gene regions of *Salmonella* phages g341 and P22, showing the presence of cargo genes (i.e. the *sieA-orf59a-mnt-arc-ant* moron) in P22.

While no logical utility can be found for the presence of Shiga, botulinum, or diphtheria toxin genes on phage genomes, problems exist with the simplified definition presented above—we still know relatively little about the roles of many bacteriophage proteins or their interactions, and about gene expression in lysogens under noninducing conditions. The simplistic view of lysogeny in which the prophage is a relatively silent parasite of the host genome, expressing only those genes necessary for its persistence in the lysogenic state (the repressor and integrase), is now known to be patently incorrect; this has been shown through both classic and "-omics" studies that reveal that a considerable number of phage genes are expressed in lysogens, many of which may affect the cell's phenotype. Recent microarray studies (Balbontin et al., 2006) have revealed higher-level expression (≥2×) of numerous prophage genes, including those of *Salmonella* prophages STM1018 (Gifsy-2), STM2622 (Gifsy-1), and STM2238, as well as cryptic prophage *oaf* genes (see below). In the case of coliphage λ, comparison of lysogens with nonlysogenic hosts revealed expression of repressor (*c*I), *rexAB,* recombination genes *xis* and *bet,* hypothetical gene *orf63,* and, interestingly, the genes that produce morphogenesis proteins gpG and gpT (Osterhout et al., 2007). Studies conducted by Chen et al. (2005) revealed that in addition to *c*I and *rexAB,* genes *lom, bor, kil, int,* and gp*E* were also being expressed in lambda lysogens. The expression data for *bor* and *lom* (see below) were confirmed by proteomic studies which indicated that the lambda lysogens express both of these membrane proteins (Barondess and Beckwith, 1990, 1995).

A closer examination of the *sieA-ant* moron in P22 and the *rexAB* region of λ illustrates this problem further. The λ *rexAB* genes are not found in other lambdoid phages—this position is empty in coliphage HK022 and contains the unrelated genes *49-48* in HK97 (Juhala et al., 2000). Nor should these genes be viewed as superfluous to the phage life cycle. RexAB have multiple potential roles including antagonism of DNA damage repair (Fix, 1993) and as a phage superinfection exclusion mechanism (Parma et al., 1992; Snyder, 1995), which must be considered beneficial to the life cycle of this virus. In the case of P22, *sieA*, with the unlinked *sieB* gene, is also part of a superinfection exclusion system (Hofer et al., 1995), while the *mnt-ant* cluster is part of a secondary immunity regulatory system (ImmI) (Sauer et al., 1983). SieA homologs are not present in other members of the "P22-like viruses" (Lavigne et al., 2008), and as illustrated above, the *mnt-ant* moron can also be totally or partially missing from relatives. The situation is additionally rendered complex by the presence of phage-related genomic islands in bacterial chromosomes.

Given that prophage genomes experience greater transcriptional activity than was previously assumed, it is quite possible that additional, probably more subtle, aspects of phage conversion await discovery. For this review, though, we will focus on "classical" examples of phage conversion, i.e., those resulting in obvious phenotypic changes that have been extensively investigated and that are now reasonably well understood.

Limits of This Review

The two best-understood categories of phage conversion involve (i) expression of exotoxins and (ii) alteration of cell surface structural components, such as lipopolysaccharide (LPS) and outer membrane proteins. In this review we will focus on the exotoxins produced by *Escherichia coli/Shigella, Staphylococcus,* and *Clostridium botulinum* phages, as well as on changes in LPS structure and membrane protein compositions brought about by *E. coli*- and *Salmonella*-specific phages; i.e., our emphasis is on phage conversion activities that affect the major food- and waterborne pathogens. For details on streptococcal phages and *Vibrio* viruses see Brüssow et al. (2004), McLeod et al. (2005), and Sánchez and Holmgren (2008).

TOXINS

Shiga Toxin

The *Shigella dysenteriae* exotoxin that was first discovered by Kiyoshi Shiga (1898) has subsequently been shown to be produced by a variety of *E. coli* (Allison, 2007), *Citrobacter freundii* (Schmidt et al., 1993), and *Enterobacter cloacae* (Paton and Paton, 1996) strains as well. This toxin is variably referred to as Shiga toxin (Stx) or verotoxin (verocytoxin; VT). There are two main types, called Stx1 and Stx2, as well as a number of variants (Persson et al., 2007). Like pertussis and cholera toxins, Stx comprises two subunits: subunit A (3.5×10^4 Da, based upon the gene sequence) is present in one copy and subunit B (ca. 9.5 kDa) is present in five copies (Color Plate 3). Subunit A is the toxin and functions as an *N*-glycosidase, cleaving the adenine base from the ribose-phosphate backbone of 28S rRNA at nucleotide residue 4324 (Endo et al., 1988). This is a critical residue of elongation factor binding, and as a result, protein synthesis is inhibited. The active site and critical residues of subunit A have been identified (Y77 in Fig. 2A), and tertiary structures of the *Shigella* and *E. coli* O157:H7 proteins have been determined (Fig. 2B) (Fraser et al., 1994, 2004). Subunit B binds to globotriaosylceramide (Gb3Cer/CD77), a glycosphingolipid found on the surface of epithelial cells. This protein possesses a signal peptide and its critical residues are indicated in Fig. 2.

In addition to their primary toxic effects, Shiga toxins may have profound secondary effects on signal transduction and immunological responses in eukaryotic cells (Creuzburg and Schmidt, 2007). Stx induces macrophages to express tumor necrosis factor α, interleukin-1β, and interleukin-6. These cytokines, along with LPS, were reported to increase the susceptibility of cells to Stx-mediated apoptosis (for discussion, see Ramamurthy, 2008).

Shiga toxin-producing *E. coli* (STEC) strains usually do not cause a disease in animals; however, the presence of the Shiga toxin in the intestine is associated with bloody diarrhea in humans (Fairbrother and Nadeau, 2006). STEC strains colonize the human large intestine by means of a characteristic attaching and effacing lesion. This lesion is induced by a bacterial type III secretion system that injects

A.

```
Phage_Nil2          MKCILFKWVLCLLLGFSSVSYSREFTIDFSTQQSYVSSLNSIRTEISTPLEHISQGTTSV
Phage_Lahn1         MKIIIFRVLTFFFVIFSVNVVAKEFTLDFSTAKTYVDSLNVIRSAIGTPLQTISSGGTSL
                    ** *:*: :  ::: **    ::***:**** ::**.*** **: *.***: **.* **:
                                                          ▼
Phage_Nil2          SVINHTPPGSYFAVDIRGLDVYQARFDHLRLIIEQNNLYVAGFVNTATNTFYRFSDFTHI
Phage_Lahn1         LMIDSGTGDNLFAVDVRGIDPEEGRFNNLRLIVERNNLYVTGFVNRTNNVFYRFADFSHV
                     :*:   . .. ****:**:*  :.**::****:*:*****:**** :.*.****:**:*:

Phage_Nil2          SVPGVTTVSMTTDSSYTTLQRVAALERSGMQISRHSLVSSYLALMEFSGNTMTRDASRAV
Phage_Lahn1         TFPGTTAVTLSGDSSYTTLQRVAGISRTGMQINRHSLTTSYLDLMSHSGTSLTQSVARAM
                    :.**.*:*::: ***********.:.*:****.****.:*** **..**.::*:..:**:

Phage_Nil2          LRFVTVTAEALRFRQIQREFRQALSETAP-VYTMTPGDVDLTLNWGRISNVLPEYRGEDG
Phage_Lahn1         LRFVTVTAEALRFRQIQRGFRTTLDDLSGRSYVMTAEDVDLTLNWGRLSSVLPDYHGQDS
                    ****************** ** :*.: :   *.**. **********:*.***:*:*:*.

Phage_Nil2          VRVGRISFNNISAILGTVAVILNCHHQGARSVRAVNEDSQPECQITGDRPVIKINNTLWE
Phage_Lahn1         VRVGRISFGSINAILGSVALILNCHHHASRVARMASDEFPSMCPADGRVRGITHNKILWD
                    ********..*.****:**:******:.:* .* ..::  . *   *    *. *: **:

Phage_Nil2          SNTAAAFLNRKSQFLYTTGK
Phage_Lahn1         SSTLGAILMRRTISS-----
                     *.* .*:* *::
```

B.

```
                                     ▽
Phage_Nil2          MKK-MFMAVLFALVSVNAMAA-DCAKGKIEFSKYNENDTFTVKVAGKEYWTSRWNLQPLL
Phage_Lahn1         MKKTLLIAASLSFFSASALATPDCVTGKVEYTKYNDDDTFTVKVGDKELFTNRWNLQSLL
                    ***.:::*. :::.*..*:*: **..**:*::***::*******..** :*.*****.**

Phage_Nil2          QSAQLTGMTVTIKSSTCESGSGFAEVQFNND
Phage_Lahn1         LSAQITGMTVTIKTNACHNGGGFSEV-----
                     ***:********:.:*..*.**:**
```

FIGURE 2 Comparison of two prophage Shiga toxin sequences (StxA and StxB subunits) showing the active site (▼) and signal peptidase cleavage sites (▽) as defined using Phobius (Kall et al., 2004).

effector proteins into the intestinal epithelial cell, resulting in profound changes in the architecture and metabolism of the host cell and intimate adherence of the bacteria (Gyles, 2007). Interestingly, STEC strains appear to use a quorum-sensing regulatory system to recognize the intestinal environment and to activate genes required for colonization (Guglielmini et al., 2008). Further production of Shiga toxin molecules causes serious changes in the host-cell metabolism, which result in the symptoms mentioned above. Moreover, 15 to 20% of patients infected with STEC progress to hemorrhagic colitis and/or hemolytic-uremic syndrome, which is a dangerous disease, especially for children (Agbodaze, 1999; Guglielmini et al., 2008). This figure may reach as much as 50% if antibiotics are used for treatment of patients (Serna and Boedeker, 2008).

Although ruminants are recognized as the main natural reservoir of STEC, water has also been documented as a way of transmission. Moreover, these bacteria may contaminate recreational, drinking, or irrigation waters, most probably through feces from humans and other animals. Indeed, the occurrence of STEC carrying *stx2* in raw municipal sewage and animal wastewater from several origins has been documented (Muniesa et al., 2006). STEC continues to be prevalent in the industrialized world because of dissemination in food products contaminated by ruminant feces (Serna and Boedeker, 2008).

The first studies indicating that *E. coli* cells produced the Shiga toxin and caused serious infections were reported by Karmali et al. (1983a, 1983b), when a severe disease caused by the *E. coli* serotype O157:H7 was described. Since the emergence of this foodborne pathogen, more than 500 different serogroups of *E. coli* have been reported to produce Shiga toxin (discussed by Allison, 2007).

The relationship of Shiga toxin to bacteriophage was discovered later. *E. coli* strains H-19 (O26:H11) and K-12 were cultivated, and it was found that a heat-labile cytotoxin (VT) could be transferred between them. This turned out to be a prophage (phage VT) that was a member of the family *Siphoviridae* (Scotland et al., 1983). Smith et al. furthered this research by isolating two converting phages, H-19A and H-19B, from strain H-19 (Smith et al., 1983, 1987). Similar results were obtained with some strains of *E. coli* O157 (Smith et al., 1984).

Further studies demonstrated that phages carrying genes for production of Shiga toxins (called Shiga toxin-converting prophages) belong to the family of lambdoid prophages. Bacteriophage λ is the best-studied member of this family, and our knowledge about the genetics, biochemistry, and biology of phage λ has been helpful in understanding processes that occur during expression of *stx* genes (Friedman and Court, 2001; Waldor and Friedman, 2005).

A considerable variability of Shiga toxin-converting phages has been demonstrated (Garcia-Aljaro et al., 2009), though there is generally more sequence variation among *stx2* genes than *stx1* (Lee et al., 2007). This variability may be enhanced by generation of new phage variants through recombination between Stx1 and Stx2 phage genomes (Asadulghani et al., 2009). Spreading of the *stx* genes may be possible not only through lysogenization of new hosts but also by transduction; however, the latter process is inefficient at low pH and low temperature (Imamovic et al., 2009). On the other hand, ability to establish multiple lysogens by at least some Shiga toxin-converting phages (e.g., $\Phi 24_B$) may enhance pathogenicity of STEC (Fogg et al., 2007).

In STEC strains the *stx* genes are located on prophages, and without prophage induction their expression is mostly repressed. Therefore, in most cases, effective production of Shiga toxins requires prophage induction and its further lytic development, including replication of the phage genome as an extrachromosomal element (Herold, et al., 2004; Schmidt, 2001; Wagner et al., 2001b, 2002; Waldor and Friedman, 2005). Shiga toxin 1 (contrary to Shiga toxin 2) may also be produced in response to low iron levels, particularly in phage H-19B (Weinstein et al., 1988), but without prophage induction and host-cell lysis this toxin is not transported outside the cell, as *E. coli* lacks an appropriate secretion system for Stx1. Synthesis of Stx2 depends exclusively on the transcriptional activity of the so-called late region of the phage genome, i.e., the region that is expressed at later stages of the phage lytic development (Fig. 3). Stx1 localizes in the periplasmic fraction, while Stx2 is found in the extracellular fraction, suggesting that STEC contains a specific Stx2 release system, contrary to Stx1. It has been demonstrated that an Stx2-encoding phage, but not an Stx1-encoding phage, is spontaneously induced at extremely low rates (Shimizu et al., 2009). These results may suggest that a specific system is involved in Stx2 release from bacterial cells.

The mechanism of λ prophage induction has been investigated in detail; however, this is true only for standard laboratory conditions and a few induction agents, like UV irradiation and mitomycin C (Ptashne, 2004; Węgrzyn and Węgrzyn, 2005). Moreover, functions of some λ genes, particularly those

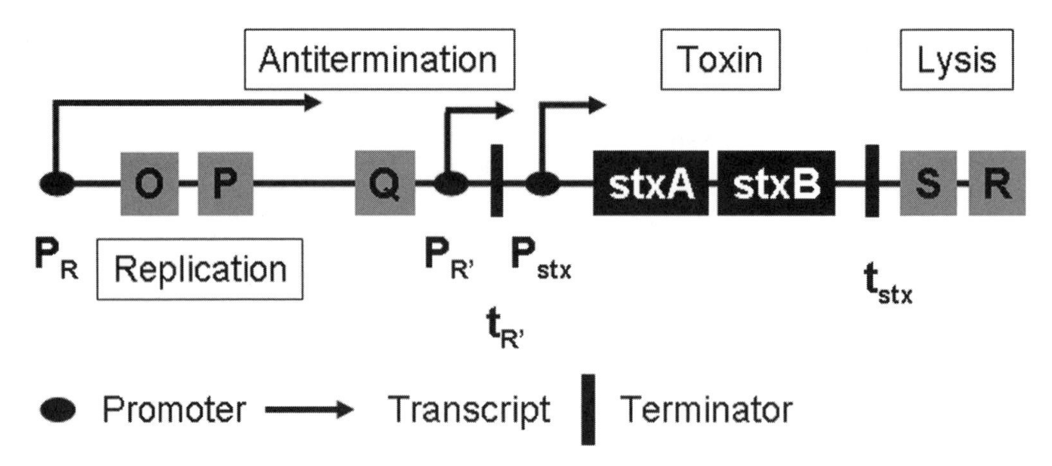

FIGURE 3 The late region of a Shiga toxin-converting lambdoid phage. Genes coding for replication proteins (O and P), a gene for antiterminator Q, the Shiga toxin genes, and genes coding for proteins causing cell lysis are marked. Promoters are shown by ovals and terminators by rectangles, and arrows indicate the directionality of transcription. (Adapted from Łoś [2009].)

located in the *b* region of the phage genome (like the λ DNA fragment between *exo* and *xis* loci), that play roles in lysogenization and prophage maintenance have been found recently (Łoś et al., 2008b), indicating that our knowledge about these processes is still incomplete. Generally, any agent that can provoke the bacterial SOS response is a potential prophage induction agent, as the first step in induction is a RecA-dependent autocleavage of the λ *c*I repressor and subsequent activation of lytic promoters (Ptashne, 2004; Węgrzyn and Węgrzyn, 2005). On the other hand, Shkilnyj and Koudelka (2007) demonstrated that an increased salt concentration (which does not induce the SOS response) caused induction of a λimm^{434} prophage, and this induction was RecA independent. Furthermore, induction of lambdoid prophages under conditions of high pressure (250 to 400 MPa) was reported (Aertsen et al., 2005). However, in contrast to UV irradiation or mitomycin C treatment, the high hydrostatic pressure that induced prophages in a standard laboratory strain (MG1655) of *E. coli* was not able to induce Stx prophages or cause an SOS response in several natural STEC isolates (Aertsen et al., 2005).

UV irradiation and mitomycin C are classical agents that can efficiently induce lambdoid prophages. However, occurrence of these agents in STEC-infected human intestine is very unlikely; thus, a search for other intestinal conditions more likely to cause lambdoid prophage induction is desirable. Recent studies demonstrated that oxidative stress conditions may occur during colonization of human intestine by enteric bacteria (Kumar et al., 2007). Moreover, earlier studies on a clinical isolate of *E. coli* O157:H7 suggested that hydrogen peroxide, produced by human neutrophils, may increase production of Stx2 (Wagner et al., 2001a). Results of recent studies indicated that hydrogen peroxide may cause Shiga toxin-converting prophage induction in both laboratory *E. coli* strains (Łoś et al., 2010) and clinical isolates of STEC (Łoś et al., 2010); however, even at optimal concentrations hydrogen peroxide caused induction of lambdoid prophages in only a fraction of the bacterial cell population (Łoś et al., 2010). These studies were possible in part due to the development of a new procedure of titrating phages that form very small plaques or no plaques in standard laboratory assays (Łoś et al., 2008a).

While oxidative stress, caused under laboratory conditions by addition of hydrogen peroxide to a bacterial culture, may be imitative of a natural condition responsible for induction of Shiga toxin-converting prophages in the intestine, there are other factors and conditions that can modulate both efficiency of prophage induction and further lytic development of the phage, which are also important in determining the intensity of the toxin production. For example, it was reported that Dam methyltransferase is essential for maintenance of lysogeny of one Shiga toxin-converting phage (Murphy et al., 2008). Furthermore, starvation conditions impair both prophage induction and lytic development of lambdoid phages (Czyz et al., 2001; Łoś et al., 2007). Recent studies indicate that replication efficiency of various lambdoid prophages, a process significantly influencing *stx* genes' copy number and thus production of Shiga toxins, may differ under various conditions (Nejman et al., 2009).

Results discussed briefly in preceding paragraphs show that the control of maintenance of the lysogenic state of STEC, as well as of prophage induction, are complicated processes whose efficiencies are difficult to predict on the basis of our current knowledge. A mathematical model for calculating the stability of Shiga toxin-converting prophages has been developed (Evans et al., 2007); while the model is useful in predicting the frequency of spontaneous prophage induction in the laboratory, its efficacy in specific clinical situations is doubtful, given the complexity of the intestinal environment.

The accurate diagnosis of STEC infection is important for medical procedures applied to patients. This is not a trivial problem, as in many medical laboratories routine tests used for identification of STEC are based on microbiological and/or serological methods, which may not always be appropriate. As mentioned above, the STEC phenotype has been initially correlated to the O157 serotype and to an inability to ferment sorbitol (Riley et al., 1983). Therefore, one of the recommended tests is the use of MacConkey sorbitol plates to detect sorbitol-negative *E. coli* strains, while another involves immunological detection of the O157 serotype. Subsequent studies, however, have shown that some *E. coli* O157 strains are sorbitol positive (Birembaux et al., 1999; Karch and Bielaszewska, 2001) and that *E. coli* serotypes other than O157 can be responsible for the STEC phenotype (Blanco et al., 2004; Prere and Fayet, 2005). Thus, other methods are needed to confirm STEC phenotype; one possibility is that genetic tests, based on detection of *stx1* and *stx2* and perhaps also other genes required for effective virulence, may be better for identification of STEC. The use of PCR and/or biochip technology to detect either phage DNA (Łoś et al., 2008c) or virions (Łoś et al., 2005) may be the solution for the current problem of adequately detecting STEC and/or Shiga toxin-converting phages in both clinical and environmental (including food) samples.

Treatment of patients with STEC infections appears to be especially complicated and difficult. This is due mainly to the mechanism of STEC pathogenicity, which depends on prophage induction and subsequent production of toxins. Because many antimicrobial agents (e.g., quinolones, trimethoprim, and furazolidone) are also potent lambdoid prophage inducers, antibiotic therapy for STEC-infected patients may increase toxin production by 100-fold or more. Thus, it is recommended that such antibiotics be avoided when treating patients with potential or confirmed STEC infections (Kimmitt et al., 2000). Current treatment recommendations are to maintain hydration to prevent thrombotic complications (Serna and Boedeker, 2008). Recent studies demonstrated that chemical or cellular sources of nitric oxide (NO) inhibit spontaneous and mitomycin C-induced *stx* mRNA expression and Stx synthesis, without altering bacterial viability (Vareille et al., 2007). The production of *stx*-bearing progeny phage is also reduced by NO. This inhibitory effect may occur through the NO-mediated sensitization of STEC because mutation of the NO sensor (nitrite-sensitive repressor) results in loss of NO-inhibiting activity on *stx* expression.

The question of why bacteria produce Shiga toxins is intriguing, since these toxins are synthesized only after prophage induction and the cells that produce them are killed by the activated lytic pathway. Results of recent studies may shed some light on this putative biological paradox. Lainhart et al. (2009) suggested that Shiga toxins may have evolved for the purpose of bacterial antipredator defense. They demonstrated that *Tetrahymena thermophila,* a bacteriovorous predator, is killed when cocultured with bacteria bearing a Shiga toxin-encoding prophage. In cocultures with *Tetrahymena,* the Stx-encoding bacteria display a growth advantage over those that do not produce Stx. *Tetrahymena* is also killed by purified Shiga toxin. Bacterially mediated *Tetrahymena* killing was blocked by mutations that prevented the bacterial SOS response or by enzymes that break down hydrogen peroxide, suggesting that the production of this compound by *Tetrahymena* signals its presence to the bacteria, leading to bacteriophage induction and production of Stx. Given the recent discovery that hydrogen peroxide induces prophages in only a fraction of cells from a population of lysogenic STEC (Łoś et al., 2010), the benefit of the prophage induction for the host cells may be easier to understand. If only some bacterial cells are lysed by bacteriophages that at the same time produce toxins that can eliminate the enemy, such a loss increases the probability of survival for the rest of the population, thereby providing a selective advantage for strains lysogenic with Shiga toxin-converting phages. The same rationale may also apply under conditions existing in the human intestine, where hydrogen peroxide-producing neutrophils are as dangerous to bacteria as are unicellular eukaryotic predators.

Staphylococcal Toxins

Staphylococcus aureus is a leading cause of gastroenteritis, resulting from the absorption of staphylococcal enterotoxins after consumption of contaminated food (reviewed in Le Loir et al., 2003). *S. aureus* is also the leading cause of mammary gland infections in dairy animals (Nickerson et al., 1995). Staphylococcal virulence is multifactorial, and staphylococci produce an array of virulence factors of three basic classes (Novick, 2006). Secreted proteins, which are assumed to facilitate attack of the structural elements of host tissues and organs, include superantigens, cytotoxins, and tissue-degrading enzymes. Cell surface-bound proteins, including fibrinogen binding protein, fibronectin binding protein, collagen binding protein, and other adhesins and antiopsonins, promote binding of the bacteria to cell surfaces, provide resistance to host defenses, and in some cases facilitate internalization by host epithelial and endothelial cells. Cell surface components include components of both the cell wall peptidoglycan and the polysaccharide capsule. While cell wall components play a role in septic shock, they are essential housekeeping genes. The remainder of these virulence factors are considered accessory factors; while important for adaptation and exploitation of the host, they are nonessential for bacterial growth. These accessory factors are generally associated with mobile genetic elements—plasmids, transposons, pathogenicity islands, and prophages.

Lysogenic conversion of staphylococci associated with expression of virulence factors was first reported in the early 1960s (Blair and Carr, 1961; Winkler et al., 1965). Subsequent molecular characterization of converting phages, discussed in detail below, reveals that these phages belong to a large family of related temperate double-stranded DNA staphylococcal phages. These phages, all members of the family *Siphoviridae,* fall into at least five different serogroups and two different morphotypes, one with an icosahedral capsid and one with an elongated head (Ackermann and DuBow, 1987). Variations in morphology notwithstanding, sequence analysis reveals that these phages share a common genetic organization and display a mosaic pattern of relatedness at the genetic level (Kwan et al., 2005), consistent with the modular theory of phage evolution. The genes for the phage-encoded virulence factors lie, in most cases, at one end of the prophage genome. This suggests that they may have been acquired by an aberrant prophage excision, similar to the generation of specialized transducing phages, although this has not been proven. It is also possible that this is one of the few places in the phage genome where insertion of foreign DNA does not interfere with expression of essential phage lytic functions.

The staphylococcal lysogenic conversion literature includes two well-documented examples of negative conversion, where staphylococcal phages (in many cases, phages carrying additional lysogenic conversion genes) insertionally inactivate chromosomal genes that encode important exoproteins. While not part of the scope of this discussion, they are mentioned here for the sake of completeness and to avoid confusion, since these are also referred to as converting phages in the literature. There are numerous examples of phages that integrate into the β-hemolysin *(hlb)* gene (Coleman et al., 1986, 1989; Dempsey et al., 2005; Kumagai et al., 2007; Sumby and Waldor, 2003; van Wamel et al., 2006; Winkler et al., 1965; Zabicka et al., 1993), and others with an attachment site in the lipase *(geh)* gene (Lee and Iandolo, 1986; Rosendal et al., 1964).

Early reports of α- and δ-hemolysin conversion (Clecner and Sonea, 1966; Diep et al., 2008) have not been confirmed by subsequent research and will not be discussed further.

EXFOLIATIVE TOXIN A

Staphylococci that produce exfoliative toxins are associated with several blistering skin diseases, ranging in severity from bullous impetigo to staphylococcal scalded-skin syndrome (Ladhani, 2001; Stanley and Amagai, 2006). Exfoliative toxins were clearly implicated in causing staphylococcal scalded-skin syndrome nearly 40 years ago, when the purified toxins were shown to be able to induce separation of the epidermal layers when injected subcutaneously into newborn mice (Melish and Glasgow, 1970; Melish et al., 1972). There are two serologically distinct isoforms, ETA and ETB, that have been implicated in human disease and cause both bullous impetigo and scalded-skin syndrome (Johnson and Bradshaw, 2001; Kondo et al., 1975). A third family member, ETD, has been found to be associated with other staphylococcal skin and soft tissue infections (Yamasaki et al., 2006). The action of ETA and ETB is characterized by a specific separation of the layers of the skin due to loss of epidermal cell adhesion at the desmosomes. Insight into the mechanism of toxin action was provided by the crystal structure of ETA (Color Plate 4) (Cavarelli et al., 1997; Papageorgiou et al., 2000; Vath et al., 1997) and its comparison to ETB (Papageorgiou and Acharya, 2000; Papageorgiou et al., 2000; Vath et al., 1999), which indicated that both proteins belong to the chymotrypsin-like family of serine proteases. Subsequent studies (reviewed in Nishifuji et al., 2008) demonstrated that the exfoliative toxins are glutamate-specific serine proteases that specifically cleave a single peptide bond between cadherin repeats in the extracellular region of desmoglein 1 (Dsg1), a desmosomal intercellular adhesion molecule (see Color Plate 4) (Amagai et al., 2000, 2002; Hanakawa et al., 2004). The resulting loss of cell-cell adhesion between keratinocytes leads to a breakdown of the cutaneous defense barrier, which facilitates bacterial invasion into the skin.

In addition to the well-documented role of the exfoliative toxins as Dsg1-specific proteases, ETA and ETB have also been reported to act as superantigens, although this remains controversial. Superantigens, which are discussed in more detail below, bind directly to the major histocompatibility complex (MHC) on the surface of antigen-presenting cells and cross-link with the variable Vβ region of the β-chain of the T-cell receptor, activating T cells in an antigen-independent fashion. Both ETA and ETB were reported to stimulate not only T-cell proliferation (Morlock et al., 1980) but also the release of tumor necrosis factor α, interleukin-6, and NO from murine macrophages (Fleming et al., 1991). Following the initial reports of superantigen activity,

Fleischer and Bailey (1992) reported that recombinant ETA purified from *S. aureus* did not have mitogenic activity and suggested that the activity reported previously was attributable to contamination of the toxin preparations with small amounts of other highly potent superantigens. Plano et al. (2000), using recombinant ETA and ETB purified from *E. coli,* also reported a lack of mitogenic activity while demonstrating that these toxins produced in *E. coli* retained exfoliating activity. In contrast, a number of studies using recombinant ETA and ETB purified from a superantigen-free strain of *S. aureus* have reported that both of these proteins can stimulate T-cell proliferation in a Vβ-dependent manner (Monday et al., 1999; Vath et al., 1997, 1999). Furthermore, substitution of the active-site serine in ETA eliminated protease activity, but did not impair superantigen activity (Vath et al., 1997). Compared to the prototypical superantigens, ETA and ETB show about 100-fold less potency in induction of T-cell proliferation, as well as reduced toxicity in a rabbit toxic shock model (Monday et al., 1999). Taken together with the demonstrated proteolytic activity of the exfoliative toxins and their lack of structural similarity to other superantigens, the data suggest that the exfoliative toxins represent a novel class of superantigens. There is little evidence to suggest that this function plays a role in the pathogenesis of scalded-skin syndrome, but it is possible that these toxins could also play a role in other diseases.

ETB and ETA share 40% sequence identity (Lee et al., 1987; O'Toole and Foster, 1987). ETB is plasmid encoded, while ETA was initially reported to be chromosomal (Jackson and Iandolo, 1986a, 1986b; Johnson et al., 1979; Rogolsky et al., 1976; Rosenblum and Tyrone, 1976). The *eta* gene encodes the mature polypeptide as well as a 38-amino-acid (aa) signal peptide that is subsequently removed (Lee et al., 1987; Sakurai et al., 1988). Based on the observation that only a limited number of *S. aureus* strains produce exfoliative toxins (Ladhani et al., 1999), it was suspected that the gene encoding ETA might also be carried on a mobile genetic element. Phage conversion to ETA production was shown by Yamaguchi et al. (2000), who determined the complete sequence of an ETA-encoding phage (ϕETA) that they isolated from *S. aureus* strain E-1. Yoshizawa et al. (2000) also described the isolation of an ETA-converting phage the same year from another ETA-producing strain. Genome sequences for two additional ETA-converting phages were reported by Sugai and Sawada: ϕETA2 from *S. aureus* strain TY94 (Entrez NC_008798) and ϕETA3 from *S. aureus* strain TY32 (Entrez NC_008799). Two ETA-converting phages have also been reported in ETA-positive bovine isolates of *S. aureus* from cows with mastitis (Endo et al., 2003). In all three sequenced phage genomes, the *eta* gene is located at the far right end of the phage genome, between the lysis genes and the phage attachment site (Fig. 4). This location is conserved in all additional *eta*-positive isolates examined thus far, as assayed by Southern blot analysis with probes to *eta* and the holin gene *(hol)* (Yamaguchi et al., 2000) or by sequence analysis of amplified regions flanking the *eta* gene (Endo et al., 2003).

FIGURE 4 Map of the right end of the prophage genome from phages ϕETA (NC_003288) (Yamaguchi et al., 2000), ϕETA2 (NC_008798), and ϕETA3 (NC_008799) showing the locations of the genes for the phage lysis functions (white) and exfoliative toxin A (gray).

The *eta* gene is expressed when cloned in *E. coli,* and putative promoter and termination sequences flanking the gene have been identified (Lee et al., 1987; O'Toole and Foster, 1987). Expression of *eta* in *S. aureus* is regulated by the accessory gene regulator *(agr)* locus (O'Toole and Foster, 1986; Sheehan et al., 1992), which encodes a global quorum-sensing system that controls expression of many staphylococcal exoproteins (Recsei et al., 1986). As expected for an *agr*-dependent gene, *eta* expression increases significantly during transition from exponential- to stationary-phase growth. Sheehan et al. (1992) also reported that *eta* expression decreased in response to increased osmolarity and was stimulated by treatment with novobiocin, which relaxes DNA supercoiling. Sakurai et al. (2004) reported that a gene upstream of *eta* in the prophage present in *S. aureus* strain ZM was involved in positive control of *eta* expression. A BLAST search reveals that the coding sequence they identified, and named ETA^{exp}, is in fact the phage amidase *(ami)* gene. It seems unlikely that this gene is involved in controlling *eta* expression, and no additional studies have been done to map transcriptional start sites or further elucidate *eta* regulation.

There is another conserved *S. aureus* protein, belonging to the C4-dicarboxylate transporter/malic acid transport protein family, that is also annotated as "exfoliative toxin A" in a number of the *S. aureus* genome sequences. This protein is completely unrelated to ETA. It is not clear where this misidentification initially occurred, but it has now unfortunately propagated itself throughout the database.

PVL LEUKOCIDIN

The Panton-Valentine leukocidin (PVL), first described by Panton et al. (1932), is a two-component, pore-forming cytotoxin that assembles into oligomeric transmembrane complexes in human polymorphonuclear leukocytes, which results in opening of calcium channels and leads to cell lysis (Finck-Barbancon et al., 1993). It is carried by about 2% of *S. aureus* isolates (Holmes et al., 2005) and was initially associated with severe skin infections and necrotizing pneumonia (Gillet et al., 2002; Lina et al., 1999; Panton et al., 1932). The role of PVL as a virulence factor remains highly controversial. Much recent interest has focused on the possible association between PVL production and successful lineages of community-associated methicillin-resistant *S. aureus* (CA-MRSA). The five predominant CA-MRSA clonal lineages that are associated with widespread disease outbreaks carry PVL, and there have been hundreds of reports over the past decade or so noting an association between PVL and outbreaks of disease caused by CA-MRSA (for recent reviews see Diep and Otto, 2008; and Kobayashi and DeLeo, 2009). However, a study by Voyich et al. (2006) of otherwise isogenic PVL^+ and PVL^- strains of two predominant CA-MRSA lineages showed that they were equally virulent in mouse abscess and sepsis models and also lysed human neutrophils at comparable rates. This led to the conclusion that PVL was not the primary virulence determinant in CA-MRSA. In contrast, Labandeira-Rey et al. (2007) reported that in a mouse acute pneumonia model of infection, PVL was critical to disease production and led to global changes in gene expression. However, similar studies found that α-hemolysin, but not PVL, made a significant contribution to pneumonia in mice (Bubeck Wardenburg et al., 2007). This apparent discrepancy was recently resolved by the discovery that the PVL^- strain used in the studies by Lanbandeira-Rey carried a point mutation in the *agr* promoter, and once this defect was repaired there were no significant differences found in gene expression or pathogenesis between the PVL^+ and PVL^- strains (Villaruz et al., 2009). Taken as a whole, these studies convincingly demonstrate that PVL is not a key virulence factor in mouse models of CA-MRSA disease. Whether the mouse models are valid models of human disease, however, remains open to question. In a bacteremia model using rabbits, in which the granulocytes are more sensitive than those of mice to the effects of PVL, the toxin was found to play a modest role in pathogenesis during early stages of infection (Diep et al., 2008). A recent study by Hongo et al. (2009) revisited the lysis of human neutrophils by PVL. They found that recombinant PVL toxins exhibited strong lytic activity against human neutrophils, as did culture supernatants from CA-MRSA strains. The lytic activity of these supernatants was neutralized by anti-

PVL monoclonal antibodies. Two findings from this study may help explain some of the conflicting results in the literature. First, the amount of PVL secreted by different CA-MRSA strains was found to be highly variable. Second, although PVL showed lytic activity against human neutrophils, murine neutrophils were unaffected by the toxin. Lysis of human polymorphonuclear cells by PVL is likely to play a role in breaching the body's defense system, but further studies will be required to clearly demonstrate a connection between PVL and specific *S. aureus* diseases.

PVL belongs to a family of β-barrel pore-forming *S. aureus* toxins that includes the homo-oligomeric α-hemolysin and three hetero-oligomeric bicomponent cytolysins: γ-hemolysin (Hlg), leukocidin (Luk), and PVL leukocidin (reviewed in Kaneko and Kamio, 2004; and Menestrina et al., 2003). The hetero-oligomeric toxins are composed of subunits of two different but related proteins of ~30 kDa that cooperatively form the active toxin: class F subunits (LukF and LukF-PV) and class S subunits (Hlg2, LukS, and LukS-PV). The proteins within each class are about 70% identical to each other, while there is about 30% identity between class S and class F subunits (Kamio et al., 2001). The class S subunit of the complex determines the cell specificity of the cytolysin. These toxins are secreted as monomers that oligomerize cooperatively and bind strongly to target cells, forming small pores with inner and outer diameters of 2 and 8 nm, respectively (Tomita and Kamio, 1997). X-ray crystallography of the homo-oligomeric α-hemolysin pore (Song et al., 1996) showed that it was a mushroom-shaped, hollow heptamer, consisting almost entirely of β-structure. The crystal structure of the monomeric LukF-PV subunit revealed a core structure very similar to that of α-hemolysin, and an additional glycine-rich stem domain (Pedelacq et al., 1999).

Modeling studies of this and the structurally related LukF monomer (Olson et al., 1999) suggest that the stem domain must undergo major rearrangement during pore formation. The precise stochiometry in the hetero-oligomeric toxins is still uncertain; there have been reports suggesting that the pore contains six, seven, or eight subunits (Jayasinghe and Bayley, 2005; Olson et al., 1999; Pedelacq et al., 1999; Sugawara-Tomita et al., 2002), with equal or approximately equal numbers of each of the two different protein subunits. The model of PVL shown in Color Plate 5 assumes a hexameric structure with three subunits each of LukF-PV and LukS-PV (Pedelacq et al., 1999).

The subunits for Hlg (LukF and Hlg2) and Luk (LukF and LukS) are encoded by the chromosomal *hlg* locus, a three-gene cluster that is present in more than 99% of *S. aureus* clinical isolates. The separate *pvl* locus, encoding PVL, was found in less than 2% of strains and contains two cotranscribed genes, *lukS*-PV and *lukF*-PV (Prevost et al., 1995a, 1995b). Lysogenic conversion of *S. aureus* to leukocidin production was first reported by van der Vijver et al. (1972). Mitomycin C induction of an icosahedral phagelike particle carrying the PVL genes on a 41-kb double-stranded DNA phage genome was first described by Kaneko et al. (1997), and the complete phage genome of this phage, ϕPVL, was reported the following year (Kaneko et al., 1998). The location of the two genes encoding PVL is between the lysis genes and *attR*, as was seen for *eta*. However, ϕPVL appears to be a defective phage; the particles do not have tails and no infection could be demonstrated. A prophage genome in *S. aureus* P83 carrying PVL, ϕPV83-pro, was reported in 2000 but this one also appeared defective. It contained insertion sequences at two positions within the phage genome, and no phage particles were induced following treatment with mitomycin C (Zou et al., 2000). The first analysis of a bona fide PVL-converting phage, ϕSLT, was reported in 2001 (Narita et al., 2001). PVL prophages have been identified in sequenced *S. aureus* genomes: ϕSa2mw in MW2 (Baba et al., 2002) and ϕSa2USA300 in USA300 (Diep et al., 2006). Others have been characterized by amplifying and sequencing the genes encoding PVL and flanking prophage DNA from otherwise unsequenced clinical isolates: ϕ108PVL (Ma et al., 2006) and ϕSa2958 (Ma et al., 2008). Of these, ϕSa2mw and ϕSa2958 have been shown to generate infective PVL phages, while ϕ108PVL is induced following treatment with mitomycin C but infective phages have not been detected (Ma et al., 2008). ϕSa2USA300 has recently been shown to be defective for induction (Wirtz et al., 2009). ϕ108PVL also contains an additional gene, homologous to a transposase gene, between the lysis gene cluster and the PVL gene cluster (Fig. 5). These phages fall into two morphologically distinct families. The ϕPVL particles have an icosahedral head about 60 nm in diameter, while ϕSLT has an elongated head of about 50 × 100 nm and a long flexible tail of about 400 nm. The genome sequences are consistent with this observed morphological difference. Outside of the PVL and lysis genes, the *int* genes are homologous but there is almost no additional sequence similarity (Fig. 5). The additional sequenced PVL-carrying prophages fall into these two different lineages (Fig. 5); ϕPV83-pro and ϕ108PVL resemble ϕPVL, while ϕSa2mw, ϕSa2958, and ϕSa2USA300 resemble ϕSLT. PCR analysis of a collection of clinical strains in Japan with probes specific to unique regions of these two different phage types indicated the presence of phages of both lineages (Ma et al., 2008). Phages belonging to the group with an elongated head predominated in the more recent isolates. Since all three current examples of the icosahedral type are defective, this led to the speculation that these phages were no longer spreading PVL in current MRSA populations. Ma et al. (2008) also demonstrated the presence of inducible PVL-converting phages in this collection of clinical isolates; these phages, by PCR analysis, were all of the elongated head type.

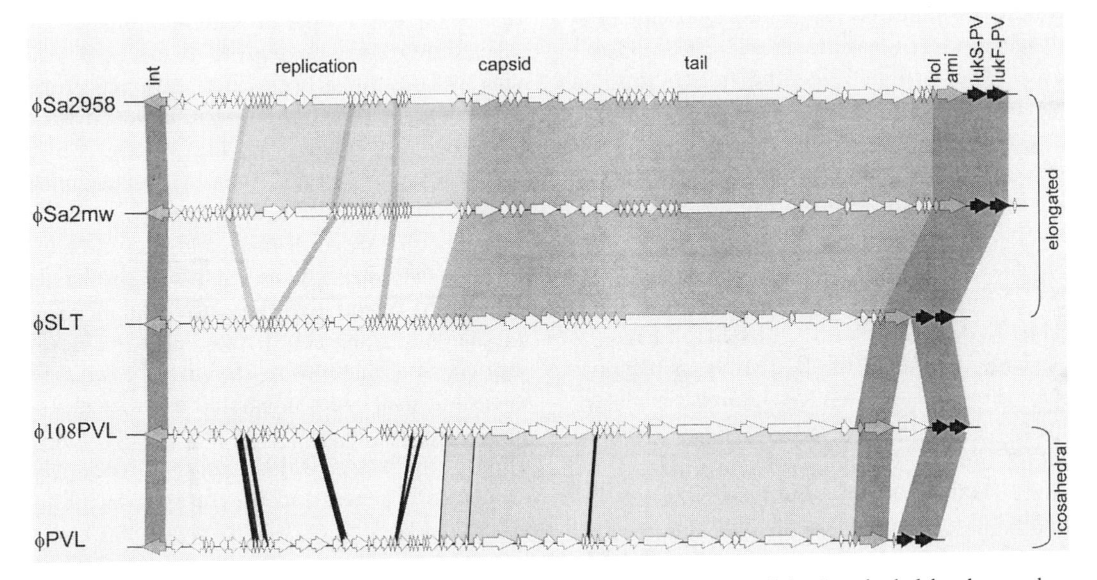

FIGURE 5 Map of PVL-converting phages. Shown are the two phages of the icosahedral head type that, although defective, produce phagelike particles and the three phages of the elongated head type that have been shown to form infective PVL phages. The *luk*-PVL genes are indicated as black arrows, the conserved phage lysis genes and integrase are gray, and all other phage genes are white. Dark gray shading between genomes indicates genes conserved among all of these phages (*int, hol, ami,* and the PVL genes *lukS*-PV and *lukF*-PV), intermediate gray shading indicates genes shared by three phages, and light gray shading indicates genes shared by two.

Like many other phage-encoded virulence factors, transcription of the *luk*-PV locus is integrated into the host regulatory circuitry. The two genes are cotranscribed as a single 2.3-kb mRNA (Wirtz et al., 2009). Transcription of *luk*-PV has been reported to depend on the activity of *agr, sar,* and *sae* (Bronner et al., 2000; Voyich et al., 2009; Wirtz et al., 2009). Despite lower expression of *luk*-PV in *agr* mutants, however, the characteristic increase in expression when cells enter late exponential phase was not observed (Wirtz et al., 2009). Treatment with subinhibitory concentrations of β-lactam antibiotics has also been reported to increase expression (Dumitrescu et al., 2007; Stevens et al., 2007). This is presumably due to induction of the SOS response by β-lactams in *S. aureus* (Maiques et al., 2006). Production of mRNA is increased following induction with mitomycin C, commensurate with the increase in gene dosage. A longer transcript initiating farther upstream was also seen after mitomycin C treatment, indicating that activation of a phage promoter also contributes to increased expression following prophage induction (Wirtz et al., 2009). However, PVL protein levels did not increase in parallel with *luk*-PV mRNA levels, suggestive of posttranscriptional regulation as well (Wirtz et al., 2009). The composition of the culture medium has also been reported to affect PVL levels (Bronner et al., 2000; Dobardzic and Sonea, 1971).

THE IMMUNE EVASION CLUSTER

A large number of β-hemolysin-negative staphylococcal isolates, containing prophages integrated in the *hlb* gene, produce one or more immune evasion molecules. These include staphylokinase, enterotoxin A (or P), staphylococcal complement inhibitor (SCIN), and chemotaxis inhibitory protein (CHIPS). Coconversion of staphylokinase and enterotoxin by *hlb*-integrating phages has been well documented over the past 2 decades (Carroll et al., 1993; Coleman et al., 1989; Dempsey et al., 2005; Kumagai et al., 2007; Sumby and Waldor, 2003; Zabicka et al., 1993). SCIN and CHIPS are more recently recognized innate immune modulators (de Haas et al., 2004; Rooijakkers et al., 2005a, 2005b, 2005c, 2006) that also lie on prophages integrated into *hlb* (van Wamel et al., 2006). The genes for all of these modulators lie in an 8-kb region at the 3′ end of the prophage genome, which has been named the immune evasion cluster (IEC) (van Wamel et al., 2006). Comparison of the sequences of six different prophages inserted into *hlb* in the genomes of sequenced *S. aureus* strains (Iandolo et al., 2002; Sumby and Waldor, 2003; van Wamel et al., 2006) revealed variants with four different combinations of these genes. Analysis of a collection of human clinical isolates by PCR and Southern blot analysis revealed that 89% of them carried prophages with an IEC cassette, representing seven different variants, or IEC types (Fig. 6) (van Wamel et al., 2006). Phages containing six of these IEC types were detected following mitomycin C induction, and stable conversion was demonstrated for four of them. The phages carrying these IEC cassettes fell into five immunity groups, and the same IEC type was found on different phages, as assayed by restriction analysis. A survey of bovine isolates from both bulk milk and from cows with mastitis showed a lower overall percentage of phage-carrying strains, 4.5%, representing five of the seven IEC types found in the human isolates. Phage particles were isolated from three of these and shown to be *Siphoviridae* with icosahedral heads similar to those seen for ϕETA (Kumagai et al., 2007). Several *hlb*$^{+}$ *sak*$^{+}$ strains have also been described, and these were shown to have *sak*-encoding prophages integrated at other chromosomal sites (Goerke et al., 2006). Thus, diverse phages carry this cluster of immune-modulatory genes, and they are prevalent in clinical isolates.

STAPHYLOKINASE

Conversion to staphylokinase production was the first convincing demonstration in *S. aureus* of a virulence factor carried on a phage (Winkler et al., 1965). Staphylokinase is a 15.5-kDa protein (136 aa) that functions as a plasminogen activator (Bokarewa et al., 2006).

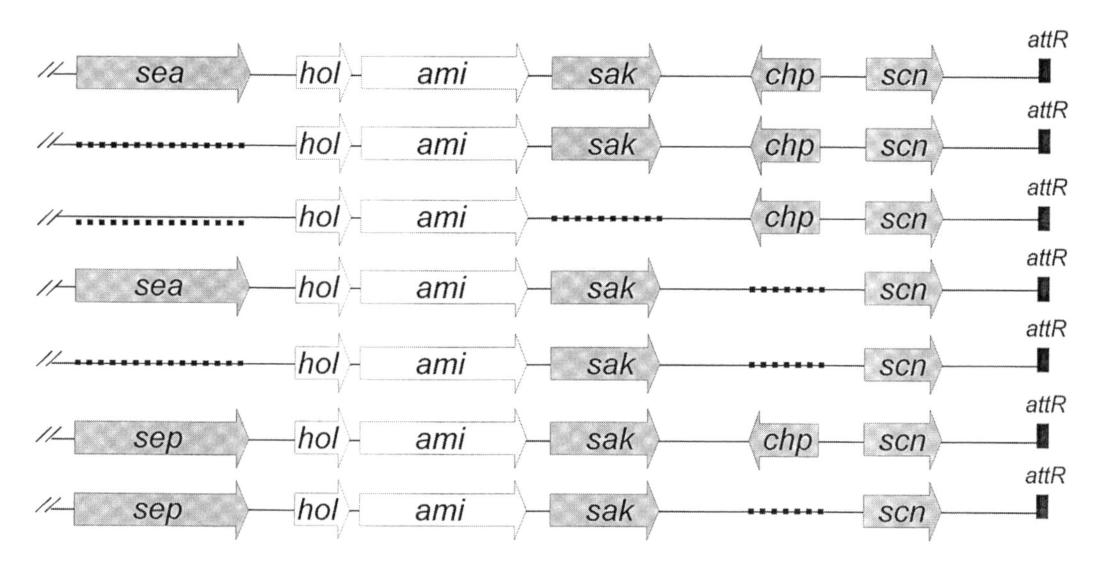

FIGURE 6 Schematic of the arrangement of genes in the IEC at the right end of the prophage genome. The seven IEC types described by van Wamel et al. (2006) are shown. The phage lysis genes are shown in white and the lysogenic conversion genes in gray. Dotted lines indicate regions where the lysogenic conversion gene is either absent, in some isolates, or replaced with open reading frames of unknown function in others.

The binding of staphylokinase to plasminogen, in a 1:1 stochiometry, results in the formation of active plasmin, which has broad-spectrum proteolytic activity against extracellular matrix proteins. It is thought that disruption of the fibrin net which often forms around an infectious focus may promote dissemination of the staphylococci into deeper host tissue. The three-dimensional structure of staphylokinase reveals a central α-helix and a five-strand β-sheet; the α-helix contains the region of main interaction between staphylokinase and plasminogen (Ohlenschlager et al., 1998; Parry et al., 1998) (Color Plate 6).

Staphylokinase may also contribute to bacterial colonization by interacting with the immune system of the host. Recent evidence suggests that staphylokinase also interacts with α-defensins (Jin et al., 2004), which may help promote bacterial survival. Human neutrophil-derived α-defensins are small (3- to 4-kDa), cationic, microbicidal peptides; they exert broad-spectrum antimicrobial activity by forming pores in bacterial membranes, which results in cell lysis (White et al., 1995). Defensins have also been reported to enhance the production of proinflammatory cytokines, enhance phagocytosis, promote recruitment of neutrophils, suppress anti-inflammatory mediators, and regulate complement activation (Yang et al., 2002). These observations suggest that defensins upregulate innate host inflammatory defenses against microbial invasion in addition to promoting direct bacterial lysis. Staphylokinase was shown to stimulate release of defensins from polymorphonuclear leukocytes and to neutralize the bactericidal effects of α-defensins by complexing with them (Jin et al., 2004). This defensin-binding activity was distinct from the plasminogen-activating activity of staphylokinase. Thus, staphylokinase appears to play multiple roles in staphylococcal virulence, including promotion of bacterial colonization/invasion of host tissue, resistance of bacteria to the microbicidal effects of defensins, and possibly interference with defensin-mediated upregulation of antimicrobial host inflammatory defenses.

The staphylokinase gene, *sak,* is 489 base pairs long and encodes a precursor protein that contains a 27-aa signal sequence at the N terminus (Sako and Tsuchida, 1983). Staphylokinase, like a number of other exoproteins, appears to be *agr* regulated. Although expressed constitutively when cloned in *E. coli* (Sako et al., 1983; Sako and Tsuchida, 1983), expression of the *sak* gene in *S. aureus* is decreased about 50-fold by disruption of the *agr* gene (Recsei et al., 1986). A similar, though smaller effect of *agr* on *sak* was reported by Rooijakkers et al. (2006). That study also examined several additional global regulators previously implicated in control of staphylococcal virulence gene expression, including the two-component response regulator SaeR/S, the staphylococcal accessory regulator SarA, and the alternative sigma factor σ^B (Chan and Foster, 1998; Giraudo et al., 1999a, 1999b; Ziebandt et al., 2001). Expression of *sak* was upregulated in a *sarA* mutant and a *sigB* mutant, while mutation of *sae* had no effect (Rooijakkers et al., 2006). Sumby and Waldor (2003) examined the effect of mitomycin C on *sak* transcription and found that, as described above for *luk*-PV, *sak* mRNA increased after induction and this increase was dependent on phage replication. Northern blot analysis revealed a 1-kb *sak* transcript produced at low levels preinduction and at elevated levels postinduction. In addition, similar to what was seen with *luk*-PV, mitomycin C treatment also led to the appearance of a longer transcript that presumably initiated upstream from a latent phage promoter activated following prophage induction. Similar results have been reported after mitomycin C induction of additional *sak*-converting phages, including some integrated into other chromosomal sites (Goerke et al., 2006; Wirtz et al., 2009), but none of these promoters have been characterized.

ENTEROTOXINS

Staphylococci encode a large family of related toxins that function as superantigens, a well-defined group of secreted proteins that bind to MHC class II molecules on the outside of antigen-presenting cells and cross-link them with the variable Vβ region of the β-chain of the T-cell receptor (Fraser and Proft, 2008) (Color Plate 7). Because interaction between the superantigen and the T-cell receptor is based on Vβ recognition rather than antigen specificity, large numbers of T cells are stimulated. This leads to rapid production of high levels of cytokines and proinflammatory mediators. In addition to well-established roles for staphylococcal superantigens in staphylococcal food poisoning (Tranter, 1990) and toxic shock (McCormick et al., 2001; Todd et al., 1978), studies have reported an association of superantigen-producing strains with atopic dermatitis, allergic rhinitis, and upper airway disease (Bachert et al., 2008; Leung et al., 1995; Rossi and Monasterolo, 2004). Over 20 serologically distinct superantigens, which share between 15 and 90% amino acid identity, have been identified in *S. aureus*. Each superantigen interacts with a defined subset of T cells. The specificity of the interaction is determined by recognition between the superantigen and the Vβ region of the T-cell receptor (Sundberg et al., 2007). Crystal structures have been reported for 10 staphylococcal superantigens (Fraser and Proft, 2008). They all share a compact, two-domain globular structure. The topology of the N-terminal domain resembles a well-known oligonucleotide binding motif, the so-called OB fold, while the C-terminal domain contains a β-grasp motif. Despite the similarities in structure, superantigens differ in the way they bind to MHC class II molecules. Some recognize the α-chain of MHC class II; this binding is mediated by the N-terminal OB domain of the superantigen. Others bind to the α-chain of MHC class II via a site in the C-terminal β-grasp domain and coordination of a zinc atom.

The staphylococcal superantigens are encoded on a variety of accessory genetic elements, including phages, plasmids, and pathogenicity islands (reviewed in Novick, 2006). Enterotoxin A (SEA) was shown to be phage encoded over 20 years ago by Betley and Mekalanos (1985). SEA was initially identified in an isolate from an outbreak of staphylococcal food poisoning (Bergdoll, 1983) and is a major cause of food-borne intoxication (Holmberg and Blake, 1984) as well as non-menstrual toxic shock syndrome (Dauwalder et al., 2006; Ferry et al., 2005). It belongs to the β-grasp family of superantigens and was the first zinc binding enterotoxin for which a structure was determined (Schad et al., 1995). Enterotoxin P, which has not been extensively characterized, was first identified by the presence of its coding sequence *(sep)* on a resident prophage in the genome of N315 (Kuroda et al., 2001). Among human clinical isolates carrying an enterotoxin gene in the IEC (see Fig. 6), *sep* was found in place of *sea* 21% of the time (van Wamel et al., 2006), and was recently found in 8 of 27 enterotoxigenic *S. aureus* food isolates (Bania et al., 2006). SEP shares 77% amino acid identity with SEA.

The *sea* (or *sep*) gene is located upstream of the phage lysis genes (Fig. 6), which argues against initial acquisition of this gene by aberrant prophage excision. Expression of *sea* was shown to be independent of *agr,* unlike a number of the other enterotoxins (Tremaine et al., 1993). Transcription of *sea* initiates 86 base pairs upstream of the translation start site, and deletion analysis indicated that there were no regulatory sequences upstream of the −35 region of the promoter that were required for *sea* expression (Bokarewa et al., 2006). Deletion of sequences between the transcription and translation starts resulted in elevated levels of SEA, suggesting that there may be post-transcriptional regulation of gene expression. Consistent with this, an increase in *sea* transcripts after mitomycin C induction was not paralleled by increased SEA production (Sumby and Waldor, 2003). Analysis of *sea* expression following mitomycin C induction of the ϕSa3ms prophage showed that, similar to what was reported for the distal *sak* gene, there is a replication-dependent increase in the 1.7-kb *sea* transcript (Sumby and Waldor, 2003). Furthermore, as was seen for *sak,* there were also longer *sea*-containing transcripts induced by mitomycin C treatment. These included a 3.0-kb transcript containing just *sea* and a 6.8-kb transcript containing both *sea* and *sak.* Thus, prophage induction appears to lead to an increase in expression of these genes, not only from their own promoters due to an increase in copy number during phage replication, but also due to activation of latent phage promoters during lytic growth.

Prophage ϕSa3ms also encodes two additional enterotoxins, SEG and SEK (Sumby and Waldor, 2003). The *seg2* and *sek2* genes lie at the opposite end of the prophage genome from the IEC cluster, between the repressor and integrase genes (Fig. 7). The prevalence of these particular toxin genes in populations of converting phages or clinical isolates has not been addressed, although a nearly identical prophage, ϕSa3mw, is present in *S. aureus* strain MW2 (Baba et al., 2002). The *sek2* gene encodes a protein that is 97% identical to that encoded by the *sek* gene present on staphylococcal pathogenicity islands carrying clusters of enterotoxin genes (Jarraud et al., 2001; Yarwood et al., 2002). However, *seg2* does not show similarity to the *seg* gene identified by Munson et al. (1998), which also lies in enterotoxin gene clusters (Jarraud et al., 2001; Monday and Bohach, 2001) and encodes the SEG that has been structurally characterized (Fernandez et al., 2007).

Expression of *seg2* and *sek2* was increased following prophage induction with mitomycin C, but to a smaller extent than *sea* or *sak* and independent of phage replication (Sumby

FIGURE 7 Map of the left end of the ϕSa3ms genome, showing the location of the two enterotoxin genes (in gray) relative to the phage genes controlling lysogeny and integration (in white).

and Waldor, 2003). Northern blot and 5′ rapid amplification of cDNA ends analysis indicated that these genes (and probably *int*) were cotranscribed with the phage repressor (*cI*) gene. Thus, regulation of *seg2* and *sek2* depends on factors that influence the expression of *cI*.

CHIPS AND SCIN

CHIPS is a recently discovered 14.1-kDa secreted protein that helps modulate immune evasion by *S. aureus*. CHIPS blocks an early step in the inflammatory response, the chemotaxis of neutrophils and monocytes toward the site of infection. CHIPS binds with high affinity to the receptor for complement component C5a (C5aR) and to the formylated peptide receptor (FPR) (de Haas et al., 2004; Postma et al., 2004), thereby preventing neutrophil activation and chemotaxis in response to these chemoattractants. The binding activities for the two receptors are discrete; the FPR binding activity is located in the first 6 amino acid residues of CHIPS, while a fragment lacking the first 31 amino acids shows normal blocking of the C5aR (Haas et al., 2004, 2005). The C5aR binding domain of CHIPS shows structural similarity to the C-terminal domain of staphylococcal superantigens (Color Plate 8) (Haas et al., 2005). CHIPS was found in over 60% of clinical *S. aureus* isolates surveyed by van Wamel et al. (2006).

SCIN is another recently discovered small secreted protein that acts to interfere with an early step of the innate immune response. The 9.3-kDa SCIN polypeptide displays a compact, triple α-helical bundle structure, with the middle helix positioned antiparallel to both the N- and C-terminal helices (Color Plate 9) (Rooijakkers et al., 2007). SCIN binds to C3 convertases on the bacterial cell surface and blocks the cleavage of the central complement protein C3 into C3a and C3b (Rooijakkers et al., 2005a). A recent study shows that SCIN traps the C3 convertase in an inactive state that is still able to bind C3 but not generate and deposit C3b (Ricklin et al., 2009). This prevents opsonization and protects the bacteria from phagocytosis. In addition, since C3b is required for formation of C5 convertases, SCIN strongly attenuates C5a-induced neutrophil activation and chemotaxis (Rooijakkers et al., 2006). Van Wamel et al. (2006) found the gene for SCIN in 90% of the clinical isolates they examined. This gene is annotated in a number of prophage genomes as "fibrinogen binding protein."

As already mentioned, the genes for both CHIPS (*chp*) and SCIN (*scn*) are located in the IEC cluster on staphylococcal prophages (van Wamel et al., 2006). They lie between *sak* and the end of the prophage genome and are divergently transcribed (Fig. 6). In the single published study of expression of these two genes, Rooijakkers et al. (2006) found that, unlike many of the other secreted *S. aureus* virulence factors, both *chp* and *scn* were expressed at maximum levels during the early exponential phase of growth. This early transcription was paralleled by a rapid appearance of both proteins in culture supernatants. The roles of several of the global regulators known to play roles in staphylococcal virulence gene expression were also examined. Mutation of *agr* had no effect on transcription of *chp*, but levels of *scn* transcription were lower in the postexponential and stationary phases of growth. Deletion of *sae* resulted in severely depressed expression from both *chp* and *scn*, while expression of both genes was elevated in a *sigB* mutant. Although there was some variation in different strain backgrounds, transcription of *chp* was upregulated in a *sarA* mutant, while *scn* expression was lower in the absence of *sarA*. Thus, several of the major *S. aureus* regulatory loci appear to control expression of these genes, but the two genes are regulated differently and also differently from *sak*, as discussed above. High expression of *chp* and *scn* during the early stages of staphylococcal growth is consistent with their proposed role in modulating the innate immune response during the initial phases of infection.

RESTRICTION-MODIFICATION

There is one additional reported example of prophage conversion genes that, like *seg2* and *sek2,* map just upstream of the *int* gene at the left end of the prophage genome (Fig. 8). These two slightly overlapping open reading frames, found on a converting phage that also carries *sak* and *sea,* encode a BcgI-like restriction-modification system (Dempsey et al., 2005). These genes confer resistance to lysis by all phages of the International Basic set of *S. aureus* typing phages. The larger open reading frame encodes the putative restriction and modification subunit (RM), while the smaller one encodes the putative specificity subunit (S). These genes differ from all of those previously discussed in that they are not secreted virulence factors. They do, however, confer an obvious selective advantage to both the phage and to the bacteria carrying it, by providing protection from lytic infection by other staphylococcal phages.

C. botulinum Neurotoxin

Seven serogroups (A to G) of botulinum neurotoxin (BoNT) are known to be produced by the soil bacterium *C. botulinum;* they act at neuromuscular junctions to block transmitter release and are exquisitely toxic, with less than 1 μg being lethal for humans (Johnson and Bradshaw, 2001). BoNTs exist in high-molecular-weight complexes (300 kDa to 1.1 MDa) (Arndt et al., 2005; Johnson and Bradshaw, 2001; Lietzow et al., 2008) that include the neurotoxin; hemagglutinins HA17 (HA2), HA34 (HA1), and HA70 (HA3); and a nontoxic and nonhemagglutinating protein (NTNH). The neurotoxin is synthesized as a 150-kDa precursor protein, which undergoes activation through proteolytic cleavage into a heavy chain (98 kDa) and a light chain (50 kDa) that remains associated through a disulfide bridge. The light chain requires a bound zinc ion for its toxicity (Brin, 1997). HA70 also undergoes posttranslational cleavage to produce HA22-23 (HA3a) and HA48-53 (HA3b). Recent work by Lietzow et al. (2008) suggests that in type A *C. botulinum,* the toxin and NTNH protein are present in a 1:1 molecular ratio, in association with approximately 22 subunits of the hemagglutinating proteins. These accessory proteins function to protect BoNT from denaturation, are involved in binding to intestinal microvilli, and facilitate uptake of toxin and transcytosis (Arndt et al., 2005; Nakamura et al., 2009). The BoNT heavy chain interacts with two receptors on the neuronal cell surface: namely, synaptotagmin and a ganglioside, often the trisialoganglioside GT1b [II^3(Neu5Ac)$_2$, IV^3(Neu5Ac)GgOse$_4$Cer] {[α-Neu5Ac-(2→3)]-β-Gal(1→3)-β-GalNAc-(1→4)-[α-Neu5Ac-(2→8)-α-Neu5Ac-(2→3)]-β-Gal-(1→4)-β-Glc-(1→1′)-Cer} (Kozaki et al., 1984). The latter was first shown by Simpson and Rapport (1971), who demonstrated that brain gangliosides inactivated BoNT. It was later confirmed by showing that ganglioside-deficient mice were less sensitive to BoNT (Kitamura et al., 1999). The three-dimensional structure of BoNT associated with these two receptors has been determined (Arndt et al., 2006; Hanson and Stevens, 2000; Swaminathan and Eswaramoorthy, 2000) (Color Plate 10). The heavy-chain portion of BoNT is also responsible for the internalization of the light-chain domain. BoNT light chain is a zinc metalloprotease (Schiavo et al.,

FIGURE 8 Map showing the location of the two genes encoding a restriction-modification system in bacteriophage ϕ42.

1992b) that cleaves SNAP-25, blocking neurotransmitter release at neuronal junctions (Schiavo et al., 1992a).

Phage conversion to toxin production in *C. botulinum* was first shown by Inoue and Iida (1968, 1970, 1971) and was later confirmed by Eklund et al. (1971, 1972). They suggested that neurotoxin production following phage infection was analogous to the *C. diphtheriae* system with an important difference: botulinum toxin production was a pseudolysogenic conversion due to the presence of phage in a carrier state. Studies have shown that growth in the presence of phage antisera or in the presence of acridine orange leads to a curing of toxigenicity (Oguma, 1976).

Two bacteriophages carrying the BoNT toxin gene have been sequenced: phage c-st (Sakaguchi et al., 2005), isolated from *C. botulinum* type C strain C-Stockholm (NCBI NC_007581); and phage D-1873 (NCBI NZ_ACSJ01000012 to NZ_ACSJ01000019) from *C. botulinum* D strain 1873. Both of these phages were inducible from their lysogens with UV and mitomycin C (Inoue and Iida, 1968). Staining with phosphotungstic acid revealed large heads (120 nm in diameter) attached to long (350- to 450-nm) contractile tails showing clear demarcation between the tail tube (15 nm in diameter) and its surrounding horizontally striated sheath (30 to 35 nm in diameter). While a tail baseplate was not observed, some of the electron micrographs suggest the presence of short tail fibers.

Phage c-st possesses a genome size of almost 186 kb, with 404-bp terminal direct repeats. Upon infection the genome circularizes, presumably through recombination between the terminal direct repeats; the prophage then persists as a plasmid within the lysogen (Sakaguchi et al., 2005). The GenBank submission information on D-1873 indicates that it, too, exists as a circular plasmid within its host bacterium. Interestingly, these two phages exhibit very little DNA sequence similarity but are serologically related (Oguma et al., 1976) and display significant similarity at the protein level. Using CoreGenes (Kropinski et al., 2009) at its default setting, the proteins of D-1873 share 98 homologs with c-st, indicating that they are most probably members of the same genus (Lavigne et al., 2008). Elegant PCR analyses by Sakaguchi et al. (2005) definitively proved that this virus does not integrate into the genome of its host, whereas data for the pseudotemperate *Listonella* phage, ϕHSIC, suggest that it persists both as a plasmid and in the integrated state (Williamson and Paul, 2006). Its plasmid state is reminiscent of coliphage P1 (Lehnherr, 2005), but there are several fundamental differences between the systems, as pointed out by Sakaguchi et al. (2005) and summarized here. (i) Partitioning of the replicated low-copy-number plasmid prophage involves CST193 (NCBI YP_398623.1), which encodes an FtsK homolog, and CST185 (NCBI YP_398615.1), producing ParB; P1 uses a phage-encoded ParAB system. (ii) While multimer resolution in P1 involves the Cre-lox system, in these clostridial phages a crossover junction endodeoxyribonuclease RusA (CST102; YP_398532.1) or XerC/D family recombinase (CST104; YP_398534.1) may be involved. The function of the gyrase A and B subunits in the life cycle of these phages is unknown. (iii) No analogs of the coliphage toxin-antitoxin control system have been found in the clostridial phages (Guglielmini et al., 2008). (iv) In P1, EcoP1res and EcoP1mod are a type III restriction-modification system, coded for by linked genes, while the clostridial phages have a putative Mrr-type restriction enzyme (CST191; YP_398621.1) (Waite-Rees et al., 1991) and an independent N^6-adenylmethylase subunit (CST012; YP_398442.1).

This begs the question of how the prophage is maintained in a repressed state in the lysogens. Protein domain analysis using the Pfam database (http://pfam.janelia.org/) (Coggill et al., 2008) reveals several proteins with homology to regulators, including protein CST150 (YP_398580.1), with a C-terminal pfam01381 (HTH_3) helix-turn-

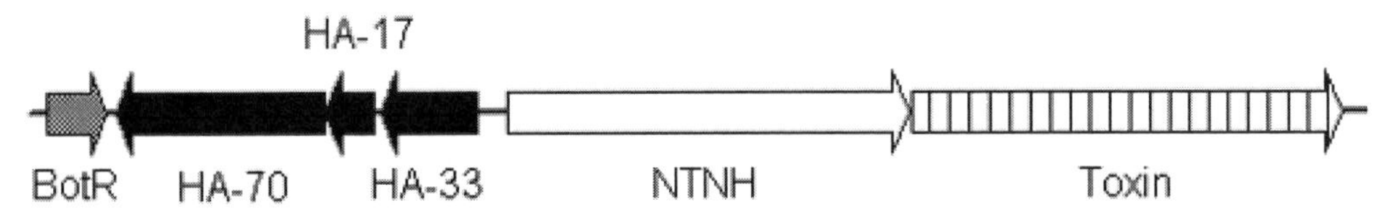

FIGURE 9 Organization of the toxin operon in *C. botulinum* bacteriophages c-st and D-1873. In the latter phage HA70 is referred to as HA3.

helix DNA binding motif; proteins CST012 (YP_398442.1) and CST170 (YP_398600.1), which contain the pfam09669 (Phage_pRha) motifs; and protein CST027 (YP_398457.1), which contains a pfam02452 (PemK) motif. None of these resemble the immunity system of any other characterized bacteriophage.

In each of these phages the genes for BoNT are clustered with NTNH and hemagglutinin-encoding genes with the addition of a regulatory protein (BotR) (Fig. 9). BotR possesses the DNA binding domains pfam04545 and cd06171 (Sigma70_r4), a region that is found at the C terminus of the sigma 70 class of accessory RNA polymerase subunits, suggesting that this is an activator of BoNT operon expression.

CELL SURFACE MACROMOLECULES

Serotype Conversion

In the case of gram-negative bacteria, lysogenization will almost always result in resistance to superinfection by the same phage due to homoimmunity. As previously mentioned, this is not regarded as an example of lysogenic conversion. At the next level are lysogenic conversion events causing changes to cell surface structures that are possibly too subtle to be serologically relevant, but which result in resistance to homo- and heterologous viruses. This is nicely illustrated by *Salmonella* phages A3 and A4, which respectively acetylate the O-2 and O-3 positions on the rhamnosyl residues in the O-antigen repeat unit of *Salmonella enterica* serovar Typhimurium (O:4,5,12) (Fig. 10) (Ellermeier and Slauch, 2006) without inducing specific antibodies (Wollin et al., 1987). Lysogenization by either one of these phages abolishes receptor activity for the other, and in addition severely reduces the efficacy with which the cellular LPS inactivates phage P22 (PhI_{50}) (Kropinski, 2009; Wollin et al., 1987).

The earliest known example of antigenic conversion involves phage ε15-mediated conversion of serotype in group E1 salmonellae, which was first demonstrated by Uetake et al. (1955). Lysogenic clones showed all the properties of true convertants (growth in the presence of antiphage serum, stability, and inducibility). Work on the mechanism of conversion progressed with the isolation of nonconverting mutants of ε15 (Uetake and Uchita, 1959) that could establish lysogeny but whose host-cell O-antigens remained, to varying degrees, unaltered. Using these mutants and host mutants defective in the synthesis of O-antigen, the structural changes brought about by conversion were elucidated (Borst and Betley, 1994; Losick and Robbins, 1967). The structure of the *S. enterica* serovar Anatum O-antigen (O:3,10) was shown (Robbins and Uchida, 1965; refined in Gajdus et al., 2008) to consist of a mannosyl-rhamnosyl-O-acetylated-galactose repeat unit, held together by α-glycosidic linkages. Upon infection by ε15, a new O-polysaccharide

D-Abe-2-OAc
| α-1→3
2)-D-Man-α(1→4)-L-Rha-α (1→3)-D-Gal-α(1→

FIGURE 10 Chemical structure of the O-antigenic repeat unit in the LPS of *Salmonella* serovar Typhimurium, where Abe, Man, Rha, and Gal are abequose (3,6-dideoxy-D-xylo-hexose), mannose, rhamnose, and galactose, respectively.

comprising nonacetylated mannosyl-rhamnosyl-galactose repeat units held together by β-glycosidic linkages is formed (Fig. 11).

Loss of the O-acetyl groups on the galactosyl residues and conversion of the galactosyl-mannosyl linkage from α- to β-configuration resulted in a new O-antigen identical to the O:3,15 antigen of *S. enterica* serovar Newington, the organism from which ε15 was originally isolated. This bacterium is now officially named *Salmonella* serovar Anatum (var. 15+). Other salmonellae with the same serotype are Weltevreden, London, and Uganda. Abundant genetic and biochemical data led Robbins and coworkers to conclude that ε15 coded for three conversion proteins: (i) a repressor of the host transacetylase gene, (ii) an inhibitor of the host α-polymerase enzyme, and (iii) a new β-polymerase. Phage mutants defective in the hypothesized transacetylase repressor *(tar),* α-polymerase inhibitor *(api),* and β-polymerase *(oap)* genes were subsequently isolated (Borst and Betley, 1994; Bray and Robbins, 1967; Robbins et al., 1965; Robbins and Uchida, 1965; Uetake and Hagiwara, 1961).

A second well-studied phage-induced conversion of group E salmonellae occurs following the lysogenization of *Salmonella* serovar Anatum (var. 15+) by the morphologically related phage ε34. The seroconvertant was originally known as *Salmonella* serovar Minneapolis, but *Salmonella* serovars Petersburg, Southbank, and Newlands also display this serotype. Phage ε34 is ordinarily unable to infect serovar Anatum because it does not recognize the O:3,10 antigenic structure. However, conversion by ε15 to the O:3,15 antigenic structure allows phage ε34 to adsorb to and infect this bacterium (Uetake and Hagiwara, 1961). Infection by ε34 causes the appearance of antigen (O:34) in the host by a mechanism involving the addition of a glucosyl side-chain residue to the galactose residue of the repeating unit (Robbins and Uchida, 1962; Wright and Barzilai, 1971).

The new O-antigen (O:3,15,34) is not recognized as a receptor by either of the converting phages. When double lysogens of *Salmonella* serovar Anatum (ε15,ε34) are cured of the ε15 prophage, the O-antigen reverts to O:3,10 rather than a possible O:3,34 (Fig. 12). Apparently, the phage ε34-encoded glucosyltransferase does not recognize the host-encoded α-linked repeating unit (Wright and Barzilai, 1971). Mutants of ε34 have been isolated which lack the ability to convert O:3,15 to the O:3,15,34 antigen. Their defects fall into two categories: (i) inability to produce the glucosyl-lipid intermediate and (ii) inability to transfer the glucose from the glucosyl-lipid intermediate onto the galactose residues of the trisaccharide repeat unit (Wright and Barzilai, 1971).

A third group E1 salmonella-specific phage known as g341 interacts with the acetyl groups of galactose residues in the O:3,10 antigen, hydrolyzing and releasing them with its enzymatically active tailspikes during the infection process. Once established as a prophage, g341 converts the host-cell O:3,10 antigen by preventing the acetylation of its galactose residues (Uetake and Hagiwara, 1969). It is interesting to note that viable hybrids can be formed between all the above

```
                                         OAc
                                         |1→6
-6)-D-Man-α(1→4)-L-Rha-α(1→3)-D-Gal-α(1→

-6)-D-Man- α(1→4)-L-Rha- α(1→3)-D-Gal-β (1→
```

FIGURE 11 The structure of the O-antigenic repeat unit in the LPS of *Salmonella* serovar Anatum before (top) and after (bottom) infection by ε15.

```
                                         D-Glc
                                         | (1→4)
-6)-D-Man- α(1→4)-L-Rha- α(1→3)-D-Gal-β (1→
```

FIGURE 12 The tetrasaccharide O-polysaccharide repeat unit of ε34 seroconvertants of *Salmonella* serovar Anatum (var. 15+).

phages that have the capability to cause conversions mimicking double infections (Hagiwara and Uetake, 1970; Uetake and Hagiwara, 1969), suggesting the relatedness of the phages.

Little more was done regarding group E1 salmonella-converting phages for almost a quarter century, but we now have the genomic sequences of ε15 (Kropinski et al., 2007), ε34 (Villafane et al., 2008), and a clear plaque mutant of g341 (c341; NC_013059). All three viruses are members of the *Podoviridae* and have genome sizes similar to that of P22 (c341, 40,975 bp; ε15, 39,671 bp; ε34, 43,016 bp; P22, 41,724 bp), but the earlier assumption that they were all related turns out to have been only partially correct. ε34 is clearly "P22-like," with c341 somewhat less so, while ε15 merits its own classification—the "ε15-like viruses" (Lavigne et al., 2008)—and is most closely related to a temperate converting phage of *E. coli* O157 (Perry et al., 2009). In the following paragraphs we will discuss how the sequence data relate to what had been previously postulated.

ε15 SEROCONVERSION GENOMICS

Over the past decade, *Salmonella* serovar Anatum strains transformed by plasmids carrying various ε15 genes have been analyzed, using a variety of biological, immunological, and biochemical assays. This work has resulted in the identification of four ε15 genes whose protein products influence the structure of serovar Anatum LPS, namely genes *21, 22, 28,* and *46* (Kropinski et al., 2007) (Fig. 13). Real-time PCR studies involving whole RNA purified from serovar Anatum cells lysogenized by ε15 confirm that these four genes are expressed in lysogens to about the same degree as gene *38,* which codes for the major ε15 repressor protein (M. R. McConnell, unpublished data). Gene *21* codes for the β-polymerase (Oap), which like most other known salmonella O-polysaccharide polymerases is just under 400 amino acids in length (390 aa), has a large number of membrane-spanning helices (10), and has a basic pI value (9.26). Gene *22* codes for the α-polymerase inhibitor (Api). Although the Api protein of ε15 was originally reported to be small, heat resistant, and water soluble (Losick, 1969), gp22 is instead a small hydrophobic membrane protein, with two membrane-spanning α-helices; investigations into its mode of action are currently under way. Forty years ago, Phil Robbins and coworkers postulated that ε15 blocked acetylation of galactose residues by producing a repressor protein (Tar) that prevented transcription of the host-cell O-polysaccharide acetyltransferase gene (Kropinski et al., 2007). Their reasoning was quite logically based on in vitro measurements of acetyltransferase activity in cell lysates as a function of time after infection with ε15, which revealed a drop in specific activity that was inversely related to the bacterial growth rate; i.e., apparently no additional acetyltransferase enzyme activity was being formed following infection, but neither was the enzyme already in existence at the time of infection adversely affected. Rather, it persisted, only to be gradually diluted out by bacterial growth and cell division. The two ε15 proteins recently discovered to prevent acetylation of galactose do not display repressor-like properties;

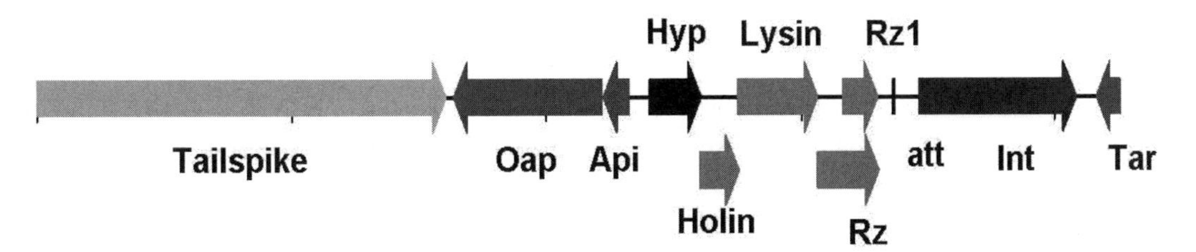

FIGURE 13 The 8,509-bp region of phage ε15 responsible for serotype conversion with genes *21* (Oap), *22* (Api), *26* (Tar), and *46* (Con) (Kropinski et al., 2007).

instead, they are small hydrophobic proteins with either one (gp28, Tar) or two (gp46, Con) inferred membrane-spanning helices. A current model (yet to be tested) is that gp28 and gp46 act within the membrane to prevent newly synthesized (but not already established) acetyltransferase enzyme from interacting productively with other proteins that produce and/or transport the undecaprenyl phosphate-linked mannosyl-rhamnosyl-galactose repeat unit, thereby preventing acetylation of its galactose residue.

GENOMICS OF SEROCONVERSION IN ε34

The genome of ε34 is 43,016 bp, circularly permuted, and presumably integrates into a host tRNA *argU* gene, since immediately downstream of the ε34 *int* gene (gp*23*) is a 43-bp sequence also found in the host tRNA *argU* gene (Villafane et al., 2008). Other than its role in seroconversion, research on ε34 has focused on the activity of its tailspike protein, which possesses, like the homologous protein in phage P22, both receptor-binding and receptor-cleaving activities (Greenberg et al., 1995; Iwashita and Kanegasaki, 1976; Zayas and Villafane, 2007). Between the tailspike- and integrase-coding genes of phage ε34 is the conversion cassette, which bears striking organizational similarity to that of P22 (see below; Fig. 14), as well as the SfV (Huan et al., 1997) and SfX (Guan et al., 1999) conversion modules. The first product, GrtA (120 aa; 13.2 kDa), is a flippase that translocates glucosylated undecaprenyl phosphate from the cytoplasmic face of the inner membrane to the periplasmic face. GtrB (305 aa; 34.3 kDa) is the enzyme responsible for glycosylating bactoprenol (a glucosyltransferase), while GtrC (551 aa; 62.5 kDa) is the glucosyltransferase responsible for modification of the LPS. The latter protein possesses 11 transmembrane domains with a long periplasmic domain (Viklund and Elofsson, 2008).

In addition, this phage codes for two proteins, gp44 (103 aa) and gp45, which are downstream of the prophage repressor gene gp*46* in a manner analogous to that of *rexA* and *rexB* in coliphage λ. gp44 possesses a C-terminal transmembrane domain with the N terminus predicted to be internal. It is identical to the g341 gp40 protein and related to hypothetical protein NCBI E2348C_2536 from *E. coli* O127:H6 strain E2348/69. Cytoplasmic gp45 (156 aa) is again related to proteins in g341 (gp41) and *E. coli* (E2348C_2535). While they are most likely expressed in lysogens, their function is currently unknown.

GENOMICS OF SEROCONVERSION WITH g341

Phage g341's only known converting function to date is its ability to block acetylation of galactose residues, resulting in production of a nonacetylated form of antigen O:3,10 (Sauer et al., 1983). Recent data (McConnell, unpublished) indicate that it does this by using the protein coded for by gene *23* (YP_003090240), which is nearly identical to ε15's gp*46* (69 out of 72 amino acids in common). The g341 genome also contains one other putative converting gene (gene *15;* YP_003090232.1) that displays strong homology with members of the acetyltransferase 3 family (pfam01757). To date, the target of g341's gp15 acetyltransferase enzyme remains unidentified, but a detailed comparative struc-

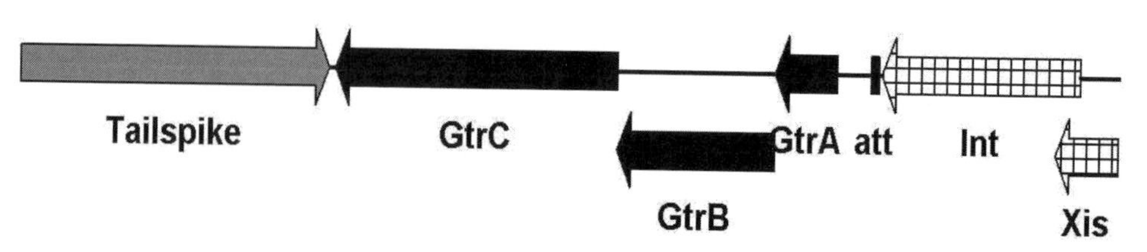

FIGURE 14 The 6,441-bp region of phage ε34 responsible for serotype conversion illustrated in black (Villafane et al., 2008).

tural analysis of O-polysaccharides purified from *Salmonella* serovars Anatum and Anatum (g341) that is currently under way should hopefully soon remedy this situation.

P22-INDUCED SEROCONVERSION

P22 has a 41,724-bp genome that is circularly permuted and terminally redundant (Pedulla et al., 2003; Vander Byl and Kropinski, 2000). P22 is one of the most extensively analyzed bacteriophages because of its activity as a generalized transducer (Thierauf et al., 2009), interest in its morphogenesis (Nemecek et al., 2008; Zheng et al., 2008), and the structure and activity of its tailspike proteins, which possess endorhamnosidase activity (Handa et al., 2008; Landstrom et al., 2008; Spatara et al., 2009). The P22 genome integrates within a tRNA gene *(thrW)* during lysogeny, and lysogenization results in superinfection exclusion mediated by the widely separated *sieA* and *sieB* genes. Lysogenization by P22 also results in the addition of an α-linked glucosyl residue (antigen O:1) to the 6 position of galactose moieties in the LPS O-antigenic tetrameric repeat. Mutational analysis by Young and his associates determined that gene *a1* was responsible for the expression of antigen O: 1 and showed its relative position within the P22 genome, adjacent to the phage attachment site *(attP)* (Young et al., 1964). With the completion of the genome analysis plus subcloning, we now know that the conversion cassette consists of GtrA (112 aa; 12.5 kDa), GtrB (310 aa; 35.1 kDa), and GtrC (486 aa; 55.4 kDa) preceded by a promoter and followed by a bidirectional Rho-independent terminator—in other words, the perfect moron. This cassette is responsible for the chemical change to the LPS O-antigen which results in a change in serotype from 4,[5],12 to 1,4,[5],12 and prevents the binding of P22 (compare Fig. 10 with Fig. 15) (Iseki and Kashiwagi, 1955; Rundell and Shuster, 1975).

Homologs to P22 *gtrC* are also found in phages ST64T (Mmolawa et al., 2003) and ST104 (Tanaka et al., 2004), and O1 is one of the most frequently found antigens among the salmonellae, being present in groups O:2 (A), O:4 (B), O:9 (D_1), O:9,46 (D_2), O: 9,46,27 (D_3), O:1,3,19 (E_4), O:13 (G), O:6,14 (H), O:40 (R), O:42 (T), O:44 (V), O:47 (X), O:51, O:53, and O:59. Oddly, P22 *gtrC* homologs are found only in genomes of *Salmonella* serovars Paratyphi A [group O:2 (A)], Dublin, and Heidelberg, while we might have expected to find homologs in the genomes of serovars Gallinarum (O:1,9,12), Saintpaul (O: 1,4,5,12), and Paratyphi B (O:1,4,5,12) but did not. This suggests that O:1 conversion can be catalyzed by a number of unrelated prophages. *Salmonella* serovars Gallinarum and Paratyphi B contain complete *gtrABC* clusters at bp positions 612634 to 615553 and 2500425 to 2500700, respectively, which are not prophage associated. It is important to note that while P22 and ε34 GtrC proteins are functionally and structurally related, they exhibit no sequence similarity.

OTHER KNOWN OR SUSPECTED LPS-CONVERTING PHAGES

There have been a myriad of conversions reported for the genus *Salmonella,* several of which are briefly reviewed by Barksdale and Arden (1974). According to the definitive *Antigenic Formulae of the* Salmonella *Serovars* (Grimont et al., 2007), in addition to antigens 1, 15, and 34, antigens 14 (groups O:7 and O: 18), 20 (groups O:8 and O:54), and sometimes 27 (Le Minor et al., 1961) are phage encoded. In *Salmonella* group 2 seroconvertants expressing the 27 antigen, the linkage between the repeat units is α1→6 rather than α1→2 (Lindberg et al., 1978; Wang et al., 2002), suggesting the presence of a glucosyltransferase inhibitor and a new polymerase.

Salmonella serovar Typhimurium *oafA* defines antigen 5, which is α-D-Abe-2-OAc (Slauch et al., 1996). Using a number of

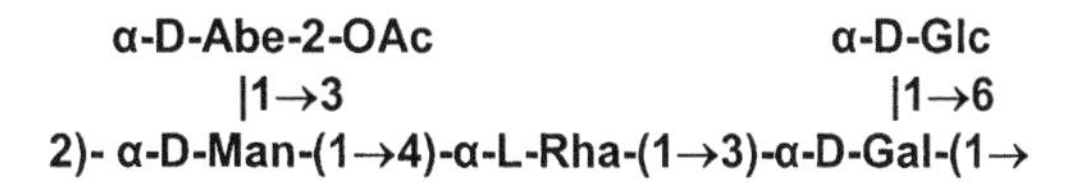

FIGURE 15 O-antigenic repeat unit from *Salmonella* serovar Typhimurium showing the impact of the P22 *gtrABC* conversion module.

Accession no.	Start	Sequence	End	Organism
	1	MKKMllatalallITGCAQQTFTVQNKPAAVAPKETITHHFFVSGIGQKKTVDAAKICGGAENVVKTETQQTFVNGLLGFITLGIYTPLEARVYCSQ	97	
X55792	24	....	314	Phage lambda Bor
CP001396	4152353	....	4152063	E. coli BW2952
CP001368	1565100	...............................T...	1564810	E. coli O157:H7 str. TW14359
CP001368	3220208	...........................QT...............................T....................................	3220498	
CP001164	1564812	...............................T...	1564522	E. coli O157:H7 str. EC4115
CP001164	3265337	...........................QT...............................T....................................	3265627	
BA000007	1628141	...............................T...	1627851	E. coli O157:H7 str. Sakai
BA000007	1275183	...........................QT...............................T....................................	1274893	
AE005174	1712116	...............................T...	1711826	E. coli O157:H7 EDL933
AE005174	1359194	...........................QT...............................T....................................	1358904	
CU928161	1200617	.....FSA...M............G...T..T...	1200327	E. coli S88
CP000468	1197095	.....FSA...M............G...T..T...	1196805	E. coli APEC O1
CP000247	1205810	.....FSA...M............G...T..T...	1205520	E. coli 536
CP000243	1259708	.....FSA...M............G...T..T...	1259418	E. coli UTI89
AE014075	1423173	.....FSA...M............G...T..T...	1422883	E. coli CFT073
AP010953	3117160	.....FSA...M............G...T..T................E..	3117450	E. coli O26:H11 str. 11368
AP010953	629937	.....FSA...M............G...T..T................E..	629647	
CP000819	550362	.....FSA...M............G...T..T................E..	550072	E. coli B str. REL606
CU928163	1429800	.....FSA...M............G...T..T................E..	1429510	E. coli UMN026
AP010958	587568	...........................QT...K	587278	E. coli O103:H2 str. 12009
CP000948	630708	...........................QT...K	630418	E. coli str. K-12 substr. DH10B
CP000948	517448	...........................QT...K	517158	
AP009048	578116	...........................QT...K	577826	E. coli str. K-12 substr. W3110
U00096	578116	...........................QT...K	577826	E. coli str. K-12 substr. MG1655
AP010960	1209169	...........................QT...............................T....................................	1208879	E. coli O111:H- str. 11128
EU311208	30045	...........................QT...............................T....................................	29755	Enterobacteria phage Min27
AB255436	7575	...........................QT...............................T....................................	7285	Stx2-converting phage 86
AP005154	61059	...........................QT...............................T....................................	60769	Stx2 converting phage II
AP005153	58219	...........................QT...............................T....................................	57929	Stx1 converting phage
AP004402	60118	...........................QT...............................T....................................	59828	Stx2 converting phage I
AP000422	29722	...........................QT...............................T....................................	29432	Enterobacteria phage VT2-Sakai
AF125520	28366	...........................QT...............................T....................................	28076	Bacteriophage 933W
FJ824853	16	.....FSA...M............G...T..T................E.................................F..............	306	E. coli strain APEC
DQ309292	16	.....FSA...M............G...T..T................E.................................F..............	306	E. coli Beijing1
DQ309291	16	.....FSA...M............G...T..T................E.................................F..............	306	E. coli Xinda
DQ309290	16	.....FSA...M............G...T..T................E.................................F..............	306	E. coli Tianda
DQ309289	16	.....FSA...M............G...T..T................E.................................F..............	306	E. coli Beijing3
DQ309288	16	.....FSA...M............G...T..T................E.................................F..............	306	E. coli Guizhou1
DQ309284	16	.....FSA...M............G...T..T................E.................................F..............	306	E. coli E18
DQ309283	16	.....FSA...M............G...T..T................E.................................F..............	306	E. coli E14
DQ309282	16	.....FSA...M............G...T..T................E.................................F..............	306	E. coli E11
DQ309281	16	.....FSA...M............G...T..T................E.................................F..............	306	E. coli E10
DQ309280	16	.....FSA...M............G...T..T................E.................................F..............	306	E. coli E9
DQ309279	16	.....FSA...M............G...T..T................E.................................F..............	306	E. coli E8
DQ299400	16	.....FSA...M............G...T..T................E.................................F..............	306	E. coli E7
DQ299399	16	.....FSA...M............G...T..T................E.................................F..............	306	E. coli E5
DQ295188	16	.....FSA...M............G...T..T................E.................................F..............	306	E. coli serotype O1
DQ295187	16	.....FSA...M............G...T..T................E.................................F..............	306	E. coli serotype O2
DQ299398	16	.....FSA...M............G...T..T..F.............K	306	E. coli E1
CP000836	101276	.....FSA...M............G...T..T................E.................................F..............	101566	E. coli chi7122 plasmid pAPEC-1
CU928160	828550	.....FSA...M............G...T..T................E...........................F....................	828260	E. coli IAI1
CU928158	2774913	.....FSA...M............G...T..T................E.................................F..............	2775203	Escherichia fergusonii ATCC 35469
CU928146	112785	.....FSA...M............G...T..T................E.................................F..............	112495	E. coli str. S88 plasmid pECOS88
CP001232	34124	.....FSA...M............G...T..T................E.................................F..............	34414	E. coli strain APEC O103 plasmid pAPEC-1
FJ416147	307	.....FSA...M............G...T..T................E.................................F..............	597	E. coli serotype O78
AP009240	628552	.....FSA...M............G...T..T................E...........................F....................	628262	E. coli SE11

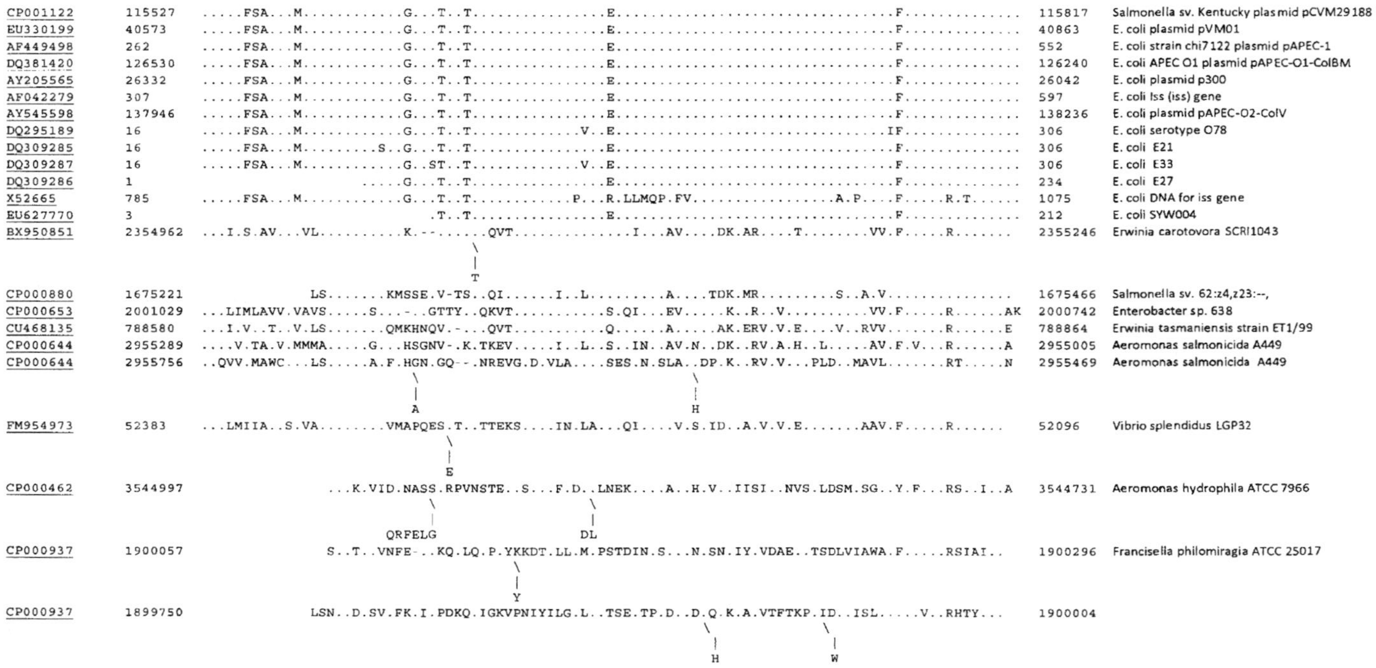

```
CP001122   115527   .....FSA...M...........G...T..T................E.............................F.............   115817   Salmonella sv. Kentucky plasmid pCVM29188
EU330199   40573    .....FSA...M...........G...T..T................E.............................F.............   40863    E. coli plasmid pVM01
AF449498   262      .....FSA...M...........G...T..T................E.............................F.............   552      E. coli strain chi7122 plasmid pAPEC-1
DQ381420   126530   .....FSA...M...........G...T..T................E.............................F.............   126240   E. coli APEC O1 plasmid pAPEC-O1-ColBM
AY205565   26332    .....FSA...M...........G...T..T................E.............................F.............   26042    E. coli plasmid p300
AF042279   307      .....FSA...M...........G...T..T................E.............................F.............   597      E. coli Iss (iss) gene
AY545598   137946   .....FSA...M...........G...T..T................E.............................F.............   138236   E. coli plasmid pAPEC-O2-ColV
DQ295189   16       .....FSA...M...........G...T..T.............V..E............................IF.............   306      E. coli serotype O78
DQ309285   16       .....FSA...M.........S..G...T..T................E.............................F.............   306      E. coli  E21
DQ309287   16       .....FSA...M...........G..ST..T.............V..E.............................F.............   306      E. coli  E33
DQ309286   1                       .....G...T..T................E.............................F.............   234      E. coli  E27
X52665     785      .....FSA...M...........G...T..T............P...R.LLMQP.FV................A.P....F.....R.T.....   1075     E. coli DNA for iss gene
EU627770   3                                .T..T................E.............................F.............   212      E. coli SYW004
BX950851   2354962  ...I.S.AV...VL..........K.--......QVT..............I...AV....DK.AR....T........VV.F.....R.......   2355246  Erwinia carotovora SCRI1043
                                                  \
                                                  |
                                                  T
CP000880   1675221            LS.......KMSSE.V-TS..QI......I..L.........A....TDK.MR.........S..A.V...............   1675466  Salmonella sv. 62:z4,z23:--,
CP000653   2001029  ...LIMLAVV.VAVS.....S...-..GTTY..QKVT...........S.QI...EV.....K..R..V..........VV.F.....R......AK   2000742  Enterobacter sp. 638
CU468135   788580   ...I.V..T..V.LS.......QMKHNQV.-...QVT..........Q......A.....AK.ERV.V.E....V..RVV.......R......E   788864   Erwinia tasmaniensis strain ET1/99
CP000644   2955289  ....V.TA.V.MMMA.....G...HSGNV-.K.TKEV.....I..L..S..IN..AV.N..DK..RV.A.H..L.....AV.F.V...R......A   2955005  Aeromonas salmonicida A449
CP000644   2955756  ..QVV.MAWC...LS.....A.F.HGN.GQ--.NREVG.D.VLA....SES.N.SLA..DP.K..RV.V...PLD..MAVL.......RT.....N   2955469  Aeromonas salmonicida  A449
                                          \                           \
                                          |                           |
                                          A                           H
FM954973   52383    ...LMIIA..S.VA........VMAPQES.T..TTEKS....IN.LA...QI....V.S.ID..A.V.V.E.......AAV.F.....R......   52096    Vibrio splendidus LGP32
                                               \
                                               |
                                               E
CP000462   3544997                ...K.VID.NASS.RPVNSTE..S...F.D..LNEK....A..H.V..IISI..NVS.LDSM.SG..Y.F...RS..I..A   3544731  Aeromonas hydrophila ATCC 7966
                                             \                \
                                             |                |
                                          QRFELG              DL
CP000937   1900057                S..T..VNFE-..KQ.LQ.P.YKKDT.LL.M.PSTDIN.S...N.SN.IY.VDAE..TSDLVIAWA.F.....RSIAI..   1900296  Francisella philomiragia ATCC 25017
                                                   \
                                                   |
                                                   Y
CP000937   1899750               LSN..D.SV.FK.I.PDKQ.IGKVPNIYILG.L..TSE.TP.D..D.Q.K.A.VTFTKP.ID..ISL.....V..RHTY...   1900004
                                                                      \          \
                                                                      |          |
                                                                      H          W
```

FIGURE 16 Diversity of phage lambda Bor homologs found in the NCBI translated nucleotide database using BLAST query. The search also indicated the occurrence of related genes in distant relatives of *E. coli* within the Gammaproteobacteria.

BLAST algorithms (Altschul et al., 1990, 1997) run against various NCBI databases, identical homologs can be identified in all of the 5-positive strains (serovars Paratyphi B strain SPB7, Saintpaul strain SARA23, Heidelberg strain SL486, and 4,[5],12:i:- strain CVM23701). This protein is characterized as possessing a pfam01757 (Acyl_transf_3, acyltransferase family) protein motif and predicted to possess 11 transmembrane domains (Kall et al., 2004) with a long periplasmic C-terminus domain. Although gene *oafA* is not indicated as phage encoded (Grimont et al., 2007), it does occur within a defective prophage (Sty-cPP2) in serovar Typhimurium LT2. Furthermore, the as-yet-uncharacterized acetyltransferase of phage g341 (gp15; YP_003090232.1) is strongly homologous to *oafA*. The 14.14-kb cryptic prophage Sty-cPP2 is inserted between bases 2330845 and 2344988 in the genome of serovar Typhimurium LT2. The proximity of a prolyl-tRNA gene (STM2229) to this chromosomal region suggested a possible insertion into that gene, and using the direct repeat finder in DNAMAN (Lynnon Corporation, Vaudreuil-Dorion, Quebec, Canada) we identified two hyphenated repeats, 55 and 56 bp, respectively, which flank this region. The consensus sequence is TGCCGACCAAAATTCCYC (N0-2) GAA (N0-1) AACCAGCCTGTYASGGCTGGTT TTTTTATGCCT. Single copies were also found adjacent to phagelike genes in serovars Choleraesuis and Typhi.

Outer Membrane Proteins Bor and Lom

Earlier sections of this chapter have nicely illustrated that some proteins encoded by phage conversion genes increase host offensive weapon repertoires, while others appear to increase host defensive capabilities. Another example of the latter category was originally reported by Muschel et al. (1968), when they showed that *E. coli* K-12 lysogenization by phage λ greatly enhanced survival of the bacterium in normal serum, relative to unlysogenized *E. coli* K-12 or *E. coli* lysogenized by ϕ80. Barondess and Beckwith (1990) later showed that the serum resistance phenomenon was attributable to the presence of a previously unidentified gene, which they named *bor*. The *bor* gene is located at 46.45 kb, immediately downstream of the lysis cassette on the coliphage λ genome, and was found to be expressed during lysogeny. It codes for a protein of 97 aa (pfam06291; Lambda_Bor), with a sequence very similar to that of the Iss protein from plasmid ColV2-K94, which also increases host serum resistance (Johnson et al., 2008). Essentially identical proteins have been found in podoviruses min27 (YP_001648941.1) and 933W (Plunkett et al., 1999), as well as numerous other prophages (Fig. 15). The amount of Bor protein molecules was estimated to be about 2,000 copies per cell, and cells carrying the protein were 20 times more resistant to serum when compared to the unlysogenized host.

Bor protein was shown to be an outer membrane lipoprotein, with an action resembling that of the Iss protein (Barondess and Beckwith, 1995). The Iss protein makes cells immune to the classical complement pathway but not to the alternative complement pathway (Chuba et al., 1986). The *iss* gene is very common among pathogenic avian strains of *E. coli*. Less frequent occurrence of *iss* and *bor* homologs in mammalian pathogenic *E. coli* strains can be attributed to the fact that colibacteriosis in mammals is mostly enteric, whereas avian infections are mostly extraintestinal; thus avian pathogens are more exposed to host defenses based on complement action (Nolan et al., 2003). Some bacterial strains, including O157:H7 strains like Sakai and EDL933, carry two copies of the *bor* gene, which are associated with separate prophages integrated to their genome. The *bor* gene is also present in the chromosome of K-12 and K-12 derivatives, where it is present as part of a DLP12 defective prophage (Fig. 16). Strains 933W and Sakai also carry a second copy of the gene on a DLP12-like prophage. Bacteriophage λ was originally isolated from the *E. coli* K-12 strain, and in the past this bacterial line may have also had two copies of the gene. It is very interesting that both pathogenic and nonpathogenic strains sometimes share some of the same virulence genes. Apparently, the *bor* gene plays only a supportive role in pathogenesis and, without additional pathogenicity genes, cannot by itself cause transformation of a nonpathogenic strain into a pathogenic one. On the other hand, the gene is widespread among prophages, which suggests that it is advantageous for phage survival. It is known that prophage induction may be associated with pathogenesis of infection of some virulent bacterial strains (Łoś et al., 2009); it could also provide for a rapid circulation of virulence factor-carrying bacteriophages between pathogenic and nonpathogenic strains. Perhaps *E. coli* K-12 was just a short stop for phage λ in its search for a host with a larger virulence gene repertoire, which could be greatly reinforced by *bor* gene addition.

The gene for Lom maps at approximately 18.96 kb on the λ genome (i.e., downstream of the gene for the tailspike protein J). Employing *E. coli* minicells, Reeve and Shaw (1979) first showed that gene *lom* codes for a 20.5-kDa outer membrane protein. The 206-aa (21.8-kDa) polypeptide possesses a 15-aa leader sequence and displays close sequence similarity to the Cig5-like protein of phage mep503 (AAQ23226.1) and, to a lesser degree, phages cdtI and 2851 (Strauch et al., 2008), and proteins from prophages YYZ-2008, BP-4795 (Creuzburg et al., 2005), 1717, and VT1-Sakai. It also possesses a pfam06316 (Ail_Lom) domain that is shared with Ail-like proteins that are known virulence factors in *Yersinia enterocolitica,* and PagC proteins that play roles in the survival of *Salmonella* in macrophages and adherence (Crago and Koronakis, 1999; Miller et al., 1992; Vica et al., 1997) (Fig. 17).

The modeled structures of Lom and Bor, shown in Color Plate 11, reveal that Lom contains a β-barrel structure.

ACKNOWLEDGMENTS

We owe a considerable debt to Dr. Le Minor (Pasteur Institute) for his work on the serology of phage conversion in salmonellae (Le Minor, 1962, 1963, 1965a, 1965b, 1965c; Le Minor et al., 1961, 1963) and to Drs. Lüderitz and Westphal (Max-Planck-Institut für Immunbiologie) for their analysis of the structures of the O-antigenic repeat units from *Salmonella* (Lüderitz, 1970; Lüderitz et al., 1966). Research conducted

```
λ Lom          MRNVCIAVAVFAALAVTVT--PARAEGGHGTFTVGYFQVKPGTLPSLSGGDTGVSHLKGI 58
CP-933C Lom    MRKLYAAILS-AAICLTVSGAPAWASEQQATLSAGYLHVST----NAPGSDN----LNGI 51
Ail            MKKTLLASSLIACLSIASVNVYAASE---SSISIGYAQSHVK-----ENGDTLDNDPKGF 52
               *::   *    *.:.::     * :.   .::: ** :          ..*.     :*:
                                                     ------LOOP1--------

λ Lom          NVKYRYELTDSVGVMASLGFAASKKSSTVMTGEDTFHYESLRGRYVSVMAGPVLQISKQV 118
CP-933C Lom    NVKYRYEFTDTLGLVTSFSYAGDRNRQITRYSDTRWHEDSVRNRWFSVMAGPSVRVNEWF 111
Ail            NLKYRYELDDNWGVIGSFAYT--HQGYDFFYGSNKFGHGDVD--YYSVTMGPSFRINEYV 108
               *:*****: *. *:: *:.::  ::      ..  :   .:   : **  ** .::.: .
                                       ------LOOP2--------

λ Lom          SAYAMAGVAHSRWSGSTMDYRKTEITPGYMKETTTARDESAMRHTSVAWSAGIQINPAAS 178
CP-933C Lom    SAYAMAGVAYSRVSTFSGDYLRVTDNKGKTHDVLTGSDDGRHSNTSLAWGAGVQFNPTES 171
Ail            SLYGLLGAAHGKVKASVFD-------------------ESISASKTSMAYGAGVQFNPLPN 150
               * *.: *.*:.: .    *                  :.    :**:*:.**:*:**  .
                         -------------LOOP3--------------

λ Lom          VVVDIAYEGSGSGDWRTDGFIVGVGYKF 206
CP-933C Lom    VAIDIAYEGSGSGDWRTDGFIVGVGYKF 199
Ail            FVIDASYEYSKLDSTKVGTWMLGAGYRF 178
               ..:* :** *  .. :.. :::*.**:*
```

FIGURE 17 ClustalW analysis of Lom and Ail proteins with the position of the periplasmic loops indicated (Miller et al., 2001).

by these individuals in the 1950s and '60s is only now being reexamined by people interested in genomics.

REFERENCES

Ackermann, H.-W., and M. S. DuBow. 1987. *Viruses of Prokaryotes*. CRC Press, Boca Raton, FL.

Aertsen, A., D. Faster, and C. W. Michiels. 2005. Induction of Shiga toxin-converting prophage in *Escherichia coli* by high hydrostatic pressure. *Appl. Environ. Microbiol.* **71:**1155–1162.

Agbodaze, D. 1999. Verocytotoxins (Shiga-like toxins) produced by *Escherichia coli:* a minireview of their classification, clinical presentations and management of a heterogeneous family of cytotoxins. *Comp. Immunol. Microbiol. Infect. Dis.* **22:**221–230.

Allison, H. E. 2007. Stx-phages: drivers and mediators of the evolution of STEC and STEC-like pathogens. *Future Microbiol.* **2:**165–174.

Altschul, S. F., W. Gish, W. Miller, E. W. Myers, and D. J. Lipman. 1990. Basic local alignment search tool. *J. Mol. Biol.* **215:**403–410.

Altschul, S. F., T. L. Madden, A. A. Schaffer, J. Zhang, Z. Zhang, W. Miller, and D. J. Lipman. 1997. Gapped BLAST and PSI-BLAST: a new generation of protein database search programs. *Nucleic Acids Res.* **25:**3389–4022.

Amagai, M., N. Matsuyoshi, Z. H. Wang, C. Andl, and J. R. Stanley. 2000. Toxin in bullous impetigo and staphylococcal scalded-skin syndrome targets desmoglein 1. *Nat. Med.* **6:**1275–1277.

Amagai, M., T. Yamaguchi, Y. Hanakawa, K. Nishifuji, M. Sugai, and J. R. Stanley. 2002. Staphylococcal exfoliative toxin B specifically cleaves desmoglein 1. *J. Invest. Dermatol.* **118:**845–850.

Arndt, J. W., Q. Chai, T. Christian, and R. C. Stevens. 2006. Structure of botulinum neurotoxin type D light chain at 1.65 Å resolution: repercussions for VAMP-2 substrate specificity. *Biochemistry* **45:**3255–3262.

Arndt, J. W., J. Gu, L. Jaroszewski, R. Schwarzenbacher, M. A. Hanson, F. J. Lebeda, and R. C. Stevens. 2005. The structure of the neurotoxin-associated protein HA33/A from *Clostridium botulinum* suggests a reoccurring β-trefoil fold in the progenitor toxin complex. *J. Mol. Biol.* **346:**1083–1093.

Asadulghani, M., Y. Ogura, T. Ooka, T. Itoh, A. Sawaguchi, A. Iguchi, K. Nakayama, and T. Hayashi. 2009. The defective prophage pool of *Escherichia coli* O157: prophage-prophage interactions potentiate horizontal transfer of virulence determinants. *PLoS Pathog.* **5:**e1000408.

Baba, T., F. Takeuchi, M. Kuroda, H. Yuzawa, K. Aoki, A. Oguchi, Y. Nagai, N. Iwama, K. Asano, T. Naimi, H. Kuroda, L. Cui, K. Yamamoto, and K. Hiramatsu. 2002. Genome and virulence determinants of high virulence community-acquired MRSA. *Lancet* **359:**1819–1827.

Bachert, C., N. Zhang, J. Patou, T. van Zele, and P. Gevaert. 2008. Role of staphylococcal superantigens in upper airway disease. *Curr. Opin. Allergy Clin. Immunol.* **8:**34–38.

Balbontin, R., G. Rowley, M. G. Pucciarelli, J. Lopez-Garrido, Y. Wormstone, S. Lucchini, F. Garcia-Del Portillo, J. C. Hinton, and J. Casadesus. 2006. DNA adenine methylation regulates virulence gene expression in *Salmonella enterica* serovar Typhimurium. *J. Bacteriol.* **188:**8160–8168.

Bania, J., A. Dabrowska, J. Bystron, K. Korzekwa, J. Chrzanowska, and J. Molenda. 2006. Distribution of newly described enterotoxin-like genes in *Staphylococcus aureus* from food. *Int. J. Food Microbiol.* **108:**36–41.

Barksdale, L., and S. B. Arden. 1974. Persisting bacteriophage infections, lysogeny, and phage conversions. *Annu. Rev. Microbiol.* **28:**265–299.

Barondess, J. J., and J. Beckwith. 1990. A bacterial virulence determinant encoded by lysogenic coliphage λ. *Nature* **346:**871–874.

Barondess, J. J., and J. Beckwith. 1995. *bor* gene of phage λ, involved in serum resistance, encodes a widely conserved outer membrane lipoprotein. *J. Bacteriol.* **177:**1247–1253.

Bergdoll, M. S. 1983. Enterotoxins, p. 559–598. *In* C. S. F. Easmon and C. Adlam (ed.), *Staphylococci and Staphylococcal Infections*. Academic Press, New York, NY.

Betley, M. J., and J. J. Mekalanos. 1985. Staphylococcal enterotoxin A is encoded by phage. *Science* **229:**185–187.

Birembaux, C., O. Fayet, and M. F. Prere. 1999. *E. coli* O157 from children with gastrointestinal symptoms. Isolation of a urease producing strain. *Acta Clin. Belg.* **46:**51–52.

Blair, J. E., and M. Carr. 1961. Lysogeny in staphylococci. *J. Bacteriol.* **82:**984–993.

Blanco, M., J. E. Blanco, A. Mora, G. Dahbi, M. P. Alonso, E. A. Gonzalez, M. I. Bernardez, and J. Blanco. 2004. Serotypes, virulence genes, and intimin types of Shiga toxin (verotoxin)-producing *Escherichia coli* isolates from cattle in Spain and identification of a new intimin variant gene (*eae*-ξ). *J. Clin. Microbiol.* **42:**645–651.

Bokarewa, M. I., T. Jin, and A. Tarkowski. 2006. *Staphylococcus aureus:* staphylokinase. *Int. J. Biochem. Cell Biol.* **38:**504–509.

Borst, D. W., and M. J. Betley. 1994. Promoter analysis of the staphylococcal enterotoxin A gene. *J. Biol. Chem.* **269:**1883–1888.

Bray, D., and P. W. Robbins. 1967. Mechanism of ε15 conversion studied with bacteriophage mutants. *J. Mol. Biol.* **30:**457–475.

Brin, M. F. 1997. Botulinum toxin: chemistry, pharmacology, toxicity, and immunology. *Muscle Nerve Suppl.* **6:**S146–S168.

Bronner, S., P. Stoessel, A. Gravet, H. Monteil, and G. Prevost. 2000. Variable expressions of *Staphylococcus aureus* bicomponent leucotoxins semiquantified by competitive reverse transcription-PCR. *Appl. Environ. Microbiol.* **66:** 3931–3938.

Brüssow, H., C. Canchaya, and W. D. Hardt. 2004. Phages and the evolution of bacterial pathogens: from genomic rearrangements to lysogenic conversion. *Microbiol. Mol. Biol. Rev.* **68:**560–602.

Bubeck Wardenburg, J., T. Bae, M. Otto, F. R. DeLeo, and O. Schneewind. 2007. Poring over pores: α-hemolysin and Panton-Valentine leukocidin in *Staphylococcus aureus* pneumonia. *Nat. Med.* **13:**1405–1406.

Carroll, J. D., M. T. Cafferkey, and D. C. Coleman. 1993. Serotype F double- and triple-converting phage insertionally inactivate the *Staphylococcus aureus* beta-toxin determinant by a common molecular mechanism. *FEMS Microbiol. Lett.* **106:**147–155.

Cavarelli, J., G. Prevost, W. Bourguet, L. Moulinier, B. Chevrier, B. Delagoutte, A. Bilwes, L. Mourey, S. Rifai, Y. Piemont, and D. Moras. 1997. The structure of *Staphylococcus aureus* epidermolytic toxin A, an atypic serine protease, at 1.7 Å resolution. *Structure* **5:**813–824.

Chan, P. F., and S. J. Foster. 1998. Role of SarA in virulence determinant production and environmental signal transduction in *Staphylococcus aureus*. *J. Bacteriol.* **180:**6232–6241.

Chen, Y., I. Golding, S. Sawai, L. Guo, and E. C. Cox. 2005. Population fitness and the regulation of *Escherichia coli* genes by bacterial viruses. *PLoS Biol.* **3:**e229.

Chuba, P. J., S. Palchaudhuri, and M. A. Leon. 1986. Contribution of *traT* and *iss* genes to the serum resistance phenotype of plasmid ColV2-K94. *FEMS Microbiol. Lett.* **37:**135–140.

Clecner, B., and S. Sonea. 1966. Acquisition of the delta type of hemolytic property by lysogenic conversion in *Staphylococcus aureus*. *Rev. Can. Biol.* **25:** 145–148.

Coggill, P., R. D. Finn, and A. Bateman. 2008. Identifying protein domains with the Pfam database. *Curr. Protoc. Bioinformatics* **Chapter 2:**Unit 2.5.

Coleman, D. C., J. P. Arbuthnott, H. M. Pomeroy, and T. H. Birkbeck. 1986. Cloning and expression in *Escherichia coli* and *Staphylococcus aureus* of the beta-lysin determinant from *Staphylococcus aureus:* evidence that bacteriophage conversion of beta-lysin activity is caused by insertional inactivation of the beta-lysin determinant. *Microb. Pathog.* **1:**549–564.

Coleman, D. C., D. J. Sullivan, R. J. Russell, J. P. Arbuthnott, B. F. Carey, and H. M. Pomeroy. 1989. *Staphylococcus aureus* bacteriophages mediating the simultaneous lysogenic conversion of β-lysin, staphylokinase and enterotoxin A: molecular mechanism of triple conversion. *J. Gen. Microbiol.* **135:**1679–1697.

Crago, A. M., and V. Koronakis. 1999. Binding of extracellular matrix laminin to *Escherichia coli* expressing the *Salmonella* outer membrane proteins Rck and PagC. *FEMS Microbiol. Lett.* **176:**495–501.

Creuzburg, K., J. Recktenwald, V. Kuhle, S. Herold, M. Hensel, and H. Schmidt. 2005. The Shiga toxin 1-converting bacteriophage BP-4795 encodes an NleA-like type III effector protein. *J. Bacteriol.* **187:**8494–8498.

Creuzburg, K., and H. Schmidt. 2007. Shiga toxin-producing *Escherichia coli* and their bacteriophages as a model for the analysis of virulence and stress response of a food-borne pathogen. *Berl. Munch. Tierarztl. Wochenschr.* **120:**288–295.

Czyz, A., M. Łoś, B. Wrobel, and G. Węgrzyn. 2001. Inhibition of spontaneous induction of lambdoid prophages in *Escherichia coli* cultures: simple procedures with possible biotechnological applications. *BMC Biotechnol.* **1:**1.

Dauwalder, O., D. Thomas, T. Ferry, A. L. Debard, C. Badiou, F. Vandenesch, J. Etienne, G. Lina, and G. Monneret. 2006. Comparative inflammatory properties of staphylococcal superantigenic enterotoxins SEA and SEG: implications for septic shock. *J. Leukoc. Biol.* **80:**753–758.

de Haas, C. J., K. E. Veldkamp, A. Peschel, F. Weerkamp, W. J. Van Wamel, E. C. Heezius, M. J. Poppelier, K. P. Van Kessel, and J. A. van Strijp. 2004. Chemotaxis inhibitory protein of *Staphylococcus aureus,* a bacterial antiinflammatory agent. *J. Exp. Med.* **199:**687–695.

Dempsey, R. M., D. Carroll, H. Kong, L. Higgins, C. T. Keane, and D. C. Coleman. 2005. *Sau*42I, a *Bcg*I-like restriction-modification system encoded by the *Staphylococcus aureus* quadruple-converting phage ϕ42. *Microbiology* **151:**1301–1311.

Diep, B. A., S. R. Gill, R. F. Chang, T. H. Phan, J. H. Chen, M. G. Davidson, F. Lin, J. Lin, H. A. Carleton, E. F. Mongodin, G. F. Sensabaugh, and F. Perdreau-Remington. 2006. Complete genome sequence of USA300, an epidemic clone of community-acquired meticillin-resistant *Staphylococcus aureus*. *Lancet* **367:**731–739.

Diep, B. A., and M. Otto. 2008. The role of virulence determinants in community-associated MRSA pathogenesis. *Trends Microbiol.* **16:**361–369.

**Diep, B. A., A. M. Palazzolo-Ballance, P. Tattevin, L. Basuino, K. R. Braughton, A. R.

Whitney, L. Chen, B. N. Kreiswirth, M. Otto, F. R. DeLeo, and H. F. Chambers. 2008. Contribution of Panton-Valentine leukocidin in community-associated methicillin-resistant *Staphylococcus aureus* pathogenesis. *PLoS One* **3:** e3198.

Dobardzic, R., and S. Sonea. 1971. Hemolysin production and lysogenic conversion in *Staphylococcus aureus. Ann. Inst. Pasteur* (Paris) **120:**42–49.

Dumitrescu, O., S. Boisset, C. Badiou, M. Bes, Y. Benito, M. E. Reverdy, F. Vandenesch, J. Etienne, and G. Lina. 2007. Effect of antibiotics on *Staphylococcus aureus* producing Panton-Valentine leukocidin. *Antimicrob. Agents Chemother.* **51:**1515–1519.

Eklund, M. W., F. T. Poysky, and S. M. Reed. 1972. Bacteriophage and the toxigenicity of *Clostridium botulinum* type D. *Nat. New Biol.* **235:**16–17.

Eklund, M. W., F. T. Poysky, S. M. Reed, and C. A. Smith. 1971. Bacteriophage and the toxigenicity of *Clostridium botulinum* type C. *Science* **172:**480–482.

Ellermeier, C. D., and J. M. Slauch. 2006. The genus *Salmonella,* p. 123–158. *In* M. Dworkin, S. Falkow, E. Rosenberg, K.-H. Schleifer, and E. Stackebrandt (ed.), *The Prokaryotes: a Handbook on the Biology of Bacteria,* 3rd ed. Springer, New York, NY.

Endo, Y., K. Tsurugi, T. Yutsudo, Y. Takeda, T. Ogasawara, and K. Igarashi. 1988. Site of action of a Vero toxin (VT2) from *Escherichia coli* O157:H7 and of Shiga toxin on eukaryotic ribosomes. RNA *N*-glycosidase activity of the toxins. *Eur. J. Biochem.* **171:**45–50.

Endo, Y., T. Yamada, K. Matsunaga, Y. Hayakawa, T. Kaidoh, and S. Takeuchi. 2003. Phage conversion of exfoliative toxin A in *Staphylococcus aureus* isolated from cows with mastitis. *Vet. Microbiol.* **96:**81–90.

Evans, T., R. G. Bowers, and M. Mortimer. 2007. Modelling the stability of Stx lysogens. *J. Theor. Biol.* **248:**241–250.

Fairbrother, J. M., and E. Nadeau. 2006. *Escherichia coli:* on-farm contamination of animals. *Rev. Sci. Tech.* **25:**555–569.

Fernandez, M. M., S. Bhattacharya, M. C. De Marzi, P. H. Brown, M. Kerzic, P. Schuck, R. A. Mariuzza, and E. L. Malchiodi. 2007. Superantigen natural affinity maturation revealed by the crystal structure of staphylococcal enterotoxin G and its binding to T-cell receptor Vβ8.2. *Proteins* **68:**389–402.

Ferry, T., D. Thomas, A. L. Genestier, M. Bes, G. Lina, F. Vandenesch, and J. Etienne. 2005. Comparative prevalence of superantigen genes in *Staphylococcus aureus* isolates causing sepsis with and without septic shock. *Clin. Infect. Dis.* **41:**771–777.

Finck-Barbancon, V., G. Duportail, O. Meunier, and D. A. Colin. 1993. Pore formation by a two-component leukocidin from *Staphylococcus aureus* within the membrane of human polymorphonuclear leukocytes. *Biochim. Biophys. Acta* **1182:**275–282.

Fix, D. 1993. The *rex* genes of lambda prophage modify ultraviolet light and *N*-methyl-*N*-nitrosourea-induced responses to DNA damage in *Escherichia coli. Mutat. Res.* **303:**143–150.

Fleischer, B., and C. J. Bailey. 1992. Recombinant epidermolytic (exfoliative) toxin A of *Staphylococcus aureus* is not a superantigen. *Med. Microbiol. Immunol.* **180:**273–278.

Fleming, S. D., J. J. Iandolo, and S. K. Chapes. 1991. Murine macrophage activation by staphylococcal exotoxins. *Infect. Immun.* **59:**4049–4055.

Fogg, P. C., S. M. Gossage, D. L. Smith, J. R. Saunders, A. J. McCarthy, and H. E. Allison. 2007. Identification of multiple integration sites for Stx-phage $\Phi 24_B$ in the *Escherichia coli* genome, description of a novel integrase and evidence for a functional anti-repressor. *Microbiology* **153:**4098–4110.

Fraser, J. D., and T. Proft. 2008. The bacterial superantigen and superantigen-like proteins. *Immunol. Rev.* **225:**226–243.

Fraser, M. E., M. M. Chernaia, Y. V. Kozlov, and M. N. James. 1994. Crystal structure of the holotoxin from *Shigella dysenteriae* at 2.5 Å resolution. *Nat. Struct. Biol.* **1:**59–64.

Fraser, M. E., M. Fujinaga, M. M. Cherney, A. R. Melton-Celsa, E. M. Twiddy, A. D. O'Brien, and M. N. James. 2004. Structure of Shiga toxin type 2 (Stx2) from *Escherichia coli* O157:H7. *J. Biol. Chem.* **279:**27511–27517.

Freeman, V. J. 1951. Studies on the virulence of bacteriophage-infected strains of *Corynebacterium diphtheriae. J. Bacteriol.* **61:**675–688.

Friedman, D. I., and D. L. Court. 2001. Bacteriophage lambda: alive and well and still doing its thing. *Curr. Opin. Microbiol.* **4:**201–207.

Gajdus, J., Z. Kaczynski, M. Czerwicka, H. Dziadziuszko, and J. Szafranek. 2008. Structure of the *Salmonella* Anatum (O:3,10) O-antigen. *Pol. J. Chem.* **82:**1393–1398.

Garcia-Aljaro, C., M. Muniesa, J. Jofre, and A. R. Blanch. 2009. Genotypic and phenotypic diversity among induced, stx_2-carrying bacteriophages from environmental *Escherichia coli* strains. *Appl. Environ. Microbiol.* **75:**329–336.

Gillet, Y., B. Issartel, P. Vanhems, J. C. Fournet, G. Lina, M. Bes, F. Vandenesch, Y. Piemont, N. Brousse, D. Floret, and J. Etienne. 2002. Association between *Staphylococcus aureus* strains carrying gene for Panton-Valentine leukocidin and highly lethal necrotising pneumonia in young immunocompetent patients. *Lancet* **359:** 753–759.

Giraudo, A. T., A. Calzolari, A. A. Cataldi, C. Bogni, and R. Nagel. 1999a. Corrigendum to: "The *sae* locus of *Staphylococcus aureus* encodes a two-component regulatory system." *FEMS Microbiol. Lett.* **180:**117.

Giraudo, A. T., A. Calzolari, A. A. Cataldi, C. Bogni, and R. Nagel. 1999b. The *sae* locus of *Staphylococcus aureus* encodes a two-component regulatory system. *FEMS Microbiol. Lett.* **177:**15–22.

Goerke, C., C. Wirtz, U. Fluckiger, and C. Wolz. 2006. Extensive phage dynamics in *Staphylococcus aureus* contributes to adaptation to the human host during infection. *Mol. Microbiol.* **61:** 1673–1685.

Greenberg, M., J. Dunlap, and R. Villafane. 1995. Identification of the tailspike protein from the *Salmonella newington* phage ε^{34} and partial characterization of its phage-associated properties. *J. Struct. Biol.* **115:**283–289.

Grimont, P. A. D., F.-X. Weill, and WHO Collaborating Center for Reference and Research on *Salmonella*. 2007. *Antigenic Formulae of the* Salmonella *Serovars,* 9th ed. Institut Pasteur, Paris, France.

Guan, S., D. A. Bastin, and N. K. Verma. 1999. Functional analysis of the O antigen glucosylation gene cluster of *Shigella flexneri* bacteriophage SfX. *Microbiology* **145:**1263–1273.

Guglielmini, J., C. Szpirer, and M. C. Milinkovitch. 2008. Automated discovery and phylogenetic analysis of new toxin-antitoxin systems. *BMC Microbiol.* **8:**104.

Gyles, C. L. 2007. Shiga toxin-producing *Escherichia coli:* an overview. *J. Anim. Sci.* **85:**E45–E62.

Haas, P. J., C. J. de Haas, W. Kleibeuker, M. J. Poppelier, K. P. van Kessel, J. A. Kruijtzer, R. M. Liskamp, and J. A. van Strijp. 2004. N-terminal residues of the chemotaxis inhibitory protein of *Staphylococcus aureus* are essential for blocking formylated peptide receptor but not C5a receptor. *J. Immunol.* **173:**5704–5711.

Haas, P. J., C. J. de Haas, M. J. Poppelier, K. P. van Kessel, J. A. van Strijp, K. Dijkstra, R. M. Scheek, H. Fan, J. A. Kruijtzer, R. M. Liskamp, and J. Kemmink. 2005. The structure of the C5a receptor-blocking domain of chemotaxis inhibitory protein of *Staphylococcus aureus* is related to a group of immune evasive molecules. *J. Mol. Biol.* **353:**859–872.

Hagiwara, S., and H. Uetake. 1970. Formation of hybrid conversion phages between heterologous *Salmonella* phages ε15 and ε34. *Virology* **41:**419–429.

Hanakawa, Y., N. M. Schechter, C. Lin, K. Nishifuji, M. Amagai, and J. R. Stanley. 2004. Enzymatic and molecular characteristics of the efficiency and specificity of exfoliative toxin cleavage of desmoglein 1. *J. Biol. Chem.* **279:**5268–5277.

Handa, H., S. Gurczynski, M. P. Jackson, G. Auner, and G. Mao. 2008. Recognition of *Salmonella* Typhimurium by immobilized phage P22 monolayers. *Surf. Sci.* **602:**1392–1400.

Hanson, M. A., and R. C. Stevens. 2000. Cocrystal structure of synaptobrevin-II bound to botulinum neurotoxin type B at 2.0 Å resolution. *Nat. Struct. Biol.* **7:**687–692.

Herold, S., H. Karch, and H. Schmidt. 2004. Shiga toxin-encoding bacteriophages—genomes in motion. *Int. J. Med. Microbiol.* **294:**115–121.

Hofer, B., M. Ruge, and B. Dreiseikelmann. 1995. The superinfection exclusion gene *(sieA)* of bacteriophage P22: identification and overexpression of the gene and localization of the gene product. *J. Bacteriol.* **177:**3080–3086.

Holmberg, S. D., and P. A. Blake. 1984. Staphylococcal food poisoning in the United States. New facts and old misconceptions. *JAMA* **251:** 487–489.

Holmes, A., M. Ganner, S. McGuane, T. L. Pitt, B. D. Cookson, and A. M. Kearns. 2005. *Staphylococcus aureus* isolates carrying Panton-Valentine leucocidin genes in England and Wales: frequency, characterization, and association with clinical disease. *J. Clin. Microbiol.* **43:**2384–2390.

Hongo, I., T. Baba, K. Oishi, Y. Morimoto, T. Ito, and K. Hiramatsu. 2009. Phenol-soluble modulin α3 enhances the human neutrophil lysis mediated by Panton-Valentine leukocidin. *J. Infect. Dis.* **200:**715–723.

Huan, P. T., D. A. Bastin, B. L. Whittle, A. A. Lindberg, and N. K. Verma. 1997. Molecular characterization of the genes involved in O-antigen modification, attachment, integration and excision in *Shigella flexneri* bacteriophage SfV. *Gene* **195:**217–227.

Iandolo, J. J., V. Worrell, K. H. Groicher, Y. Qian, R. Tian, S. Kenton, A. Dorman, H. Ji, S. Lin, P. Loh, S. Qi, H. Zhu, and B. A. Roe. 2002. Comparative analysis of the genomes of the temperate bacteriophages φ11, φ12 and φ13 of *Staphylococcus aureus* 8325. *Gene* **289:**109–118.

Imamovic, L., J. Jofre, H. Schmidt, R. Serra-Moreno, and M. Muniesa. 2009. Phage-mediated Shiga toxin 2 gene transfer in food and water. *Appl. Environ. Microbiol.* **75:**1764–1768.

Inoue, K., and H. Iida. 1968. Bacteriophages of *Clostridium botulinum. J. Virol.* **2:**537–540.

Inoue, K., and H. Iida. 1970. Conversion of toxigenicity in *Clostridium botulinum* type C. *Jpn. J. Microbiol.* **14:**87–89.

Inoue, K., and H. Iida. 1971. Phage-conversion of toxigenicity in *Clostridium botulinum* types C and D. *Jpn. J. Med. Sci. Biol.* **24:**53–56.

Ippel, J. H., C. J. de Haas, A. Bunschoten, J. A. van Strijp, J. A. Kruijtzer, R. M. Liskamp, and J. Kemmink. 2009. Structure of the tyrosine-sulfated C5a receptor N terminus in complex with chemotaxis inhibitory protein of *Staphylococcus aureus*. *J. Biol. Chem.* **284:**12363–12372.

Iseki, S., and K. Kashiwagi. 1955. Induction of somatic antigen 1 by bacteriophage in *Salmonella* group B. *Proc. Jpn. Acad.* **31:**558–564.

Iwashita, S., and S. Kanegasaki. 1976. Deacetylation reaction catalyzed by *Salmonella* phage c341 and its baseplate parts. *J. Biol. Chem.* **251:**5361–5365.

Jackson, M. P., and J. J. Iandolo. 1986a. Cloning and expression of the exfoliative toxin B gene from *Staphylococcus aureus*. *J. Bacteriol.* **166:**574–580.

Jackson, M. P., and J. J. Iandolo. 1986b. Sequence of the exfoliative toxin B gene of *Staphylococcus aureus*. *J. Bacteriol.* **167:**726–728.

Jarraud, S., M. A. Peyrat, A. Lim, A. Tristan, M. Bes, C. Mougel, J. Etienne, F. Vandenesch, M. Bonneville, and G. Lina. 2001. *egc*, a highly prevalent operon of enterotoxin gene, forms a putative nursery of superantigens in *Staphylococcus aureus*. *J. Immunol.* **166:**669–677.

Jayasinghe, L., and H. Bayley. 2005. The leukocidin pore: evidence for an octamer with four LukF subunits and four LukS subunits alternating around a central axis. *Protein Sci.* **14:**2550–2561.

Jin, T., M. Bokarewa, T. Foster, J. Mitchell, J. Higgins, and A. Tarkowski. 2004. *Staphylococcus aureus* resists human defensins by production of staphylokinase, a novel bacterial evasion mechanism. *J. Immunol.* **172:**1169–1176.

Johnson, A. D., L. Spero, J. S. Cades, and B. T. de Cicco. 1979. Purification and characterization of different types of exfoliative toxin from *Staphylococcus aureus*. *Infect. Immun.* **24:**679–684.

Johnson, E. A., and M. Bradshaw. 2001. *Clostridium botulinum* and its neurotoxins: a metabolic and cellular perspective. *Toxicon* **39:**1703–1722.

Johnson, T. J., Y. M. Wannemuehler, and L. K. Nolan. 2008. Evolution of the *iss* gene in *Escherichia coli*. *Appl. Environ. Microbiol.* **74:**2360–2369.

Juhala, R. J., M. E. Ford, R. L. Duda, A. Youlton, G. F. Hatfull, and R. W. Hendrix. 2000. Genetic sequences of bacteriophages HK97 and HK022: pervasive genetic mosaicism in the lambdoid bacteriophages. *J. Mol. Biol.* **299:**27–51.

Kall, L., A. Krogh, and E. L. Sonnhammer. 2004. A combined transmembrane topology and signal peptide prediction method. *J. Mol. Biol.* **338:**1027–1036.

Kamio, Y., T. Tomita, and J. Kaneko. 2001. Pore-forming cytolysin from *Staphylococcus aureus,* alpha- and gamma-hemolysin, and leukocidin, p. 179–212. *In* I. Friedman (ed.), *Staphylococcal Infection and Immunity*. Kluwer Academic/Plenum Publishers, New York, NY.

Kaneko, J., and Y. Kamio. 2004. Bacterial two-component and hetero-heptameric pore-forming cytolytic toxins: structures, pore-forming mechanism, and organization of the genes. *Biosci. Biotechnol. Biochem.* **68:**981–1003.

Kaneko, J., T. Kimura, Y. Kawakami, T. Tomita, and Y. Kamio. 1997. Panton-Valentine leukocidin genes in a phage-like particle isolated from mitomycin C-treated *Staphylococcus aureus* V8 (ATCC 49775). *Biosci. Biotechnol. Biochem.* **61:** 1960–1962.

Kaneko, J., T. Kimura, S. Narita, T. Tomita, and Y. Kamio. 1998. Complete nucleotide sequence and molecular characterization of the temperate staphylococcal bacteriophage φPVL carrying Panton-Valentine leukocidin genes. *Gene* **215:** 57–67.

Karch, H., and M. Bielaszewska. 2001. Sorbitol-fermenting Shiga toxin-producing *Escherichia coli* O157:H^- strains: epidemiology, phenotypic and molecular characteristics, and microbiological diagnosis. *J. Clin. Microbiol.* **39:**2043–2049.

Karmali, M. A., M. Petric, C. Lim, P. C. Fleming, and B. T. Steele. 1983a. *Escherichia coli* cytotoxin, haemolytic-uraemic syndrome, and haemorrhagic colitis. *Lancet* **2:**1299–1300.

Karmali, M. A., B. T. Steele, M. Petric, and C. Lim. 1983b. Sporadic cases of haemolytic-uraemic syndrome associated with faecal cytotoxin and cytotoxin-producing *Escherichia coli* in stools. *Lancet* **1:**619–620.

Kimmitt, P. T., C. R. Harwood, and M. R. Barer. 2000. Toxin gene expression by Shiga toxin-producing *Escherichia coli:* the role of antibiotics and the bacterial SOS response. *Emerg. Infect. Dis.* **6:**458–465.

Kitamura, M., K. Takamiya, S. Aizawa, K. Furukawa, and K. Furukawa. 1999. Gangliosides are the binding substances in neural cells for tetanus and botulinum toxins in mice. *Biochim. Biophys. Acta* **1441:**1–3.

Kobayashi, S. D., and F. R. DeLeo. 2009. An update on community-associated MRSA virulence. *Curr. Opin. Pharmacol.* **9:**545–551.

Kondo, I., S. Sakurai, Y. Sarai, and S. Futaki. 1975. Two serotypes of exfoliatin and their distribution in staphylococcal strains isolated from patients with scalded skin syndrome. *J. Clin. Microbiol.* **1:**397–400.

Kozaki, S., G. Sakaguchi, M. Nishimura, M. Iwamori, and Y. Nagai. 1984. Inhibitory effect of ganglioside GTlb on the activities of *Clostridium botulinum* toxins. *FEMS Microbiol. Lett.* **21:**219–223.

Kropinski, A. M. 2009. Measurement of the bacteriophage inactivation kinetics with purified receptors. *Methods Mol. Biol.* **501:**157–160.

Kropinski, A. M., M. Borodovsky, T. J. Carver, A. M. Cerdeno-Tarraga, A. Darling, A. Lomsadze, P. Mahadevan, P. Stothard, D. Seto, G. Van Domselaar, and D. S. Wishart. 2009. *In silico* identification of genes in bacteriophage DNA. *Methods Mol. Biol.* **502:**57–89.

Kropinski, A. M., I. V. Kovalyova, S. J. Billington, B. D. Butts, A. N. Patrick, J. A. Guichard, S. M. Hutson, A. D. Sydlaske, K. R. Day, D. R. Falk, and M. R. McConnell. 2007. The genome of ε15, a serotype-converting, Group E1 *Salmonella enterica*-specific bacteriophage. *Virology* **369:**234–244.

Kumagai, R., K. Nakatani, N. Ikeya, Y. Kito, T. Kaidoh, and S. Takeuchi. 2007. Quadruple or quintuple conversion of *hlb, sak, sea* (or *sep*), *scn,* and *chp* genes by bacteriophages in non-β-hemolysin-producing bovine isolates of *Staphylococcus aureus*. *Vet. Microbiol.* **122:**190–195.

Kumar, A., H. Wu, L. S. Collier-Hyams, J. M. Hansen, T. Li, K. Yamoah, Z. Q. Pan, D. P. Jones, and A. S. Neish. 2007. Commensal bacteria modulate cullin-dependent signaling via generation of reactive oxygen species. *EMBO J.* **26:** 4457–4466.

Kuroda, M., T. Ohta, I. Uchiyama, T. Baba, H. Yuzawa, I. Kobayashi, L. Cui, A. Oguchi, K. Aoki, Y. Nagai, J. Lian, T. Ito, M. Kanamori, H. Matsumaru, A. Maruyama, H. Murakami, A. Hosoyama, Y. Mizutani-Ui, N. K. Takahashi, T. Sawano, R. Inoue, C. Kaito, K. Sekimizu, H. Hirakawa, S. Kuhara, S. Goto, J. Yabuzaki, M. Kanehisa, A. Yamashita, K. Oshima, K. Furuya, C. Yoshino, T. Shiba, M. Hattori, N. Ogasawara, H. Hayashi, and K. Hiramatsu. 2001. Whole genome sequencing of meticillin-resistant *Staphylococcus aureus*. *Lancet* **357:**1225–1240.

Kwan, T., J. Liu, M. DuBow, P. Gros, and J. Pelletier. 2005. The complete genomes and proteomes of 27 *Staphylococcus aureus* bacteriophages. *Proc. Natl. Acad. Sci. USA* **102:**5174–5179.

Labandeira-Rey, M., F. Couzon, S. Boisset, E. L. Brown, M. Bes, Y. Benito, E. M. Barbu, V. Vazquez, M. Hook, J. Etienne, F. Vandenesch, and M. G. Bowden. 2007. *Staphylococcus aureus* Panton-Valentine leukocidin causes necrotizing pneumonia. *Science* **315:**1130–1133.

Ladhani, S. 2001. Recent developments in staphylococcal scalded skin syndrome. *Clin. Microbiol. Infect.* **7:**301–307.

Ladhani, S., C. L. Joannou, D. P. Lochrie, R. W. Evans, and S. M. Poston. 1999. Clinical, microbial, and biochemical aspects of the exfoliative toxins causing staphylococcal scalded-skin syndrome. *Clin. Microbiol. Rev.* **12:**224–242.

Lainhart, W., G. Stolfa, and G. B. Koudelka. 2009. Shiga toxin as a bacterial defense against a eukaryotic predator, *Tetrahymena thermophila*. *J. Bacteriol.* **191:**5116–5122.

Landstrom, J., E. L. Nordmark, R. Eklund, A. Weintraub, R. Seckler, and G. Widmalm. 2008. Interaction of a *Salmonella enteritidis* O-antigen octasaccharide with the phage P22 tailspike protein by NMR spectroscopy and docking studies. *Glycoconj. J.* **25:**137–143.

Lavigne, R., D. Seto, P. Mahadevan, H.-W. Ackermann, and A. M. Kropinski. 2008. Unifying classical and molecular taxonomic classification: analysis of the *Podoviridae* using BLASTP-based tools. *Res. Microbiol.* **159:**406–414.

Lee, C. Y., and J. J. Iandolo. 1986. Lysogenic conversion of staphylococcal lipase is caused by insertion of the bacteriophage L54a genome into the lipase structural gene. *J. Bacteriol.* **166:**385–391.

Lee, C. Y., J. J. Schmidt, A. D. Johnson-Winegar, L. Spero, and J. J. Iandolo. 1987. Sequence determination and comparison of the exfoliative toxin A and toxin B genes from *Staphylococcus aureus*. *J. Bacteriol.* **169:**3904–3909.

Lee, J. E., J. Reed, M. S. Shields, K. M. Spiegel, L. D. Farrell, and P. P. Sheridan. 2007. Phylogenetic analysis of Shiga toxin 1 and Shiga toxin 2 genes associated with disease outbreaks. *BMC Microbiol.* **7:**109.

Lehnherr, H. 2005. Bacteriophage P1, p. 350–364. *In* R. Calendar (ed.), *The Bacteriophages,* 2nd ed. Oxford University Press, New York, NY.

Le Loir, Y., F. Baron, and M. Gautier. 2003. *Staphylococcus aureus* and food poisoning. *Genet. Mol. Res.* **2:**63–76.

Le Minor, L. 1962. Conversion par lysogenisation de quelques sérotypes de *Salmonella* des groupes A, B et D normalement dépourvus du facteur 027 en cultures 27 positives. *Ann. Inst. Pasteur* (Paris) **103:** 684–706.

Le Minor, L. 1963. Conversion antigénique chez les *Salmonella:* IV. Acquisition du facteur 01 par les *Salmonella* des groupes R et T sous l'effet de la lysogenisation. *Ann. Inst. Pasteur* (Paris) **105:**879–896.

Le Minor, L. 1965a. Conversion antigénique chez les *Salmonella:* V. Acquisition des facteurs 15 et 34 par les *Salmonella* du sous groupe D2 sous l'effet de la lysogenisation par les phages ε15 et ε34. *Ann. Inst. Pasteur* (Paris) **109:**35–46.

Le Minor, L. 1965b. Conversions antigéniques chez les *Salmonella.* VI. Acquisition des facteurs 6, 14 par les serotypes du groupe K (018) sous l'effet de la lysogénisation. *Ann. Inst. Pasteur* (Paris) **108:** 805–811.

Le Minor, L. 1965c. Conversions antigéniques chez les *Salmonella*. VII. Acquisition du facteur 14 par les *Salmonella* du sous groupe C1 (6, 7) apres lysogénisation par un phage tempere isole des cultures du sous groupe C4 (6, (7), (14)). *Ann. Inst. Pasteur* (Paris) **109:**505–515.

Le Minor, L., H.-W. Ackermann, and P. Nicolle. 1963. Acquisition simultanée des facteurs 01 et 037 par les *Salmonella* du groupe G sous l'effet de la lysogénisation. *Ann. Inst. Pasteur* (Paris) **104:** 469–475.

Le Minor, L., S. Le Minor, and P. Nicolle. 1961. Conversion des cultures de *S. schwarzengrund* et *S. bredeney* dépourvues de l'antigène 27 en cultures 27 positives par lysogénisation. *Ann. Inst. Pasteur* (Paris) **101:**571–589.

Leung, D. Y., J. B. Travers, and D. A. Norris. 1995. The role of superantigens in skin disease. *J. Invest. Dermatol.* **105**(1 Suppl.):37S–42S.

Lietzow, M. A., E. T. Gielow, D. Le, J. Zhang, and M. F. Verhagen. 2008. Subunit stoichiometry of the *Clostridium botulinum* type A neurotoxin complex determined using denaturing capillary electrophoresis. *Protein J.* **27:**420–425.

Lina, G., Y. Piemont, F. Godail-Gamot, M. Bes, M. O. Peter, V. Gauduchon, F. Vandenesch, and J. Etienne. 1999. Involvement of Panton-Valentine leukocidin-producing *Staphylococcus aureus* in primary skin infections and pneumonia. *Clin. Infect. Dis.* **29:**1128–1132.

Lindberg, A. A., C. G. Hellerqvist, G. Bagdian-Motta, and P. H. Mäkelä. 1978. Lipopolysaccharide modification accompanying antigenic conversion by phage P27. *J. Gen. Microbiol.* **107:**279–287.

Łoś, J. M. 2009. Molecular mechanisms of development regulation in lambdoid bacteriophages. Ph.D. thesis. University of Gdansk, Gdansk, Poland.

Łoś, J. M., P. Golec, G. Węgrzyn, A. Węgrzyn, and M. Łoś. 2008a. Simple method for plating *Escherichia coli* bacteriophages forming very small plaques or no plaques under standard conditions. *Appl. Environ. Microbiol.* **74:**5113–5120.

Łoś, J. M., M. Łoś, A. Węgrzyn, and G. Węgrzyn. 2008b. Role of the bacteriophage λ *exo-xis* region in the virus development. *Folia Microbiol.* (Praha) **53:**443–450.

Łoś, J. M., M. Łoś, A. Węgrzyn, and G. Węgrzyn. 2010. Hydrogen peroxide-mediated induction of the Shiga toxin-converting lambdoid prophage ST2-8624 in *Escherichia coli* O157:H7. *FEMS Immunol. Med. Microbiol.* **58:**322–329.

Łoś, J. M., M. Łoś, G. Węgrzyn, and A. Węgrzyn. 2009. Differential efficiency of induction of various lambdoid prophages responsible for production of Shiga toxins in response to different induction agents. *Microb. Pathog.* **47:**289–298.

Łoś, M., P. Golec, J. M. Łoś, A. Weglewska-Jurkiewicz, A. Czyz, A. Węgrzyn, G. Węgrzyn, and P. Neubauer. 2007. Effective inhibition of lytic development of bacteriophages λ, P1 and T4 by starvation of their host, *Escherichia coli*. *BMC Biotechnol.* **7:**13.

Łoś, M., J. M. Łoś, L. Blohm, E. Spillner, T. Grunwald, J. Albers, R. Hintsche, and G. Węgrzyn. 2005. Rapid detection of viruses using electrical biochips and anti-virion sera. *Lett. Appl. Microbiol.* **40:**479–485.

Łoś, M., J. M. Łoś, and G. Węgrzyn. 2008c. Rapid identification of Shiga toxin-producing *Escherichia coli* (STEC) using electric biochips. *Diagn. Mol. Pathol.* **17:**179–184.

Losick, R. 1969. Isolation of a trypsin-sensitive inhibitor of O-antigen synthesis involved in lysogenic conversion by bacteriophage ε^{15}. *J. Mol. Biol.* **42:**237–246.

Losick, R., and P. W. Robbins. 1967. Mechanism of ε^{15} conversion studied with a bacterial mutant. *J. Mol. Biol.* **30:**445–455.

Lüderitz, O. 1970. Recent results on the biochemistry of the cell wall lipopolysaccharides of *Salmonella* bacteria. *Angew. Chem. Int. Ed. Engl.* **9:**649–663.

Lüderitz, O., A. M. Staub, and O. Westphal. 1966. Immunochemistry of O and R antigens of *Salmonella* and related *Enterobacteriaceae*. *Bacteriol. Rev.* **30:**192–255.

Lwoff, A. 1953. Lysogeny. *Bacteriol. Rev.* **17:**269–337.

Ma, X. X., T. Ito, P. Chongtrakool, and K. Hiramatsu. 2006. Predominance of clones carrying Panton-Valentine leukocidin genes among methicillin-resistant *Staphylococcus aureus* strains isolated in Japanese hospitals from 1979 to 1985. *J. Clin. Microbiol.* **44:**4515–4527.

Ma, X. X., T. Ito, Y. Kondo, M. Cho, Y. Yoshizawa, J. Kaneko, A. Katai, M. Higashiide, S. Li, and K. Hiramatsu. 2008. Two different Panton-Valentine leukocidin phage lineages predominate in Japan. *J. Clin. Microbiol.* **46:**3246–3258.

Maiques, E., C. Ubeda, S. Campoy, N. Salvador, I. Lasa, R. P. Novick, J. Barbe, and J. R. Penades. 2006. β-Lactam antibiotics induce the SOS response and horizontal transfer of virulence factors in *Staphylococcus aureus*. *J. Bacteriol.* **188:** 2726–2729.

McCormick, J. K., J. M. Yarwood, and P. M. Schlievert. 2001. Toxic shock syndrome and bacterial superantigens: an update. *Annu. Rev. Microbiol.* **55:**77–104.

McLeod, S. M., H. H. Kimsey, B. M. Davis, and M. K. Waldor. 2005. CTXϕ and *Vibrio cholerae:* exploring a newly recognized type of phage-host cell relationship. *Mol. Microbiol.* **57:**347–356.

Melish, M. E., and L. A. Glasgow. 1970. The staphylococcal scalded-skin syndrome. *N. Engl. J. Med.* **282:**1114–1119.

Melish, M. E., L. A. Glasgow, and M. D. Turner. 1972. The staphylococcal scalded-skin syndrome: isolation and partial characterization of the exfoliative toxin. *J. Infect. Dis.* **125:**129–140.

Menestrina, G., M. Dalla Serra, M. Comai, M. Coraiola, G. Viero, S. Werner, D. A. Colin, H. Monteil, and G. Prevost. 2003. Ion channels and bacterial infection: the case of β-barrel pore-forming protein toxins of *Staphylococcus aureus. FEBS Lett.* **552:**54–60.

Miller, V. L., K. B. Beer, G. Heusipp, B. M. Young, and M. R. Wachtel. 2001. Identification of regions of Ail required for the invasion and serum resistance phenotypes. *Mol. Microbiol.* **41:** 1053–1062.

Miller, V. L., K. B. Beer, W. P. Loomis, J. A. Olson, and S. I. Miller. 1992. An unusual *pagC*::Tn*phoA* mutation leads to an invasion- and virulence-defective phenotype in salmonellae. *Infect. Immun.* **60:**3763–3770.

Mmolawa, P. T., H. Schmieger, C. P. Tucker, and M. W. Heuzenroeder. 2003. Genomic structure of the *Salmonella enterica* serovar Typhimurium DT 64 bacteriophage ST64T: evidence for modular genetic architecture. *J. Bacteriol.* **185:** 3473–3475.

Monday, S. R., and G. A. Bohach. 2001. Genes encoding staphylococcal enterotoxins G and I are linked and separated by DNA related to other staphylococcal enterotoxins. *J. Nat. Toxins* **10:**1–8.

Monday, S. R., G. M. Vath, W. A. Ferens, C. Deobald, J. V. Rago, P. J. Gahr, D. D. Monie, J. J. Iandolo, S. K. Chapes, W. C. Davis, D. H. Ohlendorf, P. M. Schlievert, and G. A. Bohach. 1999. Unique superantigen activity of staphylococcal exfoliative toxins. *J. Immunol.* **162:**4550–4559.

Moreland, J. L., A. Gramada, O. V. Buzko, Q. Zhang, and P. E. Bourne. 2005. The Molecular Biology Toolkit (MBT): a modular platform for developing molecular visualization applications. *BMC Bioinformatics* **6:**21.

Morlock, B. A., L. Spero, and A. D. Johnson. 1980. Mitogenic activity of staphylococcal exfoliative toxin. *Infect. Immun.* **30:**381–384.

Muniesa, M., J. Jofre, C. García-Aljaro, and A. R. Blanch. 2006. Occurrence of *Escherichia coli* O157:H7 and other enterohemorrhagic *Escherichia coli* in the environment. *Environ. Sci. Technol.* **40:** 7141–7149.

Munson, S. H., M. T. Tremaine, M. J. Betley, and R. A. Welch. 1998. Identification and characterization of staphylococcal enterotoxin types G and I from *Staphylococcus aureus. Infect. Immun.* **66:** 3337–3348.

Murphy, K. C., J. M. Ritchie, M. K. Waldor, A. Lobner-Olesen, and M. G. Marinus. 2008. Dam methyltransferase is required for stable lysogeny of the Shiga toxin (Stx2)-encoding bacteriophage 933W of enterohemorrhagic *Escherichia coli* O157:H7. *J. Bacteriol.* **190:**438–441.

Muschel, L. H., L. A. Ahl, and L. S. Baron. 1968. Effect of lysogeny on serum sensitivity. *J. Bacteriol.* **96:**1912–1914.

Nakamura, T., M. Kotani, T. Tonozuka, A. Ide, K. Oguma, and A. Nishikawa. 2009. Crystal structure of the HA3 subcomponent of *Clostridium botulinum* type C progenitor toxin. *J. Mol. Biol.* **385:**1193–1206.

Narita, S., J. Kaneko, J. Chiba, Y. Piemont, S. Jarraud, J. Etienne, and Y. Kamio. 2001. Phage conversion of Panton-Valentine leukocidin in *Staphylococcus aureus:* molecular analysis of a PVL-converting phage, φSLT. *Gene* **268:**195–206.

Nejman, B., J. M. Łoś, M. Łoś, G. Węgrzyn, and A. Węgrzyn. 2009. Plasmids derived from lambdoid bacteriophages as models for studying replication of mobile genetic elements responsible for the production of Shiga toxins by pathogenic *Escherichia coli* strains. *J. Mol. Microbiol. Biotechnol.* **17:** 211–220.

Nemecek, D., G. C. Lander, J. E. Johnson, S. R. Casjens, and G. J. Thomas, Jr. 2008. Assembly architecture and DNA binding of the bacteriophage P22 terminase small subunit. *J. Mol. Biol.* **383:**494–501.

Nickerson, S. C., W. E. Owens, and R. L. Boddie. 1995. Mastitis in dairy heifers: initial studies on prevalence and control. *J. Dairy Sci.* **78:**1607–1618.

Nishifuji, K., M. Sugai, and M. Amagai. 2008. Staphylococcal exfoliative toxins: "molecular scissors" of bacteria that attack the cutaneous defense barrier in mammals. *J. Dermatol. Sci.* **49:**21–31.

Nolan, L. K., S. M. Horne, C. W. Giddings, S. L. Foley, T. J. Johnson, A. M. Lynne, and J. Skyberg. 2003. Resistance to serum complement, *iss,* and virulence of avian *Escherichia coli. Vet. Res. Commun.* **27:**101–110.

Novick, R. P. 2006. Staphylococcal pathogenesis and pathogenicity factors: genetics and regulation, p. 496–516. *In* V. A. Fischetti, R. P. Novick, J. J. Ferretti, D. Portnoy, and J. A. Rood (ed.), *Gram-Positive Pathogens.* ASM Press, Washington, DC.

Oguma, K. 1976. The stability of toxigenicity in *Clostridium botulinum* types C and D. *J. Gen. Microbiol.* **92:**67–75.

Oguma, K., H. Iida, M. Shiozaki, and K. Inoue. 1976. Antigenicity of converting phages obtained from *Clostridium botulinum* types C and D. *Infect. Immun.* **13:**855–860.

Ohlenschlager, O., R. Ramachandran, K. H. Guhrs, B. Schlott, and L. R. Brown. 1998. Nuclear magnetic resonance solution structure of the plasminogen-activator protein staphylokinase. *Biochemistry* **37:**10635–10642.

Olson, R., H. Nariya, K. Yokota, Y. Kamio, and E. Gouaux. 1999. Crystal structure of staphylococcal LukF delineates conformational changes accompanying formation of a transmembrane channel. *Nat. Struct. Biol.* **6:**134–140.

Osterhout, R. E., I. A. Figueroa, J. D. Keasling, and A. P. Arkin. 2007. Global analysis of host response to induction of a latent bacteriophage. *BMC Microbiol.* **7:**82.

O'Toole, P. W., and T. J. Foster. 1986. Molecular cloning and expression of the epidermolytic toxin A gene of *Staphylococcus aureus*. *Microb. Pathog.* **1:** 583–594.

O'Toole, P. W., and T. J. Foster. 1987. Nucleotide sequence of the epidermolytic toxin A gene of *Staphylococcus aureus*. *J. Bacteriol.* **169:**3910–3915.

Panton, P. N., M. B. Came, and F. C. O. Valentine. 1932. Staphylococcal toxin. *Lancet* **1:**506–508.

Papageorgiou, A. C., and K. R. Acharya. 2000. Microbial superantigens: from structure to function. *Trends Microbiol.* **8:**369–375.

Papageorgiou, A. C., L. R. Plano, C. M. Collins, and K. R. Acharya. 2000. Structural similarities and differences in *Staphylococcus aureus* exfoliative toxins A and B as revealed by their crystal structures. *Protein Sci.* **9:**610–618.

Parma, D. H., M. Snyder, S. Sobolevski, M. Nawroz, E. Brody, and L. Gold. 1992. The Rex system of bacteriophage λ: tolerance and altruistic cell death. *Genes Dev.* **6:**497–510.

Parry, M. A., C. Fernandez-Catalan, A. Bergner, R. Huber, K. P. Hopfner, B. Schlott, K. H. Guhrs, and W. Bode. 1998. The ternary microplasmin-staphylokinase-microplasmin complex is a proteinase-cofactor-substrate complex in action. *Nat. Struct. Biol.* **5:**917–923.

Paton, A. W., and J. C. Paton. 1996. *Enterobacter cloacae* producing a Shiga-like toxin II-related cytotoxin associated with a case of hemolytic-uremic syndrome. *J. Clin. Microbiol.* **34:**463–465.

Pedelacq, J. D., L. Maveyraud, G. Prevost, L. Baba-Moussa, A. Gonzalez, E. Courcelle, W. Shepard, H. Monteil, J. P. Samama, and L. Mourey. 1999. The structure of a *Staphylococcus aureus* leucocidin component (LukF-PV) reveals the fold of the water-soluble species of a family of transmembrane pore-forming toxins. *Structure* **7:** 277–287.

Pedulla, M. L., M. E. Ford, T. Karthikeyan, J. M. Houtz, R. W. Hendrix, G. F. Hatfull, A. R. Poteete, E. B. Gilcrease, D. A. Winn-Stapley, and S. R. Casjens. 2003. Corrected sequence of the bacteriophage P22 genome. *J. Bacteriol.* **185:**1475–1477.

Perry, L. L., P. SanMiguel, U. Minocha, A. I. Terekhov, M. L. Shroyer, L. A. Farris, N. Bright, B. L. Reuhs, and B. M. Applegate. 2009. Sequence analysis of *Escherichia coli* O157:H7 bacteriophage ϕV10 and identification of a phage-encoded immunity protein that modifies the O157 antigen. *FEMS Microbiol. Lett.* **292:**182–186.

Persson, S., K. E. Olsen, S. Ethelberg, and F. Scheutz. 2007. Subtyping method for *Escherichia coli* Shiga toxin (verocytotoxin) 2 variants and correlations to clinical manifestations. *J. Clin. Microbiol.* **45:**2020–2024.

Petersson, K., M. Thunnissen, G. Forsberg, and B. Walse. 2002. Crystal structure of a SEA variant in complex with MHC class II reveals the ability of SEA to crosslink MHC molecules. *Structure* **10:** 1619–1626.

Plano, L. R., D. M. Gutman, M. Woischnik, and C. M. Collins. 2000. Recombinant *Staphylococcus aureus* exfoliative toxins are not bacterial superantigens. *Infect. Immun.* **68:**3048–3052.

Plunkett, G., III., D. J. Rose, T. J. Durfee, and F. R. Blattner. 1999. Sequence of Shiga toxin 2 phage 933W from *Escherichia coli* O157:H7: Shiga toxin as a phage late-gene product. *J. Bacteriol.* **181:**1767–1778.

Postma, B., M. J. Poppelier, J. C. van Galen, E. R. Prossnitz, J. A. van Strijp, C. J. de Haas, and K. P. van Kessel. 2004. Chemotaxis inhibitory protein of *Staphylococcus aureus* binds specifically to the C5a and formylated peptide receptor. *J. Immunol.* **172:**6994–7001.

Prere, M. F., and O. Fayet. 2005. A new genetic test for the rapid identification of shiga-toxines producing (STEC), enteropathogenic (EPEC) *E. coli* isolates from children. *Pathol. Biol.* (Paris) **53:** 466–469.

Prevost, G., P. Couppie, P. Prevost, S. Gayet, P. Petiau, B. Cribier, H. Monteil, and Y. Piemont. 1995a. Epidemiological data on *Staphylococcus aureus* strains producing synergohymenotropic toxins. *J. Med. Microbiol.* **42:**237–245.

Prevost, G., B. Cribier, P. Couppie, P. Petiau, G. Supersac, V. Finck-Barbancon, H. Monteil, and Y. Piemont. 1995b. Panton-Valentine leucocidin and gamma-hemolysin from *Staphylococcus aureus* ATCC 49775 are encoded by distinct genetic loci and have different biological activities. *Infect. Immun.* **63:**4121–4129.

Ptashne, M. 2004. *A Genetic Switch: Phage Lambda Revisited*. Cold Spring Harbor Laboratory Press, Cold Spring Harbor, NY.

Ramamurthy, T. 2008. Shiga toxin-producing *Escherichia coli* (STEC): the bug in our backyard. *Indian J. Med. Res.* **128:**233–236.

Recsei, P., B. Kreiswirth, M. O'Reilly, P. Schlievert, A. Gruss, and R. P. Novick. 1986. Regulation of exoprotein gene expression in *Staphylococcus aureus* by agar. *Mol. Gen. Genet.* **202:**58–61.

Reeve, J. N., and J. E. Shaw. 1979. Lambda encodes an outer membrane protein: the *lom* gene. *Mol. Gen. Genet.* **172:**243–248.

Ricklin, D., A. Tzekou, B. L. Garcia, M. Hammel, W. J. McWhorter, G. Sfyroera, Y. Q. Wu, V. M. Holers, A. P. Herbert, P. N. Barlow, B. V. Geisbrecht, and J. D. Lambris. 2009. A molecular insight into complement evasion by the staphylococcal complement inhibitor protein family. *J. Immunol.* **183:**2565–2574.

Riley, L. W., R. S. Remis, S. D. Helgerson, H. B. McGee, J. G. Wells, and B. R. Davis. 1983. Serotypes, virulence genes of shigatoxin producing *Escherichia coli* isolates from human patients. *J. Clin. Microbiol.* **4:**311–319.

Robbins, P., J. M. Keller, A. Wright, and R. L. Bernstein. 1965. Enzymatic and kinetics studies on the mechanism of O-antigen conversion by bacteriophage ε^{15}. *J. Biol. Chem.* **240:**384–390.

Robbins, P., and T. Uchida. 1962. Studies on the chemical basis of the phage conversion of O-antigens in the E-group salmonellae. *Biochemistry* **1:**325–335.

Robbins, P., and T. Uchida. 1965. Chemical and macromolecular structure of O-antigens from *Salmonella anatum* strains carrying mutants of bacteriophage ε^{15}. *J. Biol. Chem.* **240:**375–383.

Rogolsky, M., B. B. Wiley, and L. A. Glasgow. 1976. Phage group II staphylococcal strains with chromosomal and extrachromosomal genes for exfoliative toxin production. *Infect. Immun.* **13:**44–52.

Rooijakkers, S. H., F. J. Milder, B. W. Bardoel, M. Ruyken, J. A. van Strijp, and P. Gros. 2007. Staphylococcal complement inhibitor: structure and active sites. *J. Immunol.* **179:**2989–2998.

Rooijakkers, S. H., M. Ruyken, A. Roos, M. R. Daha, J. S. Presanis, R. B. Sim, W. J. van Wamel, K. P. van Kessel, and J. A. van Strijp. 2005a. Immune evasion by a staphylococcal complement inhibitor that acts on C3 convertases. *Nat. Immunol.* **6:**920–927.

Rooijakkers, S. H., M. Ruyken, J. van Roon, K. P. van Kessel, J. A. van Strijp, and W. J. van Wamel. 2006. Early expression of SCIN and CHIPS drives instant immune evasion by *Staphylococcus aureus*. *Cell. Microbiol.* **8:**1282–1293.

Rooijakkers, S. H., K. P. van Kessel, and J. A. van Strijp. 2005b. Staphylococcal innate immune evasion. *Trends Microbiol.* **13:**596–601.

Rooijakkers, S. H., W. J. van Wamel, M. Ruyken, K. P. van Kessel, and J. A. van Strijp. 2005c. Anti-opsonic properties of staphylokinase. *Microbes Infect.* **7:**476–484.

Rooijakkers, S. H., J. Wu, M. Ruyken, R. van Domselaar, K. L. Planken, A. Tzekou, D. Ricklin, J. D. Lambris, B. J. Janssen, J. A. van Strijp, and P. Gros. 2009. Structural and functional implications of the alternative complement pathway C3 convertase stabilized by a staphylococcal inhibitor. *Nat. Immunol.* **10:**721–727.

Rosenblum, E. D., and S. Tyrone. 1976. Chromosomal determinants for exfoliative toxin production in two strains of staphylococci. *Infect. Immun.* **14:**1259–1260.

Rosendal, K., P. Buelow, and O. Jessen. 1964. Lysogen conversion in *Staphylococcus aureus,* to a change in the production of extracellular "Tween"-splitting enzyme. *Nature* **204:**1222–1223.

Rossi, R. E., and G. Monasterolo. 2004. Prevalence of serum IgE antibodies to the *Staphylococcus aureus* enterotoxins (SAE, SEB, SEC, SED, TSST-1) in patients with persistent allergic rhinitis. *Int. Arch. Allergy Appl. Immunol.* **133:**261–266.

Rundell, K., and C. W. Shuster. 1975. Membrane-associated nucleotide sugar reactions: influence of mutations affecting lipopolysaccharide on the first enzyme of O-antigen synthesis. *J. Bacteriol.* **123:**928–936.

Sakaguchi, Y., T. Hayashi, K. Kurokawa, K. Nakayama, K. Oshima, Y. Fujinaga, M. Ohnishi, E. Ohtsubo, M. Hattori, and K. Oguma. 2005. The genome sequence of *Clostridium botulinum* type C neurotoxin-converting phage and the molecular mechanisms of unstable lysogeny. *Proc. Natl. Acad. Sci. USA* **102:**17472–17477.

Sako, T., S. Sawaki, T. Sakurai, S. Ito, Y. Yoshizawa, and I. Kondo. 1983. Cloning and expression of the staphylokinase gene of *Staphylococcus aureus* in *Escherichia coli*. *Mol. Gen. Genet.* **190:**271–277.

Sako, T. and N. Tsuchida. 1983. Nucleotide sequence of the staphylokinase gene from *Staphylococcus aureus*. *Nucleic Acids Res.* **11:**7679–7693.

Sakurai, S., H. Suzuki, T. Hata, Y. Yoshizawa, R. Nakayama, K. Machida, S. Masuda, and T. Tsukiyama. 2004. A novel positive regulatory element for exfoliative toxin A gene expression in *Staphylococcus aureus*. *Microbiology* **150:**945–952.

Sakurai, S., H. Suzuki, and I. Kondo. 1988. DNA sequencing of the *eta* gene coding for staphylococcal exfoliative toxin serotype A. *J. Gen. Microbiol.* **134:**711–717.

Sánchez, J., and J. Holmgren. 2008. Cholera toxin structure, gene regulation and pathophysiological and immunological aspects. *Cell. Mol. Life Sci.* **65:** 1347–1360.

Sauer, R. T., W. Krovatin, J. DeAnda, P. Youderian, and M. M. Susskind. 1983. Primary structure of the *immI* immunity region of bacteriophage P22. *J. Mol. Biol.* **168:**699–713.

Saunders, J. R., H. Allison, C. E. James, A. J. McCarthy, and R. Sharp. 2001. Phage-mediated transfer of virulence genes. *J. Chem. Technol. Biotechnol.* **76:**662–666.

Schad, E. M., I. Zaitseva, V. N. Zaitsev, M. Dohlsten, T. Kalland, P. M. Schlievert, D. H. Ohlendorf, and L. A. Svensson. 1995. Crystal structure of the superantigen staphylococcal enterotoxin type A. *EMBO J.* **14:**3292–3301.

Schiavo, G., F. Benfenati, B. Poulain, O. Rossetto, P. Polverino de Laureto, B. R. DasGupta, and C. Montecucco. 1992a. Tetanus and botulinum-B neurotoxins block neurotransmitter release by proteolytic cleavage of synaptobrevin. *Nature* **359:**832–835.

Schiavo, G., O. Rossetto, A. Santucci, B. R. DasGupta, and C. Montecucco. 1992b. Botulinum neurotoxins are zinc proteins. *J. Biol. Chem.* **267:**23479–23483.

Schmidt, H. 2001. Shiga-toxin-converting bacteriophages. *Res. Microbiol.* **152:**687–695.

Schmidt, H., M. Montag, J. Bockemuhl, J. Heesemann, and H. Karch. 1993. Shiga-like toxin II-related cytotoxins in *Citrobacter freundii* strains from humans and beef samples. *Infect. Immun.* **61:**534–543.

Scotland, S. M., H. R. Smith, G. A. Willshaw, and B. Rowe. 1983. Vero cytotoxin production in strain of *Escherichia coli* is determined by genes carried on bacteriophage. *Lancet* **2:**216.

Serna, A., and E. C. Boedeker. 2008. Pathogenesis and treatment of Shiga toxin-producing *Escherichia coli* infections. *Curr. Opin. Gastroenterol.* **24:**38–47.

Sheehan, B. J., T. J. Foster, C. J. Dorman, S. Park, and G. S. Stewart. 1992. Osmotic and growth-phase dependent regulation of the *eta* gene of *Staphylococcus aureus:* a role for DNA supercoiling. *Mol. Gen. Genet.* **232:**49–57.

Shiga, K. 1898. Ueber den Dysenteric-bacillus (Bacillus dysenteriae). *Zentralbl. Bakteriol.* **24:**913–918.

Shimizu, T., Y. Ohta, and M. Noda. 2009. Shiga toxin 2 is specifically released from bacterial cells by two different mechanisms. *Infect. Immun.* **77:** 2813–2823.

Shkilnyj, P., and G. B. Koudelka. 2007. Effect of salt shock on stability of λ^{imm434} lysogens. *J. Bacteriol.* **189:**3115–3123.

Simpson, L. L., and M. M. Rapport. 1971. Ganglioside inactivation of botulinum toxin. *J. Neurochem.* **18:**1341–1343.

Slauch, J. M., A. A. Lee, M. J. Mahan, and J. J. Mekalanos. 1996. Molecular characterization of the *oafA* locus responsible for acetylation of *Salmonella typhimurium* O-antigen: *oafA* is a member of a family of integral membrane trans-acylases. *J. Bacteriol.* **178:**5904–5909.

Smith, H. R., N. P. Day, S. M. Scotland, R. J. Gross, and B. Rowe. 1984. Phage-determined production of vero cytotoxin in strains of *Escherichia coli* serogroup O157. *Lancet* **1:**1242–1243.

Smith, H. W., P. Green, and Z. Parsell. 1983. Vero cell toxins in *Escherichia coli* and related bacteria: transfer by phage and conjugation and toxic action in laboratory animals, chickens and pigs. *J. Gen. Microbiol.* **129:**3121–3137.

Smith, H. W., M. B. Huggins, and K. M. Shaw. 1987. The control of experimental *Escherichia coli* diarrhoea in calves by means of bacteriophages. *J. Gen. Microbiol.* **133:**1111–1126.

Snyder, L. 1995. Phage-exclusion enzymes: a bonanza of biochemical and cell biology reagents? *Mol. Microbiol.* **15:**415–420.

Song, L., M. R. Hobaugh, C. Shustak, S. Cheley, H. Bayley, and J. E. Gouaux. 1996. Structure of staphylococcal α-hemolysin, a heptameric transmembrane pore. *Science* **274:**1859–1866.

Spatara, M. L., C. J. Roberts, and A. S. Robinson. 2009. Kinetic folding studies of the P22 tailspike beta-helix domain reveal multiple unfolded states. *Biophys. Chem.* **141:**214–221.

Stanley, J. R., and M. Amagai. 2006. Pemphigus, bullous impetigo, and the staphylococcal scalded-skin syndrome. *N. Engl. J. Med.* **355:**1800–1810.

Stevens, D. L., Y. Ma, D. B. Salmi, E. McIndoo, R. J. Wallace, and A. E. Bryant. 2007. Impact of antibiotics on expression of virulence-associated exotoxin genes in methicillin-sensitive and methicillin-resistant *Staphylococcus aureus. J. Infect. Dis.* **195:**202–211.

Strauch, E., J. A. Hammerl, A. Konietzny, S. Schneiker-Bekel, W. Arnold, A. Goesmann, A. Puhler, and L. Beutin. 2008. Bacteriophage 2851 is a prototype phage for dissemination of the Shiga toxin variant gene 2c in *Escherichia coli* O157: H7. *Infect. Immun.* **76:**5466–5477.

Sugawara-Tomita, N., T. Tomita, and Y. Kamio. 2002. Stochastic assembly of two-component staphylococcal γ-hemolysin into heteroheptameric transmembrane pores with alternate subunit arrangements in ratios of 3:4 and 4:3. *J. Bacteriol.* **184:**4747–4756.

Sumby, P., and M. K. Waldor. 2003. Transcription of the toxin genes present within the staphylococcal phage ϕSa3ms is intimately linked with the phage's life cycle. *J. Bacteriol.* **185:**6841–6851.

Sundberg, E. J., L. Deng, and R. A. Mariuzza. 2007. TCR recognition of peptide/MHC class II complexes and superantigens. *Semin. Immunol.* **19:** 262–271.

Swaminathan, S., and S. Eswaramoorthy. 2000. Crystallization and preliminary X-ray analysis of *Clostridium botulinum* neurotoxin type B. *Acta Crystallogr. D Biol. Crystallogr.* **56:**1024–1026.

Tanaka, K., K. Nishimori, S. Makino, T. Nishimori, T. Kanno, R. Ishihara, T. Sameshima, M. Akiba, M. Nakazawa, Y. Yokomizo, and I. Uchida. 2004. Molecular characterization of a prophage of *Salmonella enterica* serotype Typhimurium DT104. *J. Clin. Microbiol.* **42:**1807–1812.

Thierauf, A., G. Perez, and A. S. Maloy. 2009. Generalized transduction. *Methods Mol. Biol.* **501:** 267–286.

Todd, J., M. Fishaut, F. Kapral, and T. Welch. 1978. Toxic-shock syndrome associated with phage-group-I staphylococci. *Lancet* **ii:**1116–1118.

Tomita, T., and Y. Kamio. 1997. Molecular biology of the pore-forming cytolysins from *Staphylococcus aureus,* alpha- and gamma-hemolysins and leukocidin. *Biosci. Biotechnol. Biochem.* **61:**565–572.

Tranter, H. S. 1990. Foodborne staphylococcal illness. *Lancet* **336:**1044–1046.

Tremaine, M. T., D. K. Brockman, and M. J. Betley. 1993. Staphylococcal enterotoxin A gene *(sea)* expression is not affected by the accessory gene regulator *(agr). Infect. Immun.* **61:**356–359.

Uetake, H., and S. Hagiwara. 1961. Genetic cooperation between unrelated phages. *Virology* **13:** 500–506.

Uetake, H., and S. Hagiwara. 1969. Transfer of conversion gene(s) between different *Salmonella* phages g341 and epsilon-15. *Virology* **37:**8–14.

Uetake, H., T. Nakagawa, and T. Akiba. 1955. The relationship of bacteriophage to antigenic changes in group E salmonellas. *J. Bacteriol.* **69:** 571–579.

Uetake, H., and T. Uchita. 1959. Mutants of *Salmonella* ε15 with abnormal conversion properties. *Virology* **9:**495–505.

Vallat, B. K., J. Pillardy, P. Majek, J. Meller, T. Blom, B. Cao, and R. Elber. 2009. Building and assessing atomic models of proteins from structural templates: learning and benchmarks. *Proteins* **76:**930–945.

Vander Byl, C., and A. M. Kropinski. 2000. Sequence of the genome of *Salmonella* bacteriophage P22. *J. Bacteriol.* **182:**6472–6481.

van der Vijver, J. C., M. van Es-Boon, and M. F. Michel. 1972. Lysogenic conversion in *Staphylococcus aureus* to leucocidin production. *J. Virol.* **10:** 318–319.

van Wamel, W. J., S. H. Rooijakkers, M. Ruyken, K. P. van Kessel, and J. A. van Strijp. 2006. The innate immune modulators staphylococcal complement inhibitor and chemotaxis inhibitory protein of *Staphylococcus aureus* are located on β-hemolysin-converting bacteriophages. *J. Bacteriol.* **188:**1310–1315.

Vareille, M., T. de Sablet, T. Hindre, C. Martin, and A. P. Gobert. 2007. Nitric oxide inhibits Shiga-toxin synthesis by enterohemorrhagic *Escherichia coli. Proc. Natl. Acad. Sci. USA* **104:**10199–10204.

Vath, G. M., C. A. Earhart, D. D. Monie, J. J. Iandolo, P. M. Schlievert, and D. H. Ohlendorf. 1999. The crystal structure of exfoliative toxin B: a superantigen with enzymatic activity. *Biochemistry* **38:**10239–10246.

Vath, G. M., C. A. Earhart, J. V. Rago, M. H. Kim, G. A. Bohach, P. M. Schlievert, and D. H. Ohlendorf. 1997. The structure of the superantigen exfoliative toxin A suggests a novel regulation as a serine protease. *Biochemistry* **36:**1559–1566.

Vica Pacheco, S., O. Garcia Gonzalez, and G. L. Paniagua Contreras. 1997. The *lom* gene of bacteriophage lambda is involved in *Escherichia coli* K12 adhesion to human buccal epithelial cells. *FEMS Microbiol. Lett.* **156:**129–132.

Viklund, H., and A. Elofsson. 2008. OCTOPUS: improving topology prediction by two-track ANN-based preference scores and an extended topological grammar. *Bioinformatics* **24:**1662–1668.

Villafane, R., M. Zayas, E. B. Gilcrease, A. M. Kropinski, and S. R. Casjens. 2008. Genomic analysis of bacteriophage ε^{34} of *Salmonella enterica* serovar Anatum (15+). *BMC Microbiol.* **8:**227.

Villaruz, A. E., J. B. Wardenburg, B. A. Khan, A. R. Whitney, D. E. Sturdevant, D. J. Gardner, F. R. DeLeo, and M. Otto. 2009. A point mutation in the *agr* locus rather than expression of the Panton-Valentine leukocidin caused previously reported phenotypes in *Staphylococcus aureus* pneumonia and gene regulation. *J. Infect. Dis.* **200:**724–734.

Voyich, J. M., M. Otto, B. Mathema, K. R. Braughton, A. R. Whitney, D. Welty, R. D. Long, D. W. Dorward, D. J. Gardner, G. Lina, B. N. Kreiswirth, and F. R. DeLeo. 2006. Is Panton-Valentine leukocidin the major virulence determinant in community-associated methicillin-resistant *Staphylococcus aureus* disease? *J. Infect. Dis.* **194:**1761–1770.

Voyich, J. M., C. Vuong, M. DeWald, T. K. Nygaard, S. Kocianova, S. Griffith, J. Jones, C. Iverson, D. E. Sturdevant, K. R. Braughton, A. R. Whitney, M. Otto, and F. R. DeLeo. 2009. The SaeR/S gene regulatory system is essential for innate immune evasion by *Staphylococcus aureus. J. Infect. Dis.* **199:**1698–1706.

Wagner, P. L., D. W. Acheson, and M. K. Waldor. 2001a. Human neutrophils and their products induce Shiga toxin production by enterohemorrhagic *Escherichia coli*. *Infect. Immun.* **69:**1934–1937.

Wagner, P. L., J. Livny, M. N. Neely, D. W. Acheson, D. I. Friedman, and M. K. Waldor. 2002. Bacteriophage control of Shiga toxin 1 production and release by *Escherichia coli*. *Mol. Microbiol.* **44:**957–970.

Wagner, P. L., M. N. Neely, X. Zhang, D. W. Acheson, M. K. Waldor, and D. I. Friedman. 2001b. Role for a phage promoter in Shiga toxin 2 expression from a pathogenic *Escherichia coli* strain. *J. Bacteriol.* **183:**2081–2085.

Waite-Rees, P. A., C. J. Keating, L. S. Moran, B. E. Slatko, L. J. Hornstra, and J. S. Benner. 1991. Characterization and expression of the *Escherichia coli* Mrr restriction system. *J. Bacteriol.* **173:** 5207–5219.

Waldor, M. K., and D. I. Friedman. 2005. Phage regulatory circuits and virulence gene expression. *Curr. Opin. Microbiol.* **8:**459–465.

Wang, L., K. Andrianopoulos, D. Liu, M. Y. Popoff, and P. R. Reeves. 2002. Extensive variation in the O-antigen gene cluster within one *Salmonella enterica* serogroup reveals an unexpected complex history. *J. Bacteriol.* **184:**1669–1677.

Węgrzyn, G., and A. Węgrzyn. 2005. Genetic switches during bacteriophage λ development. *Prog. Nucleic Acid Res. Mol. Biol.* **79:**1–48.

Weinstein, D. L., R. K. Holmes, and A. D. O'Brien. 1988. Effects of iron and temperature on Shiga-like toxin I production by *Escherichia coli*. *Infect. Immun.* **56:**106–111.

White, S. H., W. C. Wimley, and M. E. Selsted. 1995. Structure, function, and membrane integration of defensins. *Curr. Opin. Struct. Biol.* **5:**521–527.

Williamson, S. J., and J. H. Paul. 2006. Environmental factors that influence the transition from lysogenic to lytic existence in the ϕHSIC/*Listonella pelagia* marine phage-host system. *Microb. Ecol.* **52:** 217–225.

Winkler, K. C., W. J. de Waart, and C. Grootsen. 1965. Lysogenic conversion of staphylococci to loss of β-toxin. *J. Gen. Microbiol.* **39:**321–333.

Wirtz, C., W. Witte, C. Wolz, and C. Goerke. 2009. Transcription of the phage-encoded Panton-Valentine leukocidin of *Staphylococcus aureus* is dependent on the phage life-cycle and on the host background. *Microbiology* **155:**3491–3499.

Wollin, R., B. A. Stocker, and A. A. Lindberg. 1987. Lysogenic conversion of *Salmonella typhimurium* bacteriophages A3 and A4 consists of O-acetylation of rhamnose of the repeating unit of the O-antigenic polysaccharide chain. *J. Bacteriol.* **169:**1003–1009.

Wright, A., and N. Barzilai. 1971. Isolation and characterization of nonconverting mutants of bacteriophage ε^{34}. *J. Bacteriol.* **105:**937–939.

Yamaguchi, T., T. Hayashi, H. Takami, K. Nakasone, M. Ohnishi, K. Nakayama, S. Yamada, H. Komatsuzawa, and M. Sugai. 2000. Phage conversion of exfoliative toxin A production in *Staphylococcus aureus*. *Mol. Microbiol.* **38:** 694–705.

Yamasaki, O., A. Tristan, T. Yamaguchi, M. Sugai, G. Lina, M. Bes, F. Vandenesch, and J. Etienne. 2006. Distribution of the exfoliative toxin D gene in clinical *Staphylococcus aureus* isolates in France. *Clin. Microbiol. Infect.* **12:**585–588.

Yang, D., A. Biragyn, L. W. Kwak, and J. J. Oppenheim. 2002. Mammalian defensins in immunity: more than just microbicidal. *Trends Immunol.* **23:**291–296.

Yarwood, J. M., J. K. McCormick, M. L. Paustian, P. M. Orwin, V. Kapur, and P. M. Schlievert. 2002. Characterization and expression analysis of *Staphylococcus aureus* pathogenicity island 3. Implications for the evolution of staphylococcal pathogenicity islands. *J. Biol. Chem.* **277:**13138–13147.

Yoshizawa, Y., J. Sakurada, S. Sakurai, K. Machida, I. Kondo, and S. Masuda. 2000. An exfoliative toxin A-converting phage isolated from *Staphylococcus aureus* strain ZM. *Microbiol. Immunol.* **44:**189–191.

Young, B. G., Y. Fukazawa, and P. Hartman. 1964. A P22 bacteriophage mutant defective in antigen conversion. *Virology* **23:**279–283.

Zabicka, D., A. Mlynarczyk, B. Windyga, and G. Mlynarczyk. 1993. Phage-related conversion of enterotoxin A, staphylokinase and beta-toxin in *Staphylococcus aureus*. *Acta Microbiol. Pol.* **42:**235–241.

Zayas, M., and R. Villafane. 2007. Identification of the *Salmonella* phage ε^{34} tailspike gene. *Gene* **386:**211–217.

Zheng, H., A. S. Olia, M. Gonen, S. Andrews, G. Cingolani, and T. Gonen. 2008. A conformational switch in bacteriophage p22 portal protein primes genome injection. *Mol. Cell* **29:**376–383.

Ziebandt, A. K., H. Weber, J. Rudolph, R. Schmid, D. Hoper, S. Engelmann, and M. Hecker. 2001. Extracellular proteins of *Staphylococcus aureus* and the role of SarA and sigma B. *Proteomics* **1:**480–493.

Zou, D., J. Kaneko, S. Narita, and Y. Kamio. 2000. Prophage, ϕPV83-pro, carrying Panton-Valentine leukocidin genes, on the *Staphylococcus aureus* P83 chromosome: comparative analysis of the genome structures of ϕPV83-pro, ϕPVL, ϕ11, and other phages. *Biosci. Biotechnol. Biochem.* **64:** 2631–2643.

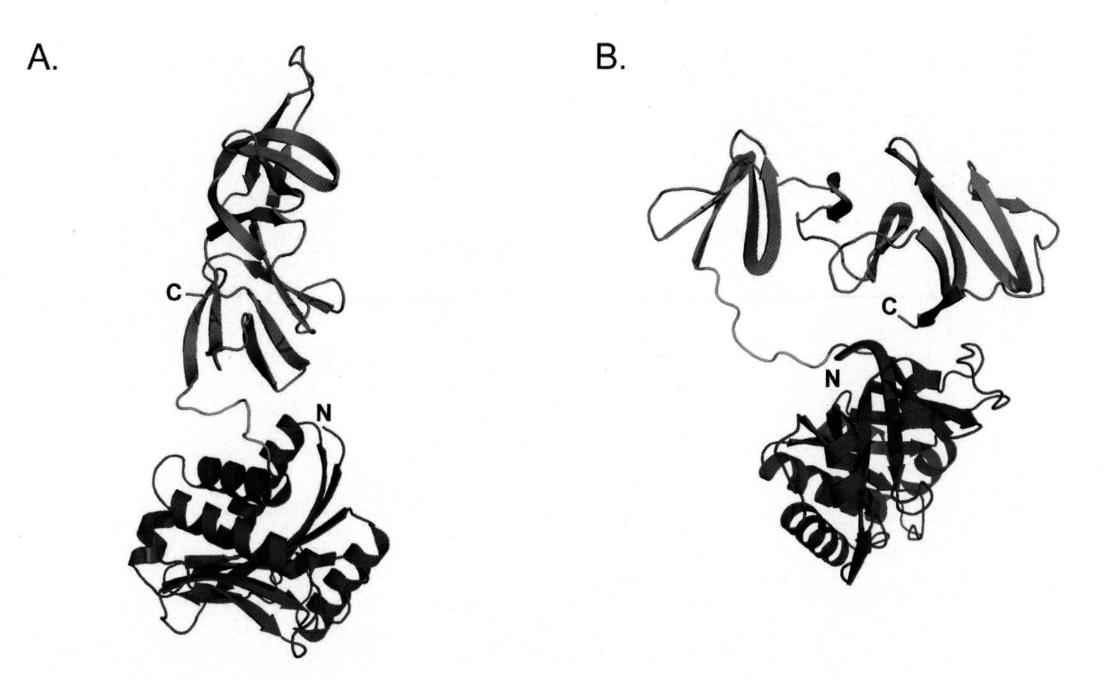

COLOR PLATE 1 [chapter 8] Common structural motifs for bacteriophage lysins. Full-length structures of (A) PlyPSA (Protein Data Bank accession number 1XOV), a *Listeria* lysin, and (B) Cpl-1 (Protein Data Bank accession number 2IXU), a pneumococcal lysin. The catalytic domains of PlyPSA (*N*-acetylmuramoyl-L-alanine amidase) and Cpl-1 (muramidase) are depicted in blue. The C-terminal CBD for each enzyme is shown in red. Flexible linker sequences between catalytic domains and CBDs are shown in green.

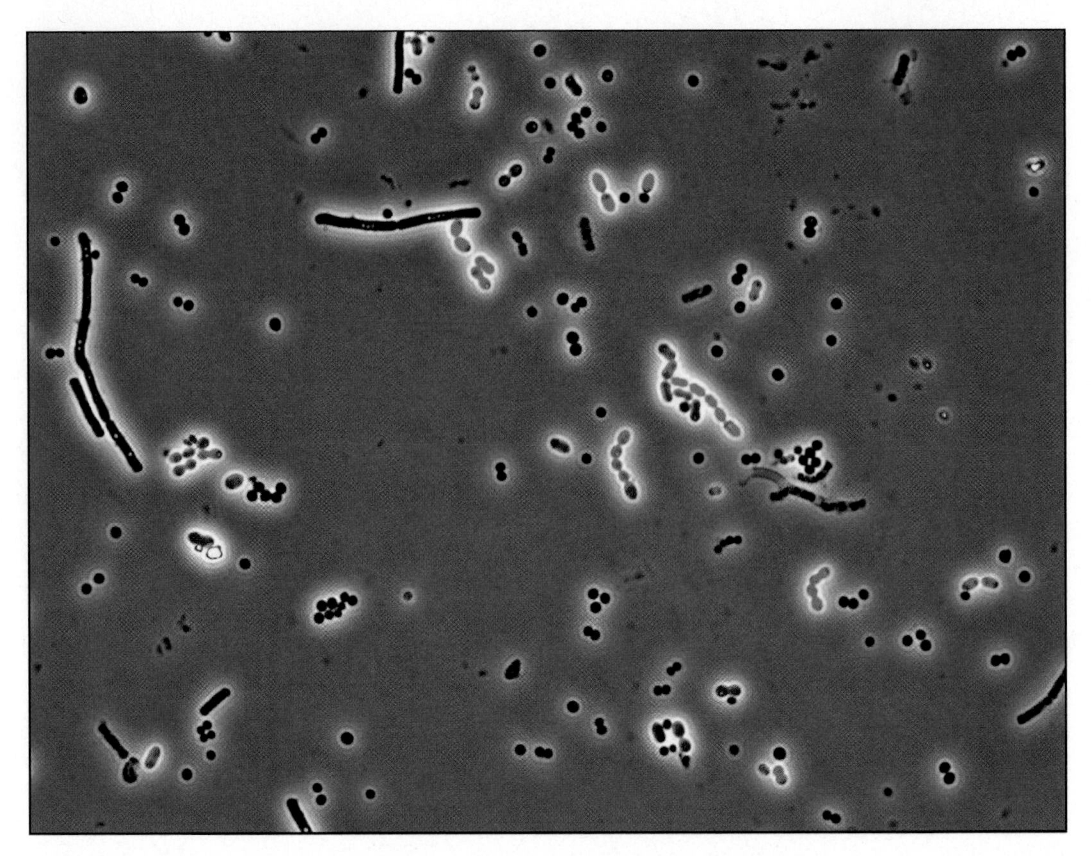

COLOR PLATE 2 [chapter 8] Specificity of lysin binding domains. Explicit labeling of *S. pyogenes* D471 cells by AlexaFluor-conjugated PlyCB in a mixture containing *S. aureus* RN4220 and *B. cereus* 4342 cells. PlyCB is the CBD of PlyC, a streptococcal lysin. The image is a merged composite of phase-contrast and fluorescent microscopy at ×1,000 magnification.

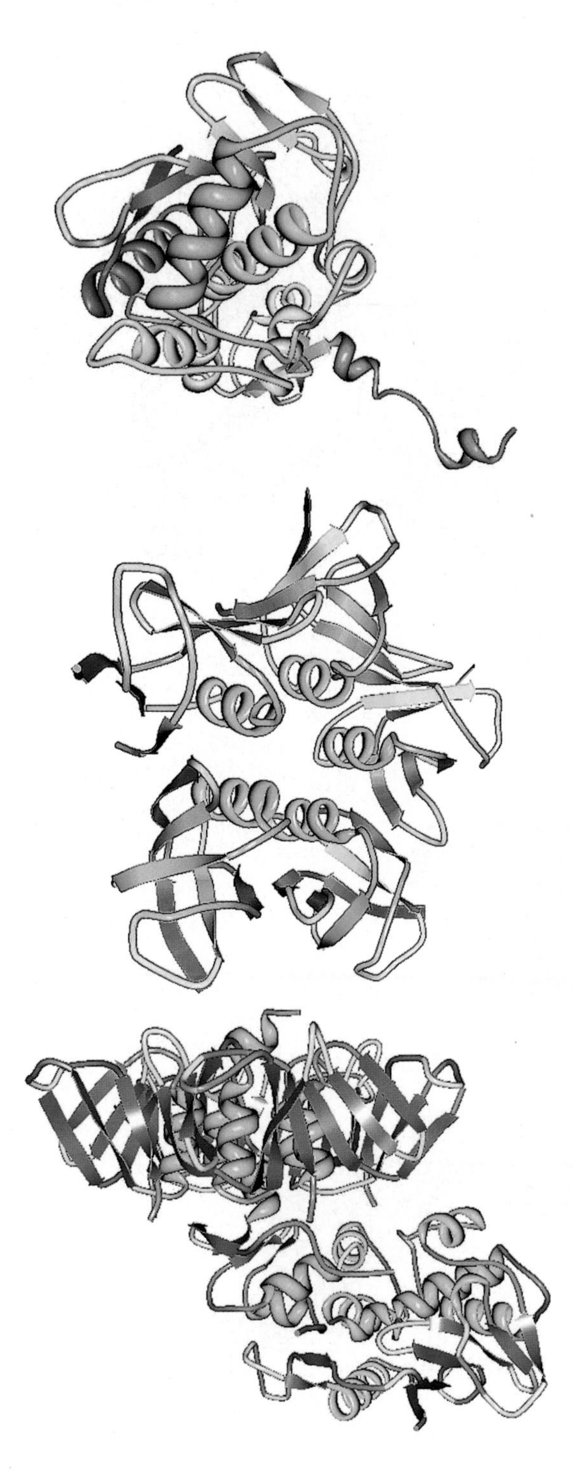

COLOR PLATE 3 [chapter 9] Three-dimensional structure of Shiga toxin showing the monomeric StxA subunit (top), StxB pentamer (middle), and complete toxin (bottom). This figure was derived from Research Collaboratory for Structural Bioinformatics Protein Data Bank (RCSB PDB) entry 2GA4, manually edited to remove hetero atoms and water molecules and visualized using Protein Workshop (Moreland et al., 2005) in conformation mode where light blue = turn, red = coil, green = α-helix, and dark blue = β-strand.

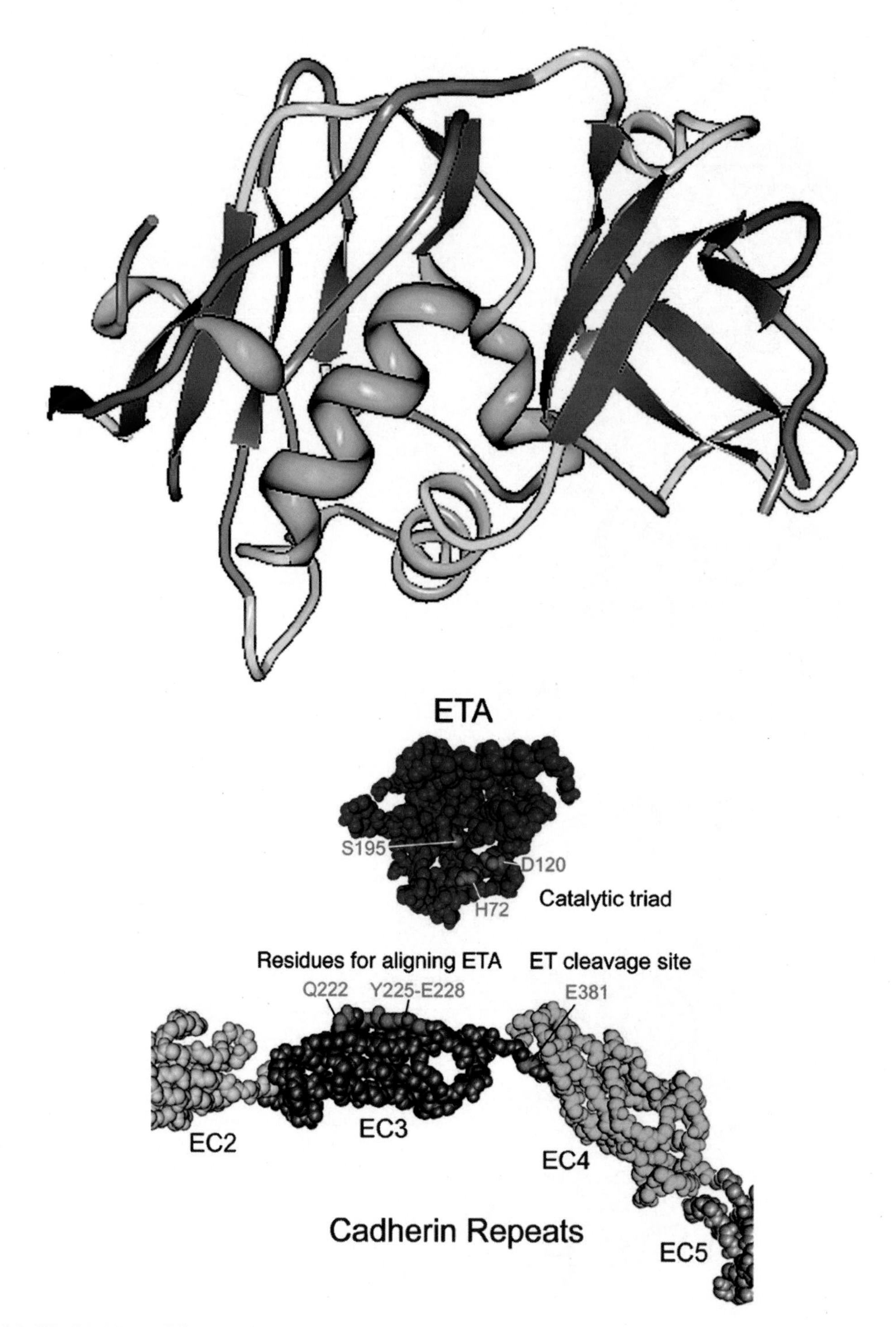

COLOR PLATE 4 [chapter 9] (Top) Three-dimensional structure of an ETA monomer derived from PDB 1ESF and visualized using Protein Workshop (Moreland et al., 2005) in conformation mode where light blue = turn, red = coil, green = α-helix, and dark blue = β-strand. (Bottom) Crystal structures of ETA and its substrate (reprinted from Nishifuji et al. [2008] with permission of the publisher). The residues comprising the catalytic triad of the serine protease active site of ETA are shown (Vath et al., 1997), as is the unique glutamate residue between the EC3 and EC4 cadherin repeats in human Dsg1 where ETA-catalyzed peptide bond cleavage occurs (Hanakawa et al., 2004). Also shown is a block of five amino acids in the EC3 domain of human Dsg1 that were implicated in alignment of ETA (Hanakawa et al., 2004).

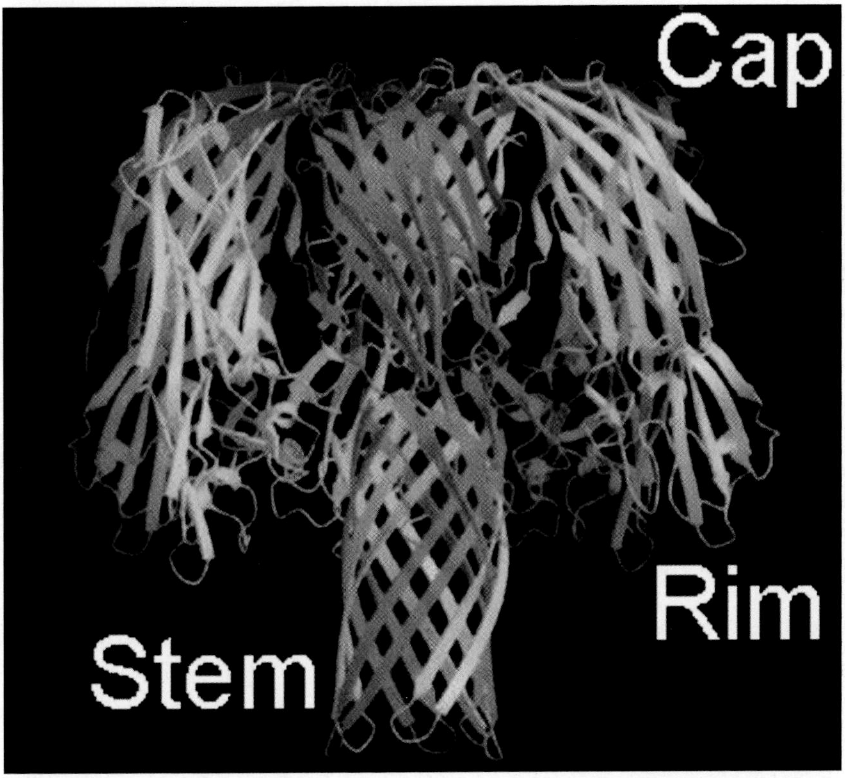

COLOR PLATE 5 [chapter 9] Molecular model of the pore formed by a heterohexameric PVL, viewed perpendicularly to the sixfold axis. The conformation of the stem region was derived from that of the α-hemolysin protomer (Song et al., 1996). The LukF and LukS subunits are shown in blue and yellow, respectively. The red regions indicate the N termini. (Adapted from Pedelacq et al. [1999] with permission of the publisher.)

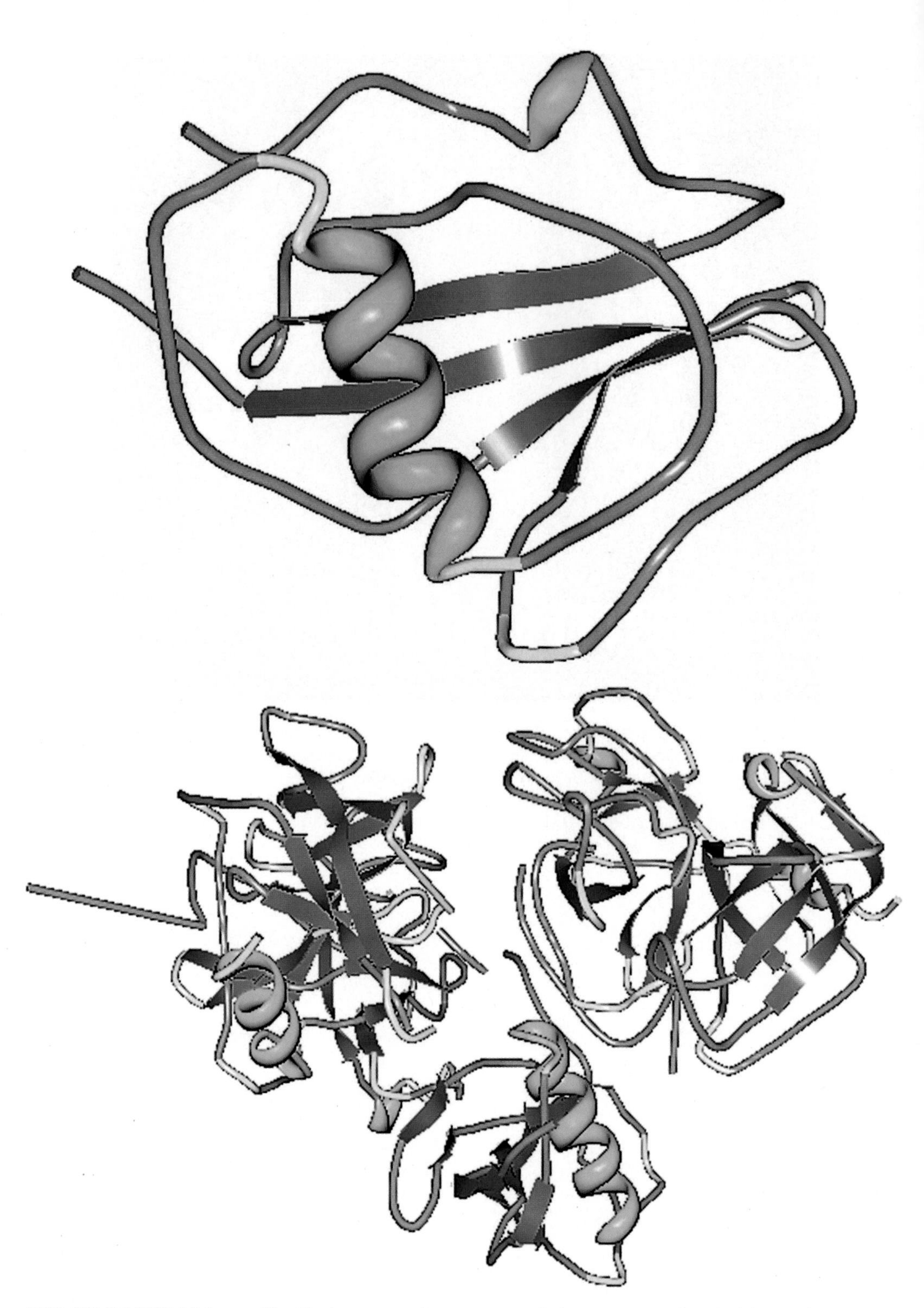

COLOR PLATE 6 [chapter 9] Nuclear magnetic resonance solution structure of staphylokinase (PDB 1SSN; Ohlenschlager et al., 1998) visualized using Protein Workshop (Moreland et al., 2005). For the ternary complex of microplasmin with staphylokinase (PDB 1BUI), see Parry et al. (1998).

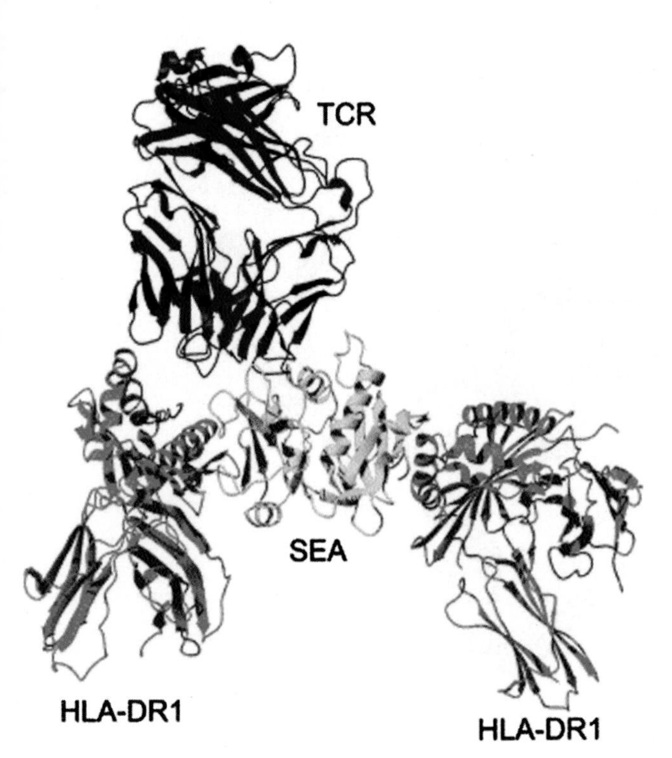

COLOR PLATE 7 [chapter 9] Illustration of MHC cross-linking by SEA. Shown is a quaternary complex of SEA (yellow) cross-linking two HLA-DR1 molecules (green) as well as interacting with the T-cell receptor (blue). (Reprinted from Petersson et al. [2002] with permission of the publisher.)

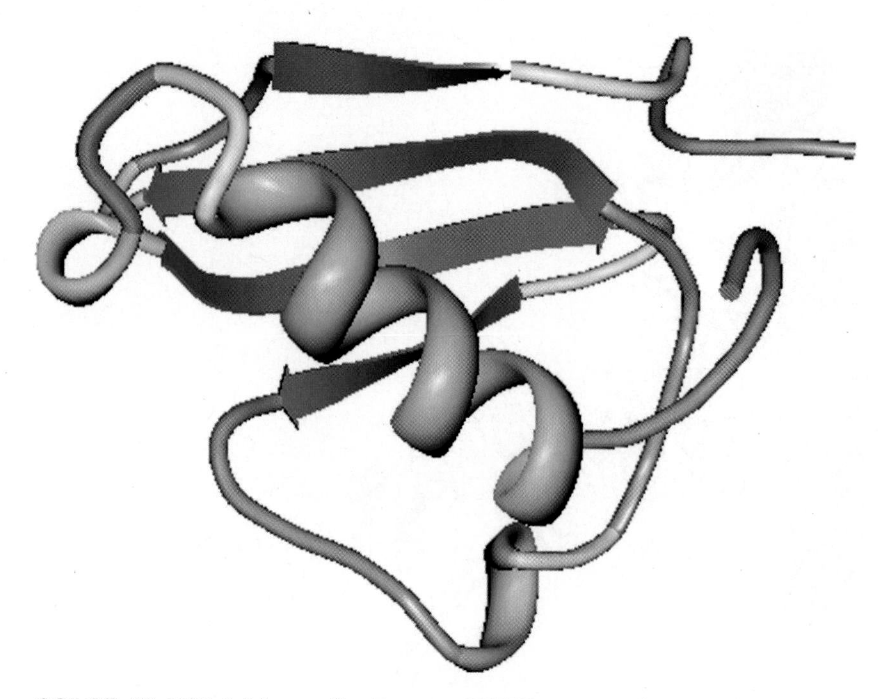

COLOR PLATE 8 [chapter 9] *S. aureus* CHIPS structure determined by solution nuclear magnetic resonance (PDB 1XEE; Haas et al., 2005). For the structure in complex with tyrosine-sulfated C5a receptor (Ippel et al., 2009) see PDB 2K3U.

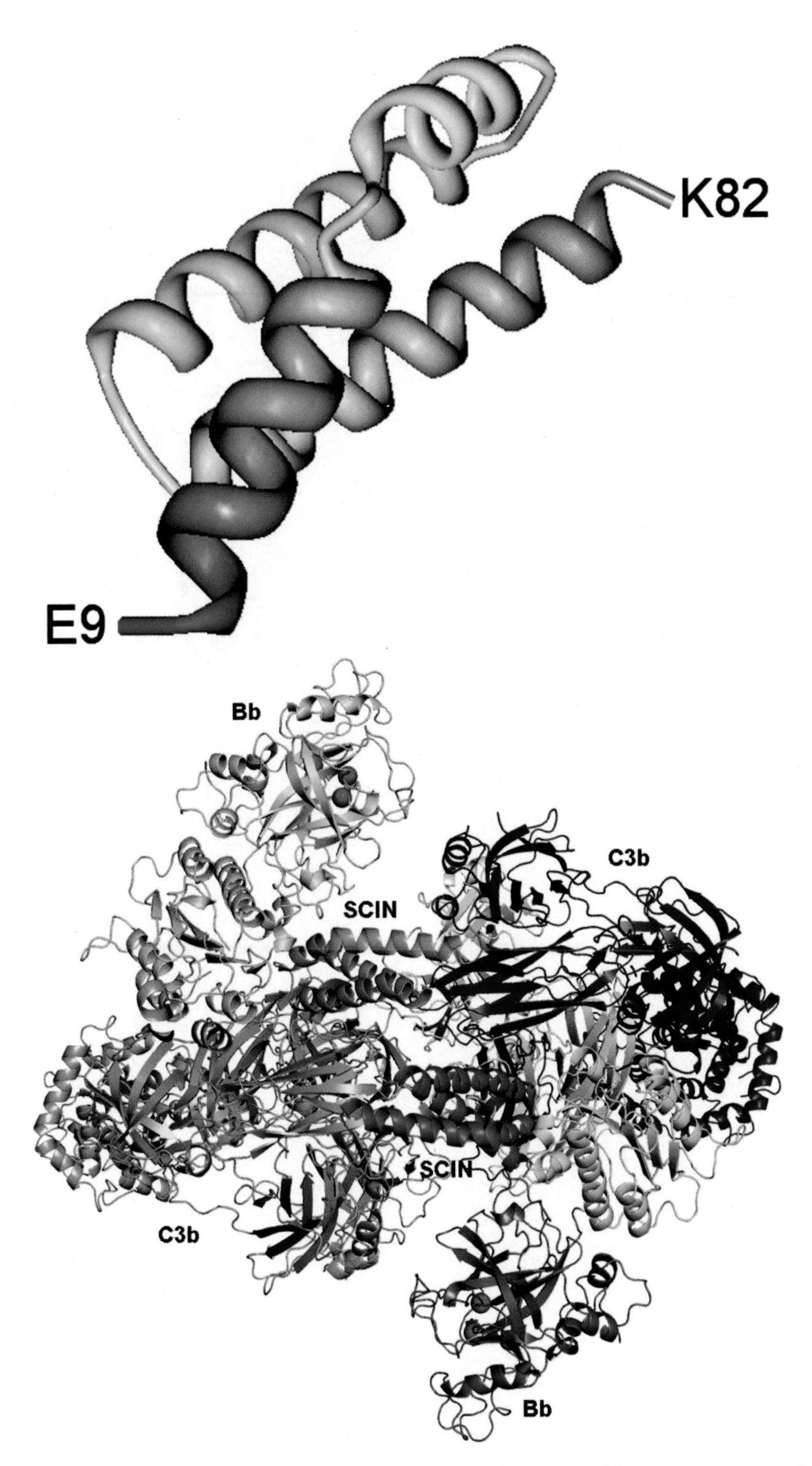

COLOR PLATE 9 [chapter 9] (Top) Structure of SCIN, shown in ribbon representation (PDB 2QFF; Rooijakkers et al., 2007). The α1-helix is colored blue, the α2-helix green, and the α3-helix orange-pink. (Bottom) The crystal structure of C3 convertase C3Bb in complex with SCIN. The C3bBb-SCIN dimeric complex is shown as a ribbon diagram with C3b in blue and turquoise, Bb in green and gold, and SCIN in purple and orange. (Reprinted from Rooijakkers et al. [2009] with permission of the publisher.)

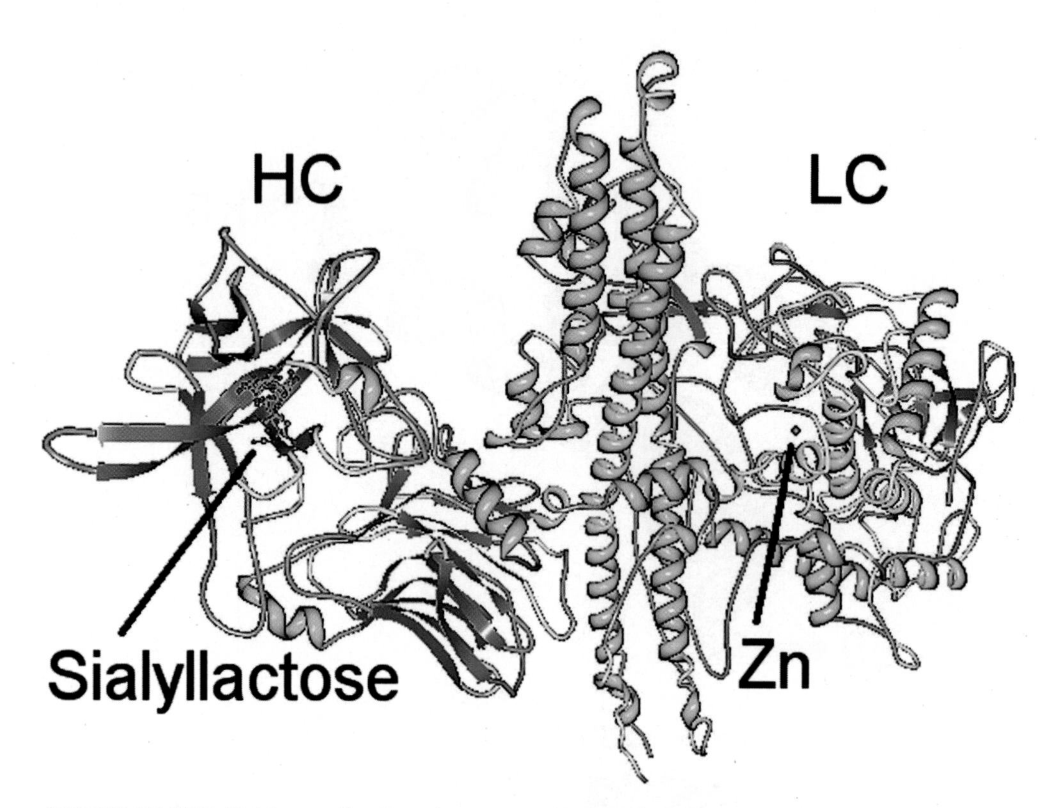

COLOR PLATE 10 [chapter 9] Crystal structure of BoNT/B (PDB 1F31) complexed to sialyllactose (Swaminathan and Eswaramoorthy, 2000) as viewed using Protein Workshop (Moreland et al., 2005). HC, heavy chain; LC, light chain; Zn, zinc cation bound to LC.

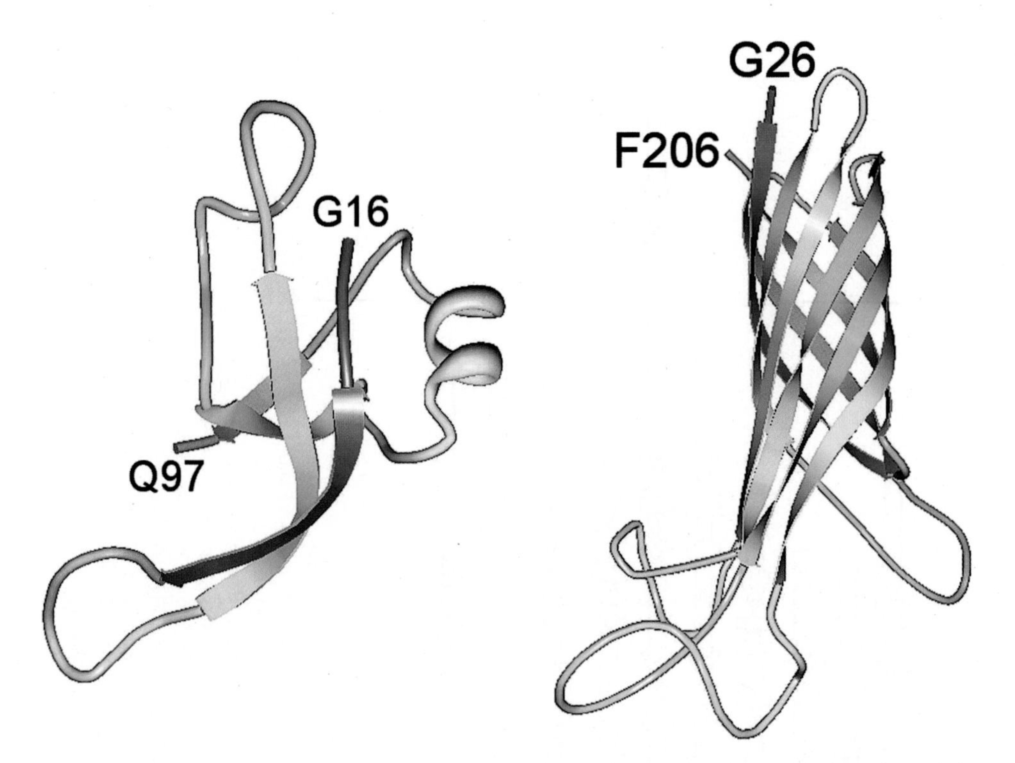

COLOR PLATE 11 [chapter 9] The LOOPP (Vallat et al., 2009) molecular models of Bor (left) and Lom (right).

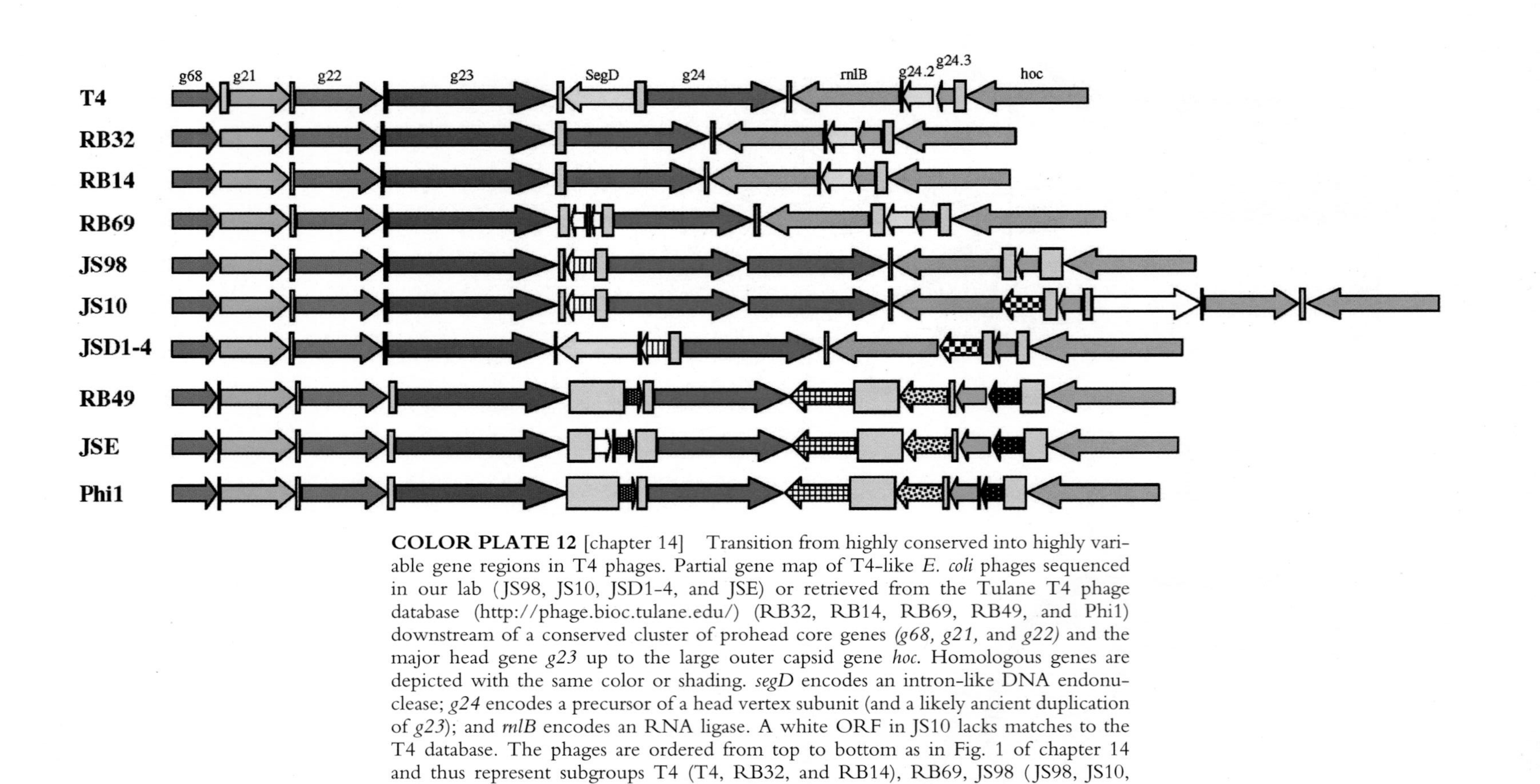

COLOR PLATE 12 [chapter 14] Transition from highly conserved into highly variable gene regions in T4 phages. Partial gene map of T4-like *E. coli* phages sequenced in our lab (JS98, JS10, JSD1-4, and JSE) or retrieved from the Tulane T4 phage database (http://phage.bioc.tulane.edu/) (RB32, RB14, RB69, RB49, and Phi1) downstream of a conserved cluster of prohead core genes *(g68, g21,* and *g22)* and the major head gene *g23* up to the large outer capsid gene *hoc*. Homologous genes are depicted with the same color or shading. *segD* encodes an intron-like DNA endonuclease; *g24* encodes a precursor of a head vertex subunit (and a likely ancient duplication of *g23*); and *rnlB* encodes an RNA ligase. A white ORF in JS10 lacks matches to the T4 database. The phages are ordered from top to bottom as in Fig. 1 of chapter 14 and thus represent subgroups T4 (T4, RB32, and RB14), RB69, JS98 (JS98, JS10, and JSD1-4), and RB49 (RB49, JSE, and Phi1). (Courtesy of E. Denou, Nestlé Research Center, Lausanne, Switzerland; reproduced with permission.)

BACTERIOPHAGES IN INDUSTRIAL FOOD PROCESSING: INCIDENCE AND CONTROL IN INDUSTRIAL FERMENTATION

Simon Labrie and Sylvain Moineau

10

INTRODUCTION

Fermentation has been used as a method of food preservation for millennia. Even today, a wide variety of food products are manufactured via microbial fermentation. These include fermented milk products such as buttermilk, cheese, sour cream, and yogurt, as well as the production of wine, soy sauce, sausages, and fermented vegetables. This short list covers only the most popular fermented products in North America (for reviews, see Hutkins, 2006; and Okafor, 2007). Not only does the fermentation process extend the shelf life of these products, but it also brings new organoleptic properties (flavor, texture, etc.) not found in the raw materials. Fermentation can also remove unwanted properties. For example, fermented soybeans contain fewer flatulence factors than unfermented soybeans (Okafor, 2007)! Fermentation can also reduce harmful compounds found in the raw product such as aflatoxins or cyanogens (Giraffa, 2004). Finally, the fermented product is likely to cause detrimental growing conditions for pathogenic bacteria. For example, the presence of organic acids, antimicrobial proteins, low pH, and limited sugar concentration are among the several hurdles limiting the growth of undesirable microbes.

Simon Labrie and Sylvain Moineau, Département de biochimie et de microbiologie, Faculté des sciences et de génie, Groupe de recherche en écologie buccale, Faculté de médecine dentaire, Félix d'Hérelle Reference Center for Bacterial Viruses, Université Laval, Québec City, Québec G1V 0A6, Canada.

FOOD FERMENTATION AND STARTER CULTURES

Some modern food fermentations are still initiated using the indigenous bacterial microflora of the raw substrate, also referred to as spontaneous fermentation. However, most industrial and large-scale food fermentations increasingly rely on concentrated bacterial cultures to control the process. These so-called starter cultures can be separated into at least two large groups: defined and mixed.

The defined starter is generally composed of two to five bacterial strains carefully selected based on precise phenotypic and genotypic criteria (Bissonnette et al., 2000). The selection of bacterial strains and the subsequent development of performing starter cultures are costly but are necessary to obtain a bacterial blend that will produce the desired characteristics in the fermented product as well as the best productivity yield. When possible, phage sensitivity is another criterion that

Bacteriophages in the Control of Food- and Waterborne Pathogens
Edited by Parviz M. Sabour and Mansel W. Griffiths, © 2010 ASM Press, Washington, DC

should be tested in order to select phage-resistant bacterial strains.

Other food fermentation processes rely on mixed starter cultures, composed of undefined numbers and proportions of bacterial strains. This type of culture is less predictable in terms of fermentation time and quality of the final product. Modern production facilities are frequently unwilling to accept this lack of consistency (Sanders, 1999). On the other hand, it has been argued that such mixed cultures can provide a unique set of characteristics to a food product, which is highly touted in artisanal food production. Mixed cultures are also often used for complex fermentation processes based on a succession of bacterial species, such as for sauerkraut (Mudgal et al., 2006), or where the bacterial strains driving the fermentation are fastidious and difficult to obtain and grow in pure culture, such as in vinegar production (Gullo and Giudici, 2008).

In the latter case, two methods are still in use for the production of vinegar, namely the traditional (slow) and the submerged (fast) fermentation process. However, both fermentation methods are initiated with the seed vinegar, also called the "mother vinegar," which is a microbiologically active undefined mixed starter obtained from previous vinegar batches (Gullo and Giudici, 2008). Sometimes mixed starter even contains bacteriophages maintaining the ecological balance of the starter. Nonetheless, whenever possible, industrial microbiologists seeking batch-to-batch consistency try to find alternative starter cultures to avoid bacteriophage infection (Boucher and Moineau, 2001; Stadhouders and Leenders, 1984).

THE BACTERIOPHAGES

Bacterial viruses are the most abundant biological entities on earth ($>10^{31}$) and are estimated to be 10-fold more abundant than bacteria (Breitbart and Rohwer, 2005; Brüssow and Hendrix, 2002; Whitman et al., 1998; see also chapter 2). Microbiologists studying bacterial viruses cannot help but be amazed by their ubiquity and their diversity. Everywhere bacteria can grow, phages will be found. From the nutrient-rich environments to extreme environments such as the human digestive tract or deep-ocean thermal vents, bacteriophages have been discovered. From a genetic perspective their diversity is also remarkable, as novel genes are often found in newly characterized phage genomes.

Phages were codiscovered by Frederick Twort and Félix d'Hérelle in 1915 and 1917, respectively. Since then, the field of phage research has broadened to focus on several different areas. In the beginning, phages were first investigated by d'Hérelle and others for their therapeutic use, but with the advent of antibiotics, phage research rapidly declined in popularity. In the last decade, however, the emergence of antibiotic-resistant pathogenic bacteria has become a serious public health concern (see chapter 1) and enthusiasm for phage therapy has been revived. In the early days of molecular biology, *Escherichia coli* bacteriophages also became the pillar of genetics research as four Nobel Prizes were attributed to researchers in that field (Norrby, 2008).

Whithead and Cox (1935) demonstrated that phages can also have an important negative impact on food fermentations. A new research area was born and numerous reports were published on phage industrial ecology. A better understanding of the industrial phage-host interactions was acquired over the years, with the ultimate objective of controlling the phage population to reduce their impact on fermentation processes. Thus, the negative effects of phage infection in fermentation are in contrast to the benefits of phage therapy discussed in many other chapters of this book (see chapters 1, 4, 5, 7, 8, 11, 13, and 15).

Bacteriophage Classification

Bacteriophages are classified according to morphology, nucleic acid composition, and homology (see chapter 2 for more details). Since the introduction of negative staining for electron microscopy in 1959 (Brenner and

Horne, 1959), more than 5,500 phages have been observed (for a review, see Ackermann, 2007). As of 2007, the vast majority (96%) of identified phages are tailed viruses belonging to the order *Caudovirales*. The *Caudovirales* are grouped into three different families: the *Siphoviridae* (61%), with a long, noncontractile tail; the *Podoviridae* (15%), with a short, noncontractile tail; and the *Myoviridae* (24%), with a long, contractile tail (see chapter 2). Pleomorphic, polyhedral, and filamentous phages represent only 4% of the observed specimens and are separated into 17 different families. When environmental microbiologists began to scrutinize extreme environments, new phages with different morphologies were discovered. *Ampullaviridae, Bicaudaviridae,* and *Globuloviridae* represent new phage families and are still awaiting classification by the International Committee on Taxonomy of Viruses (Häring et al., 2004, 2005a, 2005b; Prangishvili et al., 2006a, 2006b; Rachel et al., 2002).

Control of Phages in the Fermentation Industry

Bacteriophages are undoubtedly the greatest threat in fermented food productions, especially in the dairy industry, which has openly acknowledged this biotechnological problem. The dairy industry uses a significant amount of bacterial cells (10^7 cells/ml) in milk processing to produce cheese, yogurt, buttermilk, and other fermented products (Bogosian, 2005). In 1935, Whithead and Cox demonstrated that the milk fermentation process, driven by lactic acid bacteria, can be slowed or even stopped by virulent phages (Whithead and Cox, 1935). Over the next decades, numerous publications reported the presence of phages causing problems in industrial dairy factories (reviewed in Émond and Moineau, 2007; and Lévesque et al., 2005).

Even though the body of literature is small, phages have also been reported as a concern or a risk factor in other fermented food productions and fermentation-based bioproduction industries such as fermented feeds, organic and amino acids, alcohols, and pharmaceutical products (for a list, see Émond and Moineau, 2007). The importance of phages in bioindustries varies according to the nature of the fermentation process used. It is estimated that 0.1 to 10% of fermentations are affected in the dairy sector (Bogosian, 2005; Émond and Moineau, 2007; Moineau and Lévesque, 2005), while the occurrence of phage outbreaks in the biotechnology and pharmaceutical industries is much lower (0.1%) (Émond and Moineau, 2007). Pharmaceutical and biotechnology industries follow strict guidelines for the purity, concentration, and quality of the product. In view of that, phage contamination is not tolerated, even in low concentration, because it may lead to the complete loss of the product due to the lysis of the preferred bacterial strain. Raw substrates can also be sterilized without affecting the final product, thereby limiting phage contamination (see below).

In the dairy industry, however, milk cannot be sterilized, only pasteurized at most. Phages can enter a factory through the well-known route of raw milk. Thus, phage populations are controlled rather than eliminated in this industry. Since phages are generally recognized as safe, food fermentations compromised by phages can usually be transformed into byproducts when the amenities are available, thus reducing the economic impact of phage contamination. Moreover, a blend of non-phage-related bacterial strains is usually used for food production, thereby limiting the negative effect linked to one lysed bacterial strain.

Phage outbreaks within a production facility necessitate costly and time-consuming countermeasures. Moreover, the painstakingly timed production schedule is inevitably perturbed. First, the source of contamination must be identified. The facilities must then initiate decontamination procedures, sometimes requiring production shutdown, leading to increased raw product spoilage and loss of productivity.

THE DAIRY INDUSTRY: A CASE STUDY

The case of the dairy industry will be illustrated below to exemplify the problems caused by virulent phages infecting the starter cultures, and the corresponding strategies used to control phage population.

Starter cultures used by the dairy industry are composed of lactic acid bacteria, which represent a diverse group including, among others, the genera *Lactococcus* and *Lactobacillus* as well as the species *Streptococcus thermophilus*. The bacteriophages infecting these groups of bacteria have been extensively characterized because of their negative impact on the industry. For example, lactococcal phages are the most-studied group of bacterial viruses after the *E. coli* phages. The presence of lactococcal phages causes serious problems to the cheese industry, so their ecology and diversity have been extensively documented.

The lactococcal phages belong predominantly to the *Siphoviridae* family but some are also members of the *Podoviridae* family. *Lactococcus lactis* phages are also subgrouped into 10 so-called phage groups (Deveau et al., 2006; Labrie and Moineau, 2000) based on their morphology and nucleic acid identity (Ackermann et al., 1984, 1992; Braun et al., 1989; Jarvis, 1984; Jarvis et al., 1991; Lembke and Teuber, 1981). But three of these groups (936, c2, and P335) are by far the most predominant lactococcal phages. On the other hand, *S. thermophilus* phages are a much more homogenous group. All of them have a long, noncontractile tail (*Siphoviridae* family) and have been classified into only two groups thus far. Members of the first group possess cohesive genome extremities (*cos* type), while members of the second group include phages that package their DNA via a headful mechanism (*pac* type) and thus have redundant genomic extremities (Le Marrec et al., 1997). These two groups of phages appear to be found at the same levels in the dairy industry.

On the other hand, the reported rate of *Lactobacillus* phage infections in the dairy industry is relatively low compared to those of *S. thermophilus* and *L. lactis* (Villion and Moineau, 2009). Only phages of the *Myoviridae* and *Siphoviridae* families have been known to infect lactobacilli (Villion and Moineau, 2009). *Lactobacillus* phages are also much more genetically diverse than lactococcal and streptococcal phages, which is likely a reflection of the large number of species in the *Lactobacillus* genus.

The above studies exemplify the importance of understanding the phage population responsible for phage infections, regardless of the genus/species of bacteria used. The characterization of these bacterial viruses may sound too academic for some, but it will provide critical information to help in selecting appropriate control strategies (see below). The authors believe that it is one of the keys to an efficient phage control program.

PHAGE RESERVOIRS IN INDUSTRY

Another critical piece of information is to recognize the potential phage reservoirs within manufacturing facilities. As mentioned above, phages are ubiquitous in the environment. Each industry facing phage outbreaks should identify the sources of phage contamination as part of its phage control plan. There are different risk factors associated with the presence of phages, such as the nature of the substrate used and which treatments it can tolerate. For example, the liquid nature of milk allows effective dissemination of phages (Boucher and Moineau, 2001), while fermented sausage processes are less susceptible to phages because the meat matrix appears to curtail phage dispersion (McLoughlin and Champagne, 1994; Nes and Sorheim, 1984; Trevors et al., 1984).

Fermentation Substrate

The fermentation substrate is the first source of phage entry in a factory. Regardless of the origin of the raw material, there is a high probability of contamination by different virulent phages. In the presence of a high concentration of sensitive bacterial strains, some of these phages may rapidly multiply. Typically, milk fermentation may be compromised at the critical threshold of 10^4 PFU per ml. If the phage titer reaches that threshold, fermentation failure will occur as soon as the phage titer reaches 10^7 to 10^9 PFU per ml, with the simultaneous and significant drop of bacterial viability (Moineau and Lévesque, 2005).

Bioindustries such as biopharmaceuticals and biotechnology use bacterial media commonly found in the laboratory as the fermentation substrate. This type of substrate can easily withstand heat sterilization, eliminating the phage contamination risk. In contrast, the food industry cannot use sterile substrates since heat treatment alters essential properties of the raw product. In the dairy industry, raw milk used for cheese production generally undergoes pasteurization to reduce the population of unwanted endogenous microorganisms; however, many phages can resist pasteurization (Chopin, 1980; Madera et al., 2004). For yogurt production, milk is heated at a higher temperature (90°C) without altering the final product quality (Schmidt et al., 1985), and this treatment apparently eliminates most phages (Quiberoni et al., 1999; Schmidt et al., 1985; Suarez and Reinheimer, 2002).

Other food fermentation industries use raw materials such as vegetables (e.g., cabbage, cucumber, olive), most of which cannot withstand a heat treatment. Moreover, the manufacture of most fermented vegetables still relies on spontaneous fermentation initiated by the endogenous microflora (Hutkins, 2006), and heat treatment would kill these essential bacteria. The success of vegetable fermentation necessitates establishing good selective environmental conditions to favor growth of the wanted bacteria and inhibit the growth of others. These selective conditions rely on salt, temperature, and anaerobic conditions—none of which inhibits phages.

Interestingly, the endogenous microflora initiating the fermentation is often in equilibrium with its specific phage population; thus, these bacteria generally survive, despite the phages, and complete the fermentation. This variability means that a slower fermentation process and increased inconsistency in the final product may occur, especially when compared to fermentation driven by defined starter cultures (Émond and Moineau, 2007). Whenever possible, the vegetable fermentation industry is starting to increasingly rely on defined starter culture to initiate fermentation; thus the phage concerns will inevitably increase in scale. The use of a few selected starter lactic acid bacteria strains will allow the industry to create better consistency in the final preferred product and increase fermentation speed, but the processes will be more susceptible to attack by virulent phages (Hutkins, 2006; Lu et al., 2003; Yoon et al., 2007).

Starter Cultures as Phage Reservoirs

Many other phage reservoirs can be found, and one of the most ominous is the bacteria composing the starter. Many bacteria are lysogenic and carry prophages integrated into their genome (see chapters 2 and 9). These prophages are dormant killers waiting for the good conditions to initiate the lytic cycle. It is now well recognized that prophages play an important role in bacterial evolution and population dynamics (Casjens, 2003). Many bacterial genomes also carry more than one prophage element (Bossi et al., 2003; Canchaya et al., 2003). For example, when the genomes of *L. lactis* subsp. *lactis* IL1403 (Bolotin et al., 2001), *L. lactis* subsp. *cremoris* SK11 (Makarova et al., 2006a), and *L. lactis* subsp. *cremoris* MG1363 (Wegmann et al., 2007) were sequenced, at least six prophages were identified in the first two organisms, and five in the third. Prophages can occupy an impressive proportion of the host genome, such as for *E. coli* O157:H7 Sakai, which contains at least 18 prophages occupying 12% of the bacterial genome (Ohnishi et al., 2001). Whether functional or not, prophages often encode host fitness factors including virulence factors (Brüssow et al., 2004) or phage resistance mechanisms such as superinfection immunity. When lysogenic bacterial strains are exposed to various stress conditions, prophages can be induced, initiating their lytic cycle and causing lysis of the bacterial cells.

Any given functional prophage can also become strictly virulent due to a deletion or a mutation in its lysogeny module. Moreover, prophages constitute a gene pool that contributes to viral diversity. It has been shown that when a virulent phage infects a bacterium carrying prophages, the virulent phage can exchange parts of its genome with the resident prophages via homologous or illegitimate recombination. This process can result in rapid emergence of new virulent phages (Bouchard and Moineau, 2000; Durmaz and Klaenhammer, 2000; Labrie and Moineau, 2007).

Consequently, the risk of prophage induction and/or emergence of new virulent phages must be carefully considered when selecting bacterial strains. Some scientists advocate the use of prophage-cured strains, or the selection of bacterial strains that cannot be induced under stress conditions, to avoid cell lysis, and introduction of viral diversity within the manufacturing process (McGrath et al., 2007). Yet some lysogenic strains have been used for decades without any adverse effects on fermentation. Indeed, others believe that prophages contribute to flavor development and maturation of cheese by increasing the release of intracellular enzymes through bacterial cell lysis, following the expression of prophage-encoded endolytic enzymes. However, we believe that whenever possible, prophage-free strains should be preferred.

Fermentation By-Products

In the quest for productivity and consistency, the dairy industry may implement fermentation processes necessitating the addition of supplements such as whey proteins. Whey is a by-product of the manufacture of cheese and may contain high concentrations of phages, making it an important risk factor. Accordingly, such by-products (including dried forms) should undergo strict quality control and be tested for the presence of phages specific to the starter strains before their use in production (Émond and Moineau, 2007). Similarly, the factory design must also be carefully considered to minimize the risk of contact (in the pipelines or airflow) between raw substrate (or heat-treated substrate) and products or by-products from previous fermentations, which could potentially be contaminated with higher concentrations of phages.

Aerosols

One often overlooked source of phages is aerosol contamination. For example, although aerosols arise from all steps of cheese production, whey separation is the step producing the most aerosols (Kang and Frank, 1989; Neve et al., 1995; Sullivan, 1979). Neve et al. (2003) assessed the presence of airborne phages in a cheese factory and found higher concentrations of phages close to the whey separator (5.2×10^6 PFU/m^3 of air). Limiting airborne phage contamination requires good directional ventilation systems within the plant. Airflow must be directed from the raw product toward the final product.

Equipment and Staff

Initially, new (emerging or entering) phages are rarely present in sufficient concentration to cause fermentation problems, so they must replicate during the manufacturing process before becoming problematic. Therefore, all fermentation equipment and tools must be carefully cleaned and sanitized to lessen the phage contamination hazard. Similarly, floors, walls, and drains should be sanitized, as they may also be contaminated with undesirable bacteria and phages. The implementation of these good manufacturing procedures also requires appropriate training of the staff.

CONTROLLING PHAGES

Over the last decades a multihurdle approach has been established to control phage populations within fermentation plants. As mentioned previously, good manufacturing practices as well as sanitization, factory design, and efficient ventilation are usually sufficient to limit large-scale phage problems. In addition, establishing a good phage monitoring protocol appropriate for the industrial environment is mandatory. Furthermore, well-trained staff will aid in rapid detection of phage infections so that countermeasures can be rapidly initiated to stop phage spread within the fermentation facilities.

Phage Detection and Monitoring

When fermentation fails, the first obvious step is to determine the specific cause. There are a number of reasons why such microbial processes may not go as planned. Here we will only deal with phage infection. There are at least three different ways to confirm the presence of phages: (i) microbiological assays, (ii) biochemical detection, and (iii) electron microscopy. The industry usually chooses one of these methods according to the type of fermentation process used and accessibility of equipment.

MICROBIOLOGICAL DETECTION

This method is by far the most popular technique when the bacterial strains are available in pure culture. This is not the situation for bioprocesses using mixed starter cultures. For example, acetic acid production is driven by *Acetobacter* strains that cannot be obtained in pure culture or that are fastidious to cultivate. Thus, it is not possible to use direct microbiological detection of phages in vinegar. To detect phages using a microbiological assay, a single bacterial strain is grown to confluence in a petri dish and a serial dilution of filtered (0.45-μm), phage-containing sample is added to the bacterial stains. The development of lysis plaques confirms, without a doubt, the presence of phages. For a detailed description of three different microbiological protocols commonly used to enumerate and detect phages, readers are referred to Mazzocco et al. (2009a, 2009b). Unfortunately, if the bacterial culture is not pure, the phage-resistant or phage-insensitive strains might overgrow sensitive strains and mask phage lysis.

BIOCHEMICAL DETECTION

Biochemical methods are based primarily on immunochemical assays and molecular detection of bacteriophage genetic material (most often double-stranded DNA). The immunochemical detection requires antibodies to be raised against conserved and major structural constituents of phage particles (Chibani Azaiez et al., 1998; Ledeboer et al., 2002; Moineau et al., 1993; Schouler et al., 1992). Thus, to obtain highly sensitive antibodies detecting a wide array of phages, a detailed study of phage populations affecting the fermentation process must be undertaken before it will be possible to select a good immunological target for the antibodies. Without an appropriate antibody target, this method is subject to false-negative results.

The detection methods based on molecular detection of DNA rely on the selection of proper probes specific to phages (Binetti et al., 2005; Deveau et al., 2006; Dupont et al., 2005; Labrie and Moineau, 2000; Moineau et al., 1992). It is also important to have detailed knowledge about the phage populations occurring within the manufacturing process prior to probe selection. For instance, a multiplex PCR-based approach was developed in 2000 (Labrie and Moineau, 2000) for the detection of lactococcal phages in the dairy industry. Six years later, some of the primers had to be modified to take into account the evolving phage population within cheese factories (Deveau et al., 2008). Biochemical methods are commonly used in university and research center laboratories to detect and characterize phages, but those methods are seldom used in industry because they are time-consuming and costly (reagents and equipments) to implement. Moreover, care should be taken when biochemical methods are used in industry since phage constituents (protein or DNA) are detected whether the phage is infectious or not.

ELECTRON MICROSCOPY

When microbiological or biochemical assays are not available for a specific group of phages (or bacteria), samples from failed fermentations can be observed directly with an electron microscope (EM) (Brenner and Horne, 1959). For direct EM observation of presumed phage-containing samples, some simple purification steps (most often by filtration and centrifugation) are required to eliminate bacterial debris and other impurities, as they may affect the staining procedure. Generally, to achieve proper negative staining, either phosphotungstate or uranyl acetate is used. The phage preparation is deposited on an EM grid in the presence of the stain, and after a few washes the preparation is ready to observe. For a detailed phage purification and EM observation procedure, readers are referred to Ackermann (2009). This technique is seldom used in industry because it requires qualified personnel and expensive, specialized EM equipment. More commonly, industry contracts an academic laboratory to carry out EM detection of phages.

These detection techniques are essential in identifying an emerging phage population. A good phage-monitoring program will allow a manufacturer to quickly adapt to prevent a serious phage outbreak.

Sanitization Procedures

When a phage contamination is confirmed by one of the above techniques, one of the first steps is to clean the manufacturing facilities. Sanitizing chemicals must be selected according to the type of industry since their efficiency is affected by the nature of the contaminants found in that industrial setting. For example, sodium hypochlorite's efficiency is greatly reduced in the presence of organic material such as milk. The efficacy of commercial sanitizers against phages will vary significantly among suppliers and also depending on the active ingredients and their concentrations. Therefore, in-house testing may be required to select the appropriate sanitizers or disinfectant products. Alternatively, suppliers may be able to offer specific information on the antiviral efficacy of their chemical products.

The most efficient sanitization procedures for equipment used in the food industry are efficient against phages to different extents. These include heat, alkaline (pH 11 and higher), and acidic treatments (pH 4 and lower). Some food-grade disinfectants are also effective against phages such as sodium hypochlorite (100 ppm), quaternary ammonium compounds, and paracetic acid (0.15%, wt/vol) (Émond and Moineau, 2007). Alcohol (70%) possesses a decimal reduction rate (time required to reduce the phage population 10-fold) not suitable to the industrial environment (Bogosian, 2005), and isopropanol was demonstrated to be ineffective against *Lactobacillus* phages (Capra et al., 2004).

Ascorbic acid (vitamin C), in the presence of trace amounts of copper to stimulate oxidation, appears to be an efficient sanitizer (Bogosian, 2005). Vitamin C was demonstrated to be active against different types of phages and has a decimal reduction time suitable for industry (4 s), but care should be taken to keep the pH of the product below 5 since the efficiency starts to rapidly decrease at higher pH (Murata and Kitagawa, 1973; Murata et al., 1971, 1975; Murata and Uike, 1976). Vitamin C is not widely used in industry because it is not effective against bacteria. Thus, a second sanitizer must be used to remove bacterial flora contaminating the installations.

Starter Rotation

After (or parallel to) the cleaning procedure, the industry has to decide whether to reintroduce the same starter culture to drive the fermentation or to use a new set of bacterial strains. Based on the authors' experience, the former approach can work effectively if the source of phages has been found and corrected, as well as if the sanitizing procedures are effective. But more often than not, virulent phages will reappear, as it is extremely difficult to completely eradicate phages within large fermentation facilities.

For decades the dairy industry has been rotating different bacterial starter cultures to avoid the buildup of a specific phage population and as a control measure to reduce fermentation failure caused by phages. This simple and successful technique relies on the availability of non-phage-related bacterial strains and has a proven track record of efficiency. However, industrial microbiologists are confronted by two major difficulties when establishing a suitable starter rotation strategy. The first difficulty is the lack of bacterial strains that are phage unrelated and that also retain the same crucial metabolic properties responsible for driving their unique fermentation. The second difficulty is the inconsistency of the product between batches when using different starters.

Different tactics have been suggested to address the above problem. One approach for starter rotation is based on the use of isogenic strains encoding different phage resistance mechanisms (Sing and Klaenhammer, 1993). As a result, the bacterial strains, encoding different phage resistance mechanisms (see below), will be phage unrelated and will lead to a final product with high consistency. Nevertheless, it was demonstrated that this strategy might still lead to the rapid buildup of an emerging virulent phage mutant population that is resistant to the antiphage mechanism introduced in the bacterial strains (Labrie and Moineau, 2007).

Whenever possible, using a product rotation schedule, such as cheese varieties, would allow diverse starters to be introduced, which could dilute specific phage populations within the facilities (Moineau et al., 1996).

PHAGE-RESISTANT BACTERIAL STRAINS

The selection of bacterial strains to compose a new starter is a long and costly process. This selection process may be on a trial-and-error basis, or through a complex multifactorial selection approach. Thus, when the right strain or group of strains is found, there is an evident pressure to use them regularly. Numerous examples have shown that phage infection will inevitably occur if the same strain(s) is used for a prolonged period. However, one can also attempt to transform a phage-sensitive bacterial strain into a phage-resistant derivative.

There are different ways to make a phage-sensitive bacterial strain resistant to diverse phage isolates. The first strategy involves selecting bacteriophage-insensitive mutant strains (BIMs). A second approach relies on introducing phage resistance mechanisms into a phage-sensitive strain. These phage resistance mechanisms can be separated into two major types: natural and engineered. Natural phage resistance mechanisms are biologically found in phage-resistant bacterial strains, and some of them can be moved from one strain to another by conjugation. Knowledge acquired over the last decades on phage biology and use of molecular biology techniques makes the design of new phage resistance mechanisms possible.

Selection of BIMs

The selection of natural, spontaneous, phage-resistant bacterial stains is the easiest way to replace an industrial strain that is infected by phages. BIMs have the same genetic determinants and most likely the same desired metabolic properties as the wild-type strain. Isolation of BIMs is achieved by exposing sensitive strains to high phage titers. BIMs surviving phage exposure normally have acquired a resistant phenotype following spontaneous chromosomal mutation, usually in the gene that encodes the phage receptor. This mutation, which may result in the modification of carbohydrates, proteins, or lipoteichoic acids of the cell wall or membrane, prevents the adsorption of the phage to the bacterial cell and results in the resistance phenotype (Forde and Fitzgerald., 1999b).

Isolation of BIMs also has important disadvantages. For instance, the phage resistance phenotype may be reversible at relatively high frequency. Thus, the stability of the phage resistance phenotype must be tested through several generations. Some of the BIMs also have reduced metabolic activities. For example, it is not infrequent that *L. lactis* BIMs produce lactic acid at a slower rate than the wild-type phage-sensitive strain. In this latter case, the users must decide whether this impaired metabolic activity (but improved phage resistance) has a bearing on their production. Nevertheless, numerous BIMs have been isolated and are being used by the dairy industry.

Possibly the most effective BIMs are those obtained because of a natural deletion in the phage receptor gene. Globally, phage-host interactions are still poorly understood and very few phage receptors have been identified. But when they are known, as is the case for a few *E. coli* and *L. lactis* phages, it is possible to screen BIMs by PCR for such deletion within the gene coding for the phage receptor. This approach has been successfully used to control a group of lactococcal phages (c2) that uses the transmembrane bacterial protein PIP for efficient cell adsorption (Babu et al., 1995; Garbutt et al., 1997; Geller et al., 1993). No reversion phenotype was observed with this type of BIM, and this mutation did not influence metabolic activity. In fact, this strategy has been so successful that the number of new phages belonging to this lactococcal group is declining in cheese plants worldwide.

Natural Phage Resistance Mechanisms

Phage resistance mechanisms have been identified and characterized mainly in *L. lactis*. More than 50 of them have been found in this bacterial species and many are plasmid encoded, thereby facilitating their transfer from one strain to another. These resistance mechanisms are classified into four groups according to their general mode of action: adsorption blocking, inhibition of phage DNA ejection, restriction and modification, and abortive infection mechanisms (Forde and Fitzgerald, 1999b). The ranges and efficiencies of these phage resistance mechanisms vary according to several factors. Some mechanisms are efficient only against a narrow group of phages, while some are active against different phage groups. Measuring the phage's efficiency of plaquing (EOP) is a way to assess the efficiency of a phage resistance mechanism. The EOP is determined by dividing the phage titer for the resistant bacterial strain by the titer for the nonresistant strain. When an antiphage system provides an EOP of 10^{-7} to 10^{-9}, it is considered to be a strong antiviral system (Moineau, 1999; Moineau and Lévesque, 2005).

ADSORPTION BLOCKING

This group of resistance mechanisms blocks the first step of infection, namely phage adsorption. The genetic determinants of phage adsorption blocking are not yet well understood, but it can be attributed to different factors, including galactosyl-containing lipoteichoic acid, galactose/rhamnose or galactose/glucuronic acid polymer, and cell wall proteins (Coffey and Ross, 2002). Those compounds either mask the phage receptor or compete with the phage for adsorption to the receptor (Forde and Fitzgerald, 1999a; Lucey et al., 1992; Sijtsma et al., 1990; Tuncer and Akçelik, 2002). Most of them do not provide very efficient resistance.

BLOCKAGE OF PHAGE DNA EJECTION

Once the phage is adsorbed to the bacterial cell, another group of antiphage systems can prevent the ejection of phage DNA within the cell. These mechanisms have no effect on plasmid transformation or transfection, but transduction is impossible. Very little information is available for this type of mechanism. The genes responsible for the phage resistance phenotype are encoded by either chromosomal or plasmid DNA (Akçelik, 1998; Akçelik et al., 2000; Garvey et al., 1996). McGrath et al. (2002) have also characterized a superinfection exclusion system (Sie) encoded by the temperate lactococcal phage Tuc2009, which inhibits phage DNA ejection. Similar systems appear to be widespread in many *Lactococcus* strains.

RESTRICTION AND MODIFICATION

Bacterial restriction-modification (R-M) systems are recognized as among the first line of defense after foreign DNA entry into the cell. They are by far the best-understood and most diverse antiphage systems. They are found in most bacterial species, and their genes are either on plasmids or in the bacterial genome.

R-M systems are composed of at least two distinct enzymatic activities: a site-specific restriction endonuclease and a methylase. The host DNA remains protected by modifying its own DNA through the action of the methylase. This phage resistance mechanism has the broadest range of activity, and its efficiency is directly related to the number of specific restriction sites in the phage genome. These R-M systems are powerful defense mechanisms since they degrade phage DNA soon after its entry into the cell. Yet R-M systems are limited since the genome of the phage can sometimes be methylated and become resistant to the cognate restriction endonuclease. Once methylated, the modified progeny is impervious to the R-M, and when released, the progeny will be able to circumvent the R-M system of neighboring bacteria (Sturino and Klaenhammer, 2004b). Moreover, phages have evolved many strategies to overcome the R-M system, such as elimination of restriction sites from their genome and modification of their bases. Some phages even encode methylase genes or genes that inhibit the restriction endonuclease (Allison and Klaenhammer, 1998; Forde and Fitzgerald, 1999b). Taken altogether, R-M systems are relatively efficient phage barriers, but are even stronger when used in combination with other systems.

ABORTIVE INFECTION MECHANISMS

This class of phage resistance mechanisms is a regrouping of heterogeneous antiphage systems that generally possess a low level of amino acid identity between themselves and with databases. Abortive infection mechanisms, also referred to as Abi, inhibit phage replication after phage DNA entry into the cell and before cell lysis. Abi have been the subject of ongoing studies, but very little information about their mode of action is available. Their general impact on the phage lytic cycle is known for many Abi, but the nature of the direct interactions of Abi proteins with phage components is still unclear. For example, some Abi systems will interfere with phage DNA replication, while others will affect phage transcription or translation (Chopin et al., 2005).

Twenty-three Abi mechanisms were identified and characterized in *L. lactis,* and these are among the most efficient mechanisms. Unlike with R-M, most infected cells do not survive the infection. Thus, immature phage progeny are trapped inside the dead cell. Either the cells die from the damage induced by initiation of the phage lytic cycle or the phage infection triggers an altruistic suicide for the benefit of the neighboring bacterial population. For a complete review on Abi mechanisms in *Lactococcus,* the reader should refer to Chopin et al. (2005). Recently, an Abi-like mechanism was found in *Pectobacterium atrosepticum* (formerly *Erwinia carotovora* subsp. *atroseptica*), namely ToxIN, which functions as a toxin-antitoxin system (Fineran et al., 2009). This was the first toxin-antitoxin system based on RNA-protein interaction. The gene coding for toxic protein ToxN (Abi like) is preceded by a series of repetitions named ToxI. When transcribed, ToxI RNA antitoxin counteracts the toxin, ToxN. During phage infection, the ratio of ToxI to ToxN is changed by an unknown mechanism and ToxN is activated, killing the cell and trapping the immature phage progeny inside the cell.

Similarly to other resistance mechanisms, phages can mutate and become insensitive to the introduced Abi system. It has been reported that a single point mutation in a specific phage gene can lead to insensitivity to one Abi system (but not all) (Chopin et al., 2005). Large genomic rearrangements to bypass an Abi system have also been documented for lactococcal phages (Labrie and Moineau, 2007).

CRISPR/Cas SYSTEM

Some of the natural antiphage mechanisms discussed above have been used extensively for protection of industrial starter cultures in the dairy industry. Not surprisingly, these added selective pressures have led to the emergence of phage mutants capable of overcoming these resistance barriers. Therefore, the search is always on to isolate novel phage resistance mechanisms that can be used alone or in combination with other mechanisms to provide increased phage protection to the starter cultures. Recently, a truly novel phage resistance mechanism was discovered in the dairy bacterium *S. thermophilus* and is named CRISPR/Cas system.

Clustered regularly interspaced short palindromic repeats (CRISPRs) were also identified in a wide array of *Archaea* and *Bacteria* (Godde and Bickerton, 2006; Horvath et al., 2008; Jansen et al., 2002; Makarova et al., 2006b; Vestergaard et al., 2008). CRISPR loci are composed of 12- to 48-bp direct repeats interspersed by nonrepetitive spacers of similar length. The homology of the spacers with mobile DNA elements (such as phages) initiated the hypothesis that these genetic elements, with the CRISPR-associated gene (*cas* gene), are involved in a new foreign DNA immunity mechanism (Bolotin et al., 2005; Haft et al., 2005; Makarova et al., 2006b; Mojica et al., 2005; Pourcel et al., 2005). Nonetheless, it was only recently that the immunity function of CRISPRs was demonstrated (Barrangou et al., 2007; Deveau et al., 2008). When challenged with phages, some *S. thermophilus* cells (also named BIMs) became resistant via the acquisition of a new spacer corresponding to a region of the phage genome (Barrangou et al., 2007; Deveau et al., 2008). The newly acquired spacer provided resistance against all phages containing this DNA sequence in their genome with 100% identity.

This natural phage resistance mechanism is promising for industrial use since CRISPR BIM derivatives can be obtained relatively simply and naturally. Unfortunately, this mechanism is not the final answer against phages. Not all bacteria carry this chromosomally encoded system. Moreover, the arms race is continuing because CRISPR-resistant phage mutants have also been isolated. Virulent phages rapidly evolve resistance to specific CRISPR sequences by acquisition of single nucleotide mutations or deletion of the corresponding CRISPR/spacer targeted region (Deveau et al., 2008). The emergence of resistant phages points to the necessity of continuing to identify new phage resistance mechanisms for long-term phage resistance in important bioindustries.

Engineered Phage Resistance Mechanisms

Novel antiphage mechanisms have also been engineered for specific industrial purposes. The engineered phage resistance mechanisms are based on important phage constituents for viral replication and have been extensively reviewed recently (Coffey and Ross, 2002; Sturino and Klaenhammer, 2004b, 2006). For example, the Per system (phage-encoded resistance) is based on the presence of the phage origin of DNA replication *(ori)* cloned in *trans* on a plasmid. When phages possessing a similar origin of replication are infecting the cell containing Per, the origin of replication on the plasmid competes with the phage *ori* for enzymatic replication machinery. The plasmid rapidly replicates, leading to an augmentation of the efficiency of the mechanism while phage replication is reduced (O'Sullivan et al., 1993; Sing and Klaenhammer, 1993). Per systems were shown to be highly efficient when conjugated with other natural antiphage systems (Abi) or engineered systems such the subunit poisoning system (Sturino and Klaenhammer, 2007). The latter is based on mutated phage DNA primase expressed in *trans* from the plasmid. Since DNA primase must form oligomers to be active, the complex is inactivated by the presence of the mutated dominant allele (subunit poisoning); thus, phage DNA replication is inhibited (Sturino and Klaenhammer, 2007).

Other engineered phage resistance systems are based on gene silencing (antisense RNA), which targets important conserved phage genes (Sturino and Klaenhammer, 2002, 2004a). When those phage genes are cloned downstream of a strong promoter in an antisense orientation, silencing RNA transcripts are produced and result in a lower replication rate for the infecting phages carrying the same gene. The resistance phenotype varies according to the targeted genes, the level of expression, and the length of the antisense RNA (Sturino and Klaenhammer, 2006).

Phage-triggered suicide systems were also considered as an alternative to control phage populations. These suicide systems exploit the phage's tight genetic regulation. A toxic gene is cloned downstream from a phage-inducible promoter on a high-copy-number plasmid (Djordjevic et al., 1997). When the phage infects the bacterial cell, it induces the expression of the toxic protein, which kills the bacteria, again trapping immature phage progeny inside the cell.

All the above-mentioned engineered mechanisms resulted in a weak to medium phage resistance, so several studies have suggested combining them with other antiphage systems to increase their efficiency (Djordjevic and Klaenhammer, 1997; Sturino and Klaenhammer, 2004b, 2006, 2007). Unfortunately, engineered antiphage systems tend to have a narrow range of action since they have been designed with specific phage components, thus limiting their usefulness in the industrial context. Moreover, similarly to natural systems, phages can mutate to become insensitive to these man-made tools. Finally, it is noteworthy that bacteria encoding such engineered systems are considered to be genetically modified organisms. The European community has set strict rules to control the distribution of genetically modified organisms obtained with recombinant DNA technologies. The U.S. Food and Drug Administration allows usage of bioengineered organisms, but the product must be approved prior to its commercialization.

CONCLUSION

Almost all environments can be colonized by bacteria—from the nutrient-rich environment of the human gut to the ocean, where bacteria are slowly multiplying, as well as extreme environments with high temperature or low/high pH. For each of these ecological niches, phages infecting the local microflora have also been found. If a bacterial virus has not been found in a specific environment, one could argue that the sampling process or the isolation procedure should be revisited. Often outnumbering bacteria, phages play an important role in maintenance of the ecological balance.

On the other hand, the presence of phages is not wanted in the bacterial fermentation industry, a man-made environment where the ecological balance is not the rule. The dairy industry has been at the forefront of phage control strategies because it has been dealing with this biotechnological problem for decades. Several approaches have been tried, and many described in this chapter have been (and are still) successful to keep phage populations under control within dairy manufacturing facilities. Nonetheless, the rapidly evolving nature of bacterial viruses was experienced in several circumstances. Consequently, phage infection still poses a risk to any bacterial fermentation process.

Constant phage monitoring is one of the key factors in appropriately managing the risks associated with unwanted virulent phages. Knowledge of the phage diversity within a production unit will also provide essential information to tailor the in-house phage control program. The factors or sources facilitating the emergence of new virulent phages must also be considered as part of a long-term strategy. Albeit complete elimination of phages from an industrial food environment is not achievable at the present time, good manufacturing practices including staff formation, factory design, sanitization procedures, and strain rotations will ensure good control of the phage population and many successful fermented products for consumers.

REFERENCES

Ackermann, H.-W. 2007. 5500 phages examined in the electron microscope. *Arch. Virol.* **152:**227–243.

Ackermann, H.-W. 2009. Basic phage electron microscopy, p. 113–126. *In* M. R. J. Clokie and A. M. Kropinski (ed.), *Bacteriophages: Methods and Protocols*, vol. 1. *Isolation, Characterization, and Interactions.* Humana Press, Springer, New York, NY.

Ackermann, H.-W., E. D. Cantor, A. W. Jarvis, J. Lembke, and J. A. Mayo. 1984. New species definitions in phages of gram-positive cocci. *Intervirology* **22:**181–190.

Ackermann, H.-W., M. S. Dubow, A. W. Jarvis, L. A. Jones, V. N. Krylov, J. Maniloff, J. Rocourt, R. S. Safferman, J. Schneider, L. Seldin, T. Sozzi, P. R. Stewart, M. Werquin, and L. Wunsche. 1992. The species concept and its application to tailed phages. *Arch. Virol.* **124:** 69–82.

Akçelik, M. 1998. A phage DNA injection-blocking type resistance mechanism encoded by chromosomal DNA in *Lactococcus lactis* subsp. *lactis* PLM-18. *Milchwissenschaft* **53:**619–622.

Akçelik, M., P. Sanlibaba, and Ç. Tükel. 2000. Phage resistance in *Lactococcus lactis* subsp. *lactis* strains isolated from traditional fermented milk products in Turkey. *Int. J. Food Sci. Tech.* **35:**473–481.

Allison, G. E., and T. R. Klaenhammer. 1998. Phage resistance mechanisms in lactic acid bacteria. *Int. Dairy J.* **8:**207–226.

Babu, K. S., W. S. Spence, M. R. Monteville, and B. L. Geller. 1995. Characterization of a cloned gene *(pip)* from *Lactococcus lactis* required for phage infection. *Dev. Biol. Stand.* **85:**569–575.

Barrangou, R., C. Frémaux, H. Deveau, M. Richards, P. Boyaval, S. Moineau, D. A. Romero, and P. Horvath. 2007. CRISPR provides acquired resistance against viruses in prokaryotes. *Science* **315:**1709–1712.

Binetti, A. G., B. Del Rio, M. C. Martin, and M. A. Alvarez. 2005. Detection and characterization of *Streptococcus thermophilus* bacteriophages by use of the antireceptor gene sequence. *Appl. Environ. Microbiol.* **71:**6096–6103.

Bissonnette, F., S. Labrie, H. Deveau, M. Lamoureux, and S. Moineau. 2000. Characterization of mesophilic mixed starter cultures used for the manufacture of aged cheddar cheese. *J. Dairy Sci.* **83:**620–627.

Bogosian, G. 2005. Control of bacteriophage in commercial microbiology and fermentation facilities, p. 667–674. *In* R. Calendar (ed.), *The Bacteriophages,* 2nd ed. Oxford University Press, New York, NY.

Bolotin, A., B. Ouinquis, A. Sorokin, and S. D. Ehrlich. 2005. Clustered regularly interspaced short palindrome repeats (CRISPRs) have spacers of extrachromosomal origin. *Microbiology* **151:** 2551–2561.

Bolotin, A., P. Wincker, S. Mauger, O. Jaillon, K. Malarme, J. Weissenbach, S. D. Ehrlich, and A. Sorokin. 2001. The complete genome sequence of the lactic acid bacterium *Lactococcus lactis* ssp. *lactis* IL1403. *Genome Res.* **11:**731–753.

Bossi, L., J. A. Fuentes, G. Mora, and N. Figueroa-Bossi. 2003. Prophage contribution to bacterial population dynamics. *J. Bacteriol.* **185:** 6467–6471.

Bouchard, J. D., and S. Moineau. 2000. Homologous recombination between a lactococcal bacteriophage and the chromosome of its host strain. *Virology* **270:**65–75.

Boucher, I., and S. Moineau. 2001. Phages of *Lactococcus lactis:* an ecological and economical equilibrium. *Recent Res. Devel. Virol.* **3:**243–256.

Braun, V., Jr., S. Hertwig, H. Neve, A. Geis, and M. Teuber. 1989. Taxonomic differentiation of bacteriophages of *Lactococcus lactis* by electron microscopy, DNA-DNA hybridization, and protein profiles. *J. Gen. Microbiol.* **135:**2551–2560.

Breitbart, M., and F. Rohwer. 2005. Here a virus, there a virus, everywhere the same virus? *Trends Microbiol.* **13:**278–284.

Brenner, S., and R. W. Horne. 1959. A negative staining method for high resolution electron microscopy of viruses. *Biochim. Biophys. Acta* **34:**103–110.

Brüssow, H., C. Canchaya, and W. D. Hardt. 2004. Phages and the evolution of bacterial pathogens: from genomic rearrangements to lysogenic conversion. *Microbiol. Mol. Biol. Rev.* **68:**560–602.

Brüssow, H., and R. W. Hendrix. 2002. Phage genomics: small is beautiful. *Cell* **108:**13–16.

Canchaya, C., C. Proux, G. Fournous, A. Bruttin, and H. Brüssow. 2003. Prophage genomics. *Microbiol. Mol. Biol. Rev.* **67:**238–276.

Capra, M. L., A. Quiberoni, and J. A. Reinheimer. 2004. Thermal and chemical resistance of *Lactobacillus casei* and *Lactobacillus paracasei* bacteriophages. *Lett. Appl. Microbiol.* **38:**499–504.

Casjens, S. 2003. Prophages and bacterial genomics: what have we learned so far? *Mol. Microbiol.* **49:** 277–300.

Chibani Azaiez, S. R., I. Fliss, R. E. Simard, and S. Moineau. 1998. Monoclonal antibodies raised against native major capsid proteins of lactococcal c2-like bacteriophages. *Appl. Environ. Microbiol.* **64:**4255–4259.

Chopin, M.-C. 1980. Resistance of 17 mesophilic lactic *Streptococcus* bacteriophages to pasteurization and spray-drying. *J. Dairy Res.* **47:**131–139.

Chopin, M.-C., A. Chopin, and E. Bidnenko. 2005. Phage abortive infection in lactococci: variations on a theme. *Curr. Opin. Microbiol.* **8:**473–479.

Coffey, A., and R. P. Ross. 2002. Bacteriophage-resistance systems in dairy starter strains: molecular analysis to application. *Antonie van Leeuwenhoek* **82:** 303–321.

Deveau, H., R. Barrangou, J. E. Garneau, J. Labonte, C. Fremaux, P. Boyaval, D. A. Romero, P. Horvath, and S. Moineau. 2008. Phage response to CRISPR-encoded resistance in *Streptococcus thermophilus. J. Bacteriol.* **190:**1390–1400.

Deveau, H., S. J. Labrie, M.-C. Chopin, and S. Moineau. 2006. Biodiversity and classification of lactococcal phages. *Appl. Environ. Microbiol.* **72:** 4338–4346.

Djordjevic, G., and T. Klaenhammer. 1997. Bacteriophage-triggered defense systems: phage adaptation and design improvements. *Appl. Environ. Microbiol.* **63:**4370–4376.

Djordjevic, G. M., D. J. O'Sullivan, S. A. Walker, M. A. Conkling, and T. R. Klaenhammer. 1997. A triggered-suicide system designed as a defense against bacteriophages. *J. Bacteriol.* **179:** 6741–6748.

Dupont, K., F. K. Vogensen, and J. Josephsen. 2005. Detection of lactococcal 936-species bacteriophages in whey by magnetic capture hybridization PCR targeting a variable region of receptor-binding protein genes. *J. Appl. Microbiol.* **98:**1001–1009.

Durmaz, E., and T. R. Klaenhammer. 2000. Genetic analysis of chromosomal regions of *Lactococcus lactis* acquired by recombinant lytic phages. *Appl. Environ. Microbiol.* **66:**895–903.

Émond, É., and S. Moineau. 2007. Bacteriophages and food fermentation, p. 93–124. *In* S. McGrath and D. van Sinderen (ed.), *Bacteriophage: Genetics and Molecular Biology.* Caister Academic Press, Norfolk, United Kingdom.

Fineran, P. C., T. R. Blower, I. J. Foulds, D. P. Humphreys, K. S. Lilley, and G. P. C. Salmond. 2009. The phage abortive infection system, ToxIN, functions as a protein-RNA toxin-antitoxin pair. *Proc. Natl. Acad. Sci. USA* **106:**894–899.

Forde, A., and G. F. Fitzgerald. 1999a. Analysis of exopolysaccharide (EPS) production mediated by the bacteriophage adsorption blocking plasmid, pCI658, isolated from *Lactococcus lactis* ssp. *cremoris* HO2. *Int. Dairy J.* **9:**465–472.

Forde, A., and G. F. Fitzgerald. 1999b. Bacteriophage defence systems in lactic acid bacteria. *Antonie van Leeuwenhoek* **76:**89–113.

Garbutt, K. C., J. Kraus, and B. L. Geller. 1997. Bacteriophage resistance in *Lactococcus lactis* engineered by replacement of a gene for a bacteriophage receptor. *J. Dairy Sci.* **80:**1512–1519.

Garvey, P., C. Hill, and G. Fitzgerald. 1996. The lactococcal plasmid pNP40 encodes a third bacteriophage resistance mechanism, one which affects phage DNA penetration. *Appl. Environ. Microbiol.* **62:**676–679.

Geller, B. L., R. G. Ivey, J. E. Trempy, and B. Hettinger-Smith. 1993. Cloning of a chromosomal gene required for phage infection of *Lactococcus lactis* subsp. *lactis* C2. *J. Bacteriol.* **175:**5510–5519.

Giraffa, G. 2004. Studying the dynamics of microbial populations during food fermentation. *FEMS Microbiol. Rev.* **28:**251–260.

Godde, J. S., and A. Bickerton. 2006. The repetitive DNA elements called CRISPRs and their associated genes: evidence of horizontal transfer among prokaryotes. *J. Mol. Evol.* **62:**718–729.

Gullo, M., and P. Giudici. 2008. Acetic acid bacteria in traditional balsamic vinegar: phenotypic traits relevant for starter cultures selection. *Int. J. Food Microbiol.* **125:**46–53.

Haft, D. H., J. Selengut, E. F. Mongodin, and K. E. Nelson. 2005. A guild of 45 CRISPR-associated (Cas) protein families and multiple CRISPR/Cas subtypes exist in prokaryotic genomes. *PLoS Comp. Biol.* **1:**474–483.

Häring, M., X. Peng, K. Brügger, R. Rachel, K. O. Stetter, R. A. Garrett, and D. Prangishvili. 2004. Morphology and genome organization of the virus PSV of the hyperthermophilic archaeal genera *Pyrobaculum* and *Thermoproteus:* a novel virus family, the *Globuloviridae. Virology* **323:** 233–242.

Häring, M., R. Rachel, X. Peng, R. A. Garrett, and D. Prangishvili. 2005a. Viral diversity in hot springs of Pozzuoli, Italy, and characterization of a unique archaeal virus, acidianus bottle-shaped virus, from a new family, the *Ampullaviridae. J. Virol.* **79:**9904–9911.

Häring, M., G. Vestergaard, R. Rachel, L. M. Chen, R. A. Garrett, and D. Prangishvili. 2005b. Independent virus development outside a host. *Nature* **436:**1101–1102.

Horvath, P., D. A. Romero, A.-C. Coute-Monvoisin, M. Richards, H. Deveau, S. Moineau, P. Boyaval, C. Fremaux, and R. Barrangou. 2008. Diversity, activity, and evolution of CRISPR loci in *Streptococcus thermophilus. J. Bacteriol.* **190:**1401–1412.

Hutkins, R. W. 2006. *Microbiology and Technology of Fermented Foods.* Blackwell Publishing, Chicago, IL.

Jansen, R., J. D. A. van Embden, W. Gaastra, and L. M. Schouls. 2002. Identification of genes that are associated with DNA repeats in prokaryotes. *Mol. Microbiol.* **43:**1565–1575.

Jarvis, A. W. 1984. Differentiation of lactic streptococcal phages into phage species by DNA-DNA homology. *Appl. Environ. Microbiol.* **47:**343–349.

Jarvis, A. W., G. F. Fitzgerald, M. Mata, A. Mercenier, H. Neve, I. B. Powell, C. Ronda, M. Saxelin, and M. Teuber. 1991. Species and type phages of lactococcal bacteriophages. *Intervirology* **32:**2–9.

Kang, Y. J., and J. F. Frank. 1989. Biological aerosols—a review of airborne contamination and its measurement in dairy processing plants. *J. Food Prot.* **52:**512–524.

Labrie, S., and S. Moineau. 2000. Multiplex PCR for detection and identification of lactococcal bacteriophages. *Appl. Environ. Microbiol.* **66:**987–994.

Labrie, S. J., and S. Moineau. 2007. Abortive infection mechanisms and prophage sequences significantly influence the genetic makeup of emerging lytic lactococcal phages. *J. Bacteriol.* **189:**1482–1487.

Ledeboer, A. M., S. Bezemer, J. J. de Haard, I. M. Schäffers, C. T. Verrips, C. van Vliet, E. M. Dusterhoft, P. Zoon, S. Moineau, and L. G. Frenken. 2002. Preventing phage lysis of *Lactococcus lactis* in cheese production using a neutralizing heavy-chain antibody fragment from llama. *J. Dairy Sci.* **85:**1376–1382.

Le Marrec, C., D. van Sinderen, L. Walsh, E. Stanley, E. Vlegels, S. Moineau, P. Heinze, G. Fitzgerald, and B. Fayard. 1997. Two groups of bacteriophages infecting *Streptococcus thermophilus* can be distinguished on the basis of mode of packaging and genetic determinants for major structural proteins. *Appl. Environ. Microbiol.* **63:** 3246–3253.

Lembke, J., and M. Teuber. 1981. Serotyping of morphologically identical bacteriophages of lactic streptococci by immunoelectronmicroscopy. *Milchwissenschaft* **36:**10–12.

Lévesque, C., M. Duplessis, J. Labonté, S. Labrie, C. Fremaux, D. Tremblay, and S. Moineau. 2005. Genomic organization and molecular analysis of virulent bacteriophage 2972 infecting an exopolysaccharide-producing *Streptococcus thermophilus* strain. *Appl. Environ. Microbiol.* **71:**4057–4068.

Lu, Z., F. Breidt, H. P. Fleming, E. Altermann, and T. R. Klaenhammer. 2003. Isolation and characterization of a *Lactobacillus plantarum* bacteriophage, ϕJL-1, from a cucumber fermentation. *Int. J. Food Microbiol.* **84:**225–235.

Lucey, M., C. Daly, and G. F. Fitzgerald. 1992. Cell-surface characteristics of *Lactococcus lactis* harboring pCI528, a 46 kb plasmid encoding inhibition of bacteriophage adsorption. *J. Gen. Microbiol.* **138:**2137–2143.

Madera, C., C. Monjardin, and J. E. Suarez. 2004. Milk contamination and resistance to processing conditions determine the fate of *Lactococcus lactis* bacteriophages in dairies. *Appl. Environ. Microbiol.* **70:**7365–7371.

Makarova, K., A. Slesarev, Y. Wolf, A. Sorokin, B. Mirkin, E. Koonin, A. Pavlov, N. Pavlova, V. Karamychev, N. Polouchine, V. Shakhova, I. Grigoriev, Y. Lou, D. Rohksar, S. Lucas, K. Huang, D. M. Goodstein, T. Hawkins, V. Plengvidhya, D. Welker, J. Hughes, Y. Goh, A. Benson, K. Baldwin, J. H. Lee, I. Diaz-Muniz, B. Dosti, V. Smeianov, W. Wechter, R. Barabote, G. Lorca, E. Altermann, R. Barrangou, B. Ganesan, Y. Xie, H. Rawsthorne, D. Tamir, C. Parker, F. Breidt, J. Broadbent, R. Hutkins, D. O'Sullivan, J. Steele, G. Unlu, M. Saier, T. Klaenhammer, P. Richardson, S. Kozyavkin, B. Weimer, and D. Mills. 2006a. Comparative genomics of the lactic acid bacteria. *Proc. Natl. Acad. Sci. USA* **103:**15611–15616.

Makarova, K. S., N. V. Grishin, S. A. Shabalina, Y. I. Wolf, and E. V. Koonin. 2006b. A putative RNA-interference-based immune system in prokaryotes: computational analysis of the predicted enzymatic machinery, functional analogies with eukaryotic RNAi, and hypothetical mechanisms of action. *Biol. Direct* **1:**26.

Mazzocco, A., T. E. Waddell, E. Lingohr, and R. P. Johnson. 2009a. Enumeration of bacteriophages by the direct plating plaque assay, p. 77–80. *In* M. R. J. Clokie and A. M. Kropinski (ed.), *Bacteriophages: Methods and Protocols,* vol. 1. *Isolation, Characterization, and Interactions.* Humana Press, Springer, New York, NY.

Mazzocco, A., T. E. Waddell, E. Lingohr, and R. P. Johnson. 2009b. Enumeration of bacteriophages using the small drop plaque assay system, p. 81–85. *In* M. R. J. Clokie and A. M. Kropinski (ed.), *Bacteriophages: Methods and Protocols,* vol. 1. *Isolation, Characterization, and Interactions.* Humana Press, Springer, New York, NY.

McGrath, S., G. F. Fitzgerald, and D. van Sinderen. 2002. Identification and characterization of phage-resistance genes in temperate lactococcal bacteriophages. *Mol. Microbiol.* **43:**509–520.

McGrath, S., G. F. Fitzgerald, and D. van Sinderen. 2007. Bacteriophages in dairy products: pros and cons. *Biotechnol. J.* **2:**450–455.

McLoughlin, A. J., and C. P. Champagne. 1994. Immobilized cells in meat fermentation. *Crit. Rev. Biotechnol.* **14:**179–192.

Moineau, S. 1999. Applications of phage resistance in lactic acid bacteria. *Antonie van Leeuwenhoek* **76:** 377–382.

Moineau, S., D. Bernier, M. Jobin, J. Hebert, T. Klaenhammer, and S. Pandian. 1993. Production of monoclonal antibodies against the major capsid protein of the *Lactococcus* bacteriophage ul36 and development of an enzyme-linked immunosorbent assay for direct phage detection in whey and milk. *Appl. Environ. Microbiol.* **59:**2034–2040.

Moineau, S., M. Borkaev, B. J. Holler, S. A. Walker, J. K. Kondo, E. R. Vedamuthu, and P. A. Vandenbergh. 1996. Isolation and characterization of lactococcal bacteriophages form cultured buttermilk plants in the United States. *J. Dairy Sci.* **79:**2104–2111.

Moineau, S., J. Fortier, H.-W. Ackermann, and S. Pandian. 1992. Characterization of lactococcal bacteriophages from Québec cheese plants. *Can. J. Microbiol.* **38:**875–882.

Moineau, S., and C. Lévesque. 2005. Control of bacteriophages in industrial fermentations, p. 285–

296. *In* E. Kutter and A. Sulakvelidze (ed.), *Bacteriophages: Biology and Applications*. CRC Press, Boca Raton, FL.

Mojica, F. J. M., C. Diez-Villasenor, J. Garcia-Martinez, and E. Soria. 2005. Intervening sequences of regularly spaced prokaryotic repeats derive from foreign genetic elements. *J. Mol. Evol.* **60:**174–182.

Mudgal, P., F. Breidt, S. R. Lubkin, and K. P. Sandeep. 2006. Quantifying the significance of phage attack on starter cultures: a mechanistic model for population dynamics of phage and their hosts isolated from fermenting sauerkraut. *Appl. Environ. Microbiol.* **72:**3908–3915.

Murata, A., and K. Kitagawa. 1973. Mechanism of inactivation of bacteriophage JL by ascorbic acid. *Agric. Biol. Chem.* **37:**1145–1151.

Murata, A., K. Kitagawa, and R. Saruno. 1971. Inactivation of bacteriophages by ascorbic acid. *Agric. Biol. Chem.* **35:**294–296.

Murata, A., R. Oyadomari, T. Ohashi, and K. Kitagawa. 1975. Mechanism of inactivation of bacteriophage deltaA containing single-stranded-DNA by ascorbic acid. *J. Nutr. Sci. Vitaminol.* (Tokyo) **21:**261–269.

Murata, A., and M. Uike. 1976. Mechanism of inactivation of bacteriophage MS2 containing single-stranded RNA by ascorbic acid. *J. Nutr. Sci. Vitaminol.* (Tokyo) **22:**347–354.

Nes, I. F., and O. Sorheim. 1984. Effect of infection of a bacteriophage in a starter culture during the production of salami dry sausage—a model study. *J. Food Sci.* **49:**337–340.

Neve, H., A. Berger, and K. J. Heller. 1995. A method for detecting and enumerating airborne virulent bacteriophage of dairy starter cultures. *Kieler Milchw. Forsch.* **47:**193–207.

Neve, H., A. Laborius, and K. J. Heller. 2003. Testing of the applicability of battery-powered portable microbial air samplers for detection and enumeration of airborne *Lactococcus lactis* dairy bacteriophages. *Kieler Milchw. Forsch.* **55:**301–315.

Norrby, E. 2008. Nobel Prizes and the emerging virus concept. *Arch. Virol.* **153:**1109–1123.

Ohnishi, M., K. Kurokawa, and T. Hayashi. 2001. Diversification of *Escherichia coli* genomes: are bacteriophages the major contributors? *Trends Microbiol.* **9:**481–485.

Okafor, N. 2007. *Modern Industrial Microbiology and Biotechnology*. Science Publishers, Enfield, NH.

O'Sullivan, D. J., C. Hill, and T. R. Klaenhammer. 1993. Effect of increasing the copy number of bacteriophage origins of replication, in *trans,* on incoming-phage proliferation. *Appl. Environ. Microbiol.* **59:**2449–2456.

Pourcel, C., G. Salvignol, and G. Vergnaud. 2005. CRISPR elements in *Yersinia pestis* acquire new repeats by preferential uptake of bacteriophage DNA, and provide additional tools for evolutionary studies. *Microbiology* **151:**653–663.

Prangishvili, D., R. A. Garrett, and E. V. Koonin. 2006a. Evolutionary genomics of archaeal viruses: unique viral genomes in the third domain of life. *Virus Res.* **117:**52–67.

Prangishvili, D., G. Vestergaard, M. Haring, R. Aramayo, T. Basta, R. Rachel, and R. A. Garrett. 2006b. Structural and genomic properties of the hyperthermophilic archaeal virus ATV with an extracellular stage of the reproductive cycle. *J. Mol. Biol.* **359:**1203–1216.

Quiberoni, A., V. B. Suarez, and J. A. Reinheimer. 1999. Inactivation of *Lactobacillus helveticus* bacteriophages by thermal and chemical treatments. *J. Food Prot.* **62:**894–898.

Rachel, R., M. Bettstetter, B. P. Hedlund, M. Haring, A. Kessler, K. O. Stetter, and D. Prangishvili. 2002. Remarkable morphological diversity of viruses and virus-like particles in hot terrestrial environments. *Arch. Virol.* **147:**2419–2429.

Sanders, M. E. 1999. Phages in industrial fermentations, p. 1244–1248. *In* A. Granoff and R. G. Webster (ed.), *Encyclopedia of Virology,* 2nd ed. Academic Press, San Diego, CA.

Schmidt, R. H., M. M. Vargas, K. L. Smith, and J. J. Jezeski. 1985. The effect of ultra-high temperature milk processing on yogurt texture. *J. Food Process. Preserv.* **9:**235–240.

Schouler, C., C. Bouet, P. Ritzenthaler, X. Drouet, and M. Mata. 1992. Characterization of *Lactococcus lactis* phage antigens. *Appl. Environ. Microbiol.* **58:**2479–2484.

Sijtsma, L., N. Jansen, W. C. Hazeleger, J. T. M. Wouters, and K. J. Hellingwerf. 1990. Cell-surface characteristics of bacteriophage-resistant *Lactococcus lactis* subsp. *cremoris* SK110 and its bacteriophage-sensitive variant SK112. *Appl. Environ. Microbiol.* **56:**3230–3233.

Sing, W., and T. Klaenhammer. 1993. A strategy for rotation of different bacteriophage defenses in a lactococcal single-strain starter culture system. *Appl. Environ. Microbiol.* **59:**365–372.

Stadhouders, J., and G. J. M. Leenders. 1984. Spontaneously developed mixed-strain cheese starters: their behavior towards phages and their use in the Dutch cheese industry. *Neth. Milk Dairy J.* **38:**157–181.

Sturino, J. M., and T. R. Klaenhammer. 2002. Expression of antisense RNA targeted against *Streptococcus thermophilus* bacteriophages. *Appl. Environ. Microbiol.* **68:**588–596.

Sturino, J. M., and T. R. Klaenhammer. 2004a. Antisense RNA targeting of primase interferes with bacteriophage replication in *Streptococcus thermophilus*. *Appl. Environ. Microbiol.* **70:**1735–1743.

Sturino, J. M., and T. R. Klaenhammer. 2004b. Bacteriophage defense systems and strategies for lactic acid bacteria. *Adv. Appl. Microbiol.* **56:**331–378.

Sturino, J. M., and T. R. Klaenhammer. 2006. Engineered bacteriophage-defence systems in bioprocessing. *Nat. Rev. Microbiol.* **4:**395–404.

Sturino, J. M., and T. R. Klaenhammer. 2007. Inhibition of bacteriophage replication in *Streptococcus thermophilus* by subunit poisoning of primase. *Microbiology* **153:**3295–3302.

Suarez, V. B., and J. A. Reinheimer. 2002. Effectiveness of thermal treatments and biocides in the inactivation of Argentinian *Lactococcus lactis* phages. *J. Food Prot.* **65:**1756–1759.

Sullivan, J. J. 1979. Air microbiology and dairy processing. *Aust. J. Dairy Technol.* **34:**133–138.

Trevors, K. E., R. A. Holley, and A. G. Kempton. 1984. Effect of bacteriophage on the activity of lactic-acid starter cultures used in the production of fermented sausage. *J. Food Sci.* **49:**650–651.

Tuncer, Y., and M. Akçelik. 2002. A protein which masks galactose receptor mediated phage susceptibility in *Lactococcus lactis* subsp. *lactis* MPL56. *Int. J. Food Sci. Technol.* **37:**139–144.

Vestergaard, G., S. A. Shah, A. Bize, W. Reitberger, M. Reuter, H. Phan, A. Briegel, R. Rachel, R. A. Garrett, and D. Prangishvili. 2008. *Stygiolobus* rod-shaped virus and the interplay of crenarchaeal rudiviruses with the CRISPR antiviral system. *J. Bacteriol.* **190:**6837–6845.

Villion, M., and S. Moineau. 2009. Bacteriophages of *Lactobacillus. Front. Biosci.* **14:**1661–1683.

Wegmann, U., M. O'Connell-Motherway, A. Zomer, G. Buist, C. Shearman, C. Canchaya, M. Ventura, A. Goesmann, M. J. Gasson, O. P. Kuipers, D. van Sinderen, and J. Kok. 2007. Complete genome sequence of the prototype lactic acid bacterium *Lactococcus lactis* subsp. *cremoris* MG1363. *J. Bacteriol.* **189:**3256–3270.

Whithead, H. R., and G. A. Cox. 1935. The occurence of bacteriophage in lactic streptococci. *N. Z. J. Sci. Technol.* **16:**319.

Whitman, W. B., D. C. Coleman, and W. J. Wiebe. 1998. Prokaryotes: the unseen majority. *Proc. Natl. Acad. Sci. USA* **95:**6578–6583.

Yoon, S. S., R. Barrangou-Poueys, F. Breidt, and H. P. Fleming. 2007. Detection and characterization of a lytic *Pediococcus* bacteriophage from the fermenting cucumber brine. *J. Microbiol. Biotechnol.* **17:**262–270.

PRACTICAL AND THEORETICAL CONSIDERATIONS FOR THE USE OF BACTERIOPHAGES IN FOOD SYSTEMS

Jason J. Gill

11

INTRODUCTION

Bacteriophages represent a promising antibacterial technology that may be useful in the control of a wide variety of nuisance and pathogenic bacteria. One area that is garnering increasing interest is the use of phages for the control of bacteria—particularly food-borne pathogens—in food systems. Experimental phage treatments have been effectively employed at various points in the food processing chain, beginning as early as the production stage, such as reduction of *Campylobacter jejuni* (Loc Carrillo et al., 2005) and *Salmonella* in poultry (Atterbury et al., 2007; Hurley et al., 2008), or treatment of *Escherichia coli* O157:H7 in feedlot cattle (Callaway et al., 2008). This topic is treated thoroughly in chapter 4 of this volume. Alternatively, phages may be applied during the food processing stage, such as in the application of phages to cut fruits (Abuladze et al., 2008; Leverentz et al., 2003) or cheese (Modi et al., 2001); this topic is covered in detail in chapter 7.

The use of phages in food systems is a form of biological control. Based on certain theoretical aspects involving the kinetics of phage replication on their host-cell population, phage-based biological control strategies have been divided into two categories: passive (sometime called inundative) and active (Abedon, 2009a; Payne and Jansen, 2001). In passive phage biocontrol, the strategy is simply to introduce enough phage into the system to ensure that all of the target bacteria adsorb at least one phage in a relatively short time. In this manner, biological control is achieved without the aid of phage replication; the number of progeny phage produced during the treatment may be negligible compared to the input phage. In active biocontrol, a smaller amount of phage are added to the system, and control relies on the ability of the phage to infect, replicate within, and lyse their target bacteria over the course of several generations. This approach, while kinetically more complicated, takes advantage of the ability of phages to replicate at the target site and may allow a relatively small amount of phage to control a much larger bacterial population.

In what kinds of foods or food production and processing situations are phages most likely to be employed? In the context of food preservation and safety, the potential advantages of bacteriophages are their low toxicity (Bruttin and Brüssow, 2005; see also chapter

Jason J. Gill, Department of Biochemistry and Biophysics and Center for Phage Technology, Texas A&M University, 2128 TAMU, College Station, TX 77843.

14 of this volume), high specificity, and ability to function at ambient temperatures and neutral pH. Some limitations of phages in this context are their sensitivity to extremes of heat and pH, and the requirement for their targets to be in the form of vegetative cells (not bacterial spores) found in a hydrated environment for optimal function. Given these characteristics, phages are attractive for use in the control of bacteria during production of food animals and plants, in minimally processed foods, or as a finishing treatment for foods that have undergone a harsher processing regimen. Phages may also be of use in the targeted decontamination of surfaces that cannot be exposed to standard sanitization regimens. Even within this narrowed range of applications, there are still many variables that have the potential to strongly affect the success of a phage-based biocontrol strategy by affecting the stability of the phages, the state of the bacteria, or the ability of phages to access their hosts within the food system. This chapter will endeavor to discuss some of these factors in the context of theory, as well as empirical data from model phage-host systems and from trials of phage biocontrol in foods. While the use of phages specifically for the control of pathogens in production animals will not be discussed extensively here, many of the principles considered are applicable to both systems.

CONSIDERATIONS OF PHAGE ADSORPTION

The Principles of Phage Adsorption

Bacteriophages, like most viruses, initiate their life cycle with the adsorption of the virion to a susceptible host cell, an event that is closely followed by phage genome ejection into the host and the commencement of the viral gene expression program. Phage adsorption, under standard conditions, leads ultimately to the death of the host cell and release of progeny phage, and thus this step is a requirement for both passive and active biological control strategies. According to theoretical considerations of the kinetics of phage-mediated killing of bacteria, a rapid adsorption rate is almost always more efficacious in a passive biocontrol strategy, in which rapid infection and cell lysis are desired. In an active strategy, however, this situation is not as clear-cut; depending on the interplay of other parameters of the system, such as phage latent period, burst size, and bacterial growth rate, extremely high adsorption rates may be beneficial or detrimental. Mathematical models describing these relationships are detailed elsewhere (Abedon, 2009a; Gill, 2008; Payne and Jansen, 2001; Payne and Jansen, 2003; Stopar and Abedon, 2008).

The rate at which phage adsorb to their host is determined by second-order kinetics, as described by the relationship $-dp/dt = kPB$, where k is the phage adsorption rate constant in ml/min, P is the phage concentration, and B is the bacterial concentration. Although this process can be expressed in terms of second-order kinetics, under most conditions the behavior is pseudo-first order: during the adsorption process free phage are eliminated from the system by adsorption to a host bacterium, but the bacterium remains free in the system to adsorb additional phage. This relationship can also be expressed explicitly (here in terms of the rate constant, k) as

$$k = \frac{\ln(P_0/P_t)}{B \times t} \tag{1}$$

where P_0 is the initial concentration of free phage and P_t is the concentration of free phage at time t. One conclusion that can be drawn from this expression is that the concentration of susceptible bacteria, B, and the adsorption rate constant, k, will strongly influence the rate at which free phage are able to locate and adsorb to their hosts. A second conclusion is that, given constant parameters and the assumption that each host cell is able to adsorb an unlimited number of phage, the amount of phage adsorbed by bacteria in time period t is a constant proportion of the initial phage population. Thus, if 50% of the free phage in a given system are adsorbed during time t, the absolute number of phage adsorbed would be 50 if P_0 = 100 PFU, and 50,000 if P_0 = 100,000 PFU.

A second relationship that governs phage adsorption is the phage-to-bacteria ratio, termed the multiplicity of infection (MOI), and the random distribution of phage adsorption events across bacteria. If phage are added to bacteria in a well-mixed system at an MOI of 1 and adsorption is allowed to continue to completion, phage will randomly adsorb to susceptible cells, resulting in some cells adsorbing a single phage, some cells adsorbing multiple phage, and some cells adsorbing no phage. The probability that a bacterium will adsorb *n* number of phage when subjected to a given MOI is determined by a Poisson distribution:

$$P_r = \frac{n^r}{r!} e^{-n} \quad \textbf{(2)}$$

where P_r is the proportion of cells that will adsorb *r* phage, and *n* is the MOI. Note that this stochastic function does not take into account the rate at which phage adsorb to bacteria, but rather indicates the distribution of phage adsorption events across bacteria if the interaction were allowed to run to completion.

Note here that reporting input phage in a system in terms of simple MOI (i.e., the ratio of phage to bacteria) is not always helpful or relevant. As measured in the "optimal" conditions of in vitro culture, phage adsorption rate constants tend to be reported in the range of 10^{-8} to 10^{-11} ml/min. Given a "typical" rate of phage adsorption of 10^{-9} ml/min, it can be calculated from equation 1 that in a system containing 10^8 CFU/ml, the time required for 50% of an inoculated phage population to adsorb to a cell is 6.9 min. In contrast, if this system contained only 10^3 CFU/ml, the time required for 50% phage adsorption would increase to 6.9×10^5 min, or 479 days. We can see from this example that in such a situation the MOI, or ratio of phage to bacteria, can mean very different things depending on the absolute numbers of phage and bacteria. Given one system containing 10^3 CFU/ml and 10^5 PFU/ml, and another containing 10^8 CFU/ml and 10^{10} PFU/ml, both have an MOI of 100, but the rates at which the phage will adsorb to their host will be markedly different between the two. Kasman et al. (2002) have proposed the term MOI_{actual} to denote the number of phage adsorption events per bacterium, taking into consideration the adsorption rate, cell and phage densities, and time allowed for adsorption. In this context, MOI_{actual} is distinct from MOI_{input}, which is simply the traditional definition of MOI, that is, the ratio of phage to bacteria present in a system.

A common misconception in the use of phages as antibacterials is that phages are ineffective in systems where the bacterial density is very low, a common situation in some food systems where a pathogen may be present in numbers of 10^3 CFU/g or less. In systems in which phages greatly outnumber bacteria, even a relatively small *proportion* of phages adsorbed to hosts can represent a *numerically* overwhelming number of phage infections, resulting in a significant reduction in the bacterial population. This relationship is illustrated in Fig. 1, which shows the proportion of a phage population adsorbed to low densities of host cells (ranging from 10 to 10^6 CFU/ml) over time, assuming a binding rate constant, *k*, of 1×10^{-9} ml/min. In this situation, the proportion of phage that have adsorbed to host cells is quite low: at 10^3 CFU/ml, only 0.003% of phage will be adsorbed at 30 min, and 0.14% of phage will be adsorbed after 24 h. However, if the input phage density is 10^8 PFU/ml, even 0.003% adsorption represents 3×10^3 virions adsorbed to cells, which, given a Poisson distribution of infections, results in 95% of the bacteria in the system adsorbing at least one phage in the first 30 min of exposure and 99.75% of cells adsorbing one or more phage after 1 h. This model also illustrates the passive biocontrol strategy, as most of the bacterial killing is predicted to occur within the first round of phage

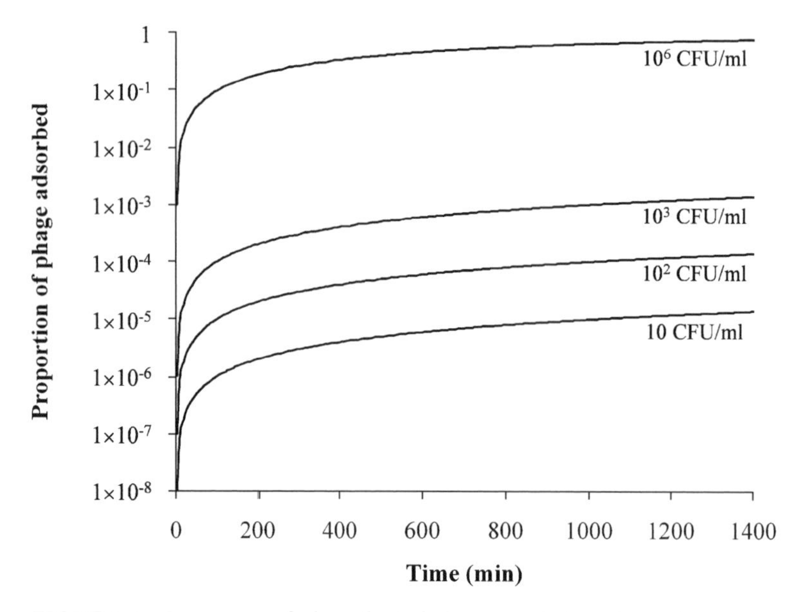

FIGURE 1 Proportion of a hypothetical phage population adsorbed to various concentrations of host bacteria over time, given an adsorption constant *(k)* of 1×10^{-9} ml/min.

infections. This strategy is expected to be the default in the case of the many food-borne pathogens that typically contaminate foods at very low levels (Abedon, 2009a).

These kinetic and stoichiometric aspects of the phage-host interaction are discussed in greater detail elsewhere (e.g., Abedon, 2009a; Kasman et al., 2002). The use of phages as antibacterial agents in any system is governed by these basic principles, and the successful application of phages is dependent, at least in part, on the proper selection of biocontrol strategy (passive or active), phage dose (ensuring that enough phage will be present in the system to control the target bacterial population), and dose timing (ensuring that the applied phage are given sufficient time to act on their hosts).

Factors Affecting Adsorption Rate

As discussed above, the phage adsorption rate constant, *k*, plays a major role in the rate at which phage adsorb to susceptible bacteria. This observable phage adsorption rate is actually the product of a number of factors that are unique to each combination of phage, host, and environment. Upon its release from a lysed host cell, the free phage enters its period of extracellular search for a new host, a process that is driven by diffusion through the medium. Upon collision of a phage with a new host, the phage receptor must irreversibly bind to its ligand on the host cell and initiate DNA ejection and subsequent infection.

Increases in the viscosity of the medium or food matrix are expected to limit the diffusion rates of both phages and bacteria and consequently lower the rate at which phages may adsorb to their hosts. Increasing medium viscosity by addition of agents such as gelatin or agar has historically been reported to inhibit the ability of phages to infect and kill their hosts in vitro (d'Herelle, 1926). Similarly, in an example of a highly defined system, increasing concentrations of agar in models of phage plaque growth limit diffusion of free phages through the medium and thereby limit their ability to access and lyse fresh hosts, as evidenced by reduced plaque sizes (Abedon and Yin, 2009). Bacterial biofilms have been shown to immobilize bacteria and also limit the rate of phage diffusion within the polysaccharide matrix (Doolittle et al., 1996), a topic that will be discussed in greater detail below.

The diffusion-driven extracellular search period of a free phage is followed by the collision of the phage with its host cell, and the probability that this collision will result in irreversible phage adsorption and infection is another important factor determining the success of a phage-based biocontrol strategy. Unlike diffusion, however, which is primarily a physical process, the affinity of the phage receptor for the host-cell surface may be modulated by a large number of factors, including the chemical environment, the temperature, and the physiology and location of the host cell. As an example, the density of the phage receptor expressed on the bacterial surface influences adsorption rate: a cell that is expressing very few receptor molecules will require numerous phage-host collisions before one occurs which is oriented in such a way that the receptor may bind. In the case of coliphage λ, a positive correlation has been observed between the number of receptor molecules expressed on the cell surface and the adsorption rate of the phage (Schwartz, 1976). The influence of growth conditions and other factors on the adsorption of phages is also key in phage-host interactions.

There has been some debate in the literature regarding the relative contributions of factors such as diffusion rate, effective radii, and the probability of irreversible adsorption upon collision to the empirically observable adsorption rate. Detailed discussions on the determination of the theoretical phage-host collision and adsorption rates may be found in the literature (e.g., Hershey, 1957; Koch, 1960; Schwartz, 1976; Tolmach, 1957). In the classical model, phage adsorption is driven by the frequency of collision between the phage and its host, which in turn is driven primarily by diffusion; the probability that a given phage-host collision results in irreversible adsorption is proposed to be high in this system (Adams, 1959; Delbruck, 1940). However, several authors (Hershey, 1957; Koch, 1960; Tolmach, 1957) have suggested that the frequency at which a phage-host collision is converted into a successful adsorption event may be quite low. Furthermore, it has been proposed that at high densities of phages and bacteria—densities that are unlikely in most food systems—diffusion ceases to be a limiting factor, and the probability that collisions convert to irreversible adsorptions becomes the primary determinant of the adsorption rate (Abedon, 2009a; Stent, 1963). This is not an issue that will be resolved here. However, where phages are to be employed for the control of undesirable bacteria in a food system, the practical significance of these relationships is that conditions that enhance the rapid diffusion of phages and their target bacterium (thus enhancing the collision rate), and/or increase the probability that a given collision will result in phage adsorption, are expected to increase the rate of phage-mediated bacterial killing. Conversely, conditions that inhibit diffusion or impede phage adsorption upon collision are expected to reduce the rate at which phages infect and kill their target cells.

CONSIDERATIONS OF PHAGE STABILITY

Like all antimicrobial agents that may be added to foods as part of a preservation strategy, phages face the practical concern of their stability in the food matrix or other treated environments. There are numerous properties of foods, both intrinsic and extrinsic, that may affect the stability and hence the efficacy of phages. Many of these factors are also able to interact with one another, resulting in either positive or negative overall effects on phage stability. Phage stability in the environment where phages are expected to operate is a requirement for both the passive and active strategies of phage biocontrol; in both cases, rapid loss of viable phages is expected to reduce overall efficacy. Strategies to protect phages from various environmental stresses, particularly those encountered in animal systems, are covered in chapter 12 of this volume.

While all of the possible parameters that affect phage stability and their interactions cannot be covered in this space, some general properties will be outlined here. When stored in "ideal" medium, such as sterile broth, phages have half-lives that have been measured in terms of days, even if held at 37°C (De Paepe and Taddei, 2006), although stability is expected to vary widely for different phage isolates and storage media. The ionic environment is also known to affect stability of phages: the presence of free divalent cations, particularly calcium (Ca^{2+}) and magnesium (Mg^{2+}), at low-millimolar concentrations tends to significantly stabilize phages (Adams, 1949, 1959).

A common extrinsic factor applied to foods is temperature, which could manifest in the form of some type of thermal preservation treatment or as prolonged storage at ambient or refrigerated temperatures, or as freezing. A number of commonly studied phages have been examined for their inactivation behavior in the presence of thermal stress: coliphage T7, for example, was found to be unstable at temperatures above 55 to 60°C, while phage T4 was inactivated at temperatures above 65 to 70°C (Adams, 1949), and these thermal limits for phage stability appear to be typical for a number of phages of mesophilic bacteria (Adams, 1959). Furthermore, when considering environmental interactions, a situation that is unfavorable for the viability of free phages will generally be exacerbated by an increase in temperature. Given that phage inactivation by many agents follows first-order kinetics, this process will be accelerated at increased temperatures (Adams, 1949). Obviously, any strategy involving the use of phages in foods that are subjected to extreme thermal processing must be designed so that the phages are added after the thermal step; some standard preservation methods, such as canning (where the food product is heated after sealing), are likely to be incompatible with the use of phages.

Phages are generally more stable at low (e.g., refrigeration) temperatures than at ambient temperatures. The effects of freezing on phages are dependent on both the phage and the medium in which the phages are frozen. Freezing of aqueous phage suspensions to low or ultralow temperatures (<-70°C) in the presence of a cryoprotectant such as glycerol has been shown to be effective at preserving various phages (Clark et al., 1962; Fortier and Moineau, 2009). A number of food matrix components, such as soluble proteins, amino acids, and sugars, may act as cryoprotectants and allow phages to be frozen with little loss of activity (Lillford and Holt, 2002). However, freezing in the absence of cryoprotectant material (e.g., freezing in water or dilute ionic solutions), and especially subjecting phages to multiple freeze-thaw cycles under these conditions, is expected to rapidly inactivate phages, as similar techniques are used to intentionally disrupt phages for laboratory study (Lavigne et al., 2009).

The sensitivity of phages to extremes of pH has been well documented. Phages tend to be stable in a neutral pH range, ranging from pH 5 to 8 (Adams, 1959). A mixture of lytic phages against *Listeria monocytogenes* was found to be relatively stable over a period of 7 days on cut honeydew melon (pH, 5.77) but was inactivated rapidly on apple slices (pH, 4.37) (Leverentz et al., 2003); however, it is not clear if this inactivation could be entirely attributable to pH or to some other factor, as some plant extracts have been shown to be virucidal against phages (de Siqueira et al., 2006). In this same study, it was reported that these phages were ineffective against *L. monocytogenes* in liquid culture at pH below 5.5 (Leverentz et al., 2003). Smith et al. (1987) found that their series of *E. coli* phages were stable for 1 h at 37°C at pH values of 3.5 and above; at pH 3 and below, the phages were inactivated, although considerable variation between phage strains was observed. As noted above, the pH tolerance of phages may be influenced by temperature; phages tend to become inactivated more slowly and may effectively tolerate more extreme pH at lower temperatures (Adams, 1959).

Phages would not be expected to function well in highly desiccated foods for several reasons, including the lack of free water available in the dried food to mediate phage diffusion and the deleterious effects of desiccation on phage stability. In general, phages tend to be unstable when desiccated in the absence of any protectant substances, although this sensitivity varies among phages (Adams, 1959). Coliphage T1, for example, is notoriously tolerant to desiccation, so much so that it can become difficult to eliminate from an environment once it becomes established. Phages may remain stable in some desiccated systems, depending on the food matrix and the means of desiccation. Lyophilization, for example, avoids exposure to high temperatures, and lyophilization of phages in a matrix of sterile nonfat milk is a common means of preserving certain phage stocks, although phage stability is highly variable in this state, with some phages experiencing multiple-log-unit reductions (Clark, 1962; Fortier and Moineau, 2009). Desiccation of phage λ-gt11 in the presence of buffer alone was found to result in an immediate drop of 95% in detectable phage titer; this loss in viability was improved to 80% by addition of protectants such as trehalose (Jepson and March, 2004). While the phages would not be expected to remain active in the desiccated environment for the reasons described above, they would be expected to regain functionality upon rehydration of the matrix.

The sequestration or inactivation of phages in the food matrix is also a significant concern. Phage particles that become immobilized by adsorption to, or entrapment within, food matrix components are often considered to be inactivated, although this is not necessarily the case: adsorbed virions may well retain their infectivity even while they are immobilized, as long as the receptor apparatus and tail remain unhindered. Phages may be nonspecifically adsorbed to surfaces on the basis of charge or hydrophobic interactions (Gerba, 1984). Raw whole and skim milk, but not pasteurized skim milk, was found to inactivate temperate *Staphylococcus aureus* phages by approximately 1 log unit over 7 h of incubation (Garcia et al., 2009). Similarly, *S. aureus* phage K was found to be inactivated by whole raw milk (J. J. Gill, unpublished observations). It is not clear from these data, however, whether the phages were truly inactivated or merely sequestered.

CONSIDERATIONS OF PHAGE HOST RANGE

The specificity of phages for a particular host species or strain is often cited as both an advantage and a disadvantage in the use of phages as antibacterials. In the case of the control of food-borne pathogens, phages present a viable option, as the diversity of pathogens requiring control is limited to relatively few bacterial species. In contrast, the use of phages as a general antimicrobial strategy in foods is severely restricted simply by the diversity of potential target bacteria in a given open environment. Therefore, phages are best suited for food systems where control of specific known bacterial species is required, rather than as a replacement for a broad-spectrum antimicrobial or sanitizer.

For example, the use of phages in the control of food spoilage has been studied by Greer and Dilts (1990), who attempted to use a phage cocktail infecting *Pseudomonas* to control spoilage of beef. While these phages were able to reduce the amount of *Pseudomonas* colonizing the treated samples, spoilage was not inhibited, presumably due to the continued growth and activity of the non-*Pseudomonas* bacteria colonizing the meat, which were unaffected by the applied phages. For the control of food-borne pathogenic bacteria, which are more likely to represent diversity at the strain rather than the species or genus level, phage cocktails have been employed experimentally to broaden the host range of the phage preparation and reduce the probability of bacterial resistance arising during treatment (Abuladze et al., 2008; Callaway et al., 2008; Leverentz et al., 2004). A commercially viable example of a phage-based antimicrobial designed for use in human medicine, targeting a relatively narrow range of human-pathogenic bacteria, contains a cocktail of over 20 phages (Markoishvili et al., 2002).

CONSIDERATIONS OF THE BACTERIAL HABITAT

The ability of a bacteriophage to infect, produce progeny, and lyse its host cell entails a complex interplay between the proteins of the host cell and those of the infecting phage, and it would be expected that alterations in host protein expression would therefore alter the reproductive program of an infecting phage. It has been known since the early days of phage biology that the state of the bacterial host can play an important role in phage infection (d'Herelle, 1926; Delbruck, 1940). However, the effects of the host physiological state or host growth conditions on phage infection are topics that have received little focused study. Therefore, only limited evidence on the effects of any type of growth conditions on phage infection may be presented here; hopefully these examples will provide the reader with some insight into the nature of these effects.

While some foods present a nutrient-rich environment that supports rapid bacterial growth, this is certainly not the case in all food environments. Food preservation methods and factors intrinsic to the food itself can induce many forms of stress and sublethal injury in bacteria (recently reviewed by Wesche et al., 2009). As a rule, significantly stressed or injured bacteria greatly reduce or arrest their rate of division. Stressing factors may be low nutrient content (starvation), low water activity (a_w), osmotic stress, pH extremes, modified atmospheres, thermal processing treatments, or the addition of other antimicrobial compounds such as organic acids or nitrates. Aside from properties that constitute a direct insult to a bacterium, intrinsic factors of the food such as the available nutrients and matrix structure may affect the state of the bacterial cell in ways that can affect the ability of a phage to successfully infect its host.

Biofilms

Biofilms are consortia of viable microorganisms attached to a surface and embedded in an extracellular polymeric matrix (Costerton et al., 1987; Kumar and Anand, 1998). Biofilms are a topic of perennial concern in the food industry, as they can cause problems in food production facilities ranging from equipment fouling to the shedding of pathogenic microbes into food products. In addition to the persistence of biofilms due to their attachment to surfaces, bacteria residing within a biofilm community tend to exhibit increased resistance to biocides (Kumar and Anand, 1998; Wong, 1998). Phages have been examined as a method for the control of bacteria in biofilms in vitro, primarily in systems involving *Pseudomonas aeruginosa* or *E. coli* and their phages. Biofilms present two potential barriers to the activity of phages: the extracellular matrix, which may act as a barrier to phage access; and the physiological state of the bacterial cell, which may be incompatible with the completion of the phage reproductive program.

The ability of phages to penetrate a biofilm and lyse the resident bacteria appears to be dependent on the bacteria and the structure of the biofilm that they form. Biofilms are, in at least some cases, able to nonspecifically trap and accumulate phages from the bulk liquid environment (Briandet et al., 2008). Once phages enter the biofilm matrix, they appear to be largely insulated from the effects of fluid shear in the bulk liquid (Doolittle et al., 1996). Thick biofilms composed of *P. aeruginosa* appeared to be largely impenetrable to phage E79 (Doolittle et al., 1996); however, some biofilms will permit enough diffusion of phages into the matrix to allow infection of the resident cells. In continuous-flow experiments using rich media, biofilms composed of

E. coli were found to be sensitive to infection by phage T4, with the majority of the phage introduced into these systems absorbed quickly into the biofilm, and biofilm populations reduced by approximately 2 to 5 log units over the course of 5 h (Doolittle et al., 1995). In *E. coli* biofilms grown under conditions of limited glucose, phage T4 was able to effect a transient reduction in bacterial populations of approximately 2 log units and to visibly lower the density of cells within the biofilms (Corbin et al., 2001). The replication of phages on cells growing within a biofilm appears to share some characteristics with phage replication in the formation of plaques in soft agar; microscopic tracking of infections by an engineered *lacZ*-expressing phage T4 strain in an *E. coli* biofilm showed progressive infections radiating in a spherical pattern from central foci, in a manner resembling plaque growth (Doolittle et al., 1996).

A number of phages have been described that express enzymes capable of degrading the bacterial exopolysaccharide capsule (Lindberg, 1973), which comprises the bulk of some biofilms. Such enzymes may be broadly referred to as capsular depolymerases, although they are sometimes referred to by more specific nomenclature based on their substrate specificity. In at least some depolymerase-encoding phages, the protein containing the depolymerase domain is assembled into the virion as a part of the phage tail and allows phage adsorption and penetration into the capsule. In the well-studied phages K29 (Bayer et al., 1979), K1E, and K1-5 (Leiman et al., 2007), the phages adsorb to bacterial exopolysaccharide and essentially "tunnel" their way through the capsule toward the cell outer membrane by processively degrading the capsular polysaccharide. In addition to being assembled into the virion, however, the depolymerase is often produced in excess and a significant amount of the free enzyme is released into the environment upon cell lysis. Phages that possess such enzymes have been shown to disrupt biofilm structure (Hanlon et al., 2001; Hughes et al., 1998), and a phage T7 strain genetically engineered to express a soluble capsule-degrading enzyme during its infection program has also been shown to effectively disrupt biofilms (Lu and Collins, 2007). These findings suggest a role for such phages in the control of biofilms in food environments.

It would appear that under ideal conditions, phages are able to significantly impact bacteria residing in biofilms; however, the heterogeneity of both the biofilm matrix and the physiological state of the bacteria appears to allow only partial elimination of the population. The metabolic activity of individual bacteria within a biofilm has been observed to be highly variable (Sternberg et al., 1999), and it has been well established that bacteria that are in stationary phase are unable to support productive infections in some phages (see below for a more detailed discussion on host physiology and phage infection). Most biofilms in natural, open environments are also highly likely to be polymicrobial, consisting of a bacterial consortium that may be composed of many unrelated species. In this case the population of nontarget bacteria in a biofilm is expected to provide some protection against phages by increasing biofilm heterogeneity (Wilkinson, 2001).

Changes to the Cell Envelope

The initial point of interaction between a phage and its host is the outer surface of the cell envelope, upon which is presented the surface receptors required for phage adsorption. As bacteria respond to their environments, however, the envelope may undergo alterations that can include the up- or down-regulation of certain membrane-associated proteins, capsule formation, or lipopolysaccharide (LPS) composition. The nature and magnitude of these adaptations are as variable as the bacteria that may express them and the environments they could inhabit, and only some will have an explicit effect on the ability of a phage to initiate infection. This author is not aware of any specific references relating to adaptive bacterial cell surface changes in response to food environments that have been shown to have an impact on phage infection. There are, however, several such changes that have been shown to have such effects in other systems, a few of which will be presented here in order to illustrate a range of potential adaptations which could be encountered.

In a case that illustrates the regulation of receptor density, phage λ utilizes the outer membrane protein LamB as its receptor (Randall-Hazelbauer and Schwartz, 1973). This porin serves as a part of the *E. coli* maltodextrin transport system, and its copy number in the outer membrane is strongly induced by the presence of maltose and cyclic AMP (Ryter et al., 1975) and to a lesser extent by trehalose (Klein and Boos, 1993). Schwartz (1976) demonstrated that cells grown in maltose or glucose + cyclic AMP adsorbed λ at a higher rate than did cells grown in glucose and that this difference correlated with the amount of LamB extracted from the outer membrane of the cell. Similarly, *E. coli* phage T5 uses FhuA (formerly TonA), a component of the iron uptake system, as its receptor. FhuA is the normal ligand for *E. coli* ferrichrome, which sequesters iron from the environment and delivers it to the cell by binding with FhuA on the cell surface (Hantke and Braun, 1978). Cells cultured under low-nutrient conditions were found to adsorb phage T5 with much lower efficiency than glucose-cultured cells in the presence of ferrichrome (Hantke and Braun, 1978). The interpretation of this result was that ferrichrome occupied FhuA for long periods under starvation conditions, which competitively inhibited the binding of T5 to FhuA. Under nutrient-rich conditions, the ferrichrome-FhuA complexes were able to turn over rapidly and therefore provided little inhibition to T5 adsorption. As a final example, the outer membrane LPS of gram-negative bacteria is able to serve as a receptor for many phages, whose tail proteins are able to recognize particular sugar residues in the LPS O-antigen or core region (Lindberg, 1973). Phage Felix-O1 of *Salmonella* utilizes the LPS core as its receptor, and its adsorption is inhibited by the presence of long O-antigen chains (Lindberg and Holme, 1969). In *Salmonella enterica* serovar Anatum, LPS O-antigen density is reduced at lower growth temperatures, and culture at 30°C has been shown to render this bacterium more susceptible to infection by Felix-O1 (McConnell and Wright, 1979). These examples of the effects of the extracellular environment on phage susceptibility are of course specialized for these phage-host systems, and a temperate phage such as λ is not likely to be used in a practical application in any food system. However, these observations provide examples of the effect of the host environment on the ability of phages to function, in this case by modulating the phage adsorption rate.

Some food-borne pathogenic bacteria are able to elaborate an extracellular polysaccharide matrix, in the form of either a bound capsule or a loosely attached slime layer, which is able to interfere with phage adsorption by occluding the cell surface (Scholl et al., 2005; Wilkinson and Holmes, 1979). In *S. aureus,* a pathogen capable of causing food-borne intoxications, capsule production has been shown to be affected by growth medium (Poutrel et al., 1997; Sutra et al., 1990), oxygen tension (Dassy and Fournier, 1996), or CO_2 levels (Herbert et al., 2001). In an observation with direct application to the microbiology of foods, *S. aureus* strains have been shown to increase CP5 capsule production when cultured on skim milk medium but not Columbia agar, although this effect was found to be variable between strains (Sutra et al., 1990). This variability of capsule production was demonstrated by Poutrel et al. (1997) to be both interstrain and intrastrain, with significant heterogeneity in capsule production by cells within the same population as measured by flow cytometry. This environmental regulation of *S. aureus* capsule production provides another example of the potential bacterial surface effects that could be observed when bacteria are present in food systems.

In the case of *S. aureus* cultured in milk, a phenomenon of surface occlusion by substances not generated by the bacterium itself has been observed (Gill et al., 2006; O'Flaherty et al., 2005). In this case, the bacterium appears to adsorb high-molecular-weight proteins and fragments of milk fat globule membrane present in the whey fraction of raw milk, which prevent phage adsorption by coating the cell surface and sterically hindering the ability of the phage to bind to its receptor (Gill et al., 2006). This inhibition was found to be highly source dependent and in several cases was so severe that the phage were essentially unable to lyse the bacteria in the presence of raw milk fractions. This phenomenon appears to be unique to *S. aureus* and may be mediated by the diverse array of mammalian tissue and serum protein-binding adhesins expressed on the cell surface, which could mediate accumulation of the inhibiting layer (Joh et al., 1999; Wann et al., 2000). Such pronounced inhibition of phage-mediated killing in raw milk is not encountered by the phages of lactic acid bacteria, which frequently proliferate in dairy cultures, resulting in fermentation failures (McGrath et al., 2007). *Salmonella enterica* serovar Enteritidis cultured in systems of both raw and pasteurized milk was likewise sensitive to phage (Modi et al., 2001).

Bacterial Growth State

Many food matrices and food processing environments present habitats that are injurious or stressful to bacteria. Such stresses could manifest in the form of pH extremes, thermal shocks, or nutrient limitation, to name only a few. Whatever the cause of the stress or injury, bacteria in such stressful environments tend to exhibit altered physiology and slowed growth rates compared to unstressed cells, with the difference proportional to the magnitude of the stress (Wesche et al., 2009).

Stationary phase, in its most general sense, refers to a bacterial population that has ceased increasing in number (Kolter et al., 1993), a state that could reflect a balance in the rate of cell division and cell death, or simply a cessation of cell division. By this criterion, the stationary-phase phenotype may be induced by any number of factors ranging from injury to starvation, but it cannot be assumed that the details of the stationary-phase cellular response will be equivalent between stresses (Huisman et al., 1996). In common practice, induction of stationary phase in a bacterial culture is most often achieved via depletion of nutrients, typically carbon, in the growth medium, by either extended culturing or physical transfer of cells into a nutrient-deficient medium. In studies of the effects of host metabolic state on phage infection, starvation is the most common condition studied and therefore will be examined here. The reader should bear in mind, however, that many other intrinsic and extrinsic factors of the food matrix, such as pH, preservatives, osmotic stress, or thermal shock, are able to induce states in bacteria that may be phenotypically described as stationary phase but metabolically are quite diverse, and the response of phage infections in these situations may be different from those observed in starvation-induced stationary-phase bacteria.

The conventional model of phage infection and replication is one in which the host is in a state of balanced growth in a nutrient-rich medium. In the case of well-studied systems such as phage T4 and *E. coli,* stationary-phase cultures, obtained by either extended culture or transfer to nutrient-poor medium, are able to adsorb the phage and permit DNA ejection, but do not support the completion of the phage lytic program (Kutter et al., 1994; Schrader et al., 1997). *E. coli* cells cultured far into stationary phase and then infected with phage T4 have been reported to be unable to produce progeny even when returned to rich culture conditions, suggesting that T4 infection impedes the ability of such stationary-phase cells to return to exponential growth (Abedon, 1990). A detailed study of host physiology and phage infection determined the effects of bacterial doubling time on the development of phage T4 (Hadas et al., 1997). In this case, host doubling time was modulated by culturing cells in media containing different carbon sources, which produced cell-doubling times ranging from 23 min (glucose) to 136 min (acetate). The authors found that as the bacterial growth rate decreased, the phage adsorption rate and burst size also decreased, and the phage latent period became extended. These observations were interpreted in the context of the relationship between cell growth rate, cell size, and intracellular DNA and RNA content. This relationship was established by Schaechter et al. (1958) in *S. enterica* serovar Typhimurium (reviewed in Bremer and Dennis, 1996) and shows that, at a given temperature, bacterial cell size and the capacity for protein synthesis increase as a function of growth rate, largely independent of the nutritional content of the medium. The implication of this work—at least as it applies to T4 infecting an *E. coli* host—is that the rate-limiting factors in phage replication are (i) the cell size (slowly growing, small cells adsorb phage more slowly) and (ii) the capacity of the cell to synthesize phage proteins (intracellular RNA and ribosome content increase with growth rate). This relationship was not absolute, however, as in one case a medium that supported rapid bacterial growth supported lower phage replication than expected, indicating that other nutritional factors may affect the ability of bacteria to support phage replication.

Under conditions of extreme nutrient limitation, phage development tends to be suppressed even further. As is the case with phage T4, *E. coli* cultures infected with phages P1*vir* or λ are unable to produce detectable levels of progeny when deprived of a carbon source (Łoś et al., 2007). In this case, phage titers were not observed to increase during the period of starvation, but the addition of glucose to the medium allowed phage reproduction to resume after up to 40 h of starvation, suggesting that the phage genome was maintained in a dormant state in at least some cells. This apparent arrest of phage multiplication under starvation conditions is not universal, however; perhaps the most notable exception to this rule is the ability of *E. coli* phage T7 to function in starved cells. This behavior is easily seen in T7 plaques grown on soft agar overlays, which, unlike plaques formed by other phages such as T4, continue to expand long after the host cells in the lawn have largely reached stationary phase (Yin, 1991). Phage T7 has also been shown to infect and lyse its *E. coli* B host under starvation conditions (Krueger et al., 1975; Schrader et al., 1997). Other phages are also able to infect and replicate within their hosts under starvation conditions: *P. aeruginosa* phages UT1 and ACQ have been demonstrated to adsorb to and replicate within hosts that had been starved for 24 h, at the expense of a 2- to 3-fold increase in latent period and 10- to 20-fold reduction in burst size (Schrader et al., 1997). In the same study, however, *P. aeruginosa* phage BLB was found to be unable to replicate in starved host cells, although it was able to adsorb. In a dramatic example of a phage overcoming the starvation barrier, phage ACQ was able to infect and lyse *P. aeruginosa* cells that had been held in starvation conditions for 5 years, exhibiting a mean burst size similar to that observed in cells starved for 24 h (Schrader et al., 1997).

Finally, phage infecting a starved host cell may enter into a state that has been described as pseudolysogeny. The term "pseudolysogeny" has been used to describe a number of possible phage-host associations (Abedon, 2009b); however, in this case it denotes an unstable association of the phage genome with its host cell during which no progeny phage are produced. Ripp and Miller (1997, 1998) have reported that the virulent phage UT1 is able to enter into a pseudolysogenic state when its *P. aeruginosa* host is grown for extended periods in nutrient-poor medium. The virulent coliphage T3 has also been reported to exhibit pseudolysogenic behavior when infecting starved cells (Krueger et al., 1975). While there are many details of the starvation-induced pseudolysogenic state that remain unresolved, the phenomenon is in practical terms somewhat equivalent to a phage infection with a significantly extended latent period; the major difference in this case is that cells in such a state may still be able to divide, thereby allowing an infected cell—or more precisely, its progeny—to escape infection by partitioning away from the infecting phage genome.

These data suggest that the requirement of vigorous host metabolism for phage infection is highly phage dependent. Some phages, such as T4, require a dividing and metabolically active host for replication, while others are able to overcome or largely ignore the stationary-phase barrier and execute their infection programs, albeit typically with lower fecundity than is observed under rapid-growth conditions. Phages such as T4, despite their high virulence and well-characterized biology, may be less attractive as biocontrol agents due to their apparent dependence on host physiology, in contrast to starvation-independent phages as represented by phage T7. A further conclusion is that, when selecting phages for use as an antimicrobial in a food system, consideration should be given to the expected physiological state of the target bacterium and how this may affect the ability of the phages to infect and lyse these cells. In many foods, nutrition is unlikely to be a limiting factor for bacterial growth, but a variety of sublethal processing procedures, intrinsic food factors, or added antimicrobial agents may induce a state that is physiologically similar to the starvation-induced stationary phase described here. It is not clear whether cells stressed by such factors would elicit a response to phage infection equivalent to that observed in experimentally starved cells.

In the case of a passive phage-based biocontrol strategy, the ability of a phage to produce progeny is secondary to its ability to simply adsorb to its host and initiate an infection. As long as the infected cell is unable to recover from the phage infection, the cell is effectively removed from the population of organisms in the treated environment. This behavior becomes practically relevant only in situations where an active control strategy, which relies on the ability of phage to replicate for its success, is implemented, or in a situation where the infection fails to be lethal to the host and the bacterium is able to resume normal growth.

Temperature

Phages, like bacteria, exhibit optimal growth temperatures for their replication. When *E. coli* cells are grown at an optimal temperature, infected with bacteriophages, and then shifted to a lower temperature, extension of the phage latent period and reduction in burst size are observed. The system of phage T4 and *E. coli,* for example, is highly sensitive to temperature. Shifting cells from 37 to 19°C immediately following infection resulted in an over sixfold increase in latent period and approximately 50% reduction of the normal burst size (Maaloe, 1950). T4-infected cells held at 16°C were observed to produce no phage progeny for at least 240 min of observation (Yanagida et al., 1984). Bacteriophage λ and a λ*vir* mutant have been reported to exhibit similar behavior upon shifting from 37 to 25°C, with total phage yield dropping by 60 to 70% and unspecified but significant extensions in latent periods on the order of three- to fourfold that observed at 37°C (Groman and Suzuki, 1962). At temperatures elevated above 37°C, phage productivity was found to drop off rapidly, with infected cells incubated at 45°C yielding only 2 to 5% of the phage observed in the 37°C control (Groman and Suzuki, 1962).

Many foods which are considered to be potential vectors for food-borne pathogens are held at temperatures well below the 20°C threshold where most work in well-defined coliphage systems has been conducted. Such foods are typically held in refrigerated storage conditions ranging from 2 to 10°C or frozen conditions as low as −20°C. At temperatures significantly below 0°C, most foods and their associated microbial flora are frozen, either partially or completely immobilized in ice. It is reasonable to assume that phages in such systems are able to diffuse only very slowly or not at all, thus limiting adsorption; their bacterial hosts—particularly mesophilic bacteria such as *E. coli* or *Salmonella*—are essentially under metabolic arrest, halting the production of progeny phage even if adsorption and DNA ejection occur successfully. Therefore the systems that primarily require consideration in terms of the behavior of phages are those in the refrigerated temperature range, held at temperatures below 10°C and above the point at which they freeze.

A number of systems involving psychrotrophic bacteria have been studied and found to support productive phage infections. Phages infecting bacteria of the genus *Pseudomonas,* many of which are psychrotrophic, have been found to infect and replicate within such hosts at low temperatures. A series of phages infecting various *Pseudomonas* species was found to successfully infect and replicate in their hosts at both 3.5 and 25°C, although the phage latent period was reduced and burst size increased at the higher incubation temperature; in this system latent periods ranged from 6 to 12 h, with burst sizes of 6 to 25 PFU at 3.5°C (Olsen et al., 1968). In contrast to this finding, a *Pseudomonas geniculata* phage was reported to exhibit a latent period of 30 min and a burst size of 100 to 150 PFU when cultured at 25°C, and a latent period of 6 h but a burst size of 300 to 350 PFU when cultured at 3.5°C (Olsen, 1967). Similar experiments in other systems have shown the ability of psychrotrophic bacteria to support phage adsorption and replication at low temperatures, as determined directly in broth culture (Greer, 1982; Whitman and Marshall, 1971) and by the ability of such phages to form plaques (Greer, 1983).

The concept that low temperatures will slow the rate at which phages are able to infect new hosts and replicate within them is, in itself, rather intuitive. While the kinetic parameters of many systems involving phages of food-borne pathogens have yet to be detailed at refrigeration temperatures, some work indicates that phages are indeed able to reduce the viable bacterial counts in foods under these conditions. In broth culture, *L. monocytogenes* phage LH7 was shown to delay the growth of its host when incubated at 7°C, indicating that this phage was able to infect the bacterium with at least some efficiency (Dykes and Moorhead, 2002). Phages A511 and P100 infecting *L. monocytogenes* were found to significantly reduce bacterial counts in seafood, cheese brine, chocolate milk, and vegetable leaf surfaces, even though these systems were held at 6°C (Guenther et al., 2009). In examinations of phages against spoilage-associated bacteria in meats, a pool of *Pseudomonas* phages was found to reduce the number of spoilage-associated pseudomonads on the surface of beef held at 8°C (Greer and Dilts, 1990), and *Brochothrix thermosphacta* phages reduced counts of this spoilage organism in pork held at 2°C (Greer and Dilts, 2002). These examples illustrate the ability of phages to infect their hosts at refrigeration temperatures, but these systems are remarkable in that they involve bacteria that are psychrotrophic and known to be able to grow at such temperatures. Among mesophilic food-borne pathogens, phages e11/2, pp01, and e4/1c of *E. coli* O157:H7 were found to be essentially inactive against this pathogen in broth culture at 12°C, even though they were able to rapidly reduce viable bacterial counts at both 30 and 37°C and could form plaques at 12°C (O'Flynn et al., 2004). Based on the details of this study, it is possible that phages were inhibited at an early step of the infection cycle, such as adsorption, as high numbers of viable cells could be recovered from these phage-inoculated cultures despite challenge at a high MOI. In contrast, Abuladze et al. (2008) have reported the ability of a three-phage cocktail to reduce *E. coli* O157:H7 contamination on the surfaces of fresh fruits, vegetables, and meat held at 10°C. This study used different phage isolates than those used by O'Flynn et al. (2004), which may possess various abilities to adsorb to and lyse their hosts under refrigerated conditions.

One conclusion that may be drawn from this work is that, in a typical refrigerated food system with otherwise adequate nutrition, the ability of a phage to replicate on a mesophilic host may be limited compared to similar conditions held closer to the bacterial growth optimum; if a mesophilic bacterium under refrigerated conditions is capable of adsorbing phage, the latent period at low temperature may be greatly extended in a manner that appears to be dependent on the phage isolates used. The effects of low temperature on the interactions between psychrotrophic bacteria and their phages may be of less concern. Given the common use of refrigerated storage in food preservation, particularly for minimally processed foods, it would therefore be advantageous to consider the behavior and efficacy of phage isolates under refrigerated conditions when selecting phages for use as a biocontrol strategy. Similar to the situation noted above in the case of starved host cells, the ability of phages to propagate in cells at refrigerated temperatures is of less consequence if a passive biocontrol strategy is being employed. As long as a bacterium has been successfully infected by phage, and the invading phage DNA is maintained by the cell until it returns to a permissive temperature, the cell would not be expected to survive when the phage resumed its lytic growth program.

CONCLUSION

The use of phages as antibacterials for the biological control of food-borne pathogens is a timely and attractive tool. Phages are nontoxic, and some specific phage products have already received regulatory approval for addition to foods as antimicrobials (Federal Register, 2006). As research in this field intensifies, careful consideration must be given to the selection of food systems that possess thermal, chemical, and physical parameters compatible with the use of phages; phages must be selected that are appropriate for a given food system in terms of their stability and ability to infect under the conditions encountered in the food environment. Highly polymicrobial systems in which the target bacteria are expected to be inaccessible to phages, highly stressed, or sublethally injured may in fact be poor candidates for the use of phages, as would foods that possess intrinsic or extrinsic properties that are incompatible with phage survival. In contrast, minimally processed foods which are considered to be at risk from contamination by a narrow spectrum of food-borne pathogens, and which are also intrinsically conducive to bacterial and phage survival, could be the most promising initial candidates for phage-based biocontrol.

ACKNOWLEDGMENTS

I thank Stephen T. Abedon for his helpful comments on the manuscript and for providing a prepublication draft of his work, and also Ry Young and Ian Molineux for their insights during the early preparation of this work. I also thank the editors of this volume for their guidance and for the opportunity to contribute. This work is supported by U.S. Public Health Service grant AI064512.

REFERENCES

Abedon, S. T. 1990. The ecology of bacteriophage T4. Ph.D. dissertation. University of Arizona, Tucson, AZ.

Abedon, S. T. 2009a. Kinetics of phage-mediated biocontrol of bacteria. *Foodborne Pathog. Dis.* **6:** 807–815.

Abedon, S. T. 2009b. Disambiguating bacteriophage pseudolysogeny: an historical analysis of lysogeny, pseudolysogeny, and the phage carrier state, p. 285–307. *In* H. T. Adams (ed.), *Contemporary Trends in Bacteriophage Research.* Nova Science Publishers, Hauppauge, NY.

Abedon, S. T., and J. Yin. 2009. Bacteriophage plaques: theory and analysis. *Methods Mol. Biol.* **501:**161–174.

Abuladze, T., M. Li, M. Y. Menetrez, T. Dean, A. Senecal, and A. Sulakvelidze. 2008. Bacteriophages reduce experimental contamination of hard surfaces, tomato, spinach, broccoli, and ground beef by *Escherichia coli* O157:H7. *Appl. Environ. Microbiol.* **74:**6230–6238.

Adams, M. H. 1949. The stability of bacterial viruses in solutions of salts. *J. Gen. Physiol.* **32:**579–594.

Adams, M. H. 1959. *Bacteriophages.* Interscience Publishers, New York, NY.

Atterbury, R. J., M. A. Van Bergen, F. Ortiz, M. A. Lovell, J. A. Harris, A. De Boer, J. A. Wagenaar, V. M. Allen, and P. A. Barrow. 2007. Bacteriophage therapy to reduce *Salmonella* colonization of broiler chickens. *Appl. Environ. Microbiol.* **73:**4543–4549.

Bayer, M. E., H. Thurow, and M. H. Bayer. 1979. Penetration of the polysaccharide capsule of *Escherichia coli* (Bi161/42) by bacteriophage K29. *Virology* **94:**95–118.

Bremer, H., and P. P. Dennis. 1996. Modulation of chemical composition and other parameters of the cell by growth rate, p. 1553–1569. *In* F. C. Neidhardt (ed.), Escherichia coli *and* Salmonella, 2nd ed., vol. 2. ASM Press, Washington, DC.

Briandet, R., P. Lacroix-Gueu, M. Renault, S. Lecart, T. Meylheuc, E. Bidnenko, K. Steenkeste, M. N. Bellon-Fontaine, and M. P. Fontaine-Aupart. 2008. Fluorescence correlation spectroscopy to study diffusion and reaction of bacteriophages inside biofilms. *Appl. Environ. Microbiol.* **74:**2135–2143.

Bruttin, A., and H. Brüssow. 2005. Human volunteers receiving *Escherichia coli* phage T4 orally: a safety test of phage therapy. *Antimicrob. Agents Chemother.* **49:**2874–2878.

Callaway, T. R., T. S. Edrington, A. D. Brabban, R. C. Anderson, M. L. Rossman, M. J. Engler, M. A. Carr, K. J. Genovese, J. E. Keen, M. L. Looper, E. M. Kutter, and D. J. Nisbet. 2008. Bacteriophage isolated from feedlot cattle can reduce *Escherichia coli* O157:H7 populations in ruminant gastrointestinal tracts. *Foodborne Pathog. Dis.* **5:**183–191.

Clark, W. A. 1962. Comparison of several methods for preserving bacteriophages. *Appl. Microbiol.* **10:** 466–471.

Clark, W. A., W. Horneland, and A. G. Klein. 1962. Attempts to freeze some bacteriophages to ultralow temperatures. *Appl. Microbiol.* **10:**463–465.

Corbin, B. D., R. J. McLean, and G. M. Aron. 2001. Bacteriophage T4 multiplication in a glucose-limited *Escherichia coli* biofilm. *Can. J. Microbiol.* **47:**680–684.

Costerton, J. W., K. J. Cheng, G. G. Geesey, T. I. Ladd, J. C. Nickel, M. Dasgupta, and T. J. Marrie. 1987. Bacterial biofilms in nature and disease. *Annu. Rev. Microbiol.* **41:**435–464.

Dassy, B., and J. M. Fournier. 1996. Respiratory activity is essential for post-exponential-phase production of type 5 capsular polysaccharide by *Staphylococcus aureus. Infect. Immun.* **64:**2408–2414.

Delbruck, M. 1940. Adsorption of bacteriophage under various physiological conditions of the host. *J. Gen. Physiol.* **23:**631–642.

De Paepe, M., and F. Taddei. 2006. Viruses' life history: towards a mechanistic basis of a trade-off between survival and reproduction among phages. *PLoS Biol.* **4:**e193.

de Siqueira, R. S., C. E. Dodd, and C. E. Rees. 2006. Evaluation of the natural virucidal activity of teas for use in the phage amplification assay. *Int. J. Food Microbiol.* **111:**259–262.

d'Herelle, F. 1926. *The Bacteriophage and Its Behavior.* Williams & Wilkins, Baltimore, MD.

Doolittle, M. M., J. J. Cooney, and D. E. Caldwell. 1995. Lytic infection of *Escherichia coli* biofilms by bacteriophage T4. *Can. J. Microbiol.* **41:** 12–18.

Doolittle, M. M., J. J. Cooney, and D. E. Caldwell. 1996. Tracing the interaction of bacteriophage with bacterial biofilms using fluorescent and chromogenic probes. *J. Ind. Microbiol.* **16:**331–341.

Dykes, G. A., and S. M. Moorhead. 2002. Combined antimicrobial effect of nisin and a listeriophage against *Listeria monocytogenes* in broth but not in buffer or on raw beef. *Int. J. Food Microbiol.* **73:** 71–81.

Federal Register. 2006. Food additives permitted for direct addition to food for human consumption; bacteriophage preparation. *Fed. Regist.* **71:** 47729–47732.

Fortier, L. C., and S. Moineau. 2009. Phage production and maintenance of stocks, including expected stock lifetimes. *Methods Mol. Biol.* **501:**203–219.

Garcia, P., C. Madera, B. Martinez, A. Rodriguez, and J. Evaristo Suarez. 2009. Prevalence of bacteriophages infecting *Staphylococcus aureus* in dairy samples and their potential as biocontrol agents. *J. Dairy Sci.* **92:**3019–3026.

Gerba, C. P. 1984. Applied and theoretical aspects of virus adsorption to surfaces. *Adv. Appl. Microbiol.* **30:**133–168.

Gill, J. J. 2008. Modeling of bacteriophage therapy, p. 439–464. *In* S. T. Abedon (ed.), *Bacteriophage Ecology: Population Growth, Evolution, and Impact of Bacterial Viruses.* Cambridge University Press, Cambridge, United Kingdom.

Gill, J. J., P. M. Sabour, K. E. Leslie, and M. W. Griffiths. 2006. Bovine whey proteins inhibit the interaction of *Staphylococcus aureus* and bacteriophage K. *J. Appl. Microbiol.* **101:**377–386.

Greer, G. G. 1982. Psychrotrophic bacteriophages for beef spoilage pseudomonads. *J. Food Prot.* **45:** 1318–1325.

Greer, G. G. 1983. Psychrotrophic *Brocothrix thermosphacta* bacteriophages isolated from beef. *Appl. Environ. Microbiol.* **46:**245–251.

Greer, G. G., and B. D. Dilts. 1990. Inability of a bacteriophage pool to control beef spoilage. *Int. J. Food Microbiol.* **10:**331–342.

Greer, G. G., and B. D. Dilts. 2002. Control of *Brochothrix thermosphacta* spoilage of pork adipose tissue using bacteriophages. *J. Food Prot.* **65:**861–863.

Groman, N. B., and G. Suzuki. 1962. Temperature and lambda phage reproduction. *J. Bacteriol.* **84:**431–437.

Guenther, S., D. Huwyler, S. Richard, and M. J. Loessner. 2009. Virulent bacteriophage for efficient biocontrol of *Listeria monocytogenes* in ready-to-eat foods. *Appl. Environ. Microbiol.* **75:**93–100.

Hadas, H., M. Einav, I. Fishov, and A. Zaritsky. 1997. Bacteriophage T4 development depends on the physiology of its host *Escherichia coli. Microbiology* **143:**179–185.

Hanlon, G. W., S. P. Denyer, C. J. Olliff, and L. J. Ibrahim. 2001. Reduction in exopolysaccharide viscosity as an aid to bacteriophage penetration through *Pseudomonas aeruginosa* biofilms. *Appl. Environ. Microbiol.* **67:**2746–2753.

Hantke, K., and V. Braun. 1978. Functional interaction of the *tonA/tonB* receptor system in *Escherichia coli*. *J. Bacteriol.* **135:**190–197.

Herbert, S., S. W. Newell, C. Lee, K. P. Wieland, B. Dassy, J. M. Fournier, C. Wolz, and G. Doring. 2001. Regulation of *Staphylococcus aureus* type 5 and type 8 capsular polysaccharides by CO_2. *J. Bacteriol.* **183:**4609–4613.

Hershey, A. D. 1957. Bacteriophages as genetic and biochemical systems. *Adv. Virus Res.* **4:**25–61.

Hughes, K. A., I. W. Sutherland, and M. V. Jones. 1998. Biofilm susceptibility to bacteriophage attack: the role of phage-borne polysaccharide depolymerase. *Microbiology* **144:**3039–3047.

Huisman, G. W., D. A. Siegele, M. M. Zambrano, and R. M. Kolter. 1996. Morphological and physiological changes during stationary phase, p. 1672–1682. *In* F. C. Neidhardt (ed.), Escherichia coli *and* Salmonella, 2nd ed., vol. 2. ASM Press, Washington, DC.

Hurley, A., J. J. Maurer, and M. D. Lee. 2008. Using bacteriophages to modulate *Salmonella* colonization of the chicken's gastrointestinal tract: lessons learned from in silico and in vivo modeling. *Avian Dis.* **52:**599–607.

Jepson, C. D., and J. B. March. 2004. Bacteriophage lambda is a highly stable DNA vaccine delivery vehicle. *Vaccine* **22:**2413–2419.

Joh, D., E. R. Wann, B. Kreikemeyer, P. Speziale, and M. Hook. 1999. Role of fibronectin-binding MSCRAMMs in bacterial adherence and entry into mammalian cells. *Matrix Biol.* **18:**211–223.

Kasman, L. M., A. Kasman, C. Westwater, J. Dolan, M. G. Schmidt, and J. S. Norris. 2002. Overcoming the phage replication threshold: a mathematical model with implications for phage therapy. *J. Virol.* **76:**5557–5564.

Klein, W., and W. Boos. 1993. Induction of the λ receptor is essential for effective uptake of trehalose in *Escherichia coli*. *J. Bacteriol.* **175:**1682–1686.

Koch, A. L. 1960. Encounter efficiency of coliphage-bacterial interaction. *Biochim. Biophys. Acta* **39:**311–318.

Kolter, R., D. A. Siegele, and A. Tormo. 1993. The stationary phase of the bacterial life cycle. *Annu. Rev. Microbiol.* **47:**855–874.

Krueger, D. H., W. Presber, S. Hansen, and H. A. Rosenthal. 1975. Biological functions of the bacteriophage T3 SAMase gene. *J. Virol.* **16:** 453–455.

Kumar, C. G., and S. K. Anand. 1998. Significance of microbial biofilms in food industry: a review. *Int. J. Food Microbiol.* **42:**9–27.

Kutter, E., E. Kellenberger, K. Carlson, S. Eddy, J. Neitzel, L. Messinger, J. North, and B. Guttman. 1994. Effects of bacterial growth conditions and physiology on T4 infection, p. 406–418. *In* J. D. Karam (ed.), *Molecular Biology of Bacteriophage T4*. ASM Press, Washington, DC.

Lavigne, R., P. J. Ceyssens, and J. Robben. 2009. Phage proteomics: applications of mass spectrometry. *Methods Mol. Biol.* **502:**239–251.

Leiman, P. G., A. J. Battisti, V. D. Bowman, K. Stummeyer, M. Muhlenhoff, R. Gerardy-Schahn, D. Scholl, and I. J. Molineux. 2007. The structures of bacteriophages K1E and K1-5 explain processive degradation of polysaccharide capsules and evolution of new host specificities. *J. Mol. Biol.* **371:**836–849.

Leverentz, B., W. S. Conway, M. J. Camp, W. J. Janisiewicz, T. Abuladze, M. Yang, R. Saftner, and A. Sulakvelidze. 2003. Biocontrol of *Listeria monocytogenes* on fresh-cut produce by treatment with lytic bacteriophages and a bacteriocin. *Appl. Environ. Microbiol.* **69:**4519–4526.

Leverentz, B., W. S. Conway, W. Janisiewicz, and M. J. Camp. 2004. Optimizing concentration and timing of a phage spray application to reduce *Listeria monocytogenes* on honeydew melon tissue. *J. Food Prot.* **67:**1682–1686.

Lillford, P. J., and C. B. Holt. 2002. In vitro uses of biological cryoprotectants. *Philos. Trans. R. Soc. Lond. B Biol. Sci.* **357:**945–951.

Lindberg, A. A. 1973. Bacteriophage receptors. *Annu. Rev. Microbiol.* **27:**205–241.

Lindberg, A. A., and T. Holme. 1969. Influence of O side chains on the attachment of the Felix O-1 bacteriophage to *Salmonella* bacteria. *J. Bacteriol.* **99:**513–519.

Loc Carrillo, C., R. J. Atterbury, A. El-Shibiny, P. L. Connerton, E. Dillon, A. Scott, and I. F. Connerton. 2005. Bacteriophage therapy to reduce *Campylobacter jejuni* colonization of broiler chickens. *Appl. Environ. Microbiol.* **71:**6554–6563.

Łoś, M., P. Golec, J. M. Łoś, A. Weglewska-Jurkiewicz, A. Czyz, A. Węgrzyn, G. Węgrzyn, and P. Neubauer. 2007. Effective inhibition of lytic development of bacteriophages λ, P1 and T4 by starvation of their host, *Escherichia coli*. *BMC Biotechnol.* **7:**13.

Lu, T. K., and J. J. Collins. 2007. Dispersing biofilms with engineered enzymatic bacteriophage. *Proc. Natl. Acad. Sci. USA* **104:**11197–11202.

Maaloe, O. 1950. Some effects of changes of temperature on intracellular growth of the bacterial virus T4r. *Acta Pathol. Microbiol. Scand.* **27:**680–694.

Markoishvili, K., G. Tsitlanadze, R. Katsarava, J. G. Morris, Jr., and A. Sulakvelidze. 2002. A novel sustained-release matrix based on biodegradable poly(ester amide)s and impregnated with bacteriophages and an antibiotic shows promise in management of infected venous stasis ulcers and other poorly healing wounds. *Int. J. Dermatol.* **41:** 453–458.

McConnell, M., and A. Wright. 1979. Variation in the structure and bacteriophage-inactivating capacity of *Salmonella anatum* lipopolysaccharide as a function of growth temperature. *J. Bacteriol.* **137:** 746–751.

McGrath, S., G. F. Fitzgerald, and D. van Sinderen. 2007. Bacteriophages in dairy products: pros and cons. *Biotechnol. J.* **2:**450–455.

Modi, R., Y. Hirvi, A. Hill, and M. W. Griffiths. 2001. Effect of phage on survival of *Salmonella enteritidis* during manufacture and storage of cheddar cheese made from raw and pasteurized milk. *J. Food Prot.* **64:**927–933.

O'Flaherty, S., A. Coffey, W. J. Meaney, G. F. Fitzgerald, and R. P. Ross. 2005. Inhibition of bacteriophage K proliferation on *Staphylococcus aureus* in raw bovine milk. *Lett. Appl. Microbiol.* **41:** 274–279.

O'Flynn, G., R. P. Ross, G. F. Fitzgerald, and A. Coffey. 2004. Evaluation of a cocktail of three bacteriophages for biocontrol of *Escherichia coli* O157:H7. *Appl. Environ. Microbiol.* **70:**3417–3424.

Olsen, R. H. 1967. Isolation and growth of psychrophilic bacteriophage. *Appl. Microbiol.* **15:**198.

Olsen, R. H., E. S. Metcalf, and J. K. Todd. 1968. Characteristics of bacteriophages attacking psychrophilic and mesophilic pseudomonads. *J. Virol.* **2:**357–364.

Payne, R. J., and V. A. Jansen. 2001. Understanding bacteriophage therapy as a density-dependent kinetic process. *J. Theor. Biol.* **208:**37–48.

Payne, R. J., and V. A. Jansen. 2003. Pharmacokinetic principles of bacteriophage therapy. *Clin. Pharmacokinet.* **42:**315–325.

Poutrel, B., P. Rainard, and P. Sarradin. 1997. Heterogeneity of cell-associated CP5 expression on *Staphylococcus aureus* strains demonstrated by flow cytometry. *Clin. Diagn. Lab. Immunol.* **4:**275–278.

Randall-Hazelbauer, L., and M. Schwartz. 1973. Isolation of the bacteriophage lambda receptor from *Escherichia coli. J. Bacteriol.* **116:**1436–1446.

Ripp, S., and R. V. Miller. 1997. The role of pseudolysogeny in bacteriophage-host interactions in a natural freshwater environment. *Microbiology* **143:** 2065–2070.

Ripp, S., and R. V. Miller. 1998. Dynamics of the pseudolysogenic response in slowly growing cells of *Pseudomonas aeruginosa. Microbiology* **144:**2225–2232.

Ryter, A., H. Shuman, and M. Schwartz. 1975. Integration of the receptor for bacteriophage lambda in the outer membrane of *Escherichia coli:* coupling with cell division. *J. Bacteriol.* **122:**295–301.

Schaechter, M., O. Maaloe, and N. O. Kjeldgaard. 1958. Dependency on medium and temperature of cell size and chemical composition during balanced growth of *Salmonella typhimurium. J. Gen. Microbiol.* **19:**592–606.

Scholl, D., S. Adhya, and C. Merril. 2005. *Escherichia coli* K1's capsule is a barrier to bacteriophage T7. *Appl. Environ. Microbiol.* **71:**4872–4874.

Schrader, H. S., J. O. Schrader, J. J. Walker, T. A. Wolf, K. W. Nickerson, and T. A. Kokjohn. 1997. Bacteriophage infection and multiplication occur in *Pseudomonas aeruginosa* starved for 5 years. *Can. J. Microbiol.* **43:**1157–1163.

Schwartz, M. 1976. The adsorption of coliphage lambda to its host: effect of variations in the surface density of receptor and in phage-receptor affinity. *J. Mol. Biol.* **103:**521–536.

Smith, H. W., M. B. Huggins, and K. M. Shaw. 1987. Factors influencing the survival and multiplication of bacteriophages in calves and in their environment. *J. Gen. Microbiol.* **133:**1127–1135.

Stent, G. S. 1963. *Molecular Biology of Bacterial Viruses.* W. H. Freeman & Co., San Francisco, CA.

Sternberg, C., B. B. Christensen, T. Johansen, A. Toftgaard Nielsen, J. B. Andersen, M. Givskov, and S. Molin. 1999. Distribution of bacterial growth activity in flow-chamber biofilms. *Appl. Environ. Microbiol.* **65:**4108–4117.

Stopar, D., and S. T. Abedon. 2008. Modeling bacteriophage population growth, p. 389–414. *In* S. T. Abedon (ed.), *Bacteriophage Ecology: Population Growth, Evolution, and Impact of Bacterial Viruses.* Cambridge University Press, Cambridge, United Kingdom.

Sutra, L., P. Rainard, and B. Poutrel. 1990. Phagocytosis of mastitis isolates of *Staphylococcus aureus* and expression of type 5 capsular polysaccharide are influenced by growth in the presence of milk. *J. Clin. Microbiol.* **28:**2253–2258.

Tolmach, L. J. 1957. Attachment and penetration of cells by viruses. *Adv. Virus Res.* **4:**63–110.

Wann, E. R., S. Gurusiddappa, and M. Hook. 2000. The fibronectin-binding MSCRAMM FnbpA of *Staphylococcus aureus* is a bifunctional protein that also binds to fibrinogen. *J. Biol. Chem.* **275:**13863–13871.

Wesche, A. M., J. B. Gurtler, B. P. Marks, and E. T. Ryser. 2009. Stress, sublethal injury, resuscitation, and virulence of bacterial foodborne pathogens. *J. Food Prot.* **72:**1121–1138.

Whitman, P. A., and R. T. Marshall. 1971. Characterization of two psychrophilic *Pseudomonas* bacteriophages isolated from ground beef. *Appl. Microbiol.* **22:**463–468.

Wilkinson, B. J., and K. M. Holmes. 1979. *Staphylococcus aureus* cell surface: capsule as a barrier to bacteriophage adsorption. *Infect. Immun.* **23:**549–552.

Wilkinson, M. H. 2001. Predation in the presence of decoys: an inhibitory factor on pathogen control by bacteriophages or bdellovibrios in dense and diverse ecosystems. *J. Theor. Biol.* **208:**27–36.

Wong, A. C. 1998. Biofilms in food processing environments. *J. Dairy Sci.* **81:**2765–2770.

Yanagida, M., Y. Suzuki, and T. Toda. 1984. Molecular organization of the head of bacteriophage T_{even}: underlying design principles. *Adv. Biophys.* **17:**97–146.

Yin, J. 1991. A quantifiable phenotype of viral propagation. *Biochem. Biophys. Res. Commun.* **174:** 1009–1014.

ENCAPSULATION AND CONTROLLED RELEASE OF BACTERIOPHAGES FOR FOOD ANIMAL PRODUCTION

Qi Wang and Parviz M. Sabour

12

INTRODUCTION

Antibiotic resistance is a microbiological food safety issue and a current worldwide public health concern. As detailed in chapter 1, overuse of chemical antibiotics has created a situation that has been referred to by some as "the decline of the antibiotic era" (Jagusztyn-Krynicka and Wyszyńska, 2008; Wright, 2007). A significant amount of antibiotics are used in the production of poultry, swine, and beef cattle for growth promotion, feed efficiency, and veterinary medicine, which in the United States amounts to about 13 million pounds (70% of the total) of antibiotics used annually (Mellon et al., 2001; Wise et al., 1998; Angulo et al., 2004; Tauxe, 1997; see also chapter 1). Livestock and poultry have the potential to become a reservoir of antibiotic resistance (McEwen and Fedorka-Cray, 2002; Partridge et al., 2009; Akwar et al., 2008; Gillings et al., 2008), which could then be transmitted to humans (Voss et al., 2005; Khanna et al., 2008). In response, the use of antibiotic growth promoters in animal feed has been banned in the European Union since 2006 (European Commission, 2005) and is under debate in many other countries. By removing antibiotic growth promoters from feeds, there has been an overall reduction in the use of antibiotics for animal production, but the amount used for therapeutic purposes has increased significantly (DANMAP 2003, 2004).

Reducing pathogens at the primary production level is essential to enhance the microbiological safety of meat products (Rajić et al., 2007). A wide range of protocols for disease prophylaxis in food animal production and growth promotion have been investigated, including vaccination, competitive exclusion (probiotics), bacteriophage therapy, and health-promoting feed additives (Aarestrup and Jenser, 2007; Poole et al., 2007; C. F. M. De Lange, J. Pluske, J. Gong, and C. M. Nyachoti, presented at the 11th International Symposium on Digestive Physiology of Pigs, Costa Daurada, Spain, 20 to 22 May 2009). To date, however, no single approach has been shown to be as effective as antibiotic growth promoters in industrial animal production.

Bacteriophages can eliminate pathogenic bacteria without disturbing the normal, beneficial intestinal microbiota, and the potential use of phage therapy in food animal production has been investigated and reviewed ex-

Qi Wang and Parviz M. Sabour, Guelph Food Research Centre, Research Branch, Ministry of Agriculture and Agri-Food Canada, 93 Stone Road West, Guelph, Ontario N1G 5C9, Canada.

Bacteriophages in the Control of Food- and Waterborne Pathogens
Edited by Parviz M. Sabour and Mansel W. Griffiths, © 2010 ASM Press, Washington, DC

tensively (for examples, see Sulakvelidze and Barrow, 2005; Johnson et al., 2008; and O'Flaherty et al., 2009; see also chapter 4, this volume). Some recent trials have demonstrated the efficacy of bacteriophages to reduce diseases caused by enteric pathogens. It has been applied successfully in chickens (Berchieri et al., 1991; Goode et al., 2003; Toro et al., 2005) and pigs (O'Flynn et al., 2006) against *Salmonella,* in chickens and calves (Barrow et al., 1998; Smith and Huggins, 1983; Smith et al., 1987; Rozema et al., 2009) against *Escherichia coli,* and in poultry against *Clostridium perfringens* (Siragusa et al., 2004). Potential benefits and questions on the experimental approaches used in some of the early trials, as well as new genomic approaches to identify phage-based antimicrobials (Liu et al., 2004) and phage enzymes (Fischetti, 2006; chapter 8, this volume), are under critical evaluation (Summers, 2001; Thiel, 2004; Fruciano and Bourne, 2007; Hanlon, 2007; Brüssow, 2007).

There is an emphasis on antibiotic use when controlling food pathogens in the food chain, particularly at the animal production and processing stages. Traditionally, once food has been contaminated by pathogens, it is disposed of at enormous economic cost to the food industry. Therefore, even though chemical antibiotics will continue to play a vital role in human and veterinary medicine, the goal is to reduce their use in food animal production and processing by replacing them with alternative approaches whenever possible.

Practical routes of administration of phages are an important limitation for acceptance in food animal production since phages could be affected by components present at the site of administration (Abedon, 2008). This was shown by the inhibition of phage-bacterial interaction in the presence of milk and whey (O'Flaherty et al., 2005; Gill et al., 2006a, 2006b). Similar interaction-influencing factors have been observed in processed foods (Greer, 2005). As reviewed in chapter 4, current administration routes include intramuscular injection (Huff et al., 2003, 2006), aerosol spray (Huff et al., 2002; Goode et al., 2003), and oral delivery. The low cost and ease of administration makes oral delivery the preferred method in the field. However, naked phages may become inactivated when exposed to different gastrointestinal (GI) fluids, reducing their efficacy. Thus, protecting phages against factors that affect their stability and infectivity during passage through the GI tract, and allowing for targeted release where most of the pathogens reside, may increase their biocontrol efficiency.

FACTORS THAT INFLUENCE PHAGE STABILITY AND INFECTIVITY

Phages are ubiquitous in nature and are able to survive in diverse environments. Nevertheless, depending on the phage type, their stability and infectivity upon oral administration could be affected by the chemical composition of the animal's GI tract, including factors such as gastric acidity, plant secondary metabolites (Atterbury et al., 2007; Topping et al., 1980; Andreatti Filho et al., 2007; Vispo and Karasov, 1997), immunological factors (Inchley, 1969; Geier et al., 1973; Merril et al., 2005), and in very young mammals, the presence of colostral antibodies in the GI tract (Adams, 1959; Ackermann and DuBow, 1987; Smith et al., 1987). There are also suggestions that the pharmacokinetic effectiveness of phage therapy to treat bacterial infection is dependent on timing of its administration (Payne and Jansen, 2003).

More information is needed, however, on the fate and infectivity of phages during their passage through the GI tract when mixed with complex feed and enzymes and under a wide range of acidity. For application in the animal production industry, long-term storage of bacteriophages in a desiccated state is another requirement for addition to animal feed. This is a tedious process that differs between phages. Because of their importance in industrial production and application, the effects of pH and temperature will be discussed further. Development of protective formulations such as encapsulation of phages could overcome some of the problems described here. Advances in the encapsulation of animal viruses for oral vaccination (Nechaeva, 2002) and live bacteria for increased viability during storage provide a good foundation of knowledge for phage encapsulation (Crittenden et al., 2006; Champagne and Fustier, 2007). Methods for encapsulation of proteins and other macromolecules in biodegradable microspheres have been developed successfully (Langer and Folkman, 1976). The technology provides an alternative means to protect phages destined for therapeutic purposes and could facilitate the direct addition of phages to animals (Puapermpoonsiri et al., 2009).

In oral application, GI fluids (such as stomach acid) may diffuse into the microcapsules, causing phage inactivation. In addition, the rate of release of phage from microspheres during GI passage could be influenced by mechanical movements of different segments of the GI tract (such as the gizzard in chickens). Thus, the microcapsule chemical composition needs to be optimized to withstand the physiological environment of the GI tract in order to be released at a specific intestinal site. Likewise, it is necessary to know where in the GI tract the specific target pathogen colonizes and to control the phage release at that specific location. Structurally, GI tracts of food animals are different (Moran, 1982). In simple monogastric animals such as swine, the phage enters the stomach with complex food, and here proteins are partially digested with gastric juice that contains pepsin and HCl, maintaining a pH at about 2.0. The gastric food mixture (chyme) is then passed into the small intestine through the duodenum, where pancreatic juice neutralizes the acidic chyme, and food is degraded by various digestive enzymes. Bile produced by the liver also enters the duodenum, emulsifying fats for digestion. In the small intestine, food digestion continues in its different segments, the jejunum and the ileum. Intestinal refuse then enters the large intestine, forming the feces containing bacteria and solids. In swine, the small cecum is located at the junction of the ileum of the small intestine and the extensive large intestine. The structure and size of different segments of the GI tract vary in mono- and digastric (ruminant) animals. In fowl, the crop serves to store food while the proventriculus and gizzard are the sites of gastric digestion and grinding, respectively (Moran, 1982; Jacob and Pescatore, 2009). Chickens have a shorter intestine than swine but increase digestion and absorption by back-and-forth peristalsis. Their large intestine consists of two large ceca, where the majority of microbial activity occurs, and a short colon. The cloaca in fowls is comparable to the swine rectum, where storage and excretion of fecal material occurs.

In ruminants such as cattle, the digestive tract is more complex (see http://edis.ifas.ufl.edu/ds061). It includes the rumen, the largest of four compartments, where feed is stored for microbial fermentation; followed by the reticulum, the omasum, and the abomasum, which produces HCl and digestive enzymes that are comparable to those found in the nonruminant stomach. As in nonruminants, this is followed by the small intestine with its various segments, a larger cecum, and the colon and rectum, where a variety of microbes, including *E. coli* H7:O157, colonize.

GI pH

Survival of phages in different pH and cultural conditions has been reviewed (see, for example, Adams, 1959; see also chapter 11 of this volume). There is a difference in acid tolerance among phages which may depend on the GI content (Smith et al., 1987). As well, acid tolerance of the host bacteria and their phages may differ. For example, the survival time of enteropathogenic *E. coli* in milk whey at pH 2.0 and 30°C ranged from 0.5 to 60 min, while survival of its lytic phage ranged from 0.5 to 5 min (Ackermann and DuBow, 1987). To design a protective formulation, pH is the most important factor to consider. Phages are usually stable in a pH range of 5 to 8; this is temperature dependent, though, so at low temperatures this range is extended to pH 4 to 9. Purified phages are generally stable in neutral, buffered solutions containing 0.1 M sodium chloride, 0.001 M magnesium chloride, and some gelatin when stored at low temperatures (Adams, 1959). Depending on the phage strain, these pH differences and changes could influence the kinetics of phage survival and infectivity. The physiological pH of the GI tract of animals is dependent on the content of the digesta, GI region, number of feedings, and time of feeding. The average pH in a chicken crop is about 6.3; in the proventriculus, 1.8; gizzard, 2.5 (ranges between 0.4 and 5.4); duodenum and jejunum, 6.5; ileum, 7.2; and cecum to cloaca, 7.0 (Farner, 1942; Moran, 1982). Likewise, in the swine stomach, the pH varies from 1.7 to 4.9, depending on feeding (Rice et al., 2002), and changes to neutral pH in the duodenum and small intestine (Moran, 1982). In ruminants, the pH in the abomasum is about 2.3, changing to near neutrality of 6.0 in the small intestine (Wheeler and Noller, 1977). Generally, due to the buffering effect of feed in stomach, pH becomes less acidic after a meal.

Rate of Passage through the GI Tract

Aside from the animal species being treated and the quality of the feed:phage dilution ratio, the efficiency of the therapy is also influenced by the length of time the feed remains in different regions of the GI tract and the rate of passage of the digesta through the GI tract. The rapid transit of ingested phage through the GI tract has been suggested as one of the problems in cholera phage therapy (Summers, 2001). Using a feed marker, it was found that the average pig needs about 20 h for the feed to move from intake through defecation (Lewis and McGlone, 2008). In laying chickens, the feed passage in the GI tract appeared in the ileum 2 h postdose and in the fecal droppings after 4 h, and by 10 h most were excreted (Dunkley et al., 2008). Similar work on broiler chickens showed that the marker appeared in the small intestine 30 min after feeding and that by 2 to 2.5 h a significant amount had passed through this segment of the tract (Svihus et al., 2002). In a study of digesta kinetics in Holstein cows, measured on the 14th day of lactation, its total mean retention time in the entire digestive tract was 55.6 h, of which 40.2 h was spent in the rumen and 15.4 h in the digestive tract posterior to the rumen (Aikman et al., 2008). The volume of the GI tract in animals, particularly in ruminants, could lead to dilution of phage, therefore affecting the kinetics of adsorption, which needs to be considered (Goodridge, 2008).

Temperature

Phage stability and bioactivity under different culture and storage temperatures has been investigated (see reviews in Adams, 1959; and Ackerman and DuBow, 1987; see also chapter 11, this volume). Under optimal laboratory conditions, most studied bacteriophages are stable as phage lysates for some time and their loss of infectivity is estimated at about 1% per day, which may differ in their natural environment. The sensitivity of naked phages to extreme temperatures is influenced by the presence of nutrient media and salts, which may have a protective effect depending on type of phage. In general, however, many tailed phages are inactivated after 30 min of exposure to 56 to 60°C heat, which is influenced by their genetic background (Sakaki and Oshima, 1975; Ackermann and Dubow, 1987).

Generally, most phages maintain their stability when stored at low and freezing temperatures, and with the addition of cryopreservatives their survival can be extended. Phages of many organisms survive freezing to −196°C (Clark, 1962; Keogh and Pettingill, 1966). However, they show a range of heat inactivation, which is influenced by the chemical composition of the medium such as the presence of other proteins, sugars, and various ionic salts, particularly calcium or magnesium. For example, phage T5 in the presence of magnesium resists heat inactivation to 70°C

(Adams, 1959; Ackermann and DuBow, 1987). Similarly, lambda particles were stable when stored in liquid suspension (in SM buffer) over a 6-month period at 4 and −70°C, with a half-life of 36 days at 20°C and 3.4 days at 37°C (which is close to the pig's body temperature of 39.2°C). Addition of trehalose significantly increased the stability of the phage, as did powdered skim milk. Stability was not affected by freeze-thawing in the presence of these protectants. Lyophilization in the presence of trehalose increased the half-life between 20 and 42°C, but not all phages remain stable with this method. Likewise, there are variations among coliphage strains in response to temperature (Prouty, 1953). Their decay rates increase exponentially with the increase of temperature when cultured in optimal broth medium at 37°C (De Paepe and Taddei, 2006). In an optimal medium most phages are stable for the expected duration of use (days) when stored at 37°C, which is within the narrow range of food animal body temperatures (38 to 41°C). However, phages ingested with food, undergoing digestion and mélange with gastric and pancreatic juices, the presence of a variety of ionic salts (Adams, 1959), and a complex microbiota (Verberkmoes et al., 2009), could experience a different environment independent of body temperature that may either protect or damage them.

During the normal passage of phages through the GI tract of most farm animals, they are exposed to and generally survive physiological temperature. The temperature manipulations that are more likely to affect phage stability and infectivity occur during industrial-scale phage preparation, purification, and drying for storage before use at the farm. The preservation and storage of phages in dry form, while maintaining their activity, are important for their practical application. Storage of phages in dry form has the advantage of reduced weight, requiring less space for storage, especially at room temperature, and is cost-effective in comparison with low-temperature storage options. The identification of proper additives and development of optimal formulations are required to protect phages during their processing and are briefly discussed below.

Some food-borne pathogens are included in the microbiota of the GI tract in some agricultural animals, and it is expected that their phages would be stable in the same environment. The success of several experiments with phage therapy suggests that phages do interact with their targeted host, possibly at lower efficiency than is observed in vitro, in the GI tract. The ability of phages to tolerate low pH in the stomach, intestinal bile acids, and proteolytic degradation by pancreatic enzymes is necessary for their use as an efficient agent in phage therapy, particularly in terms of enteric infections when antibacterials are administered orally. Changes in acidity and GI composition may affect the survival and influence the bioactivity of the phages at the site of infection. Thus, for a cost-effective phage application in food animal production, it is desirable to have large quantities of stable and active phages that can be stored for long periods under farm conditions. A variety of viruses (Scott and Woodside, 1976; Loudon and Varley, 2001) and phages (Puapermpoonsiri et al., 2009) have been tested for their stability under storage in a variety of conditions with various results.

The complex structural composition of phages makes them prone to damage from extreme physical or chemical conditions during their preparation, thus reducing their bioactivity (Adams, 1959). However, formulation of food content and the use of additives, such as the addition of a buffer to reduce stomach acidity, have been attempted (Puapermpoonsiri et al., 2009). The technology of protection and controlled release of bioactive compounds by encapsulation in a wide range of tissues and organs as well as in food systems is well developed (Lakkis, 2007). The advances in encapsulation technology for oral delivery of animal viruses provide a model for bacteriophage microencapsulation and have resulted in the submission of patented procedures. To meet the physiological challenges of the GI tract and optimal storage conditions, methods such as microencapsulation of phages with a protective agent that allows their direct addition to animal feed will be discussed.

PROTECTION AND CONTROLLED DELIVERY OF PHAGES BY MICROENCAPSULATION

Microencapsulation Technology

Microencapsulation is the technology of packaging or coating active substances with a continuous, semipermeable membrane coating into small capsules that release in a controlled manner (Langer and Folkman, 1976; Lakkis, 2007; Anal and Singh, 2007). In a broader sense, microencapsulation also includes the production of microparticles in which the active substances are buried within or throughout the polymeric matrix, leaving some of the active substance at the surface of the particles. The latter is also referred to as immobilization technology in some literature. Microencapsulation technology has been extensively studied and utilized in the pharmaceutical sector for targeted delivery of drugs and vaccines in humans and animals. During the past 2 decades, these technologies have been applied to protect biological materials such as probiotic bacteria and proteins from damage caused by food processing, storage, and digestive activity (Champagne and Fustier, 2007). However, there are only a few reports on encapsulation of phages, in particular for use in oral delivery (Gill et al., 2007). Nonetheless, methods developed for microencapsulation of animal viruses for oral delivery of vaccines provides an experimental model for phages.

Microencapsulation of Viruses for Oral Delivery

A number of live or inactivated animal viruses have been successfully encapsulated and showed bioactivity in experimental animals, a few examples of which are listed in Table 1. The common technologies used for oral delivery of viruses involve the use of emulsions with solvent evaporation or extraction, coacervation, spray drying, and extrusion complexation.

A number of factors determine the selection of polymers and associated technologies for microencapsulation of viruses and other biological materials. First, the polymers must be compatible with the substances to be encapsulated and the process should not substantially reduce their bioactivity. Extreme temperature, pH, and pressure should be avoided. Microcapsule size is also very important. Studies using animal models have shown that microcapsules smaller than ~10 μm in diameter can be taken up by the Peyer's patches in the intestine, while larger microcapsules will remain at the mucosal surface of the intestine (Eldridge et al., 1991). Therefore, for oral vaccines, the typical size range is 0.1 to 10 μm. The release of encapsulated viruses in vivo can be triggered in many ways, including through a change of pH, temperature, and pressure; through degradation of polymers by enzymes or organisms; or simply through disruption by shear forces. In general, the release rate is controlled at a low rate for sustained effects; complete release occurs by a few days up to several months (Simerska et al., 2009; Nechaeva, 2002). To illustrate some of the steps and chemical treatments that viruses have to endure before being ready for application, a few specific methods will be discussed below.

SOLVENT EVAPORATION OR EXTRACTION

The creation of microcapsules via the solvent evaporation or extraction approach is a relatively convenient technology that does not require complicated instrumentation and therefore has been a popular choice for microencapsulation of viral and bacterial antigens. For encapsulation of viruses, poly(DL-lactide-co-glycolide) (PLG) has been the polymer of choice (Mundargi et al., 2008). Because of PLG's biodegradability, biocompatibility, and safety, it has been approved by the FDA for therapeutic applications in humans. In this method, a water-in-oil-in-water (W/O/W) double emulsion is prepared. The antigens to be encapsulated are suspended in the inner aqueous phase with appropriate emulsifiers (glycerol, for example), and this

TABLE 1 Examples of microencapsulated viruses for oral vaccine delivery

Virus or antigens	Carrier system	Encapulation method(s)	Test system	Reference(s)	Encapsulation efficiency
Influenza virus, hemagglutinin antigens	Poly(lactide-co-glycolide) Poly(isobutylcyanoacrylate)	W/O/W double emulsion Anionic emulsion polymerization	Mice	Chattaraj et al., 1999	3–5%
Influenza virus, inactivated	Poly(lactide-co-glycolide)	O/W emulsion/ solvent evaporation	Mice	Moldoveanu et al., 1993	Not reported
Measles, live	Polyacrylic acid with or without polyvinyl pyrrolidone, alginate-chitosan, alginate-spermidine	Coacervation	Guinea pigs	Nechaeva et al., 2001	Not reported
Rabies virus, inactivated	Poly(lactide-co-glycolide)	W/O/W double emulsion	Mice	Ramya et al., 2009	50–55%
Reovirus, live	Alginate-spermine	Coacervation in aqueous	In vitro	Periwal et al., 1997	70%
Respiratory syncytial virus, live	Alginate-calcium	Extrusion	In vitro	Bowersock et al., 1997, 1999	Not reported
Rotavirus, inactivated	Poly(acryl starch) Poly(lactide-co-glycolide)	Polymerization in W/O emulsion W/O/W double emulsion	Mice	Sturesson and Degling Wikingsson, 2000 Sturesson et al., 1999	10% 30%
Rotavirus, live	Alginate-spermine or chondroitin sulfate-spermine	Coacervation in aqueous	Mice	Moser et al., 1998; Offit et al., 1994	14% 2–3%

mixture is emulsified in an organic phase containing PLG dissolved in methylene chloride or dichloromethane by homogenization or vortexing. The resulting primary water-in-oil emulsion (W/O) is then added to the second aqueous phase and vortexed or homogenized to form a W/O/W double emulsion. Surfactants such as polyvinyl alcohol, polyvinyl pyrrolidone, and polyethylene glycol are usually added to stabilize the emulsion. This emulsion is then left for several hours at room temperature under constant stirring, during which the organic solvent is evaporated, forming hardened PLG microcapsules that can be collected by centrifugation or filtration. The solvent in the oil phase may also be extracted by the second water phase and then removed. The PLG microcapsules are readily degraded in vivo by hydrolysis to form lactic acid and glycolic acid, thus gradually releasing the encapsulated substances. The release rate is generally very slow, from several weeks to several months. The resulting acidic environment may have a negative impact on the remaining virus or antigens.

COACERVATION METHOD

The principle of coacervation microencapsulation is the phase separation of one or more carrier polymers from the initial solution and the subsequent deposition of the newly formed coacervate phase around the active ingredient suspended or emulsified in the same reaction media. If necessary, the polymer shell is then further cross-linked using an appropriate chemical or enzymatic cross-linker. The phase separation can be a result of the addition of a low-molecular-weight nonsolvent for one polymer or the incompatibility of two polymers induced by a coacervating agent. A typical example of this method is the microencapsulation of peptide or protein drugs in poly(lactide) (PLA) and PLG microspheres (Thomasin et al., 1998). Briefly, the process involves the addition of the coacervating agent poly(dimethylsiloxane) (silicone oil) into PLA/PLG dissolved in dichloromethane; this yields the so-called coacervate droplets, which are solidified in a hardening agent, such as hexane, to produce the final microspheres.

Virus-loaded microcapsules can also be prepared in a simple water-in-oil (W/O) emulsion approach, an example of which is the preparation of poly(acryl starch) microcapsules (Sturesson and Degling Wikingsson, 2000). In this approach, acryloylated starch was dissolved in a sodium phosphate buffer solution containing EDTA and ammonium peroxydisulfate additives at pH 7.5. The virus was also added to this aqueous phase. This mixture was then homogenized in the organic phase containing toluene and chloroform with Pluronic® 68 as the surfactant. A W/O emulsion is formed with the virus entrapped in the water droplets. The formation of hardened microparticles was achieved by the addition of the cross-linker *N,N,N′,N′*-tetramethylethylenediamine, which initiates the polymerization of acryloylated starch in the water droplets. The resulting microparticles can then be collected by centrifugation or filtration.

The classical coacervation method involves the use of an organic solvent and coacervating agents, which are not suitable for encapsulation of biological materials such as bacteria or viruses. Contact of organic solvent at the interface may cause partial loss of infectivity of viruses or alter the immunogenicity of surface proteins (Moldoveanu et al., 1993; Ray et al., 1993). The organic residues may be toxic to animals.

A modified coacervation approach was developed for the microencapsulation of a live measles virus (Nechaeva et al., 2001) to be used with oral administration. Here polyacids, including polyacrylic acid and their substituted variants, were used for the encapsulation. In this method, the polyacid was added to the aqueous suspension containing the measles virus and stabilizers at a weak alkaline pH, wherein the polyacid is in an unfolded state. The phase separation was induced by lowering the pH into the mild acidic range, during which the polymer folds to form coacervates with the virus embedded in them. This material with a preformed microstructure can be further lyophilized to produce virus-containing microcapsules. Such an encapsulation method allowed the measles virus-specific activity to be preserved during the encapsulation process and the transition through the acidic stomach. The microencapsulated vaccine by oral administration induced a higher immune response than the injectable form of the live measles vaccine in guinea pigs.

SPRAY DRYING

Spray-drying encapsulation has been widely used in the food industry for the preparation of dry and stable food additives and flavors. In this method, either the component for encapsulation is suspended in the solution of carrier polymers, or it is first suspended in an aqueous solution with necessary additives (emulsifiers, stabilizers) and then further mixed with an organic phase containing carrier polymers to form a W/O emulsion. The mixture (in the form of a solution, suspension, or emulsion) is then fed into the heated chamber of the spray dryer, pumped through a nozzle or a high-speed spinning disk, and sprayed into a chamber along with warm air. During this process the mixture is atomized into millions of individual droplets, and the solvent is vigorously evaporated. The droplets are dried and harvested through the use of a cyclone current generated from the spray dryer.

Compared to the solvent evaporation method, the spray-drying method is more reliable, yields a higher loading efficiency, and allows for easier mass production. The production cost is relatively low compared to other encapsulation methods. However, the use of this application in oral delivery of vaccine has been very limited. This is due mainly to the fact that conventional spray-drying technology employs high temperature and shear force, which may inactivate the biological materials in the vaccine. Recently, a number of pH-dependent enteric polymers have been found to form microcapsules by spray drying at low temperatures (<60°C) and have been successfully applied to encapsulation of different bacterial antigens for orally administered vaccines (Lin et al., 2003; Liao et al., 2003). As an example, an orally administered vaccine has been developed based on formalin-inactivated *Mycoplasma hyopneumoniae* encapsulated in ethylcellulose microspheres by the spray-drying method (Lin et al., 2002). *M. hyopneumoniae* is a potent pathogen that causes pneumonia in swine. The bacterial antigen was suspended in an aqueous dispersion of ethylcellulose polymer with hydroxypropyl methylcellulose acetate succinate as an additive, and spray dried into microspheres in the size range of 5 to 20 μm. A high encapsulation efficiency of 95.5% was achieved. The in vitro test indicated that the ethylcellulose microspheres protected the mycoplasmal antigen from damage caused by gastric acid and released the active antigen under intestinal conditions. When orally administered to pigs challenged with the pathogen, it induced a specific immune response and reduced the severity of lung lesions, suggesting that immunization with this oral vaccine may provide effective protection against *M. hyopneumoniae* infection in pigs. Anionic polymers of methacrylic acid and ethyl acrylate have also been successfully applied to encapsulation of antigens through the spray-drying method (Lin et al., 2003).

EXTRUSION AND IONIC COMPLEXATION

An aqueous-based method of microencapsulation through complexation of anionic polymers and cationic polymers (or polyvalence ions) has been extensively studied. This method obviates concerns over the use of organic solvents and is carried out at or below room temperature, and thus is particularly attractive for oral delivery of vaccines. In this approach, the virus is suspended in a polyanionic polymer solution and the mixture is extruded as minidroplets into a solution containing divalent cations (mostly Ca^{2+}) and/or cationic polymers. Contact between the polyanions and cations in solution induces immediate interfacial ionic polymerization or complexation, thus forming the microcapsules. A large group of ionic polymers have been used in pairs of opposite charges for microencapsulation of biological materials (Offit et al., 1994). These include polyanionic alginate, pectin, chondroitin sulfate, and carboxymethyl cellulose and cationic poly-L-lysine, chitosan, and polyamines (spermine and decylamine). Among those, alginate-based microcapsules are the most commonly used carrier system for oral delivery of viruses and bacteria. Alginate is a natural complex polysaccharide composed of blocks of 1,4-β-D-mannuronic acid and α-L-guluronic acid that forms gels in the presence of divalent cations. The alginate gels swell and remain intact in acidic media but disintegrate in neutral or mild alkaline solutions. This polymer is readily available and relatively inexpensive and has great compatibility with biological materials. A number of studies have successfully encapsulated live *Pseudomonas aeruginosa* bacteria (or respiratory syncytial virus) into alginate microspheres by extruding alginate suspension containing the bacteria into calcium chloride solution, and further coating them with a cationic polymer, poly-L-lysine or chitosan (Bowersock et al., 1999). The resulting alginate microparticles, in the size range of 10 to 30 μm, were found to be taken up by the Peyer's patches of rabbits when administered orally (Bowersock et al., 1997). Also, adding cationic polymers to the gelation medium or coating the alginate microcapsules with these polymers usually enhances the mechanical strength and reduces the permeability of the microcapsules, improving the protection of the encapsulated biomaterials (Gåserød et al., 1999). Live viruses were also encapsulated in microcapsules through the complexation of alginate and spermine or chondroitin sulfate-spermine (Periwal et al., 1997; Offit et al., 1994). Orally administered alginate-based microcapsules containing herpesvirus, influenza virus, and rotaviruses, respectively, were found to resist degradation by gastric acid, penetrate well into gut-associated lymphocytes, and induce immune responses in mice (Bowersock et al., 1999; Moser et al., 1997, 1998; Nechaeva, 2002). The encapsulation efficiency of this method varied between 14 and 70% in different experiments.

ENCAPSULATION OF BACTERIOPHAGES

Literature

There are few reports on microencapsulation of phages. Bacteriophages have been encapsulated into PLG microspheres in an inhalation dosage form designed for the control of *Staphylococcus aureus* or *P. aeruginosa* (Puapermpoonsiri et al., 2009) for treatment of associated lung diseases. In this study a modified double-emulsion and solvent extraction protocol was used wherein the phages were added to the external aqueous phase to minimize the exposure of the phages at the solvent interface. The microspheres were engineered to have an appropriate size and density to facilitate inhalation via a dry-powder inhaler. It was shown that the process caused partial loss of viability of phages, which has been attributed to the exposure of the phages to the water-dichloromethane interface; the lytic activity was completely lost after 7 days' storage at 4 or 22°C. The product exhibited a burst release phase upon application during which ~60% of phages were released in the first half hour, followed by a sustained release lasting approximately 6 h. Further development is required to extend the shelf life of the formulation. Phages have also been encapsulated in biodegradable poly(esteramide) along with an antibiotic as a wound-healing preparation (PhagoBioDerm™; Intralytix, Inc., Baltimore, MD) to treat topical infections caused by antibiotic-resistant bacteria (Markoishvili et al., 2002; Jikia et al., 2005).

A recent U.S. patent application (number 20090130196; K. Murthy and R. Engelhardt) described a method for preparing encapsulated bacteriophage compositions for oral delivery. The method involves absorbing the aqueous solution of phages onto a powdered matrix such as milk or soya protein powders followed by drying and grinding. The phage-loaded particles are then coated with lipid-based substances by the use of established encapsulation technologies such as spinning-disk atomization or fluid-bed spray coating. The patent claims that the encapsulated phages are stable during the 4.5-month tested period at room temperature and for 10 months at 4°C. The phages are released from the microcapsules upon physical or chemical disruption of the capsules. There are, however, no reported in vitro or in vivo data on the stability and release profile of the phages under any physiological conditions. Such a prepared formulation is expected to gradually release the phages while carrier materials are degraded by digestive enzymes after being orally administered.

Waddell et al. (2004) have filed a patent on a pharmaceutical dosage form of encapsulated phages. In their method they use a common enteric polymer, polymethacrylate copolymer (Eudragit S100, or L100), and sucrose as a lyoprotectant. Phages are dispersed through the aqueous dispersion of polymethacrylate and sucrose, then either freeze-dried or spray dried to produce the solid form. According to the patent description, phages retained their viability during both drying processes with a slight decrease in phage titer. Phages remained in the encapsulation matrices after the acid treatments with a minimal loss of their viability. For example, when tested in 0.1 M HCl solution, the viability of phages decreased only by 1 and 3 log PFU/ml, respectively, after 120 min and overnight incubation. The dosage presents a burst release of encapsulated phages upon exposure to mildly basic conditions such as 10 mM phosphate buffer at pH 7.2; almost all encapsulated phages are released within 20 min under such conditions as a result of dissolution of the polymethacrylate copolymers.

Electrospinning is a relatively new technology used for encapsulation of bioactive materials (Salalha et al., 2006). The technique involves suspending the phages in a solution of electrospinnable polymer, such as polyvinyl alcohol. Droplets of the solution are formed by a spinneret; then an electrostatic field is applied to the droplets under electrospinning conditions, forming fibers containing the phages. Kuhn and Zussman (2009) recently applied this technology to encapsulation of three phages: T4, T7, and λ. Their results demonstrated that 1 to 6% of tested phages remained viable after the spinning process. The loss of viability is believed to be caused by the buildup of pressure within the matrix and the electronic force. Such encapsulated phages are found to be stable at −20 and −55°C for several months but gradually lose viability when stored at room temperature or at 4°C.

In Vitro Simulation Experiments

During the past few years, our research group has been working on the development of encapsulated phages for the control of intestinal colonization of pathogenic bacteria such as *Salmonella* in swine production (Ma et al., 2008). The objective is to produce encapsulated, biologically active bacteriophages that can be included in animal feed, meet environmental regulations, and be economically feasible. Two types of well-characterized model phages, Felix-O1 and phage K, were successfully encapsulated into alginate microspheres using the extrusion and ionic complexation methods (Fig. 1). Briefly, purified phages were dispersed in sodium alginate solution and extruded as fine droplets into a calcium chloride solution through a commercial encapsulator, Inotech IER-50 (Inotech Biosystems International, Inc., Rockville, MD), at room temperature. Alginate forms gels instantly at the surface of the droplets upon contact with calcium ions, enclosing the phages within the droplets. The droplets hardened when calcium ions continued to diffuse into the core of the droplets. The resulting microspheres were further coated with a layer of cationic polymers, such as chitosan and poly-lysine by suspending them in a cationic polymer solution, and then collected by filtration. The produced microspheres have a spherical shape and are fairly uniform in particle size (Fig. 1). The encapsulated phages are distributed within the cavity of the alginate network throughout the mi-

FIGURE 1 Images of wet (left) and dry (right) alginate microspheres containing phage Felix-O1.

crospheres as observed by transmission electron microscopy (Fig. 2). This mild encapsulation process had little detrimental effect on phage viability, and the encapsulation efficiency was over 90% (Ma et al., 2008).

RESISTANCE TO STOMACH ACIDS AND BILE SALTS

When tested in simulated gastric fluid, the alginate microspheres provided moderate protection to the encapsulated phages. For example, Fig. 3 illustrates that at pH 2.5, free phage K became undetectable within 2 min upon exposure to simulated gastric fluid, whereas encapsulated phage K gradually lost its viability within 1.5 h. Antacids have been shown to increase survival of *Vibrio* phages in a GI model (Koo et al., 2001). Therefore, to improve the acid resistance of encapsulated phages, the antacid agent $CaCO_3$ was added to the encapsulation matrix. It was found that the viability of encapsulated phages was completely retained after 2 h of incubation in simulated gastric fluid at pH 2.5 (Fig. 3). More recent results have shown that using heat-denatured whey protein instead of $CaCO_3$ with alginate also greatly enhanced the acid resistance of encapsulated phages. The protective effect was found to increase as the whey protein content was increased in the matrix formula. However, addition of $CaCO_3$ and whey protein to the alginate matrix affected the mechanical properties of the obtained microcapsules and these capsules were less resistant to fracture. Both free phage Felix-O1 and phage K were found to be partially inactivated when tested in simulated 1 and 2% bile solutions; phage Felix-O1 was more sensitive to bile than phage K. The above microencapsulation protocol with the use of $CaCO_3$ or whey protein provided complete protection to phages against bile.

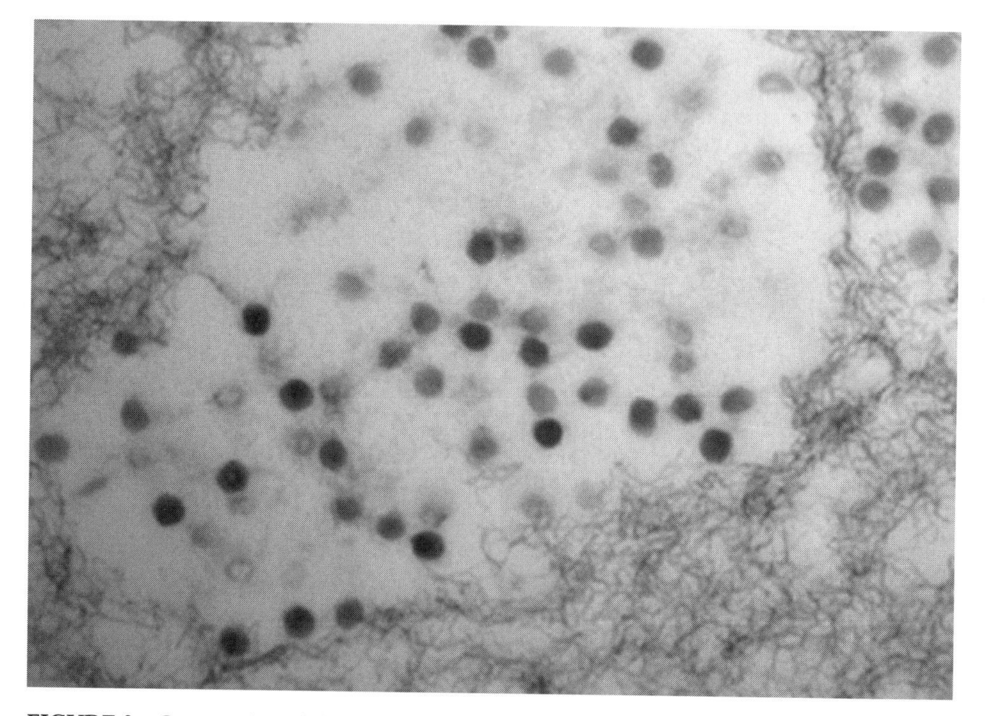

FIGURE 2 Cross section of alginate microsphere containing phage K observed under transmission electron microscopy.

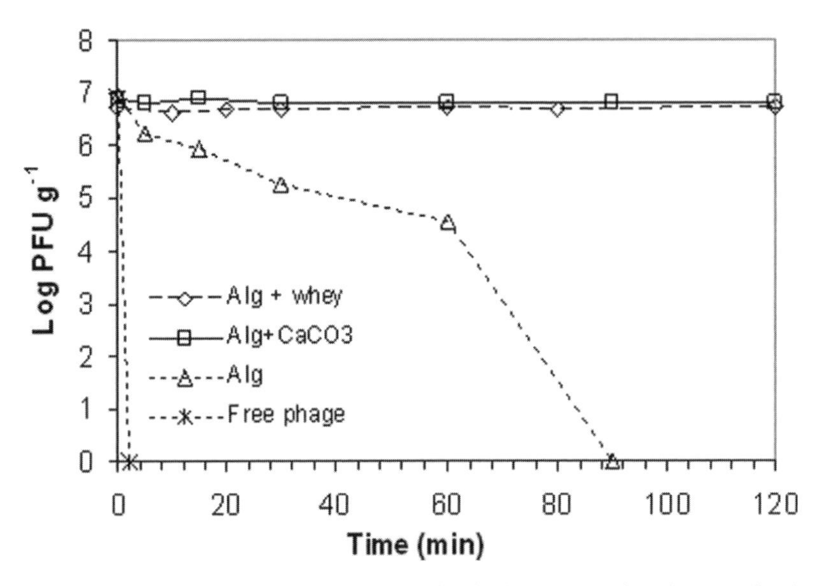

FIGURE 3 Survival of free and encapsulated phage K incubated in simulated gastric fluid at pH 2.5 and 37°C. Alg, alginate.

RELEASE OF PHAGES FROM THE MICROCAPSULES

Complete release of viable phages from the microspheres in the small intestine is an important requirement for the application and treatment of GI pathogens. In simulated intestinal fluid, 80% of the phage K particles were released from the chitosan-coated alginate microspheres within 5 h, with complete release occurring at 10 h (Fig. 4). However, the release of phages from the $CaCO_3$-incorporated microspheres was significantly retarded; fewer than 50% of phages were released in 5 h. In contrast, the addition of whey protein to the alginate matrix greatly accelerated the release of phages from the microspheres. The protection and release rate of phage microspheres can be fine-tuned by adjusting the formulation and reaction conditions, including the total polymer concentration, the ratio of polymers to other additives, reaction time, pH and temperature, and the size of the microsphere. In general, larger microspheres and higher polymer content provide better protection but lead to lower release rates.

DRYING AND STORAGE

A dry-powder form of encapsulated phages provides convenience for field applications in animal production and reduces the storage and transportation costs. The viability of phages in wet microspheres remained unchanged over a 6-week test period at 4°C; however, viability was gradually lost during the drying process and subsequent storage at room and frozen temperatures. The addition of protective additives such as sugars, proteins, and polysaccharides may stabilize phages during dehydration. In our study, several protectants were evaluated for the stabilization of phage K during air drying. The results demonstrated that sucrose, trehalose, maltodextrin, and skim milk were all able to enhance the retention of phage K viability after drying to a variable extent. The protective effect depended on the type and concentration of the additives used, as summarized in Table 2. Compared to sucrose and trehalose, maltodextrin appears to be a more appropriate protective additive because of its lower price and wide commercial availability. It is suggested that these hydrophilic substances provide their protective effect by

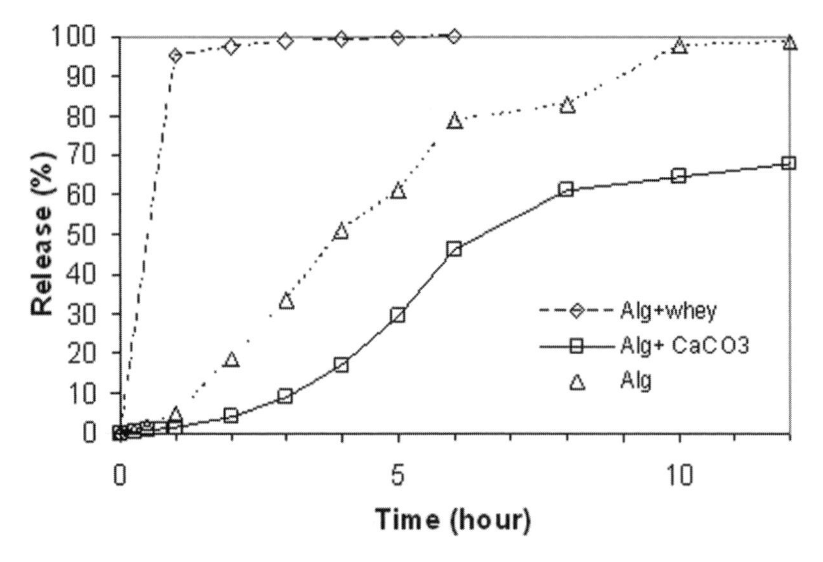

FIGURE 4 Phage release profiles from different microspheres in simulated intestinal fluid. Alg, alginate.

substituting water molecules and forming hydrogen bonds around capsid proteins, thereby stabilizing their native structure in the absence of water (Crowe et al., 1987).

OUTLOOK

Based on the data from our own investigations and in the literature, there is no doubt that microencapsulation of phages and subsequent drying allow us to preserve the viability of phages in a concentrated and solid form. Such phage preparations provide convenience for storage, transportation, and field applications of phages in animal production. In vitro simulation tests have demonstrated that appropriate encapsulation of phages may protect phages from inactivation by stomach acids and bile salts and release them under intestinal conditions. Currently, however, there is a lack of clinical animal trial data to demonstrate a substantial improvement in the efficacy of phages against targeted pathogens in vivo. A recent animal trial has demonstrated that encapsulated phages may reduce the shedding of *E. coli* in experimental animals (Murthy and Engelhardt, 2009).

Selection of highly virulent lytic and broad-spectrum phages is critical to the success of phage therapy. It has been suggested that use of a mixture of phages is more likely to be effective (Tanji et al., 2005). The phage cocktail can be encapsulated into one product, although the stability and release profile of each phage needs to be evaluated.

The increased manufacturing cost associated with encapsulation is another challenge that needs to be addressed to make it acceptable to animal producers. Some researchers have reported that application of phages with feed, or immediately after feeding, is more effective than giving phages on an empty stom-

TABLE 2 Survival of bacteriophage K after air drying with addition of different type and concentration of protective additives to the encapsulation matrix

Additive	% Survival at additive concn (%) of:				
	0	5	10	15	20
Sucrose	0	0.15	1.77	3.57	3.91
Trehalose	0	0.26	1.43	0.12	0.05
Maltodextrin	0	0.21	1.77	6.16	6.46
Skim milk	0	0.02	1.20	10.28	18.24

ach (Smith et al., 1987). This is likely due to the fact that stomach pH is less acidic after a meal because of the buffering effect from the feed. In addition, phages may be embedded in feed particles during the transition through the stomach, which provides a physical barrier to stomach acids and other digestive enzymes. Nevertheless, it is envisioned that low-cost encapsulated phages could increase efficiency of phage therapy in food animal production.

ACKNOWLEDGMENTS

Supported by Ministry of Agriculture and Agri-Food Canada research grant #1274-2009. We also thank Jason Gill for his helpful comments and suggestions regarding this chapter.

REFERENCES

Aarestrup, F. M., and L. B. Jenser. 2007. Use of antimicrobials in food animal production, p. 405–418. *In* S. Simjee (ed.), *Infectious Diseases: Foodborne Diseases.* Humana Press, Inc., Totowa, NJ.

Abedon, S. T. 2008. Phage population growth: constraints, games, adaptation, p. 64–93. *In* S. T. Abedon (ed.), *Bacteriophage Ecology: Population Growth, Evolution, and Impact of Bacterial Viruses.* Cambridge University Press, Cambridge, United Kingdom.

Ackermann, H.-W., and M. S. DuBow. 1987. *Viruses of Prokaryotes,* vol. II. *Natural Groups of Bacteriophages,* p. 26–29. CRC Press, Boca Raton, FL.

Adams, M. H. 1959. *Bacteriophages,* p. 49–62. Interscience Publishers, New York, NY.

Aikman, P. C., C. K. Reynolds, and D. E. Beever. 2008. Diet digestibility, rate of passage, and eating and rumination behaviour of Jersey and Holstein cows. *J. Dairy Sci.* **91:**1103–1114.

Akwar, H. T., C. Poppe, J. Wilson, R. J. Reid-Smith, M. Dyck, J. Waddington, D. Shang, and S. A. McEwen. 2008. Associations of antimicrobial uses with antimicrobial resistance of fecal *Escherichia coli* from pigs on 47 farrow-to-finish farms in Ontario and British Columbia. *Can. J. Vet. Res.* **72:**202–210.

Anal, A. K., and H. Singh. 2007. Recent advances in microencapsulation of probiotics for industrial applications and targeted delivery. *Trends Food Sci. Technol.* **18:**240–251.

Andreatti Filho, R. L., J. P. Higgins, S. E. Higgins, G. Gaona, A. D. Wolfenden, G. Tellez, and B. M. Hargis. 2007. Ability of bacteriophages isolated from different sources to reduce *Salmonella enterica* serovar Enteritidis in vitro and in vivo. *Poultry Sci.* **86:**1904–1909.

Angulo, F. J., N. L. Baker, S. J. Olsen, A. Anderson, and T. J. Barrett. 2004. Antimicrobial use in agriculture: controlling the transfer of antimicrobial resistance to humans. *Semin. Pediatr. Infect. Dis.* **15:**78–85.

Atterbury, R. J., M. A. Van Bergen, F. Ortiz, M. A. Lovell, J. A. Harris, A. De Boer, J. A. Wagenaar, V. M. Allen, and P. A. Barrow. 2007. Bacteriophage therapy to reduce *Salmonella* colonization of broiler chickens. *Appl. Environ. Microbiol.* **73:**4543–4549.

Barrow, P., M. Lovell, and A. Berchieri, Jr. 1998. Use of lytic bacteriophage for control of experimental *Escherichia coli* septicemia and meningitis in chickens and calves. *Clin. Diagn. Lab. Immunol.* **5:**294–298.

Berchieri, A., Jr., M. A. Lovell, and P. A. Barrow. 1991. The activity in the chicken alimentary tract of bacteriophages lytic for *Salmonella typhimurium. Res. Microbiol.* **142:**541–549.

Bowersock, T. L., H. Hogenesch, M. Suckow, P. Guimond, S. Martin, D. Borie, S. Torregrosa, H. Park, and K. Park. 1999. Oral vaccination of animals with antigens encapsulated in alginate microspheres. *Vaccine* **17:**1804–1811.

Bowersock, T. L., K. Park, and R. E. Porter, Jr. October 1997. Alginate-based vaccine compositions. U.S. patent 5,674,495.

Brüssow, H. 2007. Phage therapy: the Western perspective, p. 159–192. *In* S. McGrath and D. van Sinderen (ed.), *Bacteriophage: Genetics and Molecular Biology.* Caister Academic Press, Norfolk, United Kingdom.

Champagne, C. P., and P. Fustier. 2007. Microencapsulation for the improved delivery of bioactive compounds into foods. *Curr. Opin. Biotechnol.* **18:**184–190.

Chattaraj, S. C., A. Rathinavelu, and S. K. Das. 1999. Biodegradable microparticles of influenza viral vaccine: comparison of the effects of routes of administration on the in vivo immune response in mice. *J. Control. Release* **58:**223–232.

Clark, W. A. 1962. Comparison of several methods for preserving bacteriophage. *Appl. Microbiol.* **10:**466–471.

Crittenden, R., R. Weerakkody, L. Sanguansri, and M. Augustin. 2006. Synbiotic microcapsules that enhance microbial viability during nonrefrigerated storage and gastrointestinal transit. *Appl. Environ. Microbiol.* **72:**2280–2282.

Crowe, J. H., L. M. Crowe, J. F. Carpenter, and C. A. Wistrom. 1987. Stabilization of dry phospholipid bilayers and proteins by sugars. *Biochem. J.* **242:**1–10.

DANMAP 2003. 2004. *Use of Antimicrobial Agents and Occurrence of Antimicrobial Resistance in Bacteria from Food Animals, Food and Humans in Denmark.*

Danish Zoonoses Center, Sørborg, Denmark. (Available at http://www.dfvf.dk.)

De Paepe, M., and F. Taddei. 2006. Viruses' life history: towards a mechanistic basis of a trade-off between survival and reproduction among phages. *PLoS Biol.* **4:**e193.

Dunkley, C. S., W. K. Kim, W. D. James, W. C. Ellis, J. L. McReynolds, L. F. Kubena, D. J. Nisbet, and S. C. Ricke. 2008. Passage rates in poultry digestion using stable isotope markers and INAA. *J. Radioanal. Nucl. Chem.* **276:**35–39.

Eldridge, J. H., J. K. Staas, J. A. Meulbroek, T. R. Tice, and R. M. Gilley. 1991. Biodegradable and biocompatible poly(DL-lactide-co-glycolide) microspheres as an adjuvant for staphylococcal enterotoxin B toxoid which enhances the level of toxin-neutralizing antibodies. *Infect. Immun.* **59:**2978–2986.

European Commission. 2005. Ban on antibiotics as growth promoters in animal feed enters into effect. Reference IP/05/1687. December 22. European Commission, Brussels, Belgium. http://europa.eu/rapid/pressReleasesAction.do?reference=IP/05/1687&format=HTML&aged=0&language =EN&guiLanguage=en.

Farner, D. S. 1942. The hydrogen ion concentration in avian digestive tracts. *Poult. Sci.* **21:**445–450.

Fischetti, V. A. 2006. Using phage lytic enzymes to control pathogenic bacteria. *BMC Oral Health* **6**(Suppl. 1):S16.

Fruciano, D. E., and S. Bourne. 2007. Phage as an antimicrobial agent: d'Herelle's heretical theories and their role in the decline of phage prophylaxis in the West. *Can. J. Infect. Dis. Med. Microbiol.* **18:**19–26.

Gåserød, O., A. Sannes, and G. Skjåk-Braek. 1999. Microcapsules of alginate-chitosan. II. A study of capsule stability and permeability. *Biomaterials* **20:**773–783.

Geier, M. R., M. E. Trigg, and C. R. Merril. 1973. Fate of bacteriophage lambda in non-immune germ-free mice. *Nature* **246:**221–223.

Gill, J. J., T. Hollyer, and P. M. Sabour. 2007. Bacteriophages and phage-derived products as antibacterial therapeutics. *Expert Opin. Ther. Pat.* **17:**1341–1350.

Gill, J. J., P. M. Sabour, K. E. Leslie, and M. W. Griffiths. 2006a. Bovine whey proteins inhibit the interaction of *Staphylococcus aureus* and bacteriophage K. *J. Appl. Microbiol.* **101:**377–386.

Gill, J. J., J. C. Pacan, M. E. Carson, K. E. Leslie, M. W. Griffiths, and P. M. Sabour. 2006b. Efficacy and pharmacokinetics of bacteriophage therapy in treatment of subclinical *Staphylococcus aureus* mastitis in lactating dairy cattle. *Antimicrob. Agents Chemother.* **50:**2912–2918.

Gillings, M., Y. Boucher, M. Labbate, A. Holmes, S. Krishnan, M. Holley, and H. W. Stokes. 2008. The evolution of class 1 integrons and the rise of antibiotic resistance. *J. Bacteriol.* **190:**5095–5100.

Goode, D., V. M. Allen, and P. A. Barrow. 2003. Reduction of experimental *Salmonella* and *Campylobacter* contamination of chicken skin by application of lytic bacteriophages. *Appl. Environ. Microbiol.* **69:**5032–5036.

Goodridge, L. D. 2008. Phages, bacteria, and food, p. 302–331. *In* S. T. Abedon (ed.), *Bacteriophage Ecology: Population Growth, Evolution, and Impact of Bacterial Viruses.* Cambridge University Press, Cambridge, United Kingdom.

Greer, G. G. 2005. Bacteriophage control of foodborne bacteria. *J. Food Prot.* **68:**1102–1111.

Hanlon, G. W. 2007. Bacteriophages: an appraisal of their role in the treatment of bacterial infections. *Int. J. Antimicrob. Agents* **30:**118–128.

Huff, W. E., G. R. Huff, N. C. Rath, J. M. Balog, and A. M. Donoghue. 2002. Prevention of *Escherichia coli* infection in broiler chickens with a bacteriophage aerosol spray. *Poult. Sci.* **81:**1486–1491.

Huff, W. E., G. R. Huff, N. C. Rath, J. M. Balog, and A. M. Donoghue. 2003. Bacteriophage treatment of a severe *Escherichia coli* respiratory infection in broiler chickens. *Avian Dis.* **47:**1399–1405.

Huff, W. E., G. R. Huff, N. C. Rath, and A. M. Donoghue. 2006. Evaluation of the influence of bacteriophage titer on the treatment of colibacillosis in broiler chickens. *Poult. Sci.* **85:**1373–1377.

Inchley, C. J. 1969. The activity of Kuppfer cells following intravenous injection of T4 bacteriophage. *Clin. Exp. Immunol.* **5:**173–187.

Jacob, J., and T. Pescatore. 2009. *Chicken Anatomy and Physiology: Digestive System.* Cooperative Extension Service, University of Kentucky, Lexington, KY.

Jagusztyn-Krynicka, E. K., and A. Wyszyńska. 2008. The decline of antibiotic era—new approaches for antibacterial drug discovery. *Pol. J. Microbiol.* **57:**91–98.

Jikia, D., N. Chkhaidze, E. Imedashvili, I. Mgaloblishvili, G. Tsitlanadze, R. Katsarava, J. G. Morris, Jr., and A. Sulakvelidze. 2005. The use of a novel biodegradable preparation capable of the sustained release of bacteriophages and ciprofloxacin, in the complex treatment of multidrug-resistant *Staphylococcus aureus*-infected local radiation injuries caused by exposure to Sr90. *Clin. Exp. Dermatol.* **30:**23–26.

Johnson, R. P., C. L. Gyles, W. E. Huff, S. Ojha, G. R. Huff, N. C. Rath, and A. M. Donoghue. 2008. Bacteriophages for prophylaxis and therapy in cattle, poultry and pigs. *Anim. Health Res. Rev.* **9:**201–215.

Keogh, B. P., and G. Pettingill. 1966. Long-term storage of bacteriophages of lactic streptococci. *Appl. Microbiol.* **14:**421–424.

Khanna, T., R. Friendship, C. Dewey, and J. S. Weese. 2008. Methicillin resistant *Staphylococcus aureus* colonization in pigs and pig farmers. *Vet. Microbiol.* **128:**298–303.

Koo, J., D. L. Marshall, and A. DePaola. 2001. Antacid increases survival of *Vibrio vulnificus* and *Vibrio vulnificus* phage in gastrointestinal model. *Appl. Environ. Microbiol.* **67:**2895–2902.

Kuhn, J. C., and E. Zussman. 2009. Encapsulation of bacteria and viruses in electrospun fibers. U.S. patent 20090061496.

Lakkis, J. M. 2007. Introduction, p. 1–11. *In* J. M. Lakkis (ed.), *Encapsulation and Controlled Release Technologies in Food Systems.* Blackwell Publishing, Ames, IA.

Langer, R., and J. Folkman. 1976. Polymers for the sustained release of proteins and other macromolecules. *Nature* **263:**797–800.

Lewis, C. R. G., and J. J. McGlone. 2008. Modelling feeding behaviour, rate of feed passage and daily feeding cycles, as possible causes of fatigued pigs. *Animal* **2:**600–605.

Liao, C. W., H. Y. Chiou, K. S. Yeh, J. R. Chen, and C. N. Weng. 2003. Oral immunization using formalin-inactivated *Actinobacillus pleuropneumoniae* antigens entrapped in microspheres with aqueous dispersion polymers prepared using a co-spray drying process. *Prev. Vet. Med.* **61:**1–15.

Lin, J. H., M. J. Pan, C. W. Liao, and C. N. Weng. 2002. In vivo and in vitro comparisons of spray-drying and solvent-evaporation preparation of microencapsulated *Mycoplasma hyopneumoniae* for use as an orally administered vaccine for pigs. *Am. J. Vet. Res.* **63:**1118–1123.

Lin, J. H., C. N. Weng, C. W. Liao, K. S. Yeh, and M. J. Pan. 2003. Protective effects of oral microencapsulated *Mycoplasma hyopneumoniae* vaccine prepared by co-spray drying method. *J. Vet. Med. Sci.* **65:**69–74.

Liu, J., M. Dehbi, G. Moeck, F. Arhin, P. Bauda, D. Bergeron, M. Callejo, V. Ferretti, N. Ha, T. Kwan, J. McCarty, R. Srikumar, D. Williams, J. J. Wu, P. Gros, J. Pelletier, and M. DuBow. 2004. Antimicrobial drug discovery through bacteriophage genomics. *Nat. Biotechnol.* **22:**167–168.

Loudon, P. T., and C. A. Varley. July 2001. Stabilization of herpes virus preparations. U.S. patent 6,258,362.

Ma, Y., J. C. Pacan, Q. Wang, Y. Xu, X. Huang, A. Korenevsky, and P. M. Sabour. 2008. Microencapsulation of bacteriophage Felix O1 into chitosan-alginate microspheres for oral delivery. *Appl. Environ. Microbiol.* **74:**4799–4805.

Markoishvili, K., G. Tsitlanadze, R. Katsarava, J. G. Morris, Jr., and A. Sulakvelidze. 2002. A novel sustained-release matrix based on biodegradable poly(ester amide)s and impregnated with bacteriophages and an antibiotic shows promise in management of infected venous stasis ulcers and other poorly healing wounds. *Int. J. Dermatol.* **41:** 453–458.

McEwen, S., and P. Fedorka-Cray. 2002. Antimicrobial use and resistance in animals. *Clin. Infect. Dis.* **34:**S93–S106.

Mellon, M., C. Benbrook, and K. Lutz Benbrook. 2001. *Hogging It: Estimates of Antimicrobial Abuse in Livestock.* Union of Concerned Scientists Publications, Cambridge, MA. (Available at http://www.ucsusa.org/food_and_agriculture/science_and_impacts/impacts_industrial_agriculture/hogging-it-estimates-of.html.)

Merril, C. R., D. Scholl, and S. Adhya. 2005. Phage therapy, p. 725–741. *In* R. Calendar (ed.), *The Bacteriophages,* 2nd ed. Oxford University Press, New York, NY.

Moldoveanu, Z., M. Novak, W. Q. Huang, R. M. Gilley, J. K. Staas, D. Schafer, R. W. Compans, and J. Mestecky. 1993. Oral immunization with influenza virus in biodegradable microspheres. *J. Infect. Dis.* **167:**84–90.

Moran, E. T., Jr. 1982. *Comparative Nutrition of Fowl and Swine. The Gastrointestinal Systems,* p. 251. Office for Educational Practice, University of Guelph, Guelph, Ontario, Canada.

Moser, C. A., T. J. Speaker, and P. A. Offit. 1997. Effect of microencapsulation on immunogenicity of a bovine herpes virus glycoprotein and inactivated influenza virus in mice. *Vaccine* **15:** 1767–1772.

Moser, C. A., T. J. Speaker, and P. A. Offit. 1998. Effect of water-based microencapsulation on protection against EDIM rotavirus challenge in mice. *J. Virol.* **72:**3859–3862.

Mundargi, R. C., V. R. Babu, V. Rangaswamy, P. Patel, and T. M. Aminabhavi. 2008. Nano/micro technologies for delivering macromolecular therapeutics using poly(DL-lactide-co-glycolide) and its derivatives. *J. Control. Release* **125:**193–209.

Nechaeva, E. 2002. Development of oral microencapsulated forms for delivering viral vaccines. *Expert Rev. Vaccines* **1:**385–397.

Nechaeva, E. A., N. Varaksin, T. Ryabicheva, M. Smolina, T. Kolokoltsova, A. Vilesov, N. Aksenova, R. Stankevich, and R. Isidorov. 2001. Approaches to development of microencapsulated form of the live measles vaccine. *Ann. N. Y. Acad. Sci.* **944:**180–186.

Offit, P. A., C. A. Khoury, C. A. Moser, H. F. Clark, J. E. Kim, and T. J. Speaker. 1994. Enhancement of rotavirus immunogenicity by microencapsulation. *Virology* **203:**134–143.

O'Flaherty, S., A. Coffey, W. J. Meaney, G. F. Fitzgerald, and R. P. Ross. 2005. Inhibition of bacteriophage K proliferation on *Staphylococcus aureus* in raw bovine milk. *Lett. Appl. Microbiol.* **41:** 274–279.

O'Flaherty, S., R. P. Ross, and A. Coffey. 2009. Bacteriophage and their lysins for elimination of infectious bacteria. *FEMS Microbiol. Rev.* **33:**801–819.

O'Flynn, G., A. Coffey, G. F. Fitzgerald, and R. P. Ross. 2006. The newly isolated lytic bacteriophages st104a and st104b are highly virulent against *Salmonella enterica*. *J. Appl. Microbiol.* **101:** 251–259.

Partridge, S. R., G. Tsafnat, E. Coiera, and J. R. Iredell. 2009. Gene cassettes and cassette arrays in mobile resistance integrons: review article. *FEMS Microbiol. Rev.* **33:**757–784.

Payne, R. J. H., and V. A. A. Jansen. 2003. Pharmacokinetic principles of bacteriophage therapy. *Clin. Pharmacokinet.* **42:**315–325.

Periwal, S. B., T. J. Speaker, and J. J. Cebra. 1997. Orally administered microencapsulated reovirus can bypass suckled, neutralizing maternal antibody that inhibits active immunization of neonates. *J. Virol.* **71:**2844–2850.

Poole, T. L., T. R. Callaway, and D. J. Nisbet. 2007. Alternatives to antimicrobials, p. 419–433. *In* S. Simjee (ed.), *Infectious Diseases: Foodborne Diseases*. Humana Press, Inc., Totowa, NJ.

Prouty, C. C. 1953. Storage of the bacteriophage of the lactic acid streptococci in the desiccated state with observations on longevity. *Appl. Microbiol.* **1:** 250–251.

Puapermpoonsiri, U., J. Spencer, and C. F. van der Walle. 2009. A freeze-dried formulation of bacteriophage encapsulated in biodegradable microspheres. *Eur. J. Pharm. Biopharm.* **72:**26–33.

Rajić, A., L. A. Waddell, J. M. Sargeant, S. Read, J. Farber, M. J. Firth, and A. Chambers. 2007. An overview of microbial food safety programs in beef, pork, and poultry from farm to processing in Canada. *J. Food Prot.* **70:**1286–1294.

Ramya, R., P. C. Verma, V. K. Chaturvedi, P. K. Gupta, K. D. Pandey, M. Madhanmohan, T. R. Kannaki, R. Sridevi, and B. Anukumar. 2009. Poly(lactide-co-glycolide) microspheres: a potent oral delivery system to elicit systemic immune response against inactivated rabies virus. *Vaccine* **27:**2138–2143.

Ray, R., M. Novak, J. D. Duncan, Y. Matsuoka, and R. W. Compans. 1993. Microencapsulated human parainfluenza virus induces a protective immune response. *J. Infect. Dis.* **167:**752–755.

Rice, J. P., R. S. Pleasant, and J. S. Radcliffe. 2002. The effect of citric acid, phytase, and their interaction on gastric pH, and Ca, P, and dry matter digestibilities. Purdue University 2002 Swine Research Report, Purdue University, West Lafayette, IN. http://www.ansc.purdue.edu/swine/swineday/sday02/6.pdf.

Rozema, E. A., T. P. Stephens, S. J. Bach, E. K. Okine, R. P. Johnson, K. Stanford, and T. A. McAllister. 2009. Oral and rectal administration of bacteriophagees for control of *Escherichia coli* O157:H7 in feedlot cattle. *J. Food Prot.* **72:**241–250.

Sakaki, Y., and T. Oshima. 1975. Isolation and characterization of a bacteriophage infectious to an extreme thermophile, *Thermus thermophilus* HB8. *J. Virol.* **15:**1449–1453.

Salalha, W., J. Kuhn, Y. Dror, and E. Zussman. 2006. Encapsulation of bacteria and viruses in electrospun nanofibres. *Nanotechnology* **17:**4675–4681.

Scott, E. M., and W. Woodside. 1976. Stability of pseudorabies virus during freeze drying and storage: effect of suspending media. *J. Clin. Microbiol.* **4:**1–5.

Simerska, P., P. M. Moyle, C. Olive, and I. Toth. 2009. Oral vaccine delivery—new strategies and technologies. *Curr. Drug Deliv.* **6:**347–358.

Siragusa, G., M. Wise, and B. Seal. 2004. Lytic bacteriophage against *Clostridium perfringens,* p. 102. *Am. Soc. Microbiol. Conf. New Phage Biol.,* Key Biscayne, FL, 1 to 5 August 2004.

Smith, W. H., and M. B. Huggins. 1983. Effectiveness of phages in treating experimental *Escherichia coli* diarrhoea in calves, piglets and lambs. *J. Gen. Microbiol.* **129:**2659–2675.

Smith, W. H., M. B. Huggins, and K. M. Shaw. 1987. The control of experimental *Escherichia coli* diarrhoea in calves by means of bacteriophages. *J. Gen. Microbiol.* **133:**1111–1126.

Sturesson, C., P. Artursson, R. Ghaderi, K. Johansen, A. Mirazimi, I. Uhnoo, L. Svensson, A.-C. Albertsson, and J. Carlfors. 1999. Encapsulation of rotavirus into poly(lactide-co-glycolide) microspheres. *J. Control. Release* **59:** 377–389.

Sturesson, C., and L. Degling Wikingsson. 2000. Comparison of poly(acryl starch) and poly(lactide-co-glycolide) microspheres as drug delivery system for a rotavirus vaccine. *J. Control. Release* **68:**441–450.

Sulakvelidze, A., and P. Barrow. 2005. Phage therapy in animals and agribusiness, p. 335–380. *In* E. Kutter and A. Sulakvelidze (ed.), *Bacteriophages: Biology and Applications*. CRC Press, Boca Raton, FL.

Summers, W. C. 2001. Bacteriophage therapy. *Annu. Rev. Microbiol.* **55:**437–451.

Svihus, B., H. Hetland, M. Choct, and F. Sundby. 2002. Passage rate through the anterior digestive tract of broiler chickens fed diets with ground and whole wheat. *Br. Poultry Sci.* **43:**662–668.

Tanji, Y., T. Shimada, H. Fukudomi, K. Miyanaga, Y. Nakai, and H. Unno. 2005. Therapeutic use of phage cocktail for controlling *Escherichia coli* O157:H7 in gastrointestinal tract of mice. *J. Biosci. Bioeng.* **100:**280–287.

Tauxe, R. V. 1997. Emerging foodborne diseases: an evolving public health challenge. *Emerg. Infect. Dis.* **3:**425–434.

Thiel, K. 2004. Old dogma, new tricks—21st century phage therapy. *Nat. Biotechnol.* **22:**31–36.

Thomasin, C., H. Nam-Trân, H. P. Merkle, and B. Gander. 1998. Drug microencapsulation by PLA/PLGA coacervation in the light of thermodynamics. 1. Overview and theoretical considerations. *J. Pharm. Sci.* **87:**259–268.

Topping, D. L., G. B. Storer, G. D. Calvert, R. J. Illman, D. G. Oakenfull, and R. A. Weller. 1980. Effects of dietary saponins on fecal bile acids and neutral sterols, plasma lipids, and lipoprotein turnover in the pig. *Am. J. Clin. Nutr.* **33:** 783–786.

Toro, H., S. B. Price, S. McKee, F. J. Hoerr, J. Krehling, M. Perdue, and L. Bauermeister. 2005. Use of bacteriophages in combination with competitive exclusion to reduce *Salmonella* from infected chickens. *Avian Dis.* **49:**118–124.

Verberkmoes, N. C., A. L. Russell, M. Shah, A. Godzik, M. Rosenquist, J. Halfvarson, M. G. Lefsrud, J. Apajalahti, C. Tysk, R. L. Hettich, and J. K. Jansson. 2009. Shotgun metaproteomics of the human distal gut microbiota. *ISME J.* **3:**179–189.

Vispo, C., and H. Karasov. 1997. The interaction of avian gut microbes and their host: an elusive symbiosis, p. 116–155. *In* R. M. Makie and B. A. White (ed.), *Gastrointestinal Microbiology,* vol. 1. *Gastrointestinal Ecosystems and Fermentations.* Chapman & Hall, London, United Kingdom.

Voss, A., F. Loeffen, J. Bakker, C. Klaassen, and M. Wulf. 2005. Methicillin-resistant *Staphylococcus aureus* in pig farming. *Emerg. Infect. Dis.* **11:**1965–1966.

Waddell, T. E., R. Johnson, and A. Mazzocco. October 2009. Methods and compositions for controlled release of bioactive compounds. U.S. patent 7601347.

Wheeler, W. E., and C. H. Noller. 1977. Gastrointestinal tract pH and starch in feces of ruminants *J. Anim. Sci.* **44:**131–135.

Wise, R., T. Hart, O. Cars, M. Streulens, R. Helmuth, P. Huovinen, and M. Sprenger. 1998. Antimicrobial resistance. *Br. Med. J.* **317:** 609–610.

Wright, G. D. 2007. The antibiotic resistome: the nexus of chemical and genetic diversity. *Nat. Rev. Microbiol.* **5:**175–186.

APPLICATION OF BACTERIOPHAGES FOR CONTROL OF INFECTIOUS DISEASES IN AQUACULTURE

Toshihiro Nakai

13

Infectious diseases have posed a serious threat to aquaculture industries worldwide, as well as to terrestrial animals and humans. Although various attempts have been made to eradicate, prevent, or control these diseases in cultured fish and shellfish, these species invite a variety of viral, bacterial, or fungal infections with complex interacting factors among the host, pathogen, and environment. Particularly, being poikilothermic, fish are highly sensitive to environmental change (water temperature, dissolved oxygen, metabolites, chemical additives, etc.), including human activities such as handling, and thus environmental factors profoundly influence the susceptibility of fish to infectious agents (Plumb, 1999). This complexity in fish and aquaculture has made it difficult to establish control measures for infectious diseases. Hence different solutions are required to combat these diverse causes. For treatment of bacterial diseases encountered in aquaculture, chemotherapy has been the only method employed in practice. Chemotherapy is a rapid and effective method to treat bacterial infections, but frequent use of chemotherapeutic agents in aquaculture, as in veterinary and medical use, has allowed drug-resistant strains of bacteria to develop. This problem associated with chemotherapy is particularly serious in Japan, where 24 chemotherapeutics are licensed for fishery use (for a list, see the Japanese Ministry of Agriculture, Forestry and Fisheries website at http://www.maff.go.jp/j/syouan/suisan/suisan_yobo/index.html). Vaccination is an ideal method for prevention of infectious disease, but commercially available vaccines are still limited in the aquaculture field, due mainly to their lower marketability and to the poor effectiveness of candidate vaccines. As an alternative approach to control fish diseases, biological controls such as probiotics have been intensively studied (reviewed in Verschuere et al., 2000; and Irianto and Austin, 2002); however, probiotics involve substantial difficulties in scientific demonstration of their benefits as they relate to human use (Reid, 1999).

Apart from diseases in fish, there is an increasing concern that fish pathogens may affect the health of human beings or terrestrial animals. Although no zoonosis caused by fish-pathogenic viruses has been found in humans and terrestrial animals, probably due to extreme phylogenetic differences in their host specificities, some bacterial pathogens of fish

Toshihiro Nakai, Laboratory of Fish Pathology, Graduate School of Applied Biological Science, Hiroshima University, Higashi-Hiroshima 739-8528, Japan.

Bacteriophages in the Control of Food- and Waterborne Pathogens
Edited by Parviz M. Sabour and Mansel W. Griffiths, © 2010 ASM Press, Washington, DC

may infect humans (Plumb, 1999; Austin and Austin, 2007). Candidate zoonotic fish-pathogenic bacteria are listed in Table 1; minor pathogens are not included in the table. One of the most typical cases is infection of a human by *Vibrio vulnificus,* where it was suspected that the person was infected from purchased eels through his open wounds (Veenstra et al., 1992). *V. vulnificus* as a fish pathogen was first isolated from cultured Japanese eel *(Anguilla japonica)* in Japan (Nishibuchi et al., 1979) and was classified into a biogroup, or serovar, different from that of human isolates (Tison et al., 1982; Biosca et al., 1997). Since some fish pathogens, like *V. vulnificus, Lactococcus garvieae,* and *Edwardsiella tarda,* are also ubiquitous in natural marine or freshwater environments, the contamination of waters and fish by the pathogens raises further considerations for public and food hygiene (see chapter 7).

Bacteriophages (phages) as specific pathogen-killers are attractive agents for treating or controlling bacterial infections. After much exposure to the application of phages to human bacterial infections, scientifically well-controlled studies started with a series of *Escherichia coli* models with mice and farm animals in the 1980s (Smith and Huggins, 1982, 1983; Smith et al., 1987a, 1987b). Independently of these studies, successful clinical use of phages for drug-resistant suppurative infections in humans was described by Polish and Soviet groups. Thereafter, many successful treatments with phages have been reported using various animal models (reviewed in Alisky et al., 1998; Sulakvelidze et al., 2001; Merril et al., 2003; and Sulakvelidze and Barrow, 2005). From these studies, it is recognized that the substantial advantages of phage treatment over chemotherapy are (i) the narrow host range of phages, indicating that the phages do not harm the normal intestinal or environmental microflora; and (ii) the self-perpetuating nature of phages in the presence of susceptible bacteria, suggesting the elimination of multiple administrations of phages and possible autonomous transfer of the administered phages among nearby hosts (Smith and Huggins, 1982; Barrow and Soothill, 1997).

A pilot research project on phage therapy against *Aeromonas hydrophila* and *E. tarda* infections of Japanese eels was reported by a research group in Taiwan in the early 1980s (Wu et al., 1981; Wu and Chao, 1982). Subsequently, studies on phages of fish pathogens and aquaculture phage therapy have become more popular since the late 1990s, with an increasing interest in recent years. However, information available on phage therapy in this field is still limited (Nakai and Park, 2002). Research results, excluding the earliest studies, will be briefly reviewed here, and the potential for controlling bacterial infections in aquaculture by means of phages will be discussed. Six diseases known in freshwater or marine fish, along with one shrimp disease, have been targeted for phage therapy; these are caused by

TABLE 1 Major bacterial pathogens of fish associated with human infections

Pathogen	Host fish	Effects on humans	Reference(s)
Aeromonas hydrophila	Freshwater fishes	Diarrhea, septicemia	Goncalves et al., 1992
Edwardsiella tarda	Cultured eel, catfish, flounder, etc.	Wound infection, gastroenteritis, diarrhea	Vandepitte et al., 1983; Hargreaves and Lucey, 1990; Janda and Abbott, 1993
Vibrio vulnificus	Cultured eel	Wound infection, septicemia	Veenstra et al., 1992
Streptococcus iniae	Cultured freshwater and marine fishes	Invasive infection	Weinstein et al., 1997; Lau et al., 2006
Mycobacterium marinum	Ornamental fishes	Skin disease	Ryan and Bryant, 1997

Pseudomonas plecoglossicida, Flavobacterium psychrophilum, L. garvieae, Streptococcus iniae, E. tarda, Aeromonas salmonicida, and *Vibrio harveyi* (Table 2).

There are serial steps for research to evaluate the potential of phages in controlling these diseases and consequently to establish the practical procedures in aquaculture. These are (i) isolation of lytic phages from fish culture environments, using an enrichment method in general; (ii) culture and characterization of the phages, including phenotypic and genetic features; (iii) infectivity testing of the phages against a variety of bacterial strains (phage typing of the target bacterium); (iv) selection of phage candidates for therapeutic use; (v) evaluation of therapeutic efficacy of the phages against experimental infections in a laboratory setting followed by natural infections in a field test; (vi) determination of the presence of virulence genes or other toxic factors in the phage candidates; and (vii) establishment of large-scale culture and long-term preservation methods for the phages for commercial development.

PHAGE THERAPY OF *P. PLECOGLOSSICIDA* INFECTION IN AYU

Ayu *(Plecoglossus altivelis)* is an osmerid fish with a 1-year life cycle and is one of the most popular freshwater species for sport fishing as well as for aquaculture in Japan. Two bacterial diseases, bacterial hemorrhagic ascites caused by *P. plecoglossicida* (Nishimori et al., 2000) and bacterial cold-water disease (BCWD) caused by *F. psychrophilum* (Shotts and Starliper, 1999; Bernardet et al., 2002), have damaged the aquaculture industry and the natural resource enhancement program of this fish species since 1990. No drugs are licensed for *P. plecoglossicida* infection, while some antimicrobial agents, florfenicol and sulfisozole sodium, are used to treat BCWD. After treatment of BCWD with chemotherapeutics, however, *P. plecoglossicida* infection abruptly emerges and results in heavy mortality, often exceeding 50%. This is a typical example of microorganism substitution in fish disease. All isolates of *P. plecoglossicida* obtained from geographically and chronologically different sources are phe-

TABLE 2 Studies on phage therapy of bacterial infections in fish and shellfish

Disease	Pathogen	Host fish	Reference(s)
Bacterial hemorrhagic ascites	*Pseudomonas plecoglossicida*	Ayu *(Plecoglossus altivelis)*	Park et al., 2000 Park and Nakai, 2003
BCWD	*Flavobacterium psychrophilum*	Ayu *(Plecoglossus altivelis)*	Stenholm et al., 2008 Kim et al., 2010
Lactococcosis	*Lactococcus garvieae*	Yellowtail *(Seriola quinqueradiata)*	Park et al., 1997 Park et al., 1998 Nakai et al., 1999
Streptococcosis	*Streptococcus iniae*	Japanese flounder *(Paralichthys olivaceus)*	Matuoka et al., 2007
Furunculosis	*Aeromonas salmonicida*	Atlantic salmon *(Salmo salar)* Brook trout *(Salvelinus fontinalis)*	Verner-Jeffreys et al., 2007 Imbeault et al., 2006
Edwardsiellosis	*Edwardsiella tarda*	Japanese eel *(Anguilla japonica)*	Wu and Chao, 1982 Yamamoto and Maegawa, 2008
Eel red-fin disease	*Aeromonas hydrophila*	Loach *(Misgurnus anguillicaudatus)*	Wu et al., 1981
Luminous vibriosis	*Vibrio harveyi*	Black tiger prawn *(Penaeus monodon)*	Vinod et al., 2006 Shivu et al., 2007 Karunasagar et al., 2007

notypically homogeneous, though a nonmotile variant has been reported (Nakatsugawa and Iida, 1996). The bacterium consists of a single serotype (Park et al., 2000). At present, there are no practical procedures to control this disease other than reducing predisposing factors such as overcrowding and overfeeding.

Two types of bacteriophages specific to *P. plecoglossicida* were isolated from diseased ayu and the rearing pond water using the enrichment method (Park et al., 2000; Park and Nakai, 2003). One type of phage (PPpW-3), forming small plaques, was identified as a myovirus (of the family *Myoviridae*), and another type (PPpW-4), forming large plaques, was found to be a podovirus (the *Podoviridae*) (Fig. 1). All examined *P. plecoglossicida* strains exhibited quite similar sensitivities to phages of either type. In in vitro conditions, PPpW-4 inhibited the growth of *P. plecoglossicida* more effectively than PPpW-3, but the mixture of the two phages exhibited the highest inhibition. The lytic activities of phages were observed at temperatures from 10 to 30°C or less, which covers the entire range of rearing water temperature in ayu culture. Interestingly, in vitro-induced phage-resistant variants of a virulent *P. plecoglossicida* strain were less virulent to ayu than the parent strain. The 50% lethal dose (LD_{50}) of the in vitro variants was higher than 10^4 CFU/fish in ayu, while the phage-sensitive parent was highly virulent, with an LD_{50} of $10^{1.2}$ CFU/fish. Ultraviolet irradiation or mitomycin C induced no detectable temperate phages from any of the strains examined. Oral administration of phage-impregnated feed (pellets) to ayu increased resistance to experimental infection with *P. plecoglossicida* (Park et al., 2000; Park and Nakai, 2003). Fish were challenged by feeding live *P. plecoglossicida*-inoculated feed (estimated dose, 10^7 CFU/fish) and immediately fed with feed containing phage PPpW-3 or PPpW-4 or a mixture of PPpW-3 and PPpW-4. Cumulative mortality in the control fish groups 2 weeks after bacterial challenge was 93.3%, while mortalities of the fish receiving PPpW-3, PPpW-4, and a mixture of PPpW-3 and PPpW-4 phage-containing feed were 53.3%, 40.0%, and 20.0%, respectively, suggesting a synergistic effect by phages with different infectivities. Such protective effects of treatment with a mixture of the two phages were demonstrated in fish receiving phages 1 and 24 h after bacterial challenge. Inoculated *P. plecoglossicida* was isolated from all the kidneys of dead fish irrespective of phage treatment and from some survivors in control groups, but not from any of the fish that had received phages and survived.

P. plecoglossicida in fish receiving bacterium-inoculated feed first appeared in the kidney 3 h after feeding, and then was detected at levels of $10^{4.2}$ to $10^{5.1}$ CFU/g from all kidneys examined after 12 to 72 h. Phages were detected at $10^{5.2}$ and $10^{3.5}$ PFU/g in kidneys at 3 and 12 h, respectively, but disappeared at 24 h (Park et al., 2000). On the other hand, when fish received phage-containing feed after bacterial challenge, *P. plecoglossicida* was not detected in the kidneys of fish at 12 h or later (after a slight appearance at 3 h postchallenge), but the phages were detected in the kidneys even at 24 h postchallenge. These growth dynamics of administered bacteria and phages in fish indicate that the killing of bacteria by phages in the internal organs caused the survival of the fish.

After laboratory tests of phage therapy, a field trial was carried out using a culture pond where approximately 120,000 ayu were reared and *P. plecoglossicida* infection was prevailing, with approximately 1,000 dead fish per day for 2 weeks prior to phage treatment (Park and Nakai, 2003). The PPpW-3 and PPpW-4 phage-containing feed was administered three times on days 0, 1, and 8 to ayu using automatic feeding machines at a dose estimated at 10^7 PFU/fish. Administered phages were detected in 90% of fish kidneys 3 h after administration. Daily mortality of fish decreased at a constant rate (5% a day) to one-third of the pretreatment mortality after a 2-week period. The causal effect of phages in this treatment was verified by the appearance of administered phages in the kidneys of the fish, and by

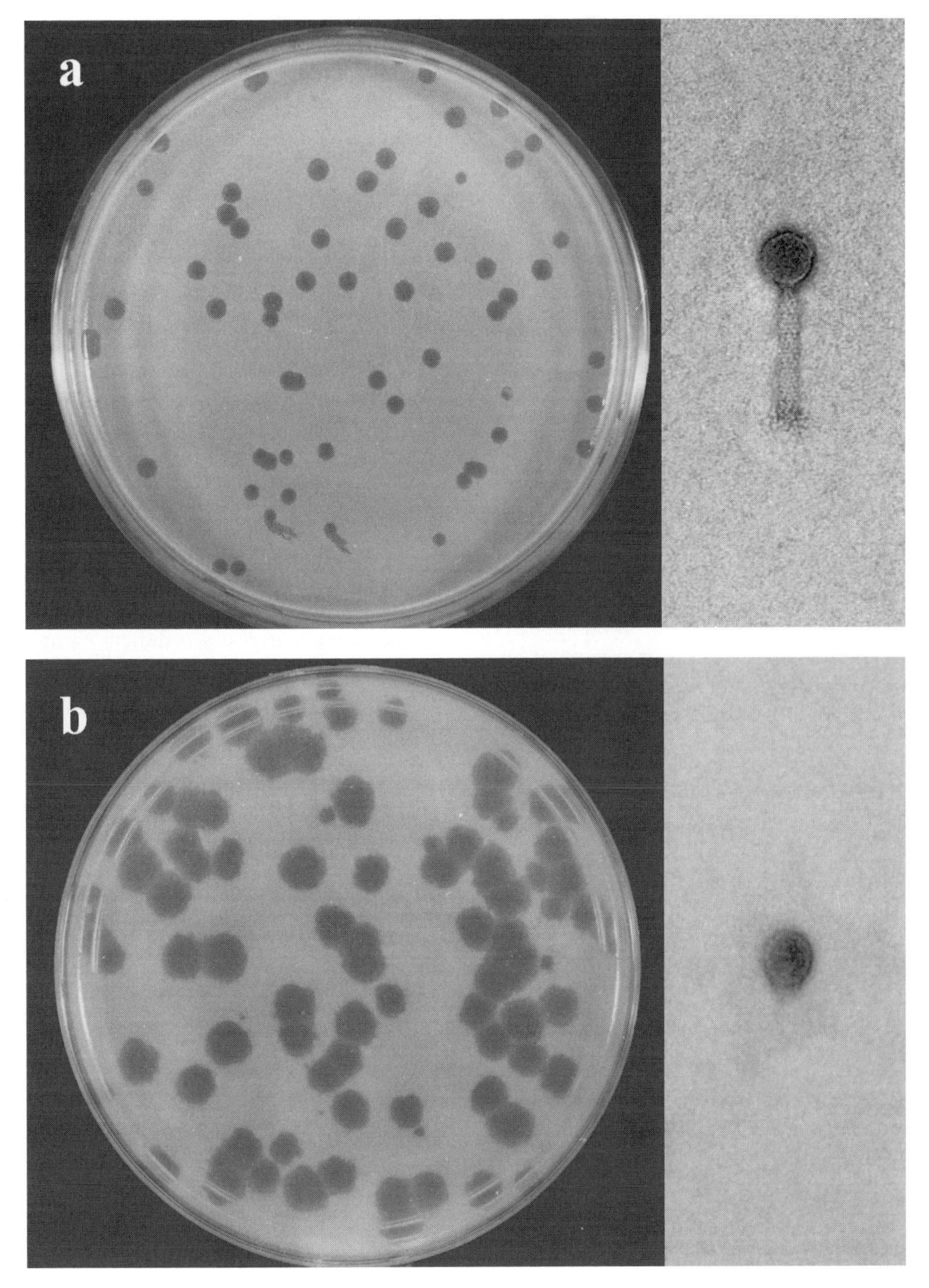

FIGURE 1 Plaques and electron micrographs of *P. plecoglossicida* phages. (a) PPpW-3. (b) PPpW-4. (Adapted from Park et al. [2000] with permission of the publisher.)

the rapid disappearance of *P. plecoglossicida* from apparently healthy fish (Fig. 2). *P. plecoglossicida*-specific phages were not isolated from the fish kidneys or the rearing water before phage treatment, but they were isolated constantly at 10^1 to 10^3 PFU/ml from the rearing water after initiation of phage treatment. Interestingly, neither phage-resistant *P. plecoglossicida* nor phage-neutralizing antibodies were detected in diseased fish or apparently healthy fish, respectively, throughout the observation period. Active immunization of ayu with live phages also demonstrated the low immunogenicity of the phages; no phage-neutralizing antibodies were detected in fish even after daily oral administration of 10^7 PFU/fish for 7 days or weekly intramuscular injections with 10^8 PFU/fish for 4 weeks. The efficacy of oral administration of phages against natural infection indicates the potential for phage control of the disease in practice. However, one problem in this field therapy was that another infection (BCWD) appeared in the late period of the phage treatment and caused mortality of fish, indicating the intertwined ecology of these diseases.

PHAGE THERAPY OF *F. PSYCHROPHILUM* INFECTION IN AYU

BCWD, also known as rainbow trout fry syndrome (RTFS), is caused by *F. psychrophilum* and has long been known as an important disease of salmonids in North America and Europe (Shotts and Starliper, 1999). The disease is characterized by erosion or ulceration of the skin and anemia. As stated previously, in Japan the disease has been one of the most serious diseases in ayu, both in cultured and wild populations, for the past 2 decades (Wakabayashi et al., 1994; Iida and Mizokami, 1996). Trials to develop efficacious vaccines have been carried out by many researchers, but commercial vaccines are still not available due to their low effectiveness (Rahman et al., 2002). Experimental infection by an immersion or bath method using bacterial suspensions of *F. psychrophilum* often fails to induce mortality in fish. The virulence factors of this pathogen remain unclear.

There are two reports on *F. psychrophilum* phages: one is from Danish rainbow trout *(Oncorhynchus mykiss)* (Stenholm et al., 2008) and another is from Japanese ayu (Kim et al., 2010). Stenholm et al. (2008) isolated 22 phages from water and sediment samples in Danish freshwater rainbow trout farms using the enrichment technique. The phages were divided into three groups based on genome size (8.5 to 12 kb, 48 kb, and 90 kb) and morphologically identified as members of the *Myoviridae, Siphoviridae,* or *Podoviridae.* Although the phages showed highly variable patterns of infectivity against the 27 tested *F. psychrophilum* strains isolated from rainbow trout

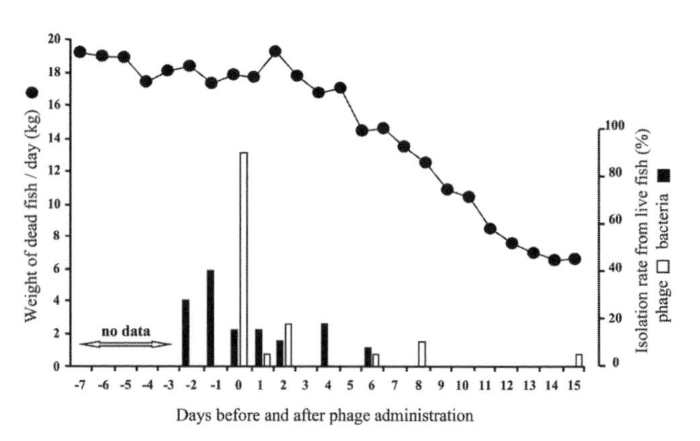

FIGURE 2 Daily mortality of ayu after phage treatment against natural infection with *P. plecoglossicida.* Fish in the *P. plecoglossicida* infection-prevailing culture pond received phage-containing feed (pellets) on days 0, 1, and 8. The presence of *P. plecoglossicida* phages and *P. plecoglossicida* was monitored by isolation from the kidneys of apparently healthy fish (n = 40) at days −2, −1, 0, 1, 2, 4, 6, 8, and 15. Fish at day 0 were examined 3 h after phage administration and fish at days 1 and 8 were examined before phage administration. (Adapted from Park and Nakai [2003] with permission of the publisher.)

in Denmark, some phage isolates with 48- or 90-kb genome size had strong infectivity with a broad host range, shown through their ability to infect between 20 and 23 of the 27 tested Danish *F. psychrophilum* strains. These results provide a foundation for future phage therapy of RTFS in rainbow trout aquaculture. On the other hand, phages with three morphotypes (myovirus, podovirus, and siphovirus) specific to *F. psychrophilum* were also isolated from rearing pond waters of ayu farms in Japan (Kim et al., 2010). Some of the myovirus and podovirus isolates formed larger plaques between 1 and 3 mm in diameter, with the others forming smaller plaques of less than 1 mm. There was considerable variation observed in the sensitivity to these phages among *F. psychrophilum* strains ($n = 72$) isolated from diseased ayu in Japan, dividing these strains into several phage sensitivity profiles. However, 77.4% of the tested *F. psychrophilum* isolates were sensitive to one or more phages. In contrast, *F. psychrophilum* strains from rainbow trout and amago trout *(Oncorhynchus rhodurus)* in Japan were not sensitive to any of the phages, indicating wide variations in phage sensitivity in this bacterium.

A preliminary phage therapy experiment was carried out using the phages isolated from ayu (T. Nakai, H. Yamashita, and S. C. Park, unpublished data). Ayu were first challenged with *F. psychrophilum* by intramuscular injection (10^5 CFU/fish) or immersion in a bacterial suspension (10^6 CFU/ml), where fish were rumpled vigorously in a net to injure the skin prior to infection. After bacterial challenge, fish were dipped in a phage (PFpW-3) suspension at 10^6 PFU/ml. During observation at 18°C, fish in the control groups apparently exhibited hemorrhage or ulceration on the skin, while such external disease signs were less severe in the phage-treated groups. In both bacterial challenges, the mortalities were significantly lower in the phage-treated groups compared with the control groups. Further detailed in vivo examinations, including effectiveness against natural infection, are required to develop phage therapy to treat BCWD in ayu or RTFS in rainbow trout.

PHAGE THERAPY OF *L. GARVIEAE* INFECTION IN YELLOWTAIL

L. garvieae, formerly *Enterococcus seriolicida,* has been known as the pathogen particularly affecting the fish of the genus *Seriola,* such as yellowtail *(S. quinqueradiata)* and amberjack *(S. dumerili)* in Japan and rainbow trout in Italy (Eldar et al., 1996; Kusada and Salati, 1999). The bacterium has also been isolated from cattle with mastitis, from clinical specimens of humans, and in plants (Eldar et al., 1996; Teixeira et al., 1996; Kawanishi et al., 2007). A cell capsule is the most crucial factor of the organism's virulence in fish, with virulent *L. garvieae* strains from fish consisting of a single serotype (Yoshida et al., 1997; Barnes and Ellis, 2004). *L. garvieae* is a typical opportunistic pathogen, as the bacterium is ubiquitous in fish and their culture environments. Several drugs including erythromycin and oxytetracycline are effective in treatment of the disease. Oral and injectable vaccines are now commercially available in Japan and other countries.

Phages specific to *L. garvieae* were isolated from fish kidneys and fish-rearing seawater and morphologically identified as members of the *Siphoviridae* (Park et al., 1997, 1998). Clinical and environmental strains ($n = 111$) of *L. garvieae* were divided into 14 phage types, with 1 major phage type containing 66% of the strains examined. However, at least 90% of *L. garvieae* strains were sensitive to some phage isolates. *L. garvieae* grows well between 17 and 41°C, but lytic activity of the phages was observed only at 29°C or lower. In in vivo conditions, a representative phage (PLgY-16) with broad host range was detected in the spleens of yellowtail up to 24 h after intraperitoneal (i.p.) injection, and the phage was recovered from the intestines of fish 3 h after the oral administration of phage-containing feed, but was undetectable after 10 h. Simultaneous administration of live *L. garvieae* and phage enhanced the survival time of the phage; in this case the phage was recovered from the spleen even 5 days after i.p. injection or from the intestine 24 h after oral administration (Nakai et al., 1999).

The protective effects of anti-*L. garvieae* phage (PLgY-16) by i.p. or oral administration were examined in experimentally infected young yellowtail (Nakai et al., 1999). After i.p. challenge with *L. garvieae,* the survival rate of fish receiving i.p. injection ($10^{7.5}$ PFU/fish) of the phage (100%) was much higher than that of the control fish that did not receive phage (10%). Similar protective effects were demonstrated even in fish that received phage treatment 24 h after bacterial challenge. In order to facilitate transfer of phage into the fish organs, the previously phage-infected bacterial cells were injected i.p. into fish after bacterial challenge. However, the present use of bacterial cells as a phage vehicle, like the "Trojan horse," did not influence the effect of phage. Protection was also obtained in yellowtail receiving phage-containing feed (approximately 10^7 PFU/fish) followed by challenge with an anal intubation with *L. garvieae*. Anally intubated *L. garvieae* was detected constantly in the spleens of the control fish for 72 h or longer, while it was detected sporadically and disappeared from the phage-treated fish after 48 h. On the other hand, orally administered phage were detected in the intestines and spleens of the phage-treated fish 3 to 48 h later, with a maximum of 10^6 PFU/g. Phage-resistant mutants are fairly common in in vitro *L. garvieae* cultures, but all *L. garvieae* isolates from dead fish obtained during the in vivo experiments were still susceptible to the phage used. No neutralizing antibodies were detected in the sera of yellowtail that repeatedly received phage-containing feed.

PHAGE THERAPY OF *S. INIAE* INFECTION IN YELLOWTAIL

S. iniae is one of the causative agents of β-hemolytic streptococcosis in a variety of cultured freshwater and marine fish species worldwide (Kusuda and Salati, 1999). The *S. iniae* isolates from diseased fish are less variable in their biochemical and serological characteristics, and the cell capsule has been shown to be a major virulence factor of this organism (Yoshida et al., 1996; Miller and Neely, 2005). The serotype of virulent *S. iniae* isolated from fish is uniform (Barnes et al., 2003; Kanai et al., 2006). In Japan, *S. iniae* has caused serious damage, particularly in cultured Japanese flounder *(Paralichthys olivaceus),* since the 1980s, due mainly to difficulties in the chemotherapeutic control of this pathogen. In 2005, an injectable vaccine for Japanese flounder was made commercially available in Japan.

Phages specific to *S. iniae* were isolated from rearing waters of Japanese flounder by the enrichment method. All phage isolates were morphologically identified as members of the *Siphoviridae*. *S. iniae* strains isolated from fish in Japan and Korea showed similar sensitivity to these phage isolates, while the *S. iniae* type strains isolated from dolphins, and *S. iniae* strains isolated from terrestrial animals or plants, were not sensitive (Matsuoka et al., 2007; Kawanishi et al., 2006, 2007).

The therapeutic effects of the *S. iniae* phages were examined using Japanese flounder (Matsuoka et al., 2007). In four trials, fish were injected i.p. with *S. iniae* (10^5 to 10^7 CFU/fish) and 1 h later injected i.p. with a mixture of phage isolates (10^8 PFU/fish). Mortalities of phage-treated fish were always significantly lower than those of untreated control fish. Phage treatment was effective even at 24 h postinfection, when cell numbers of *S. iniae* were $10^{7.7}$ and $10^{4.5}$ CFU/g in the kidneys and brains of fish, respectively. This indicates that phage injection is effective against a bacteremic condition of the brain. However, *S. iniae* strains isolated from dead fish after phage treatment were all resistant to the phages used. These phage-resistant *S. iniae* isolates (n = 11) exhibited virulence to Japanese flounder in i.p. injection at a dose of 10^5 CFU/fish. To overcome this problem, i.e., to prevent appearance of phage-resistant, virulent *S. iniae,* new phages lytic against the phage-resistant *S. iniae* strains are required. However, several attempts at isolation of such phages remain unsuccessful (E. Iwamoto, M. Ishida, S. Matsuoka, and T. Nakai, unpublished data). Another unexpected result was that oral administration of phages had no effect on experimental infection. In contrast to the cases of *P. plecoglossicida* and *L. garvieae* phages, orally administered or anally intubated *S. iniae* phages were not detected in the kidneys of inoculated fish; the ineffectiveness observed may be due partly to lower penetration of these phages via the fish intestinal wall.

PHAGE THERAPY OF OTHER BACTERIAL INFECTIONS IN FISH

A. salmonicida, a nonmotile aeromonad, was first described in 1894 and has been well-known as the causative pathogen of furunculosis of salmonids (Hiney and Olivier, 1999). The bacterium is serologically uniform and has a protein layer, designated the A-layer, on the cell surface that contributes to intracellular survival of the organism (Udey and Fryer, 1978). The previously reported phages of *A. salmonicida* (Rodgers et al., 1981) were used to examine the therapeutic effects against experimental furunculosis by Verner-Jeffreys et al. (2007). Juvenile Atlantic salmon *(Salmo salar)* were injected i.p. with a lethal dose of *A. salmonicida,* followed by i.p. injection of a mixture of three phages (10^8 PFU/fish). Fish treated with phages died at a significantly slower rate than those in the control group, but the end result was not affected. No protection was offered in fish by oral or bath treatment of phages, and phage-resistant *A. salmonicida* was recovered from dead fish in all phage treatment groups, with the conclusion that furunculosis in Atlantic salmon is not readily controllable by phages. A similar significant delay was seen in the onset of mortality of brook trout *(Salvelinus fontinalis)* that were bath-challenged with *A. salmonicida* and then immersion-treated with phages, but the authors suggested the potential of phages to prevent furunculosis in fish farms (Imbeault et al., 2006).

E. tarda, belonging to the family *Enterobacteriaceae,* has been isolated from a wide variety of animals, including mammals and humans (Sakazaki, 2005). The bacterium has long been known as a pathogen of various cultured freshwater and marine fishes worldwide (Plumb, 1999). The bacterium is resistant to killing by fish phagocytes (Iida and Wakabayashi, 1993), which makes chemotherapeutic treatment and development of vaccines difficult. Virulent strains of *E. tarda* isolated from fish are serologically uniform, consisting of a single serotype. *E. tarda* phages, which were isolated from eel (freshwater) or flounder (seawater) rearing waters, have been described (Wu and Chao, 1982; Matsuoka and Nakai, 2004; Yamamoto and Maegawa, 2008). Infectivity experiments with phages against *E. tarda* strains from fish suggest that the phage susceptibility of the bacterium is fairly variable, depending on the host fish species from which the bacterium is isolated. There are no reports on therapeutic effects of *E. tarda* phages, but the following results were obtained in a preliminary study using Japanese flounder or red sea bream *(Pagrus major)* (E. Iwamoto, K. Hanyu, and T. Nakai, unpublished data). Fish experimentally infected with *E. tarda* were treated with a phage mixture consisting of one *Myoviridae* and one *Podoviridae* phage isolate, which was administrated by various routes (injection, oral, or immersion). Although a slightly lower mortality was noticed in some of the phage-treated groups, therapeutic effects of phages were not clearly demonstrated. On the other hand, when fish were injected i.p. with the phage mixture prior to experimental infection with *E. tarda,* a significantly lower mortality was recorded in the phage-treated group. This prophylactic effect suggests that the bacterial cells were killed by phages before an intracellular population of the pathogen could be established.

In addition to their application as therapeutic or prophylactic agents, phages can be a good indicator to forecast disease outbreaks. It was reported that *E. tarda* phages were detected in culture environment of fish 1 month before the disease outbreak or prior to the detection of the bacterium (Matsuoka and Nakai, 2004).

PHAGE THERAPY OF *V. HARVEYI* INFECTION IN SHRIMP

A luminous bacterium, *V. harveyi* is a well-known shrimp pathogen worldwide, causing mass mortality in particular in the shrimp species *Penaeus monodon* at the larval stage (Lavilla-Pitogo et al., 1990). Frequent use of antibiotics to prevent bacterial infections in shrimp hatcheries has resulted in the emergence of antibiotic-resistant *V. harveyi* (Karunasagar et al., 1994). As an alternative to antibiotics, the potential of phages to control *V. harveyi* infection in a shrimp hatchery was reported by a research group in India (Vinod et al., 2006; Shivu et al., 2007; Karunasagar et al., 2007).

V. harveyi phages were isolated from shrimp farm waters using the enrichment method, and a phage isolate showing maximum lytic activity was used for further experiments (Vinod et al., 2006). All of the 50 *V. harveyi* isolates obtained from different sources were sensitive to the phage isolate. In a laboratory trial, *V. harveyi* was inoculated at 10^5 cells/ml into plastic tubs, where 18-day-old postlarvae of *P. monodon* were kept, and then the phage was inoculated at 10^9 PFU/ml. Larval survival at 48 h was 70% in the phage-treated group with a simultaneous 2-log reduction in luminous bacterial count, while survival was 25% in the control group with a 1-log increase in the bacterial count. Therapeutic or preventive effects of the phage were also demonstrated against natural infections of *V. harveyi*. Nauplii (n = 35,000) of *P. monodon* were stocked in 500-liter tanks and reared with daily addition of the phage at 10^5 PFU/ml or antibiotics (oxytetracycline at 5 ppm and kanamycin at 10 ppm) for 17 days through the stages of zoea, mysis, and postlarvae. Survival rates were 86% in the phage-treated group, 40% in the antibiotic-treated group, and 17% in the control group that received no treatment. These results suggest that phages have the potential for control of *V. harveyi* infection in hatchery settings. Other *V. harveyi* phages with different lytic activity were also isolated and characterized as candidates for hatchery use (Shivu et al., 2007), and the effects of the phages in reducing *V. harveyi* in biofilm populations were reported (Karunasagar et al., 2007).

POTENTIAL OF PHAGE THERAPY IN AQUACULTURE

Phage therapy against bacterial diseases of fish remains an inadequately studied topic, and further intensive investigations of various aspects of this practice are required to establish its practical feasibility in aquaculture. However, as shown in the studies against *P. plecoglossicida* infection of ayu (Park et al., 2000; Park and Nakai, 2003), where oral administration of phage-containing feed markedly reduced the mortality of fish due to natural infection, phage therapy is likely to find applications in aquaculture. As is the case with chemical treatments, oral administration is of practical value as a route to treat a large number of fish, which often exceeds 100,000 fish in a culture pond or net-pen in the open sea. Although it has been noted that oral administration of phages is difficult in other animals due to the low pH of the digestive tract (Smith et al., 1987b), it does not seem to be a determinant for phage treatment for fish, as pH levels in fish digestive tracts are relatively high (Nakai et al., 1999). Considering that the intestine of fish is a portal of entry or an initial multiplication site for fish pathogens in many infections, orally administered phages can directly attack resident pathogens in the intestine prior to invasion of the internal organs. It is surprising that orally administered *P. plecoglossicida* and *L. garvieae* phages quickly appeared in the kidneys of fish but *S. iniae* phages did not, suggesting that oral phage therapy depends on the ability of phages to pass through the intestinal epithelial layer or other portals. Transfer of phages from the alimentary tract to the blood circulation system has also been observed in humans and rats (Hildebrand and Wolochow, 1962; Slopek et al., 1987). Bath administration of phages may also be effective for those diseases in which infection is initiated by bacterial colonization of the skin and gills. BCWD is a typical example of this infection mode. Bath or immersion methods are particularly useful to treat larval and juvenile fish and shellfish, or eggs in hatcheries, as shown in the phage treatment of shrimp larvae (Vinod et al., 2006), with the expectation of prophylactic effects rather than treatment. Administration of phages by injection may be inconvenient to treat a large number of fish, irrespective of its accurate inoculation into the fish body, but it is tenable given the fact that a variety of injectable vaccines for fish are now commercially available (Sommerset et al., 2005). This variability in phage administration routes is very advantageous to aquaculture, where microbial infections occur at various stages from eggs to broodstocks. In addition, only limited data are now available on effective phage doses in fish. However, in contrast to chemicals and other substances, determination of a precise initial dose given to individual fish may not be essential in the aquaculture setting, because of the self-perpetuating nature of the phages in infected individuals or pathogen-contaminated water.

Previous reviews have pointed out the narrow host specificity of phages as an intrinsic obstacle to phage treatment and prevention for terrestrial animals (Barrow and Soothill, 1997; Alisky et al., 1998). Phages are strain specific rather than species specific, which leads to difficulty in preparing phages against highly diverse bacterial variants. However, the case of fish pathogens does not indicate that this is always an essential weak point of phage therapy. As previously mentioned, *P. plecoglossicida* is serologically uniform and consists of a single phage type. Similarly, a major phage type exists in *L. garvieae* isolated from fish, and some phage isolates are so broad in their infectivity that they can lyse 90% or more of strains from fish. The cell surface substance (capsule) as a virulence factor of *L. garvieae* may explain the low diversity of this organism (Yoshida et al., 1997), though such a surface substance has not been identified in *P. plecoglossicida*. Similar low diversity in phage sensitivity was also found in other fish pathogens *(S. iniae, A. salmonicida,* and *V. harveyi),* with the exceptions of *F. psychrophila* and *E. tarda*. If a given target bacterium is highly variable in its phage types or very changeable in its phage sensitivity, large collections of therapeutic phages with different lytic activities are undoubtedly required in phage therapy practices. In this point, the existence of phages having an intraspecies broad lytic activity ensures such a requirement for phage therapy. The most important consideration for phage cocktail preparation will be a combination of phages bringing synergistic effects, as shown with *P. plecoglossicida* phages.

The appearance of phage-resistant bacteria after phage treatment, like drug-resistant bacteria in chemotherapy, has been predicted as another obstacle. Phage-resistant mutants are commonly induced in bacterial culture in vitro in the presence of specific phages. However, phage-resistant variants of *P. plecoglossicida,* which were induced in vitro, lacked virulence for ayu (Park et al., 2000), and all *P. plecoglossicida* isolates from dead fish obtained during therapy experiments were still susceptible to phages used for treatment. In successful phage control against a systemic infection of mice with a K1 strain of *E. coli* or diarrhea in calves caused by enteropathogenic *E. coli* strains, only the less virulent capsule-deficient mutants emerged as phage-resistant organisms (Smith and Huggins, 1982, 1983). A surface component associated with bacterial virulence also seemed to be the receptor for phage attachment, and consequently phage-resistant variants of a virulent organism would not be pathogenic (Barrow and Soothill, 1997; Barrow et al., 1998; Nakai and Park, 2002). On the other hand, this phenomenon was not observed in the therapy of *S. iniae* infection (Matsuoka et al., 2007). In this case, the majority of phage-resistant strains isolated from dead fish after phage treatment as well as after in vitro challenges were still pathogenic to fish. The mechanism of phage resistance in this system remains unsolved, but in order for phage therapy to remain a viable option in this system new phages that have the ability to lyse these mutants must be obtained. This phenomenon resembles the story of drug resistance in chemotherapy. However, the vast number of phages in nature, and the natural coevolution of phages and their hosts, is expected to make it possible to identify new phages against phage-resistant bacteria (Sulakvelidze and Kutter, 2005).

Phage therapy targeting intracellular pathogens is thought to be far more difficult because phages cannot reach the host bacterial cells residing in phagocytes (Sulakvelidze et al., 2001; Emery and Whittington, 2004). A failure in phage therapy against *A. salmonicida* or *E. tarda* infection may be associated with their intracellular parasitic nature. This possibility was supported by a preliminary experiment showing that infection with *E. tarda* was prevented by administration of phages prior to bacterial infection. Prophylactic use of phages for intracellular pathogens is worth future consideration.

The risk that phages might mediate genetic exchange among bacteria, i.e., transduction or phage conversion, is the final obstacle to be examined. It is well known that some temperate phages contribute to bacterial virulence (Wagner and Waldor, 2002). Temperate phages with broad infectivity over species would strongly support anti-phage therapy views (Jensen et al., 1998). In fact, it was reported that some phages mediate the toxicity or virulence of *V. harveyi* in *P. monodon* (Ruangpan et al., 1999; Munro et al., 2003). Needless to say, the genome of any therapeutic phage should be examined for any known virulence or toxin genes (Wagner and Waldor, 2002) prior to its use. However, considering the importance of phages in regulating populations of planktonic bacteria (Bergh et al., 1989; Proctor and Fuhrman, 1990), it is postulated that naturally occurring phages might contribute to the killing of pathogens proliferating in naturally infected hosts (Nakai and Park, 2002). If this is true and phages isolated from natural environments are used in treatment, phage conversion and other unexpected changes in both environmental and pathogenic bacteria might not be newly added by the creation of phage therapy activities. Because lytic phages of pathogenic bacteria are generally isolated only from restricted environments where infection prevails, most of such events must have already happened in nature, as the result of endless battles between bacteria and phages (Lenski, 1984).

Phage therapy is both an old and an attractive new approach to prevent and control bacterial infections and potentially applicable to any field affected by bacterial infection, including aquaculture. However, consideration of the predisposing factors that might compromise the immune system of the host should precede every treatment, particularly in fish and shellfish diseases, the majority of which are caused by opportunistic pathogens. The essential role of phages as therapeutic agents is not to extinguish the target bacteria in the host but to reduce their number and assist the host's natural defense system. The aforementioned phage therapy studies on fish and shellfish have been carried out by only a few researchers for the past 10 years. Thus, more research is needed to establish the practical procedures and finally to gain the approval of the regulatory agencies for commercial development. Further energetic investigations are indispensable to reach the final goal.

REFERENCES

Alisky, J., K. Iczkowski, A. Rapoport, and N. Troitsky. 1998. Bacteriophages show promise as antimicrobial agents. *J. Infect.* **36:**5–15.

Austin, B., and D. Austin. 2007. Zoonoses, p. 408–411. *In* B. Austin and D. Austin (ed.), *Bacterial Fish Pathogens: Diseases of Farmed and Wild Fish,* 4th ed. Praxis Publishing, Chichester, United Kingdom.

Barnes, A. C., and A. E. Ellis. 2004. Role of capsule in serotypic differences and complement fixation by *Lactococcus garvieae. Fish Shellfish Immunol.* **16:**207–214.

Barnes, A. C., F. M. Young, M. T. Horne, and A. E. Ellis. 2003. *Streptococcus iniae:* serological differences, presence of capsule and resistance to immune serum killing. *Dis. Aquat. Org.* **53:**241–247.

Barrow, P., M. Lovell, and A. Berchieri, Jr. 1998. Use of lytic bacteriophage for control of experimental *Escherichia coli* septicemia and meningitis in chickens and calves. *Clin. Diagn. Lab. Immunol.* **5:**294–298.

Barrow, P. A., and J. S. Soothill. 1997. Bacteriophage therapy and prophylaxis: rediscovery and renewed assessment of potential. *Trends Microbiol.* **5:** 268–271.

Bergh, Ø., K. Y. Børsheim, G. Bratbak, and M. Heldal. 1989. High abundance of viruses found in aquatic environments. *Nature* **340:**467–468.

Bernardet, J.-F., Y. Nakagawa, and B. Holmes. 2002. Proposed minimal standards for describing new taxa of the family *Flavobacteriaceae* and emended description of the family. *Int. J. Syst. Evol. Microbiol.* **52:**1049–1070.

Biosca, E. G., C. Amaro, J. L. Larsen, and K. Pedersen. 1997. Phenotypic and genotypic characterization of *Vibrio vulnificus:* proposal for the substitution of the subspecific taxon biotype for serovar. *Appl. Environ. Microbiol.* **63:**1460–1466.

Eldar, A., C. Ghittino, L. Asanta, E. Bozzetta, M. Goria, M. Prearo, and H. Bercovier. 1996. *Enterococcus seriolicida* is a junior synonym of *Lactococcus garvieae,* a causative agent of septicemia and meningoencephalitis in fish. *Curr. Microbiol.* **32:** 85–88.

Emery, D. L., and R. J. Whittington. 2004. An evaluation of mycophage therapy, chemotherapy and vaccination for control of *Mycobacterium avium* subsp. *paratuberculosis* infection. *Vet. Microbiol.* **104:** 143–155.

Goncalves, J. R., G. Brum, A. Fernandes, I. Biscaia, M. J. Correia, and J. Bastardo. 1992. *Aeromonas hydrophila* fulminant pneumonia in a fit young man. *Thorax* **47:**482–483.

Hargreaves, J. E., and D. R. Lucey. 1990. Life-threatening *Edwardsiella tarda* soft-tissue infection associated with catfish puncture wound. *J. Infect. Dis.* **162:**1416–1417.

Hildebrand, G. J., and H. Wolochow. 1962. Translocation of bacteriophage across the intestinal wall of the rat. *Proc. Soc. Exp. Biol. Med.* **109:**183–185.

Hiney, M., and G. Olivier. 1999. Furunculosis *(Aeromonas salmonicida),* p. 341–425. *In* P. T. K. Woo and D. W. Bruno (ed.), *Fish Diseases and Disorders,* vol. 3. *Viral, Bacterial and Fungal Infections.* CABI Publishing, London, United Kingdom.

Iida, T., and H. Wakabayashi. 1993. Resistance of *Edwardsiella tarda* to opsonophagocytosis of eel neutrophils. *Fish Pathol.* **28:**191–192.

Iida, Y., and A. Mizokami. 1996. Outbreaks of coldwater disease in wild ayu and pale chub. *Fish Pathol.* **31:**157–164.

Imbeault, S., S. Parent, M. Lagacé, C. F. Uhland, and J.-F. Blais. 2006. Using bacteriophages to prevent furunculosis caused by *Aeromonas salmonicida* in farmed brook trout. *J. Aquat. Anim. Health* **18:**203–214.

Irianto, A., and B. Austin. 2002. Probiotics in aquaculture. *J. Fish Dis.* **25:**633–642.

Janda, J. M., and S. L. Abbott. 1993. Expression of an iron-regulated hemolysin by *Edwardsiella tarda. FEMS Microbiol. Lett.* **111:**275–280.

Jensen, E. C., H. S. Schrader, B. Rieland, T. L. Thompson, K. W. Lee, K. W. Nickerson, and T. A. Kokjohn. 1998. Prevalence of broad-host-range lytic bacteriophages of *Sphaerotilus natans, Escherichia coli,* and *Pseudomonas aeruginosa. Appl. Environ. Microbiol.* **64:**575–580.

Kanai, K., M. Notohara, T. Kato, K. Shutou, and K. Yoshikoshi. 2006. Serological characterization of *Streptococcus iniae* strains isolated from cultured fish in Japan. *Fish Pathol.* **41:**57–66.

Karunasagar, I., R. Pai, G. R. Malathi, and I. Karunasagar. 1994. Mass mortality of *Penaeus monodon* larvae due to antibiotic resistant *Vibrio harveyi* infection. *Aquaculture* **128:**203–209.

Karunasagar, I., M. M. Shivu, S. K. Girisha, G. Krohne, and I. Karunasagar. 2007. Biocontrol of pathogens in shrimp hatcheries using bacteriophages. *Aquaculture* **268:**288–292.

Kawanishi, M., T. Yoshida, M. Kijima, K. Yagyu, T. Nakai, S. Okada, A. Endo, M. Murakami, S. Suzuki, and H. Morita. 2007. Characterization of *Lactococcus garvieae* isolated from radish and broccoli sprouts that exhibited a KG^{+} phenotype, lack of virulence and absence of a capsule. *Lett. Appl. Microbiol.* **44:**481–487.

Kawanishi, M., T. Yoshida, S. Yagashiro, M. Kijima, K. Yagyu, T. Nakai, M. Murakami, H. Morita., and S. Suzuki. 2006. Differences between *Lactococcus garvieae* isolated from the genus *Seriola* in Japan and those isolated from other animals (trout, terrestrial animals from Europe) with regard to pathogenicity, phage susceptibility and genetic characterization. *J. Appl. Microbiol.* **101:** 496–504.

Kim, J. H., D. K. Gomez, T. Nakai, and S. C. Park. 2010. Isolation and identification of bacteriophages infecting ayu *Plecoglossus altivelis altivelis* specific *Flavobacterium psychrophilum. Vet. Microbiol.* **140:**109–115.

Kusuda, R., and F. Salati. 1999. *Enterococcus seriolicida* and *Streptococcus iniae,* p. 303–317. *In* P. T. K. Woo and D. W. Bruno (ed.), *Fish Diseases and Disorders,* vol. 3. *Viral, Bacterial and Fungal Infections.* CABI Publishing, London, United Kingdom.

Lau, S. K. P., P. C. Y. Woo, W.-K. Luk, A. M. Y. Fung, W.-T. Hui, A. H. C. Fong,

C.-W. Chow, S. S. Y. Wong, and K.-Y. Yuen. 2006. Clinical isolates of *Streptococcus iniae* from Asia are more mucoid and β-hemolytic than those from North America. *Diagn. Microbiol. Infect. Dis.* **54:**177–181.

Lavilla-Pitogo, C. R., M. C. L. Baticados, E. R. Cruz-Lacierda, and L. D. de la Pena. 1990. Occurrence of luminous bacterial disease of *Penaeus monodon* larvae in the Philippines. *Aquaculture* **9:**1–13.

Lenski, R. E. 1984. Coevolution of bacteria and phage: are there endless cycles of bacterial defenses and phage counterdefenses? *J. Theor. Biol.* **108:** 319–325.

Matsuoka, S., T. Hashizume, H. Kanzaki, E. Iwamoto, S. C. Park, T. Yoshida, and T. Nakai. 2007. Phage therapy against β-hemolytic streptococcosis of Japanese flounder *Paralichthys olivaceus. Fish Pathol.* **42:**181–189.

Matsuoka, S., and T. Nakai. 2004. Seasonal appearance of *Edwardsiella tarda* and its bacteriophages in the culture farms of Japanese flounder. *Fish Pathol.* **39:**145–152.

Merril, C. R., D. Scholl, and S. L. Adhya. 2003. The prospect for bacteriophage therapy in Western medicine. *Nat. Rev. Drug Discov.* **2:**489–497.

Miller, J. D., and M. N. Neely. 2005. Large-scale screen highlights the importance of capsule for virulence in the zoonotic pathogen *Streptococcus iniae. Infect. Immun.* **73:**921–934.

Munro, J., J. Oakey, E. Bromage, and L. Owens. 2003. Experimental bacteriophage-mediated virulence in strains of *Vibrio harveyi. Dis. Aquat. Org.* **54:**187–194.

Nakai, T., and S. C. Park. 2002. Bacteriophage therapy of infectious diseases in aquaculture. *Res. Microbiol.* **153:**13–18.

Nakai, T., R. Sugimoto, K.-H. Park, S. Matsuoka, K. Mori, T. Nishioka, and K. Maruyama. 1999. Protective effects of bacteriophage on experimental *Lactococcus garvieae* infection in yellowtail. *Dis. Aquat. Org.* **37:**33–41.

Nakatsugawa, T., and Y. Iida. 1996. *Pseudomonas* sp. isolated from diseased ayu, *Plecoglossus altivelis. Fish Pathol.* **31:**221–227.

Nishibuchi, M., K. Muroga, R. J. Seidler, and J. L. Fryer. 1979. Pathogenic *Vibrio* isolated from cultured eels. IV. Deoxyribonucleic acid studies. *Bull. Jpn. Soc. Sci. Fish.* **45:**1469–1473.

Nishimori, E., K. Kita-Tsukamoto, and H. Wakabayashi. 2000. *Pseudomonas plecoglossicida* sp. nov., the causative agent of bacterial hemorrhagic ascites of ayu, *Plecoglossus altivelis. Int. J. Syst. Evol. Microbiol.* **50:**83–89.

Park, K.-H., H. Kato, T. Nakai, and K. Muroga. 1998. Phage typing of *Lactococcus garvieae* (formerly *Enterococcus seriolicida*), a pathogen of cultured yellowtail. *Fish. Sci.* **64:**62–64.

Park, K.-H., S. Matsuoka, T. Nakai, and K. Muroga. 1997. A virulent bacteriophage of *Lactococcus garvieae* (formerly *Enterococcus seriolicida*) isolated from yellowtail *Seriola quinqueradiata. Dis. Aquat. Org.* **29:**145–149.

Park, S. C., and T. Nakai. 2003. Bacteriophage control of *Pseudomonas plecoglossicida* infection in ayu *Plecoglossus altivelis. Dis. Aquat. Org.* **53:**33–39.

Park, S. C., I. Shimamura, M. Fukunaga, K. Mori, and T. Nakai. 2000. Isolation of bacteriophages specific to a fish pathogen, *Pseudomonas plecoglossicida,* as a candidate for disease control. *Appl. Environ. Microbiol.* **66:**1416–1422.

Plumb, J. A. 1999. *Health Maintenance and Principal Microbial Diseases of Cultured Fishes.* Iowa State University Press, Ames, IA.

Proctor, L. M., and J. A. Fuhrman. 1990. Viral mortality of marine bacteria and cyanobacteria. *Nature* **343:**60–62.

Rahman, M. H., A. Kuroda, J. M. Dijkstra, I. Kiryu, T. Nakanishi, and M. Ototake. 2002. The outer membrane fraction of *Flavobacterium psychrophilum* induces protective immunity in rainbow trout and ayu. *Fish Shellfish Immunol.* **12:**169–179.

Reid, G. 1999. Testing the efficacy of probiotics, p. 129–140. *In* G. W. Tannock (ed.), *Probiotics: a Critical Review.* Horizon Scientific Press, Wymondham, United Kingdom.

Rodgers, C. J., J. H. Pringle, D. H. McCarthy, and B. Austin. 1981. Quantitative and qualitative studies of *Aeromonas salmonicida* bacteriophage. *J. Gen. Microbiol.* **125:**335–345.

Ruangpan, L., Y. Danayadol, S. Direkbusarakom, S. Siurairatana, and T. W. Flegel. 1999. Lethal toxicity of *Vibrio harveyi* to cultivated *Penaeus monodon* induced by a bacteriophage. *Dis. Aquat. Org.* **35:**195–201.

Ryan, J. M., and G. D. R. Bryant. 1997. Fish tank granuloma—a frequently misdiagnosed infection of the upper limb. *J. Accid. Emerg. Med.* **14:**398–399.

Sakazaki, R. 2005. Genus XI. *Edwardsiella* Ewing and McWhorter 1965, 37^{AL}, p. 657–661. *In* D. J. Brenner, N. R. Krieg, J. T. Staley, and G. M. Garrity (ed.), *Bergey's Manual of Systematic Bacteriology,* 2nd ed., vol. 2, part B. Springer, New York, NY.

Shivu, M. M., B. C. Rajeeva, S. K. Girisha, I. Karunasagar, G. Krohne, and I. Karunasagar. 2007. Molecular characterization of *Vibrio harveyi* bacteriophages isolated from aquaculture environment along the coast of India. *Environ. Microbiol.* **9:**322–331.

Shotts, E. B., Jr., and C. E. Starliper. 1999. Flavobacterial diseases: columnaris disease, cold-water disease and bacterial gill disease, p. 559–576. *In* P. T. K. Woo and D. W. Bruno (ed.), *Fish Diseases and Disorders,* vol. 3. *Viral, Bacterial and Fungal Infections.* CABI Publishing, London, United Kingdom.

Slopek, S., B. Weber-Dabrowska, M. Dabrowski, and A. Kucharewicz-Krukowska. 1987. Results of bacteriophage treatment of suppurative bacterial infections in the years 1981–1986. *Arch. Immunol. Ther. Exp.* (Warsz.) **35:**569–583.

Smith, H. W., and M. B. Huggins. 1982. Successful treatment of experimental *Escherichia coli* infections in mice using phage: its general superiority over antibiotics. *J. Gen. Microbiol.* **128:**307–318.

Smith, H. W., and M. B. Huggins. 1983. Effectiveness of phages in treating experimental *Escherichia coli* diarrhoea in calves, piglets and lambs. *J. Gen. Microbiol.* **129:**2659–2675.

Smith, H. W., M. B. Huggins, and K. M. Shaw. 1987a. The control of experimental *Escherichia coli* diarrhoea in calves by means of bacteriophages. *J. Gen. Microbiol.* **133:**1111–1126.

Smith, H. W., M. B. Huggins, and K. M. Shaw. 1987b. Factors influencing the survival and multiplication of bacteriophages in calves and in their environment. *J. Gen. Microbiol.* **133:**1127–1135.

Sommerset, I., B. Krossøy, E. Biering, and P. Frost. 2005. Vaccines for fish in aquaculture. *Expert Rev. Vaccines* **4:**89–101.

Stenholm, A. R., I. Dalsgaard, and M. Middelboe. 2008. Isolation and characterization of bacteriophages infecting the fish pathogen *Flavobacterium psychrophilum. Appl. Environ. Microbiol.* **74:** 4070–4078.

Sulakvelidze, A., Z. Alavidze, and J. G. Morris, Jr. 2001. Bacteriophage therapy. *Antimicrob. Agents Chemother.* **45:**649–659.

Sulakvelidze, A., and P. Barrow. 2005. Phage therapy in animals and agribusiness, p. 335–380. *In* E. Kutter and A. Sulakvelidze (ed.), *Bacteriophages: Biology and Applications,* CRC Press, Boca Raton, FL.

Sulakvelidze, A., and E. Kutter. 2005. Bacteriophage therapy in humans, p. 381–436. *In* E. Kutter and A. Sulakvelidze (ed.), *Bacteriophages: Biology and Applications,* CRC Press, Boca Raton, FL.

Teixeira, L. M., V. L. Merquior, M. C. Vianni, M. G. Carvalho, S. E. Fracalanzza, A. G. Steigerwalt, D. J. Brenner, and R. R. Facklam. 1996. Phenotypic and genotypic characterization of atypical *Lactococcus garvieae* strains isolated from water buffalos with subclinical mastitis and confirmation of *L. garvieae* as a senior subjective synonym of *Enterococcus seriolicida. Int. J. Syst. Bacteriol.* **46:**664–668.

Tison, D. L., M. Nishibuchi, J. D. Greenwood, and R. J. Seidler. 1982. *Vibrio vulnificus* biogroup 2: new biogroup pathogenic for eels. *Appl. Environ. Microbiol.* **44:**640–646.

Udey, L. R., and J. L. Fryer. 1978. Immunization of fish with bacterins of *Aeromonas salmonicida. Mar. Fish. Rev.* **40:**12–17.

Vandepitte, J., P. Lemmens, and L. de Swert. 1983. Human edwardsiellosis traced to ornamental fish. *J. Clin. Microbiol.* **17:**165–167.

Veenstra, J., P. J. Rietra, C. P. Stoutenbeek, J. M. Coster, H. H. de Gier, and S. Dirks-Go. 1992. Infection by an indole-negative variant of *Vibrio vulnificus* transmitted by eels. *J. Infect. Dis.* **166:**209–210.

Verner-Jeffreys, D. W., M. Algoet, M. J. Pond, H. K. Virdee, N. J. Bagwell, and E. G. Roberts. 2007. Furunculosis in Atlantic salmon (*Salmo salar* L.) is not readily controllable by bacteriophage therapy. *Aquaculture* **270:**475–484.

Verschuere, L., G. Rombaut, P. Sorgeloos, and W. Verstraete. 2000. Probiotic bacteria as biological control agents in aquaculture. *Microbiol. Mol. Biol. Rev.* **64:**655–671.

Vinod, M. G., M. M. Shivu, K. R. Umesha, B. C. Rajeeva, G. Krohne, I. Karunasagar, and I. Karunasagar. 2006. Isolation of *Vibrio harveyi* bacteriophage with a potential for biocontrol of luminous vibriosis in hatchery environments. *Aquaculture* **255:**117–124.

Wagner, P. L., and M. K. Waldor. 2002. Bacteriophage control of bacterial virulence. *Infect. Immun.* **70:**3985–3993.

Wakabayashi, H., T. Toyama, and T. Iida. 1994. A study on serotyping of *Cytophaga psychrophila* isolated from fishes in Japan. *Fish Pathol.* **29:**101–104.

Weinstein, M. R., M. Litt, D. A. Kertesz, P. Wyper, D. Rose, M. Coulter, A. McGeer, R. Facklam, C. Ostach, B. M. Willey, A. Borczyk, and D. E. Low. 1997. Invasive infections due to a fish pathogen, *Streptococcus iniae. N. Engl. J. Med.* **337:**589–594.

Wu, J.-L., and W.-J. Chao. 1982. Isolation and application of a new bacteriophage, φET-1, which infect *Edwardsiella tarda,* the pathogen of edwardsiellosis, p. 8–17. *In* Council for Agricultural Planning and Development Fisheries Series no. 8, *Fish Disease Research* (IV). Council for Agricultural Planning and Development, Taipei, Taiwan.

Wu, J.-L., H.-M. Lin, L. Jan, Y.-L. Hsu, and L.-H. Chang. 1981. Biological control of fish bacterial pathogen, *Aeromonas hydrophila,* by bacteriophage AH1. *Fish Pathol.* **15:**271–276.

Yamamoto, A., and T. Maegawa. 2008. Phage typing of *Edwardsiella tarda* from eel farm and diseased eel. *Aquac. Sci.* **56:**611–612.

Yoshida, T., M. Endo, M. Sakai, and V. Inglis. 1997. A cell capsule with possible involvement in resistance to opsonophagocytosis in *Enterococcus seriolicida* isolated from yellowtail *Seriola quinqueradiata. Dis. Aquat. Org.* **29:**233–235.

Yoshida, T., Y. Yamada, M. Sakai, V. Inglis, X. J. Xie, S.-C. Chen, and R. Kruger. 1996. Association of the cell capsule with anti-opsonophagocytosis in β-hemolytic *Streptococcus* spp. isolated from rainbow trout. *J. Aquat. Anim. Health* **8:**223–228.

CONTROL OF BACTERIAL DIARRHEA WITH PHAGES: COVERAGE AND SAFETY ISSUES IN BACTERIOPHAGE THERAPY

Harald Brüssow

14

FROM D'HÉRELLE TO COMMERCIAL PHAGE PREPARATIONS IN RUSSIA: THE LONG SAFE USE OF PHAGES

Félix d'Hérelle, the Franco-Canadian codiscoverer of bacteriophages (Summers, 1999), launched the first clinical applications of bacteriophages soon after their first scientific description (Sulakvelidze and Kutter, 2005). The nature of the phages was at that time not well known, and the early clinical application of phage work suffered from this limitation. However, d'Hérelle was a visionary scientist and a modern biotech entrepreneur. With his biological approaches against infectious diseases and agronomical pests, he was clearly ahead of his time and was therefore treated as a maverick by the scientific establishment (Häusler, 2006). When he faced difficulties at the Pasteur Institute in Paris, where he made his phage discovery and where he also initiated the earliest medical application of phages, he traveled with his ideas throughout the world. One of his destinations was Tbilisi in Georgia, then a part of the Soviet Union. There he met G. Eliava, a former collaborator at the Pasteur Institute. He helped Eliava to found an impressive phage research institute, which today carries the name of this Georgian microbiologist. Soon afterwards this institute became the cradle of the Soviet phage therapy approaches. Over time, the Tbilisi phage institute gained a large phage collection, tested these phages against major bacterial pathogens, and developed two main applications: an "intestiphage" preparation targeting bacterial diarrheal diseases and a "pyophage" preparation aimed at wound infections (Sulakvelidze and Kutter, 2005). The approach was largely empirical but very systematic. The soldiers of the Soviet Red Army were the major target of early phage therapy. Dysentery and wound infections of the soldiers were treated with bacteriophages produced in Tbilisi. In fact, from the Second World War to the Afghanistan War, phages were a standard tool of Soviet military medicine (Häusler, 2006). These phage preparations were also developed for the civil sector, and phages became a commonplace alternative treatment to antibiotics in the Soviet Union.

With respect to diarrheal diseases, a large randomized, placebo-controlled trial was conducted with an anti-*Shigella* phage preparation during the 1960s. This test came close to current standards of controlled clinical trials.

Harald Brüssow, Nutrition and Health Department, Nestlé Research Center, Vers-chez-les-Blanc, CH-1000 Lausanne 26, Switzerland.

Bacteriophages in the Control of Food- and Waterborne Pathogens
Edited by Parviz M. Sabour and Mansel W. Griffiths, © 2010 ASM Press, Washington, DC

However, few results of the early microbiological and clinical work were published in scientific journals. For example, the mentioned clinical trial, which enrolled more than 30,000 children, was documented in a single publication written in Russian, consisting of just 75 lines and a single table (Babalova et al., 1968). Yet the clinical trial seems to reflect very careful work. The randomization was well done: 15,000 children from one side of the streets in Tbilisi received the anti-*Shigella* phage preparation pressed into tablet form, while 15,000 children from the opposite street sides received a placebo tablet. The enrolled children were closely followed over 3 months for the occurrence of diarrhea. Stool samples were analyzed in the laboratory for enteric pathogens. The trial showed highly significant effects of phage treatment on *Shigella*-associated dysentery, still significant effects on *Escherichia coli* diarrhea, and to a lesser extent, an impact on undifferentiated diarrhea (Sulakvelidze et al., 2001). Based on these and other clinical trials, the phage cocktails targeting diarrhea and wound infections became a registered medicine in Russia during the 1980s. Numerous phage preparations are still sold in pharmacies in Russia and a few other member states of the former Soviet Union. They seem to represent follow-up products based on phage cocktails developed in the Soviet era. According to discussions with scientists at the Eliava Institute, adverse effects of phage therapy were not observed. Also, the Russian supplier of the commercial phage preparations, Microgen/Biomed, reports in its product information that no significant adverse events of phage therapy were observed. With the orally and anally applied *E. coli/Proteus* phage preparation, they rarely observed benign skin reactions in newborn and infant recipients. These reactions quickly disappeared when the oral dose of the phages was halved. For the "sextaphage" preparation (containing phages against six pathogens including *Staphylococcus, Streptococcus, Enterococcus, Klebsiella,* and *Proteus*), more-invasive application forms were prescribed. The sextaphage preparation was applied orally but also given into abscess cavities after the removal of pus or into the pleural cavity, the joints, the vagina, the uterus, the bladder, the kidneys, and the eyes. Adults and even premature infants were treated with these phages, and according to the product information of Microgen/Biomed, no adverse events were observed during their application.

According to electron microscopy, the *E. coli/Proteus* phage preparation from Microgen/Biomed consists of a cocktail containing phages belonging to different morphological classes. The exact composition of this phage cocktail cannot be inferred from the product information in the package or the website of the producer (http://home.sinn.ru/~imbio/Bakteriofag.htm). According to information obtained from Georgian scientists currently working at the Eliava Institute, who do not have contacts with the Russian phage producers, the corresponding Georgian Intestiphage and Pyo-phage preparations are regularly updated to match the current epidemiological situation (M. Kutateladze, presented at the Phage Biology, Ecology and Therapy Meeting, Tbilisi, Georgia, 12 to 15 June 2008). As the phages are propagated in bulk, the actual number of phages in the Georgian cocktails might be substantial. However, the Eliava Institute is currently putting forth efforts to explore the possibility of phage therapy with a single, particularly broad-host-range *Staphylococcus aureus* phage strain (Merabishvili et al., 2009).

On the basis of the long and apparently safe use of complex phage cocktails in the former Soviet Union and now in Russia, one might ask whether phage safety is an issue. However, in view of the lack of published safety and efficacy reports on these Eastern phages, part of the Western scientific community has remained unimpressed by the claims for safe and efficient use of phage therapy in Eastern Europe. Use of bacteriophages for medical or human food applications in Western countries will therefore critically depend on careful safety evaluations, documented in detailed publications in peer-reviewed scientific journals, before a medical approval of phage therapy can be expected from Western health authorities (Merril et al., 2003). Also, at the Eliava Institute in Tbilisi (which is now financially supported by U.S. and European Union funds), the phage scientists are starting to characterize the safety of their phages for therapy according to current scientific standards, even if this means reinventing the Georgian phage therapy practice.

As the focus of the current book is bacteriophage control of food-borne pathogens, this chapter concentrates on *E. coli* diarrhea (Brüssow, 2005), a major cause of childhood morbidity and mortality. As other chapters of this book deal with pathogen control in food animal production (chapters 4, 12, and 13), food plant production (chapter 5), and food processing (chapters 7 and 15), this restriction in focus seems justified, particularly since these topics also reflect the personal experience of the author. This chapter concentrates on technical and scientific aspects, since regulatory and industrial issues are discussed in chapter 15.

BACTERIAL DIARRHEA: THE EXTENT OF THE PROBLEM

After respiratory infections, diarrhea is the second-most-important cause of morbidity and mortality worldwide (Walsh and Warren, 1979). The toll of diarrhea is particularly high in children from developing countries. Of the 15 million children younger than 5 years old who die each year, 96% die in developing countries (Gwatkin, 1980). About 30% of these children die from pneumonia (World Health Organization, 1983). Estimates from Latin America specify diarrhea as the cause for 20% of all recorded early childhood deaths (American Health Organization, 1980). Despite substantial progress in research and medical practice (e.g., oral rehydration solution), the overall figures remain very high. The Child Health Epidemiology Reference Group of the World Health Organization reported that in 2000 to 2003 an estimated 10.6 million children younger than 5 years old died. The leading causes were pneumonia (19%) and diarrhea (18%). More than 70% of the childhood death cases occurred in just 15 developing countries. Undernutrition was an underlying cause of half of all deaths in children (Boschi-Pinto et al., 2008; Bryce et al. 2005).

Diarrhea has dire nutritional consequences. Each diarrheal episode causes a flattening in the weight increase curve of the affected infant, and children from developing countries experience multiple episodes of diarrhea per year. Prospective clinical and epidemiological studies in urban slum areas of Chile (Brunser et al., 1992) and Bangladesh (Qadri et al., 2007) describe about two episodes of diarrhea per child per year. These are conservative estimates from populations that were receiving regular visits of health workers during the study. It is well known that these visits heighten the attention of the parents to personal and food hygiene, causing a decrease in diarrhea incidence compared to the time before the study (a type of epidemiological Heisenberg uncertainty principle where the fact of observation interferes with data acquisition). The diarrhea rates in developing countries under unobserved conditions are higher. In fact, in rural Guatemala up to eight episodes/child/year were reported (Mata et al., 1983). The duration of diarrhea in the community averages about 3 days (Qadri et al., 2007). Thus, Guatemalan children experience diarrhea for nearly a month per year. Under these conditions, it is apparent that diarrhea will, in addition to problems of low-birth-weight deliveries and food shortage, further aggravate malnutrition in underprivileged societies. Effective interventions against diarrhea would thus have a substantial impact on child health. Oral rehydration therapy has dramatically reduced diarrhea-induced mortality. However, it does not prevent the disease nor does it treat the causes of the disease. It simply treats the symptom, electrolyte disturbances, that can lead to fatalities. Antibiotics are frequently and uncritically used against diarrheal diseases in developing countries. While antibiotics are indicated against *Shigella* dysentery, antibiotics have only limited efficacy against a number of bacterial pathogens, including *E. coli*. High rates of resistance against common and cheap antibiotics have been reported in bacterial pathogens isolated from diarrhea patients in developing countries (Djie-Maletz et al., 2008). Vaccines against many bacterial enteropathogens are still in the development stage, a drastic contrast to highly efficient and safe second-generation rotavirus vaccines. Improvement in clean water supply and the hygienic removal of fecal material would surely reduce the incidence of diarrhea, but these interventions require substantial investments into infrastructures, which are well beyond the financial possibilities of countries most affected by diarrhea. The application of oral phages for the treatment and prevention of bacterial diarrhea—if proven to be effective—represents a theoretically attractive and relatively inexpensive possibility of intervention.

BACTERIAL CHILDHOOD DIARRHEA: DEFINING A TARGET FOR PHAGE THERAPY

With respect to their bacterial hosts, the vast majority of bacteriophages are species specific. On one side this fact is a clear advantage of phages over antibiotics, which target broad classes of bacteria and therefore kill not only the pathogen but also harmless or even beneficial commensals. Antibiotic-induced diarrhea is a clinically well-defined disease entity. One cannot simply dismiss this as an undesired adverse side effect of antibiotic treatment. Under some conditions, antibiotic treatment favors the outgrowth of *Clostridium difficile*, causing hard-to-treat infections which are second only to *S. aureus* as nosocomial infections. However, in infectious diseases displaying complex etiologies, the target specificity of phages becomes a liability. Large phage cocktails might be needed to achieve a good coverage of bacterial enteropathogens. It is thus pertinent to quickly review the etiology of diarrheal diseases with a few examples from countries displaying different social levels.

The Netherlands is a wealthy, densely populated country with a very good public health system. Gastroenteritis in children <4 years of age, as reported by general practitioners, showed rotavirus and norovirus infections as main forms of diarrhea, followed by adenovirus infections. Of secondary importance are bacterial infections with *Salmonella, Campylobacter,* and *E. coli* (de Wit et al., 2001a). Prospective epidemiological studies from The Netherlands confirmed the predominance of viral infections (norovirus > rotavirus > adenovirus) over bacterial infections (de Wit et al., 2001b). In epidemiological parlance, these viral infections are classified as "democratic" infections: rotavirus affects children the same way, whether they are living in The Netherlands or in Bangladesh. The epidemiological reason for this "democratic" behavior of viral enteropathogens is their transmission mode. Rotaviruses show up in epidemiological and seroepidemiological analysis as a respiratory infection-type transmission, not a typical fecal-oral transmission route typical of infections like *E. coli* diarrhea. Norovirus and rotavirus infections are characterized by a high frequency of vomiting and (usually) by relatively low infectious doses. Clean water supplies and water toilets will not protect against viral diarrhea, while they will efficiently curtail bacterial diarrhea showing fecal-oral transmission. In conclusion, bacteriophages targeting bacterial enteropathogens will be of minimal impact when given to children from industrialized countries.

The epidemiological situation is different for threshold countries like China, another densely populated country. Here the major pathogens identified in children hospitalized with diarrhea were bacteria (enterotoxigenic *E. coli* [ETEC] > *Salmonella* > enteroinvasive *E. coli* [EIEC] and then enterohemorrhagic *E. coli* [EHEC]), followed by rotavirus (Kain et al., 1991). A dominance of *E. coli* (enteropathogenic *E. coli* [EPEC] infections were the primary cause) over rotavirus infection then followed by *Campylobacter* and *Shigella* infections was also seen in children with diarrhea from Uruguay. In threshold countries, phages, particularly phages directed against *E. coli,* might have a sizable impact on the diarrheal disease burden.

Since our laboratory has a long-term scientific collaboration with the International Centre for Diarrheal Diseases Research, Bangladesh (ICDDR,B), I will illustrate the situation of a developing country with data from Dhaka in Bangladesh. ICDDR,B is the world's largest diarrhea hospital, which treats more than 100,000 patients per year. In hospitalized diarrhea patients a wide range of enteropathogens was identified (Albert et al., 1999). However, the largest share was contributed by pathogenic *E. coli* (Qadri et al., 2000), which was detected in more than 40% of the pediatric patients. The next most frequent pathogen group was rotavirus (20%), followed by *Campylobacter, Aeromonas,* and *Shigella.* All other enteropathogens made only minor contributions. In a birth cohort study from urban Dhaka, pathogenic *E. coli* was also the primary agent associated with diarrhea in the community (20% of all cases), followed by rotavirus (10%). *Campylobacter* and *Vibrio* made only minor contributions (Qadri et al., 2007). From these data one would conclude that phage cocktails for the prevention or treatment of pediatric diarrhea in developing countries should contain mainly *E. coli* phages.

Data from ICDDR,B are a valuable source of epidemiological information on diarrhea for a developing country. A surveillance system was established 20 years ago in which a 4% systematic sample of feces from all hospitalized patients is studied for enteropathogens (Stoll et al., 1982). These data allow for a dynamic survey of the enteropathogens over time, particularly the seasonal changes of the pathogens in each year (*E. coli* peaks between April and August; rotavirus is high between November and March). These surveys also documented the shifting prevalence of the major pathogens during diarrheal epidemics following floods, when *Vibrio cholerae* becomes the dominant pathogen (Harris et al., 2008).

One point observed in case-control studies from Bangladesh and other developing and threshold countries is noteworthy. Epidemiological studies with intensive laboratory support identify a potential pathogen in up to 75% of the diarrhea cases. However, enteropathogens are also identified in nonsymptomatic controls. Healthy carriage of rotavirus is rare, but bacterial pathogens are also found in control children, albeit with significantly lower frequency in the case of *E. coli.* However, *Campylobacter* and enteric parasites were found with higher frequency in nondiarrheal than in diarrheal children (Qadri et al., 2007), making them less likely targets for an antidiarrhea approach, at least in this epidemiological setting.

If one settles for a single pathogen, *E. coli* is the best target, promising a sizable impact on overall diarrhea rates in developing and threshold countries. The Russian pharmaceutical industry offers different phage cocktails against diarrheal diseases targeting distinct pathogen groups. One phage product targets *E. coli* and *Proteus,* another targets *Shigella,* and a third targets *Salmonella.* This phage product diversification reflects the long experience of Georgian/Russian phage cocktail design tailored to specific clinical situations. In extremis, it was not unusual to practice an individualized approach, which has recently been reinvented in the Western world and termed personalized medicine. When the patients did not respond to the standard phage cocktails, the clinicians isolated the pathogen from the patient and sent it to the centralized phage institute in Tbilisi. The microbiologists screened the Georgian phage collection against this particular pathogen isolate or isolated a new phage from the environment (this was frequently done at the river crossing Tbilisi). This phage was then grown at the institute and then sent to the clinic for the treatment of the individual patient. This approach was naturally applicable only in patients suffering from chronic wound infections and not for acute diarrhea patients.

TRAVELER'S DIARRHEA

Traveler's diarrhea (turista) has a major impact on the 400 million persons traveling each year (Steffen et al., 1987). The incidence varies from 8% in travelers to highly developed countries to 50% in travelers to threshold or developing countries (Steffen, 1986). These figures have not changed during the last 2 decades (Steffen, 2005). The etiology of turista was initially an enigma, but now it is well established that ETEC represents the causative agent in more than 50% of traveler's diarrhea cases in Latin America, Africa, and Asia (Wolfe, 1997). Other types of *E. coli, Shigella, Salmonella,* and *Campylobacter* each contribute about 5% of the remaining diarrhea cases. Parasites are rare causes of turista, and viruses (norovirus, rotavirus) are mostly associated with outbreaks on cruise ships. Due to side effects, antimicrobial prophylaxis is generally not recommended, and its usefulness is further compromised by increasing antibiotic resistance in major pathogens associated with turista. Phages are therefore a potentially attractive prophylactic treatment, especially since half of all turista cases are induced by a single enteric pathotype of *E. coli,* which contrasts favorably with the more complex *E. coli* epidemiology of childhood diarrhea reviewed in the next section.

E. COLI DIARRHEA: FINE-TUNING OF A TARGET

Many phages are not only species specific, but also strain specific. This is also the case for *E. coli* phages despite studies reporting that multiple genera of *Enterobacteriaceae* are lysed by coliphages. This observation reflects more the genetically flawed historical definition of genera in this group of bacteria than true polyvalency of coliphages. Bacteria of the genus *Shigella,* for example, do not even qualify as a subspecies of *E. coli.* In our phage surveys we have detected only exceptional phages that infect different enterobacterial species (Zuber et al., 2008). Broad-host-range phages like Twort-like *S. aureus* phages, where a single phage isolate can lyse nearly 50% of large clinical collections of *S. aureus* isolates (Pantucek et al., 1998), were not observed in *E. coli.* This difference might not reflect distinct characteristics of T4 or Twort-like *Myoviridae,* but rather highlights the fact that *E. coli* is genetically more diverse than *S. aureus. E. coli* is in fact a versatile bacterial pathogen comprising enteropathogens, uropathogens, and strains involved in sepsis and meningitis (Donnenberg, 2002). Even diarrhea-associated *E. coli* isolates are differentiated into many pathotypes (Nataro and Kaper, 1998). In this category EPEC strains were the most prevalent strains at ICDDR,B (found in 16% of the stools from hospitalized children with diarrhea), quickly followed by ETEC (12%), enteroaggregative *E. coli* (10%), and diffusely adherent *E. coli* strains (8%). *Shigella* was seen in another 10% of the stools (Albert et al., 1995, 1999). Enteroinvasive *E. coli* and EHEC strains are emerging but still relatively rare isolates. *E. coli* diarrhea investigated in the community of urban Dhaka showed a lesser variety of *E. coli* pathotypes. In this context, the primary agent of diarrhea was ETEC strains (37% of the subjects), as other *E. coli* pathotypes were rarely isolated in this study (Qadri et al., 2007). However, even ETEC strains are serologically heterogeneous, and many O serotypes (based on lipopolysaccharide [LPS] chemistry), H serogroups (based on flagella antigens), and colonizing factor antigens (based on fimbrial adhesins) were identified (Qadri et al., 2005; Wolf, 1997).

With this background it is clear that phage therapy approaches against *E. coli* diarrhea have to rely on a cocktail of phages. Possibly due to the precarious hygiene situation in Dhaka, further compounded by regular floods (Harris et al., 2008), ETEC serotypes show a dynamic state and changes in serotype distribution have been documented over 2-year periods (Stoll et al., 1983). It is currently unknown if phage therapy approaches against *E. coli* diarrhea have to rely on dynamically defined phage cocktails.

About 10 different O serotypes represent the majority of ETEC strains from an international database (O6, 8, 25, 27, 78, 115, 128, 148, 153, and 159) (Wolf, 1997). A similar association of O serotypes was seen in ETEC strains from hospitalized patients in Dhaka, and in environmental water (Begum et al., 2007). The major O serotypes associated with the EPEC strains from the children hospitalized in Dhaka belonged to the traditional serogroups. The dominant EPEC O serotypes in Dhaka were O114 and O127 (Albert et al., 1995), suggesting nevertheless a limited pathogen diversity that can be matched with phages even if they are serotype specific.

PATHOGENIC *E. COLI* COVERAGE WITH PHAGES

The serological diversity of pathogenic *E. coli* is relevant for phage therapy since LPS, the physical determinant of the O serotype, is a preferred bacterial receptor structure recognized by T4 phages (Wilson et al., 1970), a major group of professional virulent coliphages. Since part of the arms race between phages and their host bacteria occurs at the level of the bacterial receptor/phage antireceptor interaction, one could expect a coevolution of these structures over evolutionary time scales. Bacteria might be expected to vary the phage receptor by mutation. In parallel, phages are predicted to adapt to these receptor variations by antireceptor variations. This is indeed observed: one major hot spot of genetic variability in *E. coli* is the LPS locus (Milkman, 1999), and a hot spot of T4 coliphage variation is the distal tail fiber region encoding the phage antireceptor (Tétart et al., 1996, 1998). Intriguingly, the second hot spot of *E. coli* genetic variation is restriction-modification genes (Milkman, 1999), which are also directly involved in the defense of the bacterial cell against bacteriophage attack. From the phage side, LPS is a good target molecule since the basic structure of LPS is essential for *E. coli* and can thus not be discarded. However, LPS is also a difficult target since it is chemically fairly complex. This phosphorylated heteropolysaccharide is divided into a peripheral, highly variable O-antigen chain (defining the immunodominant O serotype of *E. coli*), an internal core region showing a less variable carbohydrate structure, and an innermost unique lipid A. Mutations interfering with the biosynthesis of the O side chain and even part of the core structure are known (causing various rough colony phenotypes), but loss of the inner part of the LPS structure is lethal for *E. coli*.

When constituting a phage cocktail for diarrhea-associated *E. coli,* one could consider two strategies. In one approach, one would first constitute a collection of *E. coli* strains representing the major O serotypes of ETEC and EPEC in the disease and epidemiological setting targeted for phage therapy. These pathogenic strains are then used to isolate the corresponding phages from environmental sources known to contain coliphages. Standard sources are sewage, fecally polluted environmental water, hospital water discharge, and stools from healthy subjects and diarrhea patients. The most complex source is sewage since it represents the fecal output of a large number of individuals. In our experience the stools of diarrhea patients are another rich source for coliphages (Chibani-Chennoufi et al., 2004b). Older ecological studies have suggested that the stools from healthy subjects contain a smaller number of mainly lambdoid *Siphoviridae* and RNA phages (Dhillon et al., 1976), while the stools from diarrhea patients contain higher phage titers and are enriched for T4-like *Myoviridae* (Chibani-Chennoufi et al., 2004b; Furuse, 1987).

When applying this phage search strategy, we isolated mainly phages that were specific to the O serotype on which the phage was isolated (Chibani-Chennoufi et al., 2004b). The consequence is that relatively complex phage cocktails are needed to achieve a reasonable coverage of the pathogenic *E. coli* strains that might be encountered in the field.

We reasoned that this O-serotype specificity of the isolated phages might be the consequence of the distinct LPS receptor structure on the strains used for the isolation of the phages. To circumvent this difficulty, we used a single propagating *E. coli* strain with a severely truncated LPS devoid of the O side chain, which exposes only the relatively conserved core structure. This is the case for the well-known laboratory *E. coli* strain K-12. On this propagating strain, we now obtained phages that infected not only the K-12 strain but also to various extents pathogenic *E. coli* isolates belonging to different O serotypes. In addition to the extended host range, the use of the K-12 strain offers other advantages. We observed that when using the K-12 strain for phage isolation, we isolated nearly exclusively T4-like phages (Chibani-Chennoufi et al., 2004b), while when using other *E. coli* indicator strains like WG-5, commonly used by environmental microbiologists, we isolated a much larger morphological diversity of coliphages. Since T4 phages are the safest representative of the professional virulent coliphages, this selectivity of K-12 represents a definitive advantage. There are further advantages when using K-12 as a propagating strain. K-12 is a nonpathogenic *E. coli* strain and thus can be produced under standard biotechnological working conditions. In contrast, phage production on pathogenic strains necessitates working under biohazard containment levels, which makes phage propagation more expensive. In addition, when producing phages on pathogenic cells, carryover of virulence factors from the lysed pathogen into the phage preparation is a possibility. K-12 has been sequenced, and the genome analysis demonstrated that it is devoid of virulence genes, which adds substantially to the safety of the phage product. Finally, when producing phages on relatively uncharacterized pathogenic *E. coli* isolates, one might also induce an endogenous prophage from a lysogenic strain. This leads to a coproduction of two phage strains, and as the prophage was not characterized, it may contain prophage-encoded virulence genes. This was also the case for the K-12 strain, which initially contained phage lambda as a prophage. Lambda, despite being a workhorse of molecular biology, contains two virulence genes, *bor* and *lom,* conferring serum resistance to the *E. coli* lysogen (Boyd and Brüssow, 2002). The common laboratory derivative of K-12 was cured from prophage lambda, and it contains only a few, highly defective prophage remnants (Casjens, 2003). There is therefore no possibility of contamination of the phage propagated on K-12 with endogenous prophages.

In view of these advantages, we constituted first a large collection of T4 coliphages derived from various environmental sources. The classification of these phage isolates was made by electron microscopy and diagnostic mass spectrometric analysis. The attribution to the T4 subgroups was done through the sequencing of the major head gene obtained by diagnostic PCR (Tétart et al., 2001). This effort led to a collection comprising more than 100 different isolates of T4 phages. From this collection we now select the pools of T4 phages that achieve the widest coverage on a pathogenic *E. coli* collection representative for a given epidemiological setting with the lowest number of constituent phage isolates.

WHY RESTRICT PHAGE THERAPY TO PROFESSIONAL VIRULENT PHAGES?

One poorly explored aspect of the Soviet phage therapy approach is the safety of the utilized phages with respect to their gene content. We believe that certain classes of coliphages are intrinsically unsuitable for phage therapy. One of these unsuitable classes is lambdoid coliphages. Phage lambda is the prototype of temperate phages that can establish a lysogenic state. In this condition the infecting phage DNA does not initiate a new phage multiplication cycle but is incorporated into the bacterial DNA as prophage DNA (see chapters 2 and 9). The bacterial cell survives the infection and carries the viral DNA passively in its chromosome. In contrast, T4 phage uses a very different strategy (Karam, 1994), destroying the bacterial DNA early in the infection cycle. This nucleolytic attack seals the fate of the cell and removes a genetic competitor by releasing nucleotides that are necessary for the massive phage DNA synthesis task in the infected cell, which equates to five times the DNA synthesis load in exponentially growing *E. coli* cells.

The fact that the target cell can survive phage lambda infection is only part of the problem. Of course, survival of the pathogen is not a desired event in phage therapy, but even for phage lambda, lysogenization of the target cell is a rare event. In most cases, phage lambda infection leads to a lytic amplification of the infecting phage. The problem with temperate phages lies at the level of their genomes and gene content. It is the very fact that the phage DNA can be integrated into the bacterial DNA that creates a problem. When the prophage DNA becomes part of the bacterial host, selection works on the survival of the lysogen. When applying simple evolutionary reasoning, it should not be a surprise that the prophage contributes genes to the new genetic system that further the survival of the lysogen (Brüssow et al., 2004; Canchaya et al., 2003b). If the infected bacterium is a pathogen, the phage is likely to contribute genes that ameliorate the pathogenicity of the pathogenic bacterium. This conclusion is not idle armchair speculation. The fact that prophages encode important bacterial toxins has been known for a long time—the earliest examples were *Streptococcus pyogenes* strains causing scarlet fever. In fact, one can argue that *S. pyogenes* is a pathogen made by its prophages (Banks et al., 2002). However, what surprised researchers was the extent of the genetic contribution of temperate phages to bacterial pathogenicity. This became particularly apparent by comparative genome analysis of the sequenced bacterial pathogens (Brüssow et al., 2004; Canchaya et al., 2003b). The contribution of prophages to bacterial pathogenicity was also demonstrated for *E. coli*. The recently emerged EHEC strain O157:H7 is a particularly impressive example. The sequenced Sakai strain from an outbreak in Japan contains no less than 18 prophages (Ohnishi et al., 2001). This mobile DNA amounts to 16% of the bacterial genome content and about half of the 0.5 Mb of strain-specific DNA. Furthermore, if one considers that bioinformatic and biological analyses (Tobe et al., 2006) have revealed potential virulence genes (including the pathognomic Shiga-like toxin) in many of these O157:H7 prophages, one can appreciate the contribution of prophages to the pathogenic potential of this particular *E. coli* strain. Many Sakai prophages share extensive DNA sequence similarities that are probably the result of prophage-prophage recombination events. These observations fit very well with the hypothesis of a modular evolution, which was formulated 30 years ago for this group of phages (Botstein, 1980). This hypothesis was so influential that it was believed to apply to all phages. If this were true, phages would represent very volatile genetic entities that can pick up genes from nearly every DNA they meet, whether it is from other phages or bacterial hosts. In fact, these conclusions apply to lambdoid phages. For example, lambdoid prophages from gram-positive bacterial pathogens carry the lysogenic conversion genes (which are frequently virulence genes) near the right attachment site of the prophage DNA (Canchaya et al., 2003a). Since this is the transition point for the phage into bacterial DNA, their origin is most likely from the imprecise excision of the prophage, which leads to the incorporation of a bacterial gene into the phage genome. As phages can, at least at a global level, travel relatively freely between bacteria, temperate phages apparently became a very useful tool for bacteria to explore the DNA sequence space for genes that could be of selective advantage for them. With prophages, lysogenic bacteria achieve short-term adaptation to a quickly changing environment, which in the case of bacterial pathogens is the human being, with its high population density. In this theoretical framework it should not be surprising that temperate phages became a motor for bacterial evolution and particularly for bacterial pathogens that need to adapt to the human host. From these considerations it is clear that lambdoid coliphages are unsuitable tools for the phage therapy of *E. coli* diarrhea.

SAFETY EVALUATION OF T4-TYPE PHAGES: GENOMICS

The genetic risk associated with phage therapy is much smaller when working only with professional virulent phages (i.e., phages that do not yield lysogenic bacteria). Phage T4 is the classical example for this group of phages. By virological standards, phage T4 has a complex genome: nearly 300 open reading frames (ORFs) are encoded on its 169-kb double-stranded DNA. However, decades of research using T4 made this phage one of the best-characterized biological materials. Genetic approaches demonstrated that only 62 T4 genes are essential under laboratory conditions of phage growth. Biochemical approaches have attributed functions to many nonessential genes. Despite these efforts, about half of the ORFs from phage T4 still do not have an assigned function (Miller et al., 2003), but none has links to pathogenicity factors.

However, the reference T4 phage represents only one type of T4-like coliphages. To better define the genetic variability in this phage group, either to understand the evolution of the group or to better define the genetic risk associated with the oral application of T4-like coliphages, different groups have sequenced a number of T4-like coliphages. The sequencing of T4-like coliphages defined additional subgroups of T4-like coliphages in addition to T4 proper (for a visual display, see the dot plot matrix shown in Fig. 1). New subgroups were represented by phage isolates RB49 (Comeau et al., 2007; Desplats et al., 2002), JS98 (Zuber et al., 2007), and RB43 (Comeau et al., 2007; Nolan et al., 2006).

A phylogenetic tree analysis based on the sequence of the major head gene *g23* defined the same subgroups (Filée et al., 2005; Tétart

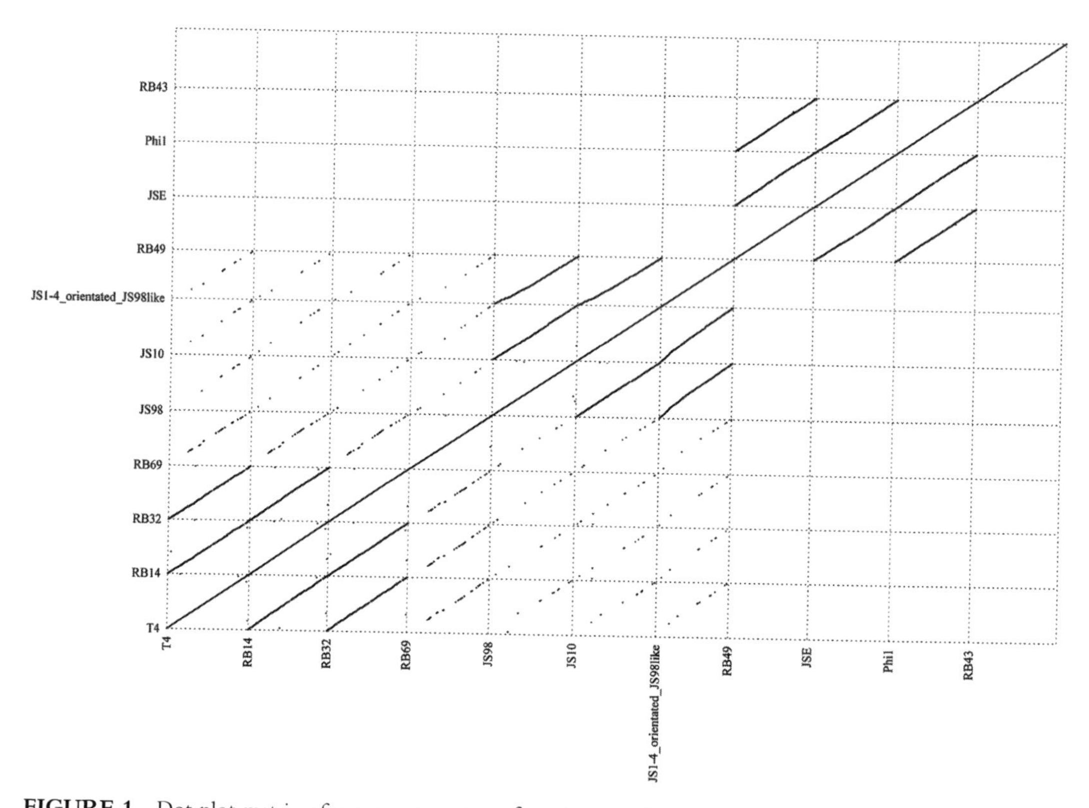

FIGURE 1 Dot plot matrix of genome sequences from T4 coliphages. The alignment criterion for a dot was identity over a 40-bp DNA segment. With the exception of JSD1-4, all sequences represent complete genomes retrieved from the NCBI database. The dot plot matrix attributes in the following order the phages T4, RB14, and RB32 to a T4 subgroup; RB69 is a more distant relative of this subgroup; phages JS98, JS10, and JSD1-4 are members of the JS98 subgroup; RB49, JSE, and Phi1 belong to the RB49 subgroup; and RB43 is currently the only member of a further subgroup of T4-like coliphages. (Courtesy of S. Zuber, C. Barretto, and C. Ngom-Bru, Nestlé Research Center, Lausanne, Switzerland; reproduced with permission.)

et al., 2001; Zuber et al., 2007). The *g23* sequence is therefore a surprisingly good indicator of the degree of genetic diversity within T4 coliphages. In fact, when more than 20 conserved T4 genes were concatenated, similar trees were obtained as with *g23* alone (Filée et al., 2005). In striking contrast, the temperate lambdoid phages showed genomes composed of about a dozen distinct modules, each encoding a specific function and represented by multiple sequence-unrelated alleles (Casjens et al., 1992; Highton et al., 1990). The modular mode of evolution so characteristic for lambdoid phages is thus not a general model for the evolution of all phage genomes.

T4-like phages are highly conserved not only in *E. coli*. Protein sequence identity was maintained between virion proteins from T4 phages infecting hosts as different as *E. coli* and cyanobacteria (Hambly et al., 2001). T4 phages thus share a conserved core genome probably derived from a very ancient ancestor phage via vertical evolution. The reason for the distinct modes of evolution in T4 phages and lambdoid coliphages (i.e., predominantly vertical versus predominantly horizontal evolution) is not clear. Some researchers have postulated a selective barrier to horizontal gene transfer in T4 phages (Filée et al., 2006). Nevertheless, T4 phage has several recombination-dependent replication pathways and homologous recombination across 20 bp, so shared sequences could mediate DNA exchanges between T4-like phages. The acquisition of T4 *g56* was quoted in this context (Mosig et al., 2001). Furthermore, a 20-bp sequence, TTGATAGGTCTATAGTATCA, was detected in 14 distinct genome regions of JS98 and a related 20-bp repeat, AGTTGATAGTTGTATAGTAC, was found in 6 genome regions of phage T4 (bases shared between JS98 and T4 repeats are underscored). Mechanistically, gene exchange between different T4-like phages should therefore be possible during double infections, and indeed they were observed in the laboratory (Abe et al., 2007). However, modular exchanges of variable genome segments between different subgroups of T4-like coliphages have up to now not been documented, possibly due to strong superinfection exclusion mechanisms.

Comparative genomics of T4-like coliphages revealed a core of conserved genes represented by three contiguous blocks of structural genes and DNA replication genes (referred to as "viral self") (Bamford et al., 2002). The morphogenesis module, which is located in the central part of the aligned genomes, represents the most conserved part of the T4-like *E. coli* phages (Chibani-Chennoufi et al., 2004a; Comeau et al., 2007; Nolan et al., 2006; Zuber et al., 2007).

The three conserved regions are separated by hyperplastic regions containing mostly novel genes of unknown function and origin (Comeau et al., 2007; Nolan et al., 2006; Petrov et al., 2006). The intersubgroup comparison of T4 coliphages demonstrated some pertinent features for the unassigned T4 genes: they are mostly clustered in a few genome regions, they are encoded on one strand, they represent particularly small ORFs, the direction of transcription is the same as that of the flanking conserved genes, most are found among early and middle genes, and many of them encode lethal functions for *E. coli* (Chibani-Chennoufi et al., 2004a; Comeau et al., 2007; Miller et al., 2003; Zuber et al., 2007). Some researchers suspect luxury functions in these unassigned genes that further ameliorate basic functions provided by the essential T4 or host proteins (Nolan et al., 2006). Other unassigned T4 genes might confer adaptation of the phage to a particular host or environmental niche (Comeau et al., 2007; Desplats et al., 2002). Still other genome segments represent selfish DNA elements like introns and homing endonuclease genes (Quirk et al., 1989; Sandegren and Sjöberg, 2004).

The comparison of T4-like coliphage genomes belonging to the same subgroup allowed the definition of isolate-specific T4 genes. In the following I provide a relatively detailed genomics analysis for three subgroups of T4 coliphages, which demonstrates only few isolate-specific ORFs, none of which are linked to any virulence genes. In fact, the sequencing of the *g23* head gene alone does not only define the subtype to which this T4 isolate belongs, but it also predicts the vast majority of genes that will be encountered in this new phage isolate.

Intrasubgroup RB49 Diversity

In a dot plot analysis (Fig. 2), the JSE phage genome sequence could be aligned at the DNA sequence level with the genome of phage RB49, the type strain of this subgroup (Comeau et al., 2007; Desplats et al., 2002). Notably, the alignment extends over the entire genome length, showing that the vast majority of the subgroup-specific genes are actually shared. This observation is remarkable since these two phages represent independent ecological isolates of phages, which were isolated more than 40 years apart from two different continents. Phages JSE and RB49 differed only by a total of 17 insertions/deletions (Fig. 3). Most differences represented small hypothetical ORFs, two encoded homing endonucleases and two intergenic alterations. In addition, a number of regions showed DNA sequence variability. Most variable regions covered a single ORF. One larger variable region encoded the homologues of the T4 baseplate wedge protein gp11, the short tail fiber gp12, and the neck whisker gpwac. gpwac forms the six fibers that radiate from the phage neck, which interact with the long tail fibers, preventing phage adsorption as an environmental sensing device (Miller et al., 2003). Genetic data suggested gpwac interaction with gp36 (Letarov et al., 2005). However, we ob-

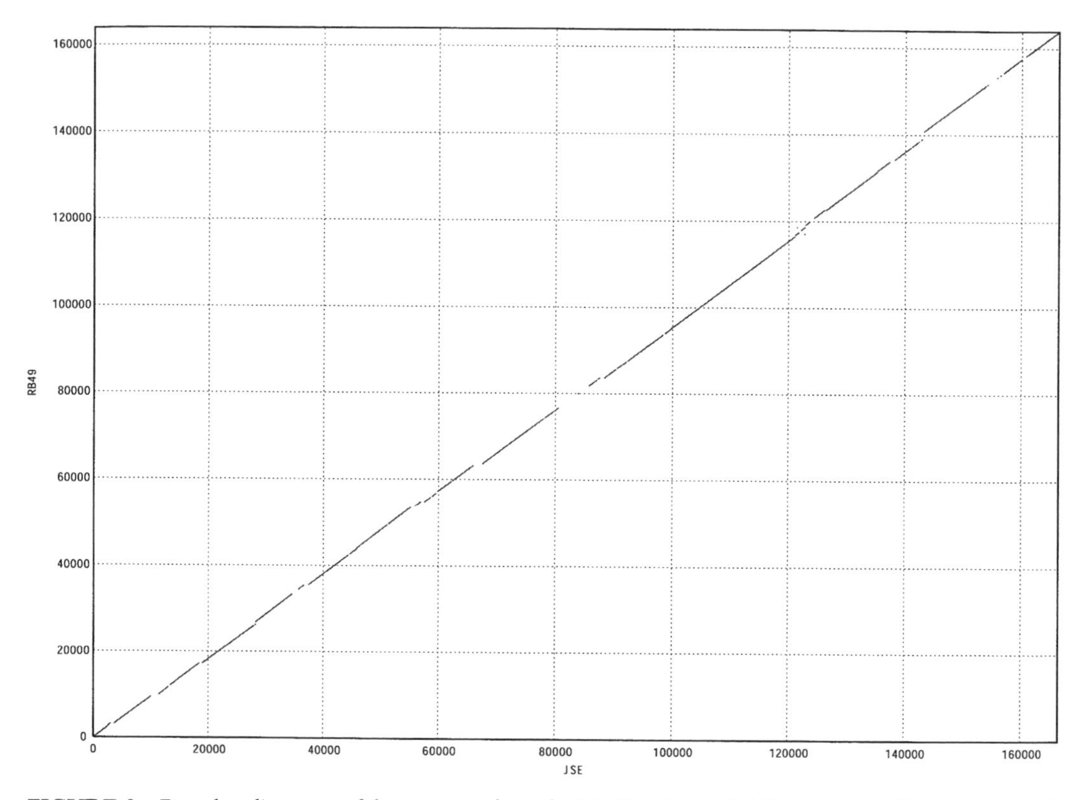

FIGURE 2 Dot plot alignment of the genomes from the T4-like phages RB49 (*y* axis) and JSE (*x* axis). Each point indicates an exact match of 40 consecutive nucleotides between the two compared genomes using the NUCmer program. (Courtesy of S. Zuber, C. Barretto, and C. Ngom-Bru, Nestlé Research Center, Lausanne, Switzerland; reproduced with permission.)

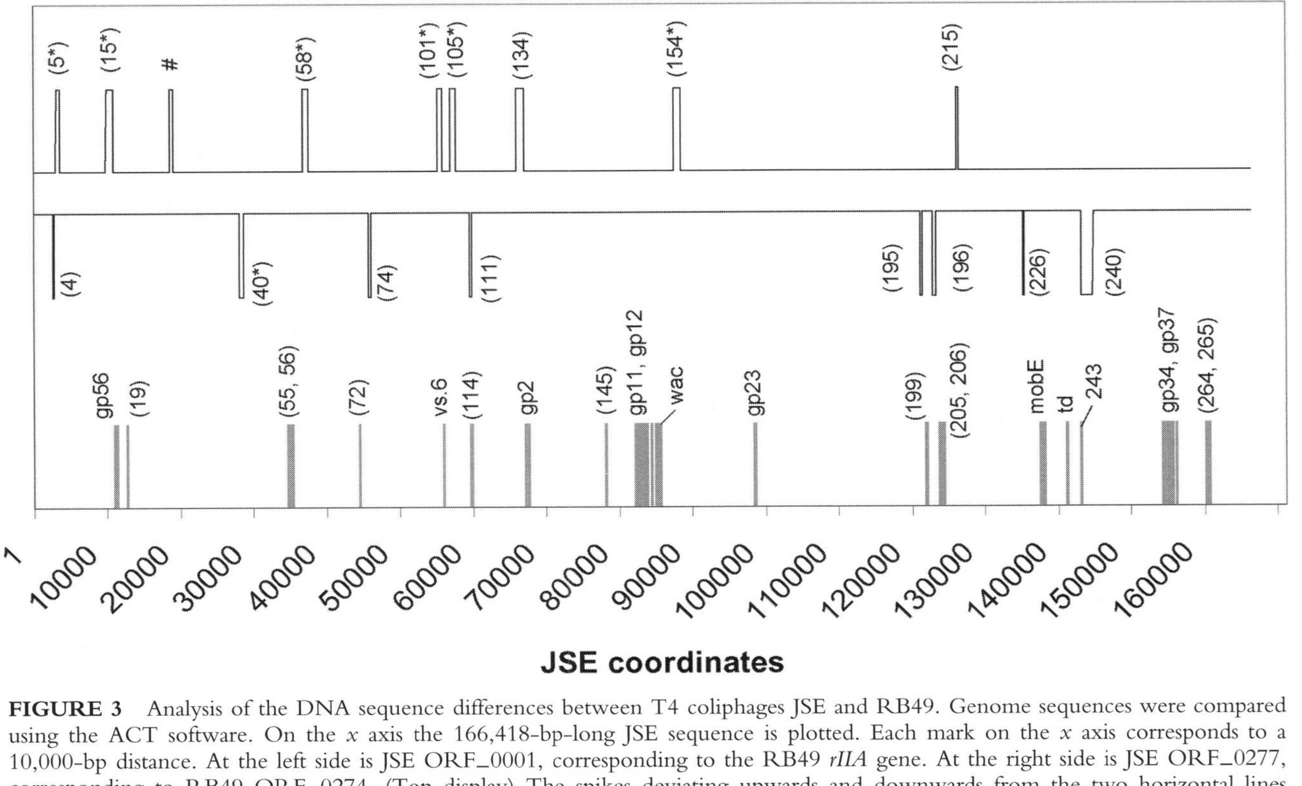

FIGURE 3 Analysis of the DNA sequence differences between T4 coliphages JSE and RB49. Genome sequences were compared using the ACT software. On the x axis the 166,418-bp-long JSE sequence is plotted. Each mark on the x axis corresponds to a 10,000-bp distance. At the left side is JSE ORF_0001, corresponding to the RB49 *rIIA* gene. At the right side is JSE ORF_0277, corresponding to RB49 ORF_0274. (Top display) The spikes deviating upwards and downwards from the two horizontal lines represent insertions/deletions at the indicated genome position in JSE and RB49, respectively, i.e., gaps in the sequence alignment. The width of the spike indicates the length of the indel. The location of the indel is further indicated by the number of the JSE or RB49 ORF. ORFs that lack a complement in T4 have their number indicated in parentheses. If the ORF number is followed by * it identifies a putative homing endonuclease. If the spike is marked by # it indicates an intergenic region. The height of the downward or upward spikes has no meaning. (Bottom display) The bars on the bottom x axis locate regions of DNA sequence diversity between JSE and RB49. The width of the bars indicates the length of the region showing sequence diversity. The regions of sequence diversity are further identified by the JSE ORF number; if the JSE ORF shares protein sequence identity with a T4 gene product, it is identified by the T4 gene symbol (*gp56, vs.6, wac,* and so forth). The height of the bar has no meaning. (Courtesy of S. Zuber, C. Barretto, and C. Ngom-Bru, Nestlé Research Center, Lausanne, Switzerland; reproduced with permission.)

served covariance of gpwac with gp37 and not gp36 in the JSE-RB49 comparison. A second large region with substantial sequence variability covered the tail fiber genes *g34* to *g37*.

Two split genes were observed in JSE. The first was JSE ORF 27 and 28, encoding the N-terminal and the C-terminal halves of gp43 T4 DNA polymerase, respectively. The gene was split between amino acid positions 408 and 409 of the DNA polymerase. The two gene segments were separated by 480 bp without database match. A bicistronic DNA polymerase was reported for T4-like phages from *Aeromonas* and *Acinetobacter* (Petrov et al., 2006), yet not for T4 coliphages. This split was not observed in RB49 or Phi1. The second split gene was ORF 57 and 59, corresponding to the T4 *nrdD* gene, encoding an anaerobic ribonucleotide reductase. The intervening ORF 58 encodes an endonuclease gene. In T4 the intron-split *nrdD* gene showed a 10-fold higher expression than the undisrupted gene and thereby increased the number of phages produced under anaerobic conditions (Young et al., 1994).

For all but 10 of the 277 predicted JSE proteins, the closest database match was with phage RB49; 6 of them showed the best match with another T4-like *E. coli* phage. Two proteins showed a match with a *Lactobacillus* phage, and two JSE proteins lacked any database matches. Strikingly, half of the 10 proteins were annotated as homing endonucleases, suggesting an invasion of the JSE genome by selfish DNA elements, as was previously observed in T4 phage (Quirk et al., 1989).

We screened the JSE sequence for possible gene transfer from its host *E. coli*. Nineteen JSE ORFs showed similarity to *E. coli* proteins. All but 2 of these 19 ORFs also had a match with T4 phage genes, and none showed higher amino acid identity with the corresponding *E. coli* protein than 60%. No DNA sequence identity was detected between these viral and *E. coli* proteins, arguing against recent horizontal gene transfer events between both biological systems.

Intrasubgroup JS98 Diversity

Phage JS10 was isolated from the feces of a pediatric diarrhea patient in Bangladesh (Denou et al., 2009). A dot plot analysis (Fig. 4) showed that the JS10 sequence was closely related to that of phage JS98, displaying only a few small gaps. Nearly all JS10 ORFs had homologues in other T4-like coliphages; in most cases the best match was with phage JS98. Only nine JS10 ORFs lacked links to JS98; three of them did not match entries of the NCBI database or of the Tulane T4 phage database (http://phage.bioc.tulane.edu/). The JS10/JS98 alignment showed 12 insertions/deletions. Most of them represented small ORFs, with seven ORFs being matched with the Tulane database (Denou et al., 2009).

Phages JS10 and JS98 differed in 14 DNA regions (Fig. 5). One large region of variability covered the adjacent tail fiber genes *g37* and *g38*, gene *t* (encoding the phage holin), and *asiA* (encoding a protein that facilitates interaction of the MotA activator with T4 phage middle promoters). The other regions showing diversification over single genes included the large outer capsid protein gene *hoc;* the *alt* gene homologue, encoding an adenosylribosyltransferase; and *motB,* whose gene product interacts with MotA.

Fifteen JS10 ORFs showed links with *E. coli* proteins, with the best match demonstrating 54% amino acid identity. However, all these JS10 ORFs displayed better alignments with T4 or T4-like phages than with *E. coli.* As for JSE, no DNA sequence identity was detected between the phage and *E. coli.*

Intrasubgroup T4 Diversity

Closely related genome sequences were also observed within the T4-like subgroup of T4 coliphages. The genome from phage RB32 could be aligned with the genome from the reference T4 phage over essentially the entire T4 genome length, but the alignment showed more gaps than observed in the two other intrasubgroup alignments, suggesting some greater variability in the T4 subgroup (Fig. 6). However, when compared to RB32 and

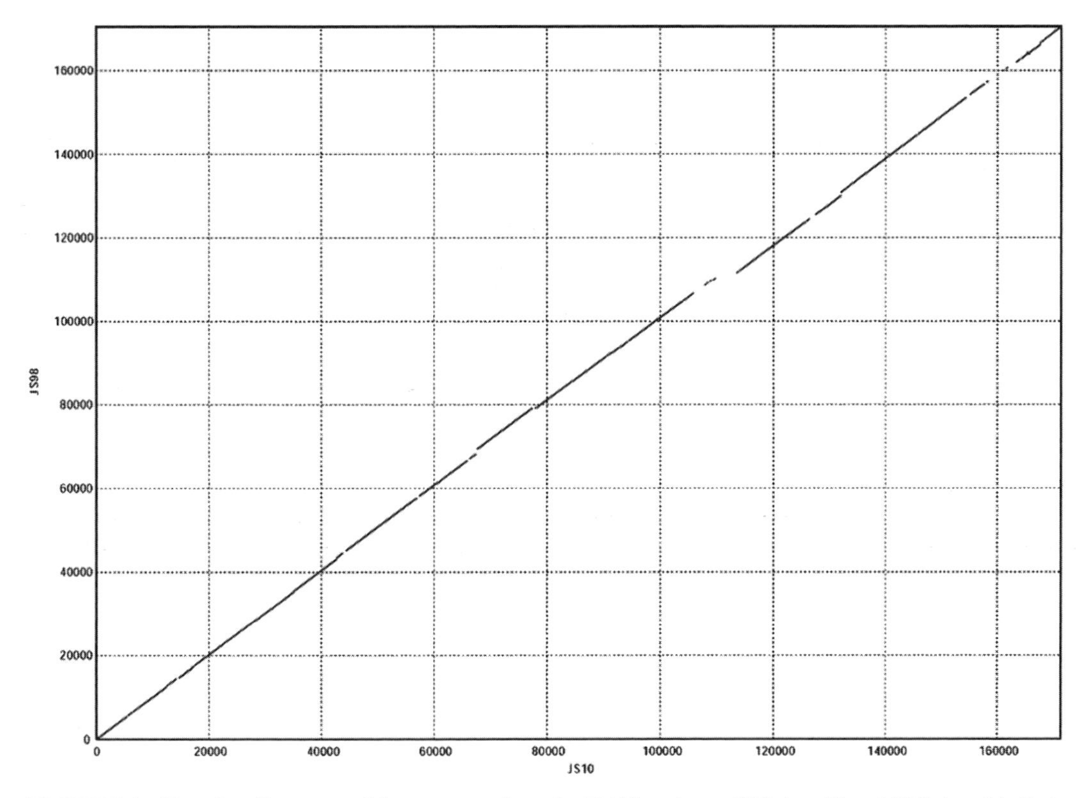

FIGURE 4 Dot plot alignment of the genomes from the T4-like phages JS98 (*y* axis) and JS10 (*x* axis). Each point indicates an exact match of 40 consecutive nucleotides between the two compared genomes using the NUCmer program. (Courtesy of S. Zuber, C. Barretto, and C. Ngom-Bru, Nestlé Research Center, Lausanne, Switzerland; reproduced with permission.)

RB14 (Tulane database: http://phage.bioc.tulane.edu/), the reference T4 phage showed only a few strain-specific genes. Most of them were mobile DNA elements. In addition to selfish endonuclease genes, only two genes were specific for the reference T4 phage: these were *g56,* encoding a dCTPase, and the hypothetical gene *e.7,* located next to genes *ipII* and *ipIII,* a known hot spot of diversity in T4 phages (Repoila et al., 1994).

T4 protein sequence identity with *E. coli* proteins was, with one exception, <55%. The exception was the phage T4 gene *vs.6,* which shared 98% DNA sequence identity with the 3′ end of a pyruvate formate lyase gene. This enzyme catalyzes a key step in anaerobic glycolysis, namely the conversion of pyruvate and coenzyme A to formate and acetyl coenzyme A by an unusual radical cleavage of pyruvate. Gene *vs.6* is found in several T4-like phages (RB43, RB49, RB69, and Phi1, but not JS98, JS10, and JSE); it is located within a cluster of hypothetical T4 genes *(vs.3* to *vs.8)* that are widely conserved in T4-like phages but lack bacterial homologues.

Diversification of T4 Genomes: Possible Mechanisms

Within T4 coliphages the prohead core protein genes *g68, g21,* and *g22* and the major head gene *g23* represent the most conserved genome region of these phages, differing only in the length of the intergenic regions (Chibani-Chennoufi et al., 2004a; Comeau et al., 2007; Zuber et al., 2007). However, downstream of *g23* a variable gene map was

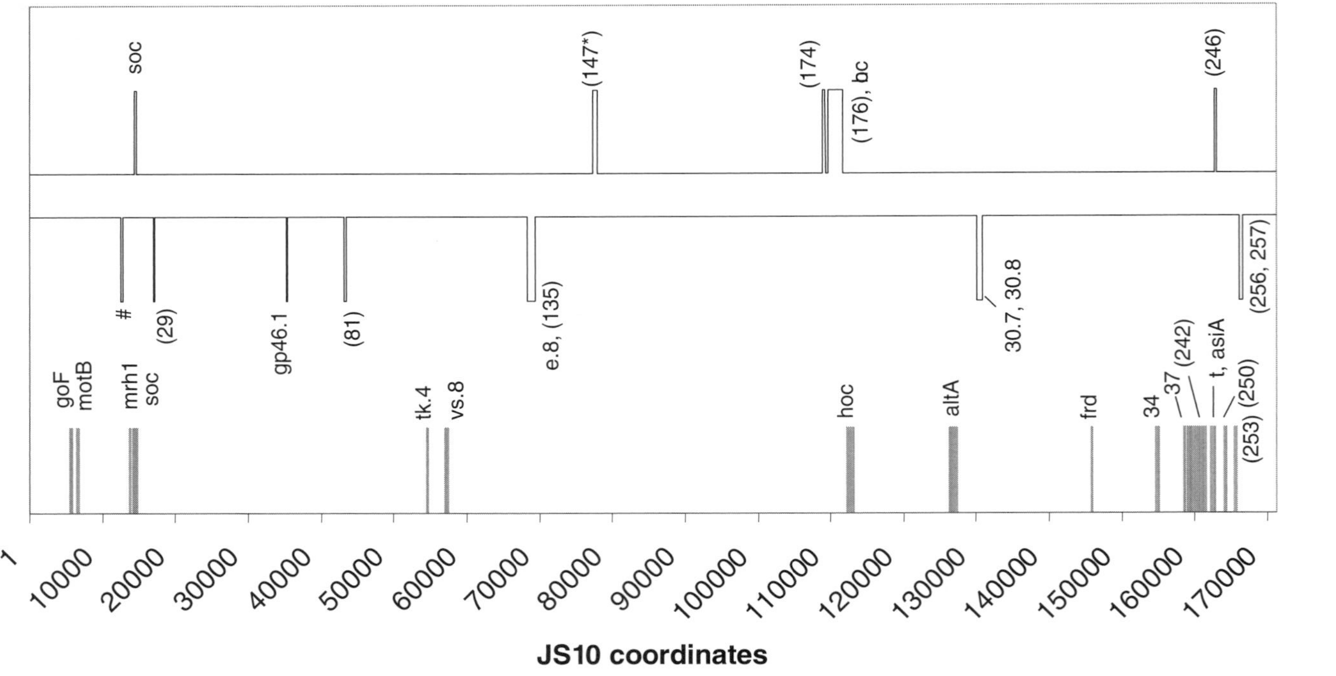

FIGURE 5 Analysis of the DNA sequence differences between T4 coliphages JS10 and JS98. Genome sequences were compared using the ACT software. On the x axis the 171,451-bp-long JS10 sequence is plotted. Each mark on the x axis corresponds to a 10,000-bp distance. At the left side is JS10 ORF_0001, corresponding to the *rIIA* gene. At the right side is JS10 ORF_0265, corresponding to JS98 ORF_0266, the *rIIB* gene. (Top display) The spikes deviating upwards and downwards from the two horizontal lines represent insertions/deletions at the indicated genome position in JS10 and JS98, respectively, i.e., gaps in the sequence alignment. The width of the spike indicates the length of the indel. The location of the indel is further indicated by the number of the JS10 or JS98 ORF. ORFs that lack a complement in T4 are indicated in parentheses. If the ORF number is followed by * it identifies a putative homing endonuclease. If the spike is marked by # it indicates an intergenic region. If the ORFs share protein sequence identity with T4 phage they are identified by the T4 gene symbol (*soc, hoc, gp49.1,* and so forth). The height of the downward or upward spikes has no meaning. (Bottom display) The bars on the bottom x axis locate regions of DNA sequence diversity between JS10 and JS98. The width of the bars indicates the length of the region showing sequence diversity. The regions of sequence diversity are further identified by the JS10 ORF number; if the JSE ORF shares protein sequence identity with a T4 gene product, it is identified by the T4 gene symbol (*motB, altA,* and so forth). The height of the bar has no meaning. (Courtesy of S. Zuber, C. Barretto, and C. Ngom-Bru, Nestlé Research Center, Lausanne, Switzerland; reproduced with permission.)

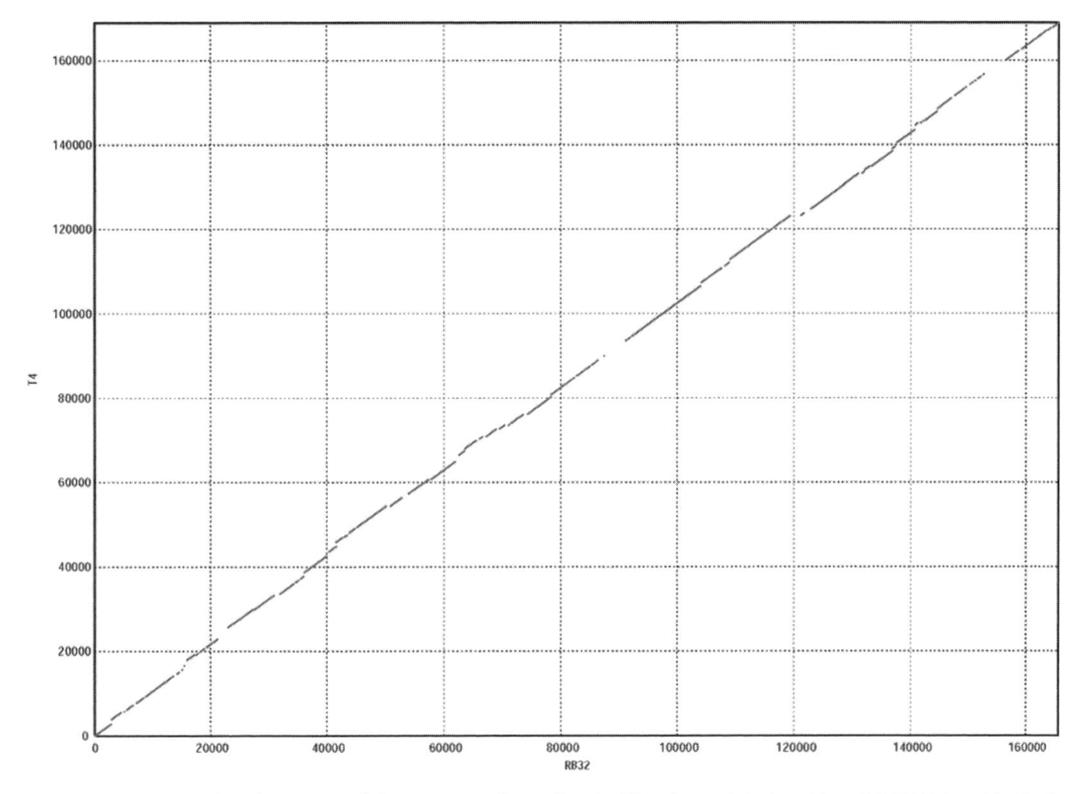

FIGURE 6 Dot plot alignment of the genomes from the T4-like phages T4 (*y* axis) and RB32 (*x* axis). Each point indicates an exact match of 40 consecutive nucleotides between the two compared genomes using the NUCmer program. (Courtesy of S. Zuber, C. Barretto, and C. Ngom-Bru, Nestlé Research Center, Lausanne, Switzerland; reproduced with permission.)

observed (Color Plate 12). Only three genes were conserved across all sequenced T4 coliphages. These conserved genes were *g24,* which encodes a head vertex subunit; *hoc,* which codes for an auxiliary large outer capsid protein; and the hypothetical gene *g24.3.* Phages from the same subgroup shared a relatively conserved gene map, except for subgroup JS98, where phages JS10 and JS98 showed a duplicated *g24,* but JSD1-4 did not. Interestingly, *g24* was probably already the result of a more ancient duplication of *g23* (Haynes and Eiserling, 1996). In addition, JS10 contained a duplicated *hoc.* Genome plasticity suggestive of domain duplication has also been described for the tail fiber genes (Comeau et al., 2007; Tétart et al., 1998). Thus, gene duplication followed by sequence diversification might be a mechanism of T4 genome diversification.

Evidence for horizontal gene transfer is limited to selfish DNA, as already described for the T4 phage genome. In all except one case (*segF* and *g56* in T4) we did not observe that the mobile DNA was accompanied by strain-specific genes. Mobile DNA is therefore not a motor for new gene acquisition in T4 coliphages. Furthermore, a few tail fiber genes from T4 phages shared distant protein sequence identity with the *Siphoviridae* λ (T4 gp37 and gp38) and T5 (RB49 gp36) (Comeau et al., 2007; Letarov et al., 2005; Tétart et al., 1996). If this is evidence for gene exchange, it reflects very ancient events.

Overall, T4 coliphages possess unique sequences and describe a secluded DNA sequence space. This diagnosis is underlined by a GC content of T4 phages very distinct from that of *E. coli*. A number of T4 phage proteins shared sequence similarity with *E. coli* proteins, but this cannot be quoted as evidence for horizontal gene transfer. In fact, the alignment and phylogenetic tree analysis of the T4 thymidylate synthase, dihydrofolate reductase, and three topoisomerase components (gp39, -52, and -60) suggest a divergence from the *E. coli* proteins before the separation of prokaryotes and eukaryotes (Miller et al., 2003). As mentioned above, a single case of a recent horizontal gene transfer between T4 phages and *E. coli* was identified with the C-terminal domain of pyruvate formate lyase, a glycolytic enzyme. It is currently unknown what selective advantage the acquisition of this cellular gene confers to T4 phages. As all of subgroup JS98 and some of subgroup RB49 phages lack this gene, it does not encode an essential function for T4 phages.

At the time of this writing, about 10 T4-like coliphages have been sequenced. Their genomes were run against a database of undesired genes containing virulence and antibiotic resistance functions, and no suspect matches were observed. Likewise, matches of T4 proteins with a food allergen database were not found (Zuber et al., 2007). In conclusion, in silico genome analysis makes T4 phages a safe candidate for phage therapy in humans.

Safety of T4 Phages: Animal Experiments

In silico analysis, albeit necessary, cannot replace in vivo safety evaluations of T4 phages. Before testing phages in human volunteers, it is appropriate to explore whether T4 phages are innocuous in animals. Conventional mice received T4 phages with their drinking water either individually or as cocktails. In initial experiments phages were applied with titers ranging from 10^3 to 10^6 PFU/ml in drinking water for 3 days. Phages appeared in the feces within half a day and reached peak titers between 10^3 and 10^7 PFU/ml proportional to the oral dose (Chibani-Chennoufi et al., 2004c). Upon stopping phage feeding with the drinking water, the phages disappeared from the feces (<100 PFU/g feces) within a day. No adverse effects were observed upon these short-term phage applications. Subsequently, T4 phages were given to mice at a high dose of 10^9 PFU/ml in the drinking water over 1 month. During that time period, no adverse effects were observed. The mice showed normal behavior, and no changes in appearance (e.g., fur changes) or food intake were observed. Weight gain of 6-week-old female C3H mice was comparable in phage-exposed and control mice. After the experiment, phage-exposed mice were searched for phages that had transgressed the gut. We did not find phages in blood, liver, spleen, or mesenteric lymph nodes. Mice also tested negative for serum antibodies to T4 phages. Tissue samples from phage-exposed and control mice were investigated with standard histological techniques. For the gut samples, both longitudinal and cross-sectional histological cuts were analyzed. No histological changes were observed after phage exposure (Chibani-Chennoufi et al., 2004c; Denou et al., 2009).

A special concern in the phage therapy of *E. coli* diarrhea is the fact that commensal *E. coli* is part of the normal gut microbiota in humans. One might therefore expect changes in commensal microbial population by the administration of oral phages. While such an effect is not a major concern for therapeutic coliphages (diarrhea patients already have a disturbed gut microbiota), it could compromise the prophylactic use of coliphages. Not all conventional mice colonies contain fecal *E. coli* strains. In *E. coli*-devoid mice, no negative effect of oral phages was observed on the fecal bacterial population as determined by colony counting on a number of media (Denou et al., 2009). Comparatively, mice that contained *E. coli* titers of $\geq 10^5$ CFU/g in feces samples showed only a very small titer decrease of this microflora upon oral phage exposure. Notably, the majority of *E. coli* colonies isolated from the feces of the phage-treated mice were susceptible in vitro to the orally applied T4 phages (Chibani-Chennoufi et al., 2004c). Commensal *E. coli* strains colonizing the large intestine of mice are thus spared from the lytic action of oral phages by current poorly defined effects like the physiological state of the bacterial cell in the colon, lack of accessibility (bacterial microcolonies in the mucus layer), or threshold effects (Hadas et al., 1997; Kasman et al., 2002, Krogfelt et al., 1993; Kutter et al., 1994; Poulsen et al., 1995; Tanji et al., 2005, Weiss et al., 2009). Recent reports indicated that different morphological groups of coliphages showed a strikingly distinct in vivo replication potential on commensal *E. coli* strains in the gut of mice (Weiss et al., 2009).

Safety of T4 Phages: Human Safety Test

After completion of the animal trials, phage T4 was given at doses of 10^3 and 10^5 PFU/ml in mineral water to healthy adult human volunteers (Bruttin and Brüssow, 2005). All volunteers who received the higher phage dose showed fecal phage 1 day after phage exposure, while feces before the intervention tested negative for phages. The mean peak titer of the fecal phage was 3×10^4 PFU/g stool. Two days after the phage exposure, fecal phage titers had returned to very low or negative values (<100 PFU/g feces). Volunteers who received the lower T4 phage dose displayed phage in only half of the fecal samples. The 15 volunteers enrolled into this safety study received, in a randomized way, the higher phage dose, the lower phage dose, or a placebo. Five mild adverse events were reported to the study physician, namely stomach pain, nausea, increased peristalsis, and sore throat. These events were not associated with the application of the high phage dose, and the study physician concluded that they were unrelated to the phage intervention. Two serum transaminases were determined before and after the study. No titer increases were observed, excluding a toxic effect of phage on liver cell integrity. Phage were not detected in the bloodstream, and no antibodies to T4 phage developed in the volunteers. Commensal *E. coli* was detected in the feces of all 15 volunteers before the phage intervention. The development of the fecal *E. coli* population was followed in daily stool samples. Phage application was not linked with a decrease of the total fecal *E. coli* counts, but a nonsignificant decrease in T4-sensitive *E. coli* strains was observed.

In conclusion, investigations ranging from genome analysis to the biology of T4 phages, and from animal experiments to human volunteer studies, have so far underlined that phage therapy approaches against *E. coli* diarrhea are safe. It is now important for the safety and efficacy of T4 phages to be demonstrated in carefully evaluated clinical trials.

REFERENCES

Abe, M., Y. Izumoji, and Y. Tanji. 2007. Phenotypic transformation including host-range transition through superinfection of T-even phages. *FEMS Microbiol. Lett.* **269:**145–152.

Albert, M. J., A. S. G. Faruque, S. M. Faruque, P. K. B. Neogi, M. Ansaruzzaman, N. A. Bhuiyan, K. Alam, and M. S. Akbar. 1995. Controlled study of *Escherichia coli* diarrheal infections in Bangladeshi children. *J. Clin. Microbiol.* **33:** 973–977.

Albert, M. J., A. S. G. Faruque, S. M. Faruque, R. B. Sack, and D. Mahalanabis. 1999. Case-control study of enteropathogens associated with childhood diarrhea in Dhaka, Bangladesh. *J. Clin. Microbiol.* **37:**3458–3464.

American Health Organization. 1980. Diarrhoeal diseases in the Americas. *Epidemiol. Bull.* **1:**1–4.

Babalova, E. G., K. T. Katsitadze, L. A. Sakvarelidze, N. S. Imnaishvili, T. G. Sharashidze, V. A. Badashvili, G. P. Kiknadze, A. N. Meipariani, N. D. Gendzekhadze, E. V. Machavariani, K. L. Gogoberidze, E. I. Gozalov, and N. G. Dekanosidze. 1968. Preventive value of dried dysentery bacteriophage. *Zh. Mikrobiol. Epidemiol. Immunobiol.* **2:**143–145.

Bamford, D. H., R. M. Burnett, and D. I. Stuart. 2002. Evolution of viral structure. *Theor. Popul. Biol.* **61:**461–470.

Banks, D. J., S. B. Beres, and J. M. Musser. 2002. The fundamental contribution of phages to GAS evolution, genome diversification and strain emergence. *Trends Microbiol.* **10:**515–521.

Begum, Y. A., K. A. Talukder, G. B. Nair, S. I. Khan, A.-M. Svennerholm, R. B. Sack, and F. Qadri. 2007. Comparison of enterotoxigenic *Escherichia coli* isolated from surface water and diarrhoeal stool samples in Bangladesh. *Can. J. Microbiol.* **53:**19–26.

Boschi-Pinto, C., L. Velebit, and K. Shibuya. 2008. Estimating child mortality due to diarrhoea in developing countries. *Bull. W.H.O.* **86:**710–717.

Botstein, D. 1980. A theory of modular evolution for bacteriophages. *Ann. N. Y. Acad. Sci.* **354:**484–491.

Boyd, E. F., and H. Brüssow. 2002. Common themes among bacteriophage-encoded virulence factors and diversity among the bacteriophages involved. *Trends Microbiol.* **10:**521–529.

Brunser, O., J. Espinoza, G. Figueroa, M. Araya, E. Spencer, H. Hilpert, H. Link-Amster, and H. Brüssow. 1992. Field trial of an infant formula containing anti-rotavirus and anti-*Escherichia coli* milk antibodies from hyperimmunized cows. *J. Pediatr. Gastroenterol. Nutr.* **15:**63–72.

Brüssow, H. 2005. Phage therapy: the *Escherichia coli* experience. *Microbiology* **151:**2133–2140.

Brüssow, H., C. Canchaya, and W.-D. Hardt. 2004. Phages and the evolution of bacterial pathogens: from genomic rearrangements to lysogenic conversion. *Microbiol. Mol. Biol. Rev.* **68:**560–602.

Bruttin, A., and H. Brüssow. 2005. Human volunteers receiving *Escherichia coli* phage T4 orally: a safety test of phage therapy. *Antimicrob. Agents Chemother.* **49:**2874–2878.

Bryce, J., C. Boschi-Pinto, K. Shibuya, and R. E. Black. 2005. WHO estimates of the causes of death in children. *Lancet* **365:**1147–1152.

Canchaya, C., G. Fournous, S. Chibani-Chennoufi, M.-L. Dillmann, and H. Brüssow. 2003a. Phage as agents of lateral gene transfer. *Curr. Opin. Microbiol.* **6:**417–424.

Canchaya, C., C. Proux, G. Fournous, A. Bruttin, and H. Brüssow. 2003b. Prophage genomics. *Microbiol. Mol. Biol. Rev.* **67:**238–276.

Casjens, S. 2003. Prophages and bacterial genomics: what have we learned so far? *Mol. Microbiol.* **49:** 277–300.

Casjens, S., G. Hatfull, and R. Hendrix. 1992. Evolution of dsDNA tailed-bacteriophage genomes. *Semin. Virol.* **3:**383–397.

Chibani-Chennoufi, S., C. Canchaya, A. Bruttin, and H. Brüssow. 2004a. Comparative genomics of the T4-like *Escherichia coli* phage JS98: implications for the evolution of T4 phages. *J. Bacteriol.* **186:**8276–8286.

Chibani-Chennoufi, S., J. Sidoti, A. Bruttin, M. L. Dillmann, E. Kutter, F. Qadri, S. A. Sarker, and H. Brüssow. 2004b. Isolation of *Escherichia coli* bacteriophages from the stool of pediatric diarrhea patients in Bangladesh. *J. Bacteriol.* **186:**8287–8294.

Chibani-Chennoufi, S., J. Sidoti, A. Bruttin, E. Kutter, S. Sarker, and H. Brüssow. 2004c. In vitro and in vivo bacteriolytic activities of *Escherichia coli* phages: implications for phage therapy. *Antimicrob. Agents Chemother.* **48:**2558–2569.

Comeau, A. M., C. Bertrand, A. Letarov, F. Tétart, and H. M. Krisch. 2007. Modular architecture of the T4 phage superfamily: a conserved core genome and a plastic periphery. *Virology* **362:** 384–396.

Denou, E., A. Bruttin, C. Barretto, C. Ngom-Bru, H. Brüssow, and S. Zuber. 2009. T4 phages against *Escherichia coli:* potential and problems. *Virology* **388:**21–30.

Desplats, C., C. Dez, F. Tétart, H. Eleaume, and H. M. Krisch. 2002. Snapshot of the genome of the pseudo-T-even bacteriophage RB49. *J. Bacteriol.* **184:**2789–2804.

de Wit, M. A., M. P. Koopmans, L. M. Kortbeek, N. J. van Leeuwen, J. Vinjé, and Y. T. van Duynhoven. 2001a. Etiology of gastroenteritis in sentinel general practices in The Netherlands. *Clin. Infect. Dis.* **33:**280–288.

de Wit, M. A., M. P. Koopmans, L. M. Kortbeek, W. J. Wannet, J. Vinjé, F. van Leusden, A. I. Bartelds, and Y. T. van Duynhoven. 2001b. Sensor, a population based cohort study on gastroenteritis in the Netherlands: incidence and etiology. *Am. J. Epidemiol.* **154:**666–674.

Dhillon, T. S., E. K. Dhillon, H. C. Chau, W. K. Li, and A. H. Tsang. 1976. Studies on bacteriophage distribution: virulent and temperate bacteriophage content of mammalian feces. *Appl. Environ. Microbiol.* **32:**68–74.

Djie-Maletz, A., K. Reither, S. Danour, L. Anyidoho, E. Saad, F. Danikuu, P. Ziniel, T. Weitzel, J. Wagner, U. Bienzle, K. Stark, A. Seidu-Kokor, F. P. Mockenhaupt, and R. Ignatius. 2008. High rate of resistance to locally used antibiotics among enteric bacteria from children in Northern Ghana. *J. Antimicrob. Chemother.* **61:**1315–1318.

Donnenberg, M. S. 2002. Escherichia coli: *Virulence Mechanisms of a Versatile Pathogen.* Academic Press, San Diego, CA.

Filée, J., E. Bapteste, E. Susko, and H. M. Krisch. 2006. A selective barrier to horizontal gene transfer in the T4-type bacteriophages that has preserved a core genome with the viral replication and structural genes. *Mol. Biol. Evol.* **23:** 1688–1696.

Filée, J., F. Tétart, C. A. Suttle, and H. M. Krisch. 2005. Marine T4-type bacteriophages, a ubiquitous component of the dark matter of the biosphere. *Proc. Natl. Acad. Sci. USA* **102:**12471–12476.

Furuse, K. 1987. Distribution of coliphages in the general environment: general considerations, p. 87–124. *In* S. M. Goyal, C. P. Gerba, and G. Bitton (ed.), *Phage Ecology*. John Wiley & Sons, New York, NY.

Gwatkin, D. R. 1980. How many die? A set of demographic estimates of the annual number of infant and child deaths in the world. *Am. J. Public Health* **70:**1286–1289.

Hadas, H., M. Einav, I. Fishov, and A. Zaritsky. 1997. Bacteriophage T4 development depends on the physiology of its host *Escherichia coli*. *Microbiology* **143:**179–185.

Hambly, E., F. Tétart, C. Desplats, W. H. Wilson, H. M. Krisch, and N. H. Mann. 2001. A conserved genetic module that encodes the major virion components in both the coliphage T4 and the marine cyanophage S-PM2. *Proc. Natl. Acad. Sci. USA* **98:**11411–11416.

Harris, A. M., F. Chowdhury, Y. A. Begum, A. I. Khan, A. S. G. Faruque, A.-M. Svennerholm, J. B. Harris, E. T. Ryan, A. Cravioto, S. B. Calderwood, and F. Qadri. 2008. Shifting prevalence of major diarrheal pathogens in patients seeking hospital care during floods in 1998, 2004, and 2007 in Dhaka, Bangladesh. *Am. J. Trop. Med. Hyg.* **79:**708–714.

Häusler, T. 2006. *Viruses vs. Superbugs: a Solution to the Antibiotics Crisis?* Macmillan, Basingstoke, United Kingdom.

Haynes, J. A., and F. A. Eiserling. 1996. Modulation of bacteriophage T4 capsid size. *Virology* **221:**67–77.

Highton, P. J., Y. Chang, and R. J. Myers. 1990. Evidence for the exchange of segments between genomes during the evolution of lambdoid bacteriophages. *Mol. Microbiol.* **4:**1329–1340.

Kain, K. C., R. L. Barteluk, M. T. Kelly, X. He, G. de Hua, Y. A. Ge, E. M. Proctor, S. Byrne, and H. G. Stiver. 1991. Etiology of childhood diarrhea in Beijing, China. *J. Clin. Microbiol.* **29:**90–95.

Karam, J. D., et al. (ed.). 1994. *Molecular Biology of Bacteriophage T4.* ASM Press, Washington, DC.

Kasman, L. M., A. Kasman, C. Westwater, J. Dolan, M. G. Schmidt, and J. S. Norris. 2002. Overcoming the phage replication threshold: a mathematical model with implications for phage therapy. *J. Virol.* **76:**5557–5564.

Krogfelt, K. A., L. K. Poulsen, and S. Molin. 1993. Identification of coccoid *Escherichia coli* BJ4 cells in the large intestine of streptomycin-treated mice. *Infect. Immun.* **61:**5029–5034.

Kutter, E., E. Kellenberger, K. Carlson, S. Eddy, J. Neitzel, L. Messinger, J. North, and B. Guttman. 1994. Effects of bacterial growth conditions and physiology on T4 infection, p. 406–418. *In* J. D. Karam et al. (ed.), *Molecular Biology of Bacteriophage T4.* ASM Press, Washington, DC.

Letarov, A., X. Manival, C. Desplats, and H. M. Krisch. 2005. gpwac of the T4-type bacteriophages: structure, function, and evolution of a segmented coiled-coli protein that controls viral infectivity. *J. Bacteriol.* **187:**1055–1066.

Mata, L., A. Simhon, J. J. Urrutia, R. A. Kronmal, R. Fernandez, and B. Garcia. 1983. Epidemiology of rotaviruses in a cohort of 45 Guatemalan Mayan Indian children observed from birth to the age of three years. *J. Infect. Dis.* **148:** 452–461.

Merabishvili, M., J.-P. Pirnay, G. Verbeken, N. Chanishvili, M. Tediashvili, N. Lashki, T. Glonti, V. Krylov, J. Mast, L. van Parys, R. Lavigne, G. Volckaert, W. Mattheus, G. Verween, P. de Corte, T. Rose, S. Jennes, M. Zizi, D. de Vos, and M. Vaneechoutte. 2009. Quality-controlled small-scale production of a well-defined bacteriophage cocktail for use in human clinical trials. *PLoS ONE* **4:**e4944.

Merril, C. R., D. Scholl, and S. L. Adhya. 2003. The prospect for bacteriophage therapy in Western medicine. *Nat. Rev. Drug Discov.* **2:**489–497.

Milkman, R. 1999. Gene transfer in *Escherichia coli,* p. 291–309. *In* R.L. Charlebois (ed.), *Organization of the Prokaryotic Genome.* ASM Press, Washington, DC.

Miller, E. S., E. Kutter, G. Mosig, F. Arisaka, T. Kunisawa, and W. Ruger. 2003. Bacteriophage T4 genome. *Microbiol. Mol. Biol. Rev.* **67:** 86–156.

Mosig, G., J. Gewin, A. Luder, N. Colowick, and D. Vo. 2001. Two recombination-dependent DNA replication pathways of bacteriophage T4, and their roles in mutagenesis and horizontal gene transfer. *Proc. Natl. Acad. Sci. USA* **98:** 8306–8311.

Nataro, J. P., and J. B. Kaper. 1998. Diarrheagenic *Escherichia coli. Clin. Microbiol. Rev.* **11:**142–201.

Nolan, J. M., V. Petrov, C. Bertrand, H. M. Krisch, and J. D. Karam. 2006. Genetic diversity among five T4-like bacteriophages. *Virol. J.* **3:** e30.

Ohnishi, M., K. Kurokawa, and T. Hayashi. 2001. Diversification of *Escherichia coli* genomes: are bacteriophages the major contributor? *Trends Microbiol.* **9:**481–485.

Pantucek, R., A. Rosypalova, J. Doskar, J. Kruzickova, P. Borecka, S. Snopkova, R. Horvath, F. Götz, and S. Rosypal. 1998. The polyvalent staphylococcal phage φ812: its host-range mutants and related phages. *Virology* **246:** 241–252.

Petrov, V. M., J. M. Nolan, C. Bertrand, D. Levy, C. Desplats, H. M. Krisch, and J. D. Karam. 2006. Plasticity of the gene functions for

DNA replication in the T4-like phages. *J. Mol. Biol.* **361:**46–68.

Poulsen, L. K., T. R. Licht, C. Rang, K. A. Krogfelt, and S. Molin. 1995. Physiological state of *Escherichia coli* BJ4 growing in the large intestines of streptomycin-treated mice. *J. Bacteriol.* **177:** 5840–5845.

Qadri, F., S. K. Das, A. S. G. Faruque, G. J. Fuchs, M. J. Albert, R. B. Sack, and A.-M. Svennerholm. 2000. Prevalence of toxin types and colonization factors in enterotoxigenic *Escherichia coli* isolated during a 2-year period from diarrheal patients in Bangladesh. *J. Clin. Microbiol.* **38:**27–31.

Qadri, F., A. Saha, T. Ahmed, A. A. Tarique, Y. A. Begum, and A.-M. Svennerholm. 2007. Disease burden due to enterotoxigenic *Escherichia coli* in the first 2 years of life in an urban community in Bangladesh. *Infect. Immun.* **75:**3961–3968.

Qadri, F., A.-M. Svennerholm, A. S. G. Faruque, and R. B. Sack. 2005. Enterotoxigenic *Escherichia coli* in developing countries: epidemiology, microbiology, clinical features, treatment, and prevention. *Clin. Microbiol. Rev.* **18:**465–483.

Quirk, S. M., D. Bell-Pedersen, and M. Belford. 1989. Intron mobility in the T-even phages: high frequency inheritance of group I introns promoted by intron open reading frames. *Cell* **56:**455–465.

Repoila, F., F. Tétart, J.-Y. Bouet, and H. M. Krisch. 1994. Genomic polymorphism in the T-even bacteriophages. *EMBO J.* **13:**4181–4192.

Sandegren, L., and B.-M. Sjöberg. 2004. Distribution, sequence homology, and homing of group I introns among T-even-like bacteriophages: evidence for recent transfer of old introns. *J. Biol. Chem.* **279:**22218–22227.

Steffen, R. 1986. Epidemiological studies of travelers' diarrhea, severe gastrointestinal infections and cholera. *Rev. Infect. Dis.* **8**(Suppl. 2)**:**122–130.

Steffen, R. 2005. Epidemiology of traveler's diarrhea. *Clin. Infect. Dis.* **41:** S536–S540.

Steffen, R., M. Rickernbach, and U. Wilhelm. 1987. Health problems after travel to developing countries. *J. Infect. Dis.* **156:**84–91.

Stoll, B. J., R. I. Glass, M. I. Huq, M. U. Khan, J. E. Holt, and H. Banu. 1982. Surveillance of patients attending a diarrhoeal disease hospital in Bangladesh. *Br. Med. J.* **285:**1185–1188.

Stoll, B. J., B. Rowe, R. I. Glass, R. J. Gross, and I. Huq. 1983. Changes in serotypes of enterotoxigenic *Escherichia coli* in Dhaka over time: usefulness of polyvalent antisera. *J. Clin. Microbiol.* **18:**935–937.

Sulakvelidze, A., Z. Alavidze, and J. G. Morris. 2001. Bacteriophage therapy. *Antimicrob. Agents Chemother.* **45:**649–659.

Sulakvelidze, A., and E. Kutter. 2005. Bacteriophage therapy in humans, p. 381–436. *In* E. Kutter and A. Sulakvelidze (ed.), *Bacteriophages: Biology and Applications.* CRC Press, Boca Raton, FL.

Summers, W. C. 1999. *Felix d'Herelle and the Origins of Molecular Biology.* Yale University Press, New Haven, CT.

Tanji, Y., T. Shimada, H. Fukudomi, Y. Nakai, and H. Unno. 2005. Therapeutic use of phage cocktail for controlling *Escherichia coli* O157:H7 in gastrointestinal tract of mice. *J. Biosci. Bioeng.* **100:** 280–287.

Tétart, F., C. Desplats, and H. M. Krisch. 1998. Genome plasticity in the distal tail fiber locus of the T-even bacteriophage: recombination between conserved motifs swaps adhesin specificity. *J. Mol. Biol.* **282:**543–556.

Tétart, F., C. Desplats, M. Kutateladze, C. Monod, H. W. Ackermann, and H. M. Krisch. 2001. Phylogeny of the major head and tail genes of the wide-ranging T4-type bacteriophages. *J. Bacteriol.* **183:**358–366.

Tétart, F., F. Repoila, C. Monod, and H. M. Krisch. 1996. Bacteriophage T4 host range is expanded by duplications of a small domain of the tail fibre adhesion. *J. Mol. Biol.* **258:**726–731.

Tobe, T., S. A. Beatson, H. Taniguchi, H. Abe, C. M. Bailey, A. Fivian, R. Younis, S. Matthews, O. Marches, G. Frankel, T. Hayashi, and M. J. Pallen. 2006. An extensive repertoire of type III secretion effectors in *Escherichia coli* O157 and the role of lambdoid phages in their dissemination. *Proc. Natl. Acad. Sci. USA* **103:** 14941–14946.

Walsh, J. A., and K. S. Warren. 1979. Selective primary health care: an interim strategy for disease control in developing countries. *N. Engl. J. Med.* **301:**967–974.

Weiss, M., E. Denou, A. Bruttin, R. Serra-Moreno, M.-L. Dillmann, and H. Brüssow. 2009. *In vivo* replication of T4 and T7 bacteriophages in germ-free mice colonized with *Escherichia coli. Virology* **393:**16–23.

Wilson, J. H., R. B. Luftig, and W. B. Wood. 1970. Interaction of bacteriophage T4 tail fiber components with a lipopolysaccharide fraction from *Escherichia coli. J. Mol. Biol.* **51:**423–434.

Wolf, M. K. 1997. Occurrence, distribution, and associations of O and H serogroups, colonization factor antigens, and toxins of enterotoxigenic *Escherichia coli. Clin. Microbiol. Rev.* **10:**569–584.

Wolfe, M. S. 1997. Protection of travelers. *Clin. Infect. Dis.* **25:**177–184.

World Health Organization. 1983. Global medium term programme, programme 13.7, acute respiratory infections. Document TRI/ARI/MTP/83.1. World Health Organization, Geneva, Switzerland.

Young, P., M. Ohman, M. Q. Xu, D. A. Shub, and B.-M. Sjöberg. 1994. Intron-containing T4 bacteriophage gene *sunY* encodes an anaerobic ribonucleotide reductase. *J. Biol. Chem.* **269:**20229–20232.

Zuber, S., C. Boissin-Delaporte, L. Michot, C. Iversen, B. Diep, H. Brüssow, and P. Breeuwer. 2008. Decreasing *Enterobacter sakazakii* (*Cronobacter* spp.) food contamination level with bacteriophages: prospects and problems. *Microb. Biotechnol.* **1:**532–543.

Zuber, S., C. Ngom-Bru, C. Barretto, A. Bruttin, H. Brüssow, and E. Denou. 2007. Genome analysis of phage JS98 defines a fourth major subgroup of T4-like phages in *Escherichia coli*. *J. Bacteriol.* **189:**8206–8214.

INDUSTRIAL AND REGULATORY ISSUES IN BACTERIOPHAGE APPLICATIONS IN FOOD PRODUCTION AND PROCESSING

Alexander Sulakvelidze and Gary R. Pasternack

15

BACTERIOPHAGE APPLICATIONS FOR IMPROVING FOOD PRODUCTION AND PROCESSING

General Overview

Naturally occurring bacteriophages were used to prevent and treat infectious diseases of bacterial origin almost immediately after their discovery during 1915 to 1917 (Summers, 1999). Their therapeutic applications have been described in numerous publications and were recently summarized in several articles and books (Merril et al., 2003; Sulakvelidze et al., 2001; Sulakvelidze and Barrow, 2005; Sulakvelidze and Kutter, 2005; Summers, 2001). However, despite the long history of using bacteriophages therapeutically in animals and humans, their potential applications for improving food safety constitute a relatively recent trend that appears to be gaining increased recognition. For example, several phage preparations recently were approved for the first time by the Food and Drug Administration (FDA), U.S. Department of Agriculture (USDA), and Environmental Protection Agency (EPA). Also, the number of peer-reviewed publications in which the value of using various bacteriophages to reduce or eliminate bacterial contamination of various foods was examined has been steadily increasing lately, as has been the number of patent applications dealing with that subject. Phage-based food safety applications have also attracted increased media attention and some prestigious international awards. Two examples of such recognition are the 2006 *Popular Science* "Best of What's New" award to Intralytix, Inc. (Baltimore, MD; www.intralytix.com) for its food safety product LMP-102™, and the 2007 F1 "Best Innovation in Food Industry" Gold award for the LISTEX™ phage preparation developed by EBI Food Safety (Wageningen, The Netherlands; www.ebifoodsafety.com). Both preparations target *Listeria monocytogenes* and can be directly added to various foods, including ready-to-eat (RTE) foods.

A requirement that must be met before bacteriophage preparations can be used in real-life food processing environments is that the companies that develop them must be able to sell the products at a reasonable profit. This requires that (i) the governmental regulatory approvals necessary for commercial sales can be obtained and (ii) patent protection for their products and/or technology is sufficient to

Alexander Sulakvelidze and Gary R. Pasternack, Intralytix, Inc., The Columbus Centre, 701 East Pratt Street, Baltimore, MD 21202.

Bacteriophages in the Control of Food- and Waterborne Pathogens
Edited by Parviz M. Sabour and Mansel W. Griffiths, © 2010 ASM Press, Washington, DC

make producers (and investors) feel comfortable investing the funds necessary to develop and commercialize them. Both these issues are reviewed at some length in this chapter, along with other considerations that can play an important role in the eventual acceptance of this novel and promising technology. Noteworthy, bacteriophages can also be used to prevent or treat bacterial infections of animals (including humans), aquacultured fish (see chapter 13), and plants (chapter 5). However, strategies for commercializing such products are not reviewed in this chapter, which is primarily focused on food safety applications of bacteriophages. Readers interested in learning more about other possible applications for phages are directed to several review articles and general mass media publications available on that subject (Gill et al., 2007; Merril et al., 2003; Stone, 2002; Sulakvelidze and Barrow, 2005; Sulakvelidze and Kutter, 2005; Thiel, 2004).

Food Safety Interventions Using Bacteriophages

There are three broad areas of potential applications using phage-based products to improve food safety. First, treating live, agriculturally important animals with phages may be used to reduce their normal intestinal colonization with bacteria known to cause food-borne diseases (see chapter 4). The rationale for this approach is that reducing the number of pathogenic bacteria brought into food processing facilities by the animals may improve the existing decontamination protocols' efficacy, which may reduce or eliminate specific bacterial pathogens in foods prepared in the facilities. Second, applying phages to environmental surfaces in food processing facilities may be used to reduce their contamination with food-borne pathogens and, subsequently, to reduce the risk of the bacteria contaminating foods processed in the facilities. Third, directly adding phages to raw or RTE foods may be used to reduce their content of pathogenic bacteria (see chapter 7). Additional approaches and combined approaches (e.g., treating domesticated livestock with bacteriophages and adding the phages to foods prepared from the animals) also may be utilized. The former method (i.e., treating live animals with phages) may be considered to be an "indirect food safety" intervention strategy; therefore, it is only briefly discussed in this chapter. The chapter's main focus is on "direct food safety" applications for phages, i.e., using them to treat food processing facilities and directly adding them to foods prepared and packaged in them (see also chapter 7).

Factors Enhancing the Development of Phage-Based Intervention Strategies

Numerous factors may improve the speed with which phage-based preparations for improving food safety can be introduced into the marketplace and the degree to which they ultimately will be integrated into hazard analysis and critical control point programs by various food producers. Developing an effective product that has been approved by appropriate regulatory agencies is only one part of the equation. In order for the product to be used by food producers, it also must pass the effectiveness/potential benefits-versus-cost analysis test, the results of which will dictate whether food processors will buy it and the degree to which it will be used in real-life food processing facilities. Most of the published studies dealing with that subject, and the recent regulatory approvals of at least two phage-based products for food safety applications, suggest that phages can significantly reduce the levels of their targeted bacteria in foods; however, transferring those findings to real-life food processing environments presents some challenges.

Some of the key requirements for optimally integrating phage-based products into real-life food processing environments is a thorough understanding of food processors' operations along with an in-depth knowledge of the epidemiology of the pathogenic bacteria being targeted, so that effective phage intervention strategies can be designed. For example, while some bacteria (e.g., *L. monocytogenes*) are "environmental" pathogens that usually contaminate foods in food processing/packaging plants, other bacteria (e.g., *Salmonella*) are part of the normal intestinal microflora of many animal species, and they contaminate foods during the slaughter or carcass-processing cycle. Thus, this difference must be considered when developing phage intervention strategies because the optimal intervention times, doses, and delivery mechanisms for phage-based products may vary. Also, identifying the most effective intervention times and treatment regimens is important for optimal cost-effectiveness. Although phage-based preparations are not necessarily expensive, some of the industries that are likely to become first customers for phage-based food safety products (e.g., the poultry industry) operate on very narrow profit margins and, therefore, will be very cautious about the products' pricing. Finally, the difficulty of obtaining regulatory approvals for the anticipated applications will significantly affect the products' commercial viability. For example, some regulatory approvals (i) may be more difficult and time-consuming to obtain than others, which may cause the final product to be more expensive; and (ii) may require that phage-treated foods must be labeled as such, which may be met with caution by food processors, at least initially until the general public is educated about the benefits of phage-based methods for improving food safety. Some of these issues, and the pros and cons of some of the possible regulatory strategies for commercializing phage-based preparations for food safety applications, are discussed in greater detail below.

REGULATORY APPROVALS

One of the main challenges in obtaining regulatory approvals of bacteriophage-based preparations has been, as implausible as it may seem, their novelty. Bacteriophages are the most ubiquitous microorganisms on earth (Brüssow and Hendrix, 2002; Rohwer, 2003), their evolutionary history dates back an estimated 3 billion years (Brüssow et al., 2004), and they have been used to prevent and treat bacterial infections of domesticated animals and humans since 1919 (Summers, 1999). Yet, except for their brief safety review by the FDA in 1973 (after phages were found to be present in many vaccines used in the United States), which determined that they did not pose a health hazard, their presence in products developed to prevent or treat bacterial diseases, or to improve food safety, had never been approved by any governmental regulatory agency in the United States or Western Europe until 2005. Thus, the strategy for obtaining governmental approval of phage-based food safety products was not highly developed; i.e., since phages are not chemicals, antimicrobials, or other compounds previously approved by Western regulatory agencies, the existing regulatory guidelines were not immediately applicable. The ice was broken in 2005, with the EPA's approval of a phage-based preparation developed by OmniLytics, Inc. (Salt Lake City, UT). On December 9, 2005, the EPA registered (registration #67986-1) the first biopesticide product (designated AgriPhage™) containing lytic bacteriophages against *Xanthomonas campestris* pv. *vesicatoria* and *Pseudomonas syringae* pv. *tomato* as its active component. The product was approved for use in the fields where tomatoes are grown. In this context, several publications suggest that application of bacteriophages may help control various plant diseases (Civerolo, 1973; Civerolo and Keil, 1969; Flaherty et al., 2000; Jones et al., 2007; Moore, 1926; Thomas, 1935), although additional research in this area is clearly indicated (Goodridge, 2004; see also http://www.apsnet.org/online/feature/phages/). AgriPhage™ was developed to prevent and treat disease of certain plants, rather than to enhance food safety. However, its approval in the United States was followed approximately 1 year later (August 18, 2006) (21 CFR part 172.785) by that for the first phage-based product designed to improve food safety: LMP-102™ (ListShield™), an Intralytix-developed phage cocktail for reducing or eliminating contamination of RTE foods, which represented the first approved use within food processing plants. After its approval, several other phage-based food safety products developed by other companies were approved by other worldwide governmental agencies. Some examples include (i) the GRAS ("Generally Recognized as Safe") approval of LISTEX™ P100 (developed by EBI Food Safety) in October 2006 and (ii) the issuance of a series of "No Objection" letters (by the USDA's Food Safety Inspection Service [FSIS]) for phage products developed by OmniLytics, including anti-*Escherichia coli* O157:H7 and anti-*Salmonella* products (the letters were issued in January 2007 and March 2007, respectively) for externally treating (in the form of a mist, spray, or wash) domesticated livestock prior to the slaughtering process.

Phage-based products developed to improve food safety may be used in a variety of ways, and the regulatory requirements and approval processes will vary, sometimes dramatically, depending on the applications for which they were designed. Some of the products' possible applications and the strategies that may be used to obtain their regulatory approval are reviewed in this chapter. However, irrespective of the strategies chosen, one key issue is that the products must be *safe*. In that regard, although the safety of lytic bacteriophages is usually not an issue, other factors (e.g., the composition of the carrier solution for the phages) must be considered. These factors are also discussed at some length in this chapter.

Areas of Applications

IN VIVO INTERVENTIONS

One possible way to use bacteriophages in order to improve food safety is to treat domesticated livestock during specific stages of their development, e.g., immediately prior to slaughter. The goal of that approach is to eliminate or significantly reduce colonization with pathogenic bacteria that may be transmitted to the food product during the animals' slaughter and/or subsequent carcass processing. The approach's rationale is based on the fact that phages regulate the levels of their specific bacterial hosts in every ecosystem in which they are found. Studies of the efficacy of phage administration for reducing or preventing intestinal colonization and/or fecal shedding of the targeted bacteria have yielded somewhat inconclusive results. For example, some authors (Hurley et al., 2008) reported that *Salmonella* phages did not reduce the levels of *Salmonella* shed by treated chickens. On the other hand, the likelihood of fecal shedding of *E. coli* O157:H7 was recently shown (Niu et al., 2009) to be reduced if penned cattle harbored *E. coli* O157:H7-specific phages. Also, several investigators (Berchieri et al., 1991; Smith and Huggins, 1983; Smith et al., 1987a, 1987b) have reported that oral administration of *E. coli* phages significantly reduced that bacterium's concentration in the intestines of the treated animals, and several authors (Brabban et al., 2003; Kudva et al., 1999; Raya et al., 2003) recently suggested that phage-mediated reduction of intestinal colonization with pathogenic bacteria and/or of shedding of pathogenic bacteria will improve food safety. The appropriate strategy for having such phage-based products approved by regulatory agencies will largely depend upon how they are used and the desired marketing claims, e.g., claims that their use improves the treated animals' health, reduces the number of pathogenic bacteria they shed, improves food safety, etc.

A possible strategy for increasing the speed with which regulatory approval may be obtained for an in vivo-administered phage-based preparation is to designate the candidate preparation a "bacterin" (i.e., a suspension of killed or attenuated bacteria used as a vaccine) and to seek its approval by the USDA. This strategy derives from the Viruses, Serums, Toxins, Antitoxins, and Analogous Products Act (21 U.S. Code, sections 151 to 159) enacted in 1913, known commonly as the Virus-Serum-Toxin Act. The USDA's Animal and Plant Health Inspection Service (APHIS) and the FDA have agreed that the jurisdiction for animal vaccines designed to reduce or eliminate the carrier state for bacteria that can infect other animals (even if the infection is only rarely associated with significant disease in animals) will reside with the APHIS. Their joint agreement requires that (i) the products are only for administration to animals, and they act primarily by directly stimulating or enhancing immune responses; (ii) their labels' claims and advertising contain only factual statements supported by data (e.g., the products are effective in reducing bacterial colonization and/or shedding); and (iii) the products show significant and clinically relevant efficacy as defined by the APHIS. Thus, it may be possible to designate phage-based preparations as bacterins. For example, the results of several studies have suggested that when used prophylactically (i.e., before the onset of disease), bacteriophage lysates protect by eliciting active immunity; i.e., because they contain released bacterial antigens that elicit the production of protective antibodies. The first investigation into using phages to prepare bacterial lysate vaccines occurred in 1919 when Tamezo Kabeshima successfully used a phage-lysed *Shigella* preparation (called "bactériolysat") to immunize small laboratory animals (reviewed in Summers, 1999). Several subsequent publications reported that preparations of phage-lysed bacteria are excellent immunizing agents whose protective effect may be stronger than that of "regular" vaccines, e.g., vaccines prepared by heat or chemical inactivation of bacteria (reviewed in Krueger and Scribner, 1941). The proposed superior immunogenicity reported for phage-generated bacterial lysates may be due to the fact that bacteriophage-mediated lysis is a more effective and gentler method for exposing protective antigens of bacteria than are other approaches used to prepare "dead-cell vaccines." For example, the methods commonly used to inactivate bacterial pathogens for dead-cell vaccines (e.g., heat treatment, irradiation, and chemical treatment) may reduce the vaccines' efficacy by reducing the antigenicity of relevant immunological epitopes (Holt et al., 1990; Lauvau et al., 2001; Melamed et al., 1991). Using phages to prepare lysed bacterial vaccines has not been actively pursued since the 1930s and 1940s; however, a similar approach recently has been gaining increased attention, particularly in Western Europe. The approach uses lysis gene *e* of bacteriophage φX174 to lyse various gram-negative bacteria, in order to prepare dead-cell vaccines called "ghost vaccines." Some of the vaccines were tested in vivo and subsequently reported to prevent infections caused by various gram-negative bacteria (Jalava et al., 2002, 2003; Szostak et al., 1996). All ghost vaccines clearly fall into the "veterinary biologic" classification because of their mode of action. Thus, since phage lysate bacterins are very similar to ghost vaccines, when used prophylactically they are likely to elicit protection via the same active immunization mechanism as the latter, and it may be possible to designate phage-based preparations as bacterins and have them approved by the APHIS. This route for obtaining regulatory approval has the potential advantage of being more straightforward and less time-consuming than the typical FDA "drug" application route. However, it may be complicated by the need to demonstrate that most of a phage lysate's protective ability is a function of its acting "primarily through the direct stimulation, supplementation, enhancement, or modulation of the immune system or immune response" (Center for Veterinary Biologics, 2005). Since the live phages in such bacterins also are likely to contribute strongly to successful prophylaxis, it may be difficult to demonstrate clearly that the *primary* mechanism for the vaccine's in vivo activity meets the above requirement. Furthermore, food safety or human health claims, either implicit or explicit, are not allowed by the APHIS, which may complicate the marketing efforts for such phage-based products.

An alternative strategy for obtaining regulatory agency approval of phage-based products designed for use in animals may be designating them as GRAS substances or as dietary supplements. The pros and cons of such approvals are discussed elsewhere in this chapter. However, given the marketing limitations of such approvals, it is likely that any phage-based product developed for live-animal administration, with the goal of reducing or eliminating bacterial colonization and/or shedding (and the claim that it enhances food safety), will need to be regulated as an "animal drug" by the Center for Veterinary Medicine of the FDA (http://www.fda.gov/cvm/default.html). In addition, the FDA is required, by the general safety provisions of sections 409, 512, and 706 of the Federal Food, Drug, and Cosmetic Act (FFDCA), to determine whether each food additive, new animal drug, or color additive proposed for use in food-producing animals is safe for those animals and whether the edible products derived from treated animals are safe. The relevant regulations are found in 21 CFR parts 70, 514.1, and 570, and they are also discussed in detail in "Published Guidance Documents" on the FDA's website (http://www.fda.gov/cvm/Guidance/published.htm). Through its website, the Center for Veterinary Medicine also informs the public about all animal drug products approved for safety and efficacy. The list, known as the Green Book, has been updated monthly since January 1989. At the present time (October 2009), the Green Book does not contain any phage-based product.

ENVIRONMENTAL DECONTAMINATION

Another potential use for phage-based food safety products is to apply them to surfaces (walls, floors, drains, gratings, non-food-contact and food-contact equipment, etc.) in food processing facilities and food handling establishments, in order to eliminate or significantly reduce the levels of specifically targeted bacteria on those surfaces. Decontamination of food processing facilities presents considerable challenges because many potentially pathogenic bacteria are resistant to traditional sanitizers (Davidson and Harrison, 2002) *and* because many chemical sanitizers are corrosive and toxic, and therefore are unacceptable for treating foods or surfaces that come into direct contact with foods. Thus, bacteriophage-based preparations may be useful additional tools for reducing contamination of such facilities with various food-borne bacterial pathogens. This type of intervention is especially relevant for so-called environmental pathogens (e.g., *L. monocytogenes*), which typically get into foods (including RTE foods) from contaminated food processing environments, rather than from domesticated livestock before and during the slaughtering process, e.g., *Salmonella* and *Campylobacter* species. The intervention's approach is to apply phages regularly to all problem locations in the food processing facility, in order to significantly reduce the concentration of, or eradicate, specifically targeted pathogenic bacteria from the facility.

Using bacteriophages to decontaminate environmental surfaces of food processing facilities is a relatively novel approach that has not been studied rigorously, although the possibility was suggested in a 1929 patent proposing to use bacteriophages to purify polluted water (as briefly discussed later in this chapter). However, the results of the few available studies suggest that a phage-based approach has merit for that purpose. For example, Roy et al. (1993) reported that applying a mixture of three anti-*Listeria* phages (ca. 4×10^8 PFU/ml) onto surfaces experimentally contaminated with *L. monocytogenes* was as effective as a 20-ppm solution of a quaternary ammonium compound (QUATAL) in reducing the surfaces' contamination with *L. monocytogenes*. Also, *E. coli* O157:H7-specific phages were more recently reported to significantly reduce the concentrations of *E. coli* O157:H7 on stainless steel coupons (Sharma et al., 2005) and on other inanimate surfaces mimicking those of various types of building materials in food processing facilities (Abuladze et al., 2008).

The market for environmentally active antimicrobial products is currently ca. $1 billion/year, and >5,000 of them are currently registered with the EPA (http://www.epa.gov/pesticides/factsheets/antimic.htm). The EPA currently recognizes three classes of antimicrobials for public health applications: (i) sterilizers (sporacides), which destroy and eliminate all microbial life (bacteria, fungi, viruses, etc.), including spores; (ii) disinfectants, which destroy or irreversibly inactivate infectious microorganisms (but not necessarily spores) and are available for general use and for use in hospitals; and (iii) sanitizers, which reduce the concentrations of (but do not necessarily eliminate) microorganisms to levels considered safe by public health codes and regulations. A fourth class of antimicrobials, antiseptics/germicides, prevent infections and decay by inhibiting microbial growth. However, since they usually are used to treat living animals (including humans), they are considered to be drugs and are regulated by the FDA.

In order to be placed in one of the classes of antimicrobial products described above, a product's active ingredient must meet specific criteria for efficacy, usually expressed as its "log kill" during a specific time period. For example, in order to be classified as a sanitizer, a product must elicit a 5-log (99.999%) reduction in the number of microorganisms within 30 seconds. However, given the timing of a phage-mediated lysis cycle (each cycle requires 20 to 40 minutes) (Mathews, 1994), it is likely that phages will exhibit maximum efficacy only if they are allowed to stay in contact with their targeted host bacteria for longer time periods (i.e., a few hours or more), in order for several lytic cycles to be completed. Shorter periods of contact time also may cause bacterial lysis, due primarily to the "lysis from without" mechanism; however, the outcome is not likely to be as effective as it could be during the longer contact time. Also, bacteriophages are very specific and will lyse only their targeted bacteria. These factors create some specific challenges for commercializing phage-based products, e.g., determining their most appropriate regulatory designation, special labeling requirements, special instructions for use, etc. These issues are briefly reviewed below.

In the United States, bacteriophages do not readily fall into any of the existing EPA designations for antimicrobials for public health applications. Also, in Europe, their designation as "decontaminants" was recently considered, but it was thought not to be appropriate (von Jagow and Teufer, 2007). The EPA has decided to regulate bacteriophage-based products for environmental applications as "biopesticides," or more specifically, as "microbial pesticides." The first phage-based product to be designated a microbial pesticide was AgriPhage™, which was developed and commercialized by OmniLytics. The product, consisting of *X. campestris* pv. *vesicatoria*- and *P. syringae* pv. *tomato*-specific phages (chemical PC codes 006449 and 006521, respectively), was the first phage-based product to receive EPA registration (registration #67986-1) as a microbial pesticide for preventing or treating infections of tomatoes and peppers. Another product that recently (June 18, 2008) received EPA registration (registration #74234-1) as a microbial pesticide is LMP-102™, an *L. monocytogenes*-specific phage cocktail developed by Intralytix and intended as a tool for controlling *L. monocytogenes* contamination of food processing plants and food handling establishments.

Phage-based microbial pesticides require special labeling because of their high specificity; e.g., a typical EPA label stating that the product is effective in killing a certain percentage of all bacteria cannot be used because it may mislead the customer. Instead, the specific bacterial species targeted by the product must be indicated on the label, to make it clear that it will not kill any other bacteria. Also, using phage-based microbial pesticides must be part of an integrated sanitization/microbial control program with EPA-approved sanitizers or disinfectants possessing broad-spectrum activity against bacterial pathogens not affected by the phages. However, that combined approach requires careful design of the intervention strategy. For example, although some phages are known (Roy et al., 1993) to be fairly stable when exposed to some quaternary ammonium compounds (e.g., 50 ppm QUATAL, 4 hours), bacteriophages with lipid envelopes are the most susceptible microorganisms (followed by vegetative bacteria, rickettsiae, chlamydia, fungi, coccidia, bacterial spores, and prions, in that order) to disinfectants commonly used in food processing facilities. Thus, phages may be inactivated as soon as food processors use the chemical disinfectants that they normally employ in their facilities. Therefore, appropriately timing the application of phage-based microbial pesticides may be critical for their successful use in food processing plants and food service establishments. The results of recent efficacy studies of various experimentally contaminated hard surfaces (Abuladze et al., 2008) indicated that 5 minutes of exposure to phages (the shortest time examined during the studies) significantly reduced contamination with their targeted bacteria. Thus, an appropriate protocol may be to apply phage-based biopesticides at least 5 minutes before treatment with a sanitizer or disinfectant, although as mentioned above, longer phage contact times are likely to increase the phage intervention's efficacy. In conclusion, the optimal treatment parameters for a phage-based microbial pesticide will need to be determined by each food processor, in close collaboration with the protocol that the product's manufacturer has developed for environmental applications.

Changes in governmental regulations required for food processors to test for the presence of food-borne bacterial pathogens in their facilities may result in the increased use of phage products for improving food safety. For example, the Canadian Food Inspection Agency recently (April 2009) instituted new rules for *L. monocytogenes* testing (Canadian Food Inspection Agency, 2009). In addition to other changes, it now requires that companies producing RTE foods and hot dogs test their food-contact surfaces at least once a week per production line and that all positive results must be reported to the agency, which will undertake additional tests in that facility. Also, food processors must test their products for possible *L. monocytogenes* contamination up to 12 times/year. Although the new program has been criticized as not being stringent enough for large companies (http://www.foodproductiondaily.com/Quality-Safety/New-Canadian-listeria-rules-said-to-lack-bite-for-big-companies), it does require more-rigorous detection protocols than did the previous regulations; therefore, it is likely to prompt additional efforts by food processors to reduce their environmental *L. monocytogenes* loads. Thus, since bacteriophages may, in contrast to harsh chemical disinfectants, be applied directly to food-contact surfaces during food processing, using phage-based products may become an important additional intervention step in that effort.

DIRECT APPLICATION ONTO FOODS

One of the most, if not the most, effective ways to reduce the levels of food-borne bacterial pathogens in various foods may be to apply phages directly onto raw or RTE foods before the product is packaged. At that point, contamination with the usual food-borne bacterial pathogens usually is minimal, as a result of current food safety intervention practices employed by all major food producers. Thus, applying phages at that time may provide a final means for product cleanup. Also, applying the phages as close as possible to the final product's packaging will minimize their back-carryover into the processing facility, which in turn will reduce the possible emergence of phage-resistant bacterial mutants. Several studies suggest that applying phages onto raw or RTE foods significantly reduces their contamination with bacterial pathogens. For example, Goode et al. (2003) reported that applying *Campylobacter jejuni*- and *Salmonella*-specific phages onto raw chicken reduced *C. jejuni* and *Salmonella* contamination by 10- to 100-fold and by 99%, respectively. Also, when the initial, pretreatment concentration of *Salmonella* contamination was low, the phage treatment reduced it to an undetectable level. Leverentz et al. (2001) were the first to examine the value of using bacteriophages to eliminate, or to reduce the number of, food-borne pathogens in various RTE foods. They reported that *Salmonella* phages reduced the concentration of viable *Salmonella* by ca. 3.5 log units on experimentally contaminated honeydew melon slices stored at 5 and 10°C, and by ca. 2.5 log units on slices stored at 20°C. Several additional studies (reviewed in Greer, 2005; and Sulakvelidze and Barrow, 2005) revealed that applying bacteriophages to RTE foods significantly reduced their experimental contamination with the phages' specifically targeted *L. monocytogenes, E. coli* O157:H7, and *Salmonella.* The first phage-based product for improving food safety (i.e., LMP-102™, also known as ListShield™, a phage cocktail targeting *L. monocytogenes*) was approved by the FDA in August 2006 as a direct food additive.

Safety of Bacteriophages

The main goal of an FDA review and approval process is to ensure the safety of the product under the conditions of its intended use. While evaluating the safety of a phage-based food safety product, the FDA considers not only its active component (i.e., bacteriophages), but also other substances that may be present in the final product, e.g., salts and residual amounts of bacterial toxins and/or of the bacterial culture medium's constituents.

Regarding the active component of phage-based products, numerous observations and a large amount of data strongly indicate that bacteriophages are very safe (see also chapter 14). For example:

- Bacteriophages are the most ubiquitous microorganisms on earth. For example, 1 ml of nonpolluted stream water may contain as many as ca. 2×10^8 PFU of phages/ml (Bergh et al., 1989), and the total number of phages on earth has been estimated to be in the range of 10^{30} to 10^{32} (Brüssow and Hendrix, 2002; Rohwer, 2003). Also, based on the amount of phages estimated to be present in sewage (Araujo et al., 1997; Lasobras et al., 1999; Puig et al., 1999) and current U.S. Census data, ca. 8×10^{17} coliphages are shed each day in the United States (and coliphages account for only a small percentage of all bacteriophages in the environment).
- An estimated 10^{25} phage infections are initiated in nature every second (Pedulla et al., 2003) (see also chapter 9).
- Phages are common in the human mouth, where they are harbored in dental plaque and saliva (reviewed in Sulakvelidze and Kutter, 2005).
- Feeds containing bacteriophages are commonly consumed by animals (including domesticated livestock), with no apparent side effects ever reported. For example, in a recent study from Texas A&M University (Maciorowski et al., 2001), male-specific and somatic coliphages were detected in all of the animal feeds, feed ingredients, and poultry diets examined, even after they were stored for 14 months at −20°C.
- Phages have been administered to animals in doses significantly exceeding those used therapeutically, with no apparent side effects. For example, Bogovazova et al. (1991) injected mice and guinea pigs intramuscularly, intravenously, and intraperitoneally with a dose of phages that was ca. 3,500-fold larger/g of body weight than the projected human dose, with no signs of acute toxicity or gross or histological changes observed in any of the treated animals. Also, during long-term holding experiments (Milstien et al., 1977), a variety of animals (including nonhuman primates) were treated for 2 years with a V1 phage, and adverse effects were not observed.
- Bacteriophages often are ingested by humans via their drinking water (Armon et al., 1997; Armon and Kott, 1993; Grabow and Coubrough, 1986; Lucena et al., 1995) and various foods. In this regard, bacteriophages have been commonly isolated from a wide range of food products, including ground beef, pork sausage, chicken, farmed freshwater fish, common carp and marine fish, oil sardines, raw skim milk, cheese, and many "healthy foods," including yogurt (Atterbury et al., 2003; Brüssow et al., 1998; Gautier et al., 1995; Greer, 2005; Hsu et al., 2002; Kennedy et al., 1984, 1986; Whitman and Marshall, 1971).

- Bacteriophages are common commensals in the human intestinal tract, and they are believed to play an important role in regulating the diversity of bacteria in that part of the human body (Calci et al., 1998; Furuse et al., 1983; Havelaar et al., 1986). Also, recent data obtained from metagenomic analyses (using partial shotgun sequencing) of an uncultured viral community from human feces suggested that bacteriophages are, after bacteria, the largest category of microorganisms in the uncultured fecal library (Breitbart et al., 2003).
- Orally ingested bacteriophages do not adversely affect the gastrointestinal tract's microflora, presumably because of their highly specific activity. For example, oral administration of the *E. coli*-specific phage T4 to healthy adult volunteers did not decrease the total number of viable *E. coli* in their feces (Bruttin and Brüssow, 2005). In addition, the phage did not replicate significantly in the commensal *E. coli* population, and adverse side effects were not observed in the volunteers.
- Phages have been used for human therapy, without any serious side effects, since 1919, when they were first used to treat children with shigellosis (Summers, 1999). During 1930 to the 1940s, several major companies, e.g., Eli Lilly and Co., E. R. Squibb & Sons, and Swan-Myers (Abbot Laboratories), manufactured therapeutic phage products. Also, the Russian and German armies routinely used phage preparations to treat bacterial infections (reviewed in Sulakvelidze and Kutter, 2005; and Summers, 2001).
- During the long history of using phages as therapeutic agents in Eastern Europe and the former Soviet Union (and before the "antibiotic era" in the United States, France, Australia, and other countries), phages were administered to humans (i) orally, in tablet or liquid formulations; (ii) rectally; (iii) locally (skin, eye, ear, nasal mucosa, etc.), in tampons, rinses, and creams; (iv) via aerosols and intrapleural injections; and, albeit less commonly, (v) intravenously. Also, despite those multiple uses, there have been virtually no reports of serious complications associated with phage therapy. Recent reviews have summarized the results of some of the human therapy studies involving bacteriophages (Alisky et al., 1998; Sulakvelidze et al., 2001; Summers, 2001).
- Phages have been administered to humans for nontherapeutic purposes, without recorded illnesses or deaths. For example, during the 1970s to 1990s, phage ϕX174 was used extensively to monitor humoral immune function of patients in the United States, including patients with Down syndrome, Wiskott-Aldrich syndrome, human immunodeficiency virus infection, and other immunodeficiencies (Lopez et al., 1975; Ochs et al., 1982, 1992, 1993).
- Phages have been administered to humans via various sera and FDA-approved vaccines commercially available in the United States (Merril et al., 1972; Milch and Fornosi, 1975; Moody et al., 1975; Petricciani et al., 1973). The presence of phages in vaccines was deemed acceptable by the FDA commissioner (Petricciani et al., 1978) in 1973, which probably was the first indirect review of phage safety ever conducted by that agency. The first full review of a phage-based product for direct application onto food was completed by the FDA in August 2006, for LMP-102™ (ListShield™), a phage cocktail of six monophages, which was designed (by Intralytix) to reduce or eliminate contamination of RTE foods with *L. monocytogenes* (21 CFR part 172.785).

Phages do not replicate in (or integrate their DNA into the DNA of) mammals, even when concentrations as high as 10^{12} PFU are administered intravenously to rhesus monkeys (Milstien et al., 1977). However, transducing phages can transfer genetic material between bacterial strains of the same species (Boyd,

2005) and, perhaps, between strains of different species (Chen and Novick, 2009). Thus, it is prudent to select lytic phages rather than transducing phages for inclusion in therapeutic phage preparations, irrespective of whether they are designed for human clinical applications or for agricultural uses, including food safety applications.

Several criteria have been used to differentiate lytic phages from transducing phages. For example, Adams (1959) suggested that plaque morphology and activity spectrum differentiate lytic and transducing phages: clear plaques and broad-spectrum activity for lytic phages, and turbid plaques and a narrow target range for transducing phages. However, while clearly of some merit, those criteria may be considered to be somewhat subjective because the two properties are influenced by multiple experimental parameters. That is to say, a phage's target range and the clarity of the plaques it produces are strongly influenced by multiple, variable factors, including the nature of the bacterial host strain and its physiological condition at the time of testing, and the temperature and pH of the media in which the testing is conducted. Therefore, those criteria may not always provide accurate information about the transducing ability of a given phage, and it is prudent to obtain additional information for every monophage (single phage strain) considered for inclusion in a commercial phage-based product. This could be achieved by fully sequencing all component monophages. After their full genome sequences are available, bioinformatics analysis can be used to determine whether any of the phages' genomes contain "undesirable genes" (e.g., the genes listed in 40 CFR part 725.421 and bacterial toxin-encoding genes) and/or bacterial 16S rRNA genes, which would suggest prior transduction events. The analysis is very robust, and using an appropriate cutoff value will ensure that no undesirable genes are missed during the analysis. For example, with a cutoff *e* value of 1×10^{-4}, the chance that the gene in question is not an undesirable gene is 1 in 10,000 (an *e* value of 0 indicates a perfect match, i.e., absolute identity). However, in practice, significant matches are considered to be those with *e* values of $\leq 10^{-5}$ (Miller et al., 2003). Therefore, a cutoff value 1×10^{-4} will provide very strong assurance that undesirable genes are not missed during the analysis.

Typically, the "inactive" components (e.g., NaCl) of phage-based products are GRAS; therefore, their presence is not a safety concern. However, the concentration of NaCl and/or other salts in phage-based products may affect the commercial sale of foods treated with them, and it may need to be adjusted in order to satisfy marketing requirements. Also, the concentrations of various bacterial toxins in the commercial products will need to be determined, to ensure that the bacteria used to propagate the products' phages have not released dangerous amounts of toxins into them. For example, for the first phage-based product approved by the FDA, LMP-102™, it was necessary to show that listeriolysin O—the major toxin produced by *L. monocytogenes*—was either not present in the final phage preparation or was present at a concentration of ≤5 hemolytic units/ml, which is not of toxicological concern (21 CFR part 172.785). Similar analyses for other toxic bacterial byproducts will need to be performed for other phage-based preparations active against other food-borne bacterial pathogens. The risk assessment must be based on the (i) concentration of a given toxin in the phage preparation, (ii) estimated application rate of the phage preparation onto/into foods (to estimate the amount of toxin that may be added to foods treated with the phage preparation), (iii) maximum dose of toxin that may be consumed by humans eating the treated foods, and (iv) potential impact/safety margins. Such an assessment is certainly possible; however, it may be simplified if a nonpathogenic bacterium or strain that does not release toxin(s) during its growth and/or lysis is used to propagate the phages for the proposed commercial products. This approach was recently employed for the P100 phage (a phage active against *L. mono-*

cytogenes), which was propagated in *L. innocua*—a nonpathogenic *Listeria* species. The FDA issued a "No Objection" ruling in June 2007 (Center for Food Safety and Applied Nutrition, 2007) re the GRAS designation for P100 grown in a nonpathogenic *L. innocua* strain.

Environmental Impact

One part of the regulatory evaluation process for bacteriophage-based products (as well as for other products destined for use in food processing environments) is determining whether or not the product's intended use will affect the environment and, if it does, performing a rigorous environmental assessment analysis. As discussed elsewhere in this book and also briefly in this chapter, phages are the most ubiquitous microorganisms on earth, with their total numbers estimated to be 10^{30} to 10^{32}. For virtually all imaginable, technically feasible applications of bacteriophages in food processing environments, the amount of phages released into the environment (compared to the amount of phages already and naturally present there) will be less than negligible, which provides a strong basis for obtaining a categorical exclusion pursuant to 21 CFR part 25.32(r). This exclusion would exempt phage preparations and their intended uses from the need for an environmental assessment. The following example may help to illustrate this point.

In 2006 the FDA approved LMP-102™ as a food additive for reducing the levels of *L. monocytogenes* on various RTE foods (21 CFR part 172.785). The amount of *L. monocytogenes* phages released and propagated in the environment as a result of LMP-102™'s being used in food processing facilities was estimated, based on several "worst-case scenarios," to be ca. 6×10^{11} PFU. This number represents ca. 0.00000000000000000006% of the estimated 10^{32} phages naturally present in the environment, a truly negligible amount. Even if the number of LMP-102™ phages released into the environment was underestimated by 1,000-fold and, *at the same time,* the number of phages naturally present in the environment overestimated by 1,000-fold, the concentration of LMP-102™ phages released into the environment as a result of LMP-102™'s use would still be negligible, and it would amount to only ca. 0.00000000000006% of the phages naturally present in the environment. Based on these calculations, and on other information contained in the food additive petition for LMP-102™, the FDA determined that "there is no significant impact on the human environment and that an environmental impact statement is not required" (21 CFR part 172.785).

A conclusion essentially identical to that mentioned in the above paragraph was reached by the EPA after reviewing AgriPhage™, another phage-based product (developed by OmniLytics). AgriPhage™ is used "to optimize prevention of bacterial disease," namely plant pathogen-elicited lesions (see chapter 5), by spraying the plants growing in fields or in greenhouses. The product contains $>4 \times 10^9$ phages/ml, and the application regimens range from ca. twice a week, at a rate of ca. 1 liter/acre (field applications), to approximately 200 ml (approximately 6.8 fluid ounces) per 9,600 ft^2 of area (greenhouse applications). The EPA conducted a review of the product and its proposed applications, and it performed an environmental risk assessment analysis based on the literature citations provided by the registrant. After performing its studies, the EPA concluded that the "proposed uses of AgriPhage™ will have no adverse effects on avian species, wild mammals, fish, aquatic invertebrates, and insects, including the honeybee." Also, the EPA concluded that AgriPhage™ does not pose a hazard to wildlife at the proposed use rates, and it issued a "No Effect" finding in connection with any endangered/threatened species listed by the U.S. Fish and Wildlife Service (http://www.epa.gov/pesticides/biopesticides/ingredients/tech_docs/brad_006449-006521.pdf). Thus, it seems likely that other phage-based products, irrespective of whether they are for food safety, in vivo, or environmental applications, will not require an environmental impact or full environmental assessment analysis. Indeed, phage-based food safety products may be one of the most, if not the most, environmentally friendly/"green" approaches available today for managing specific bacterial pathogens in a variety of settings.

Efficacy of Bacteriophages

Efficacy is another criterion that is required for obtaining regulatory approval for a commercially viable food safety modality. The efficacy criterion may be based on the requirements of a specific regulatory agency, or market requirements, or both—and it may vary in different countries. For example, the FDA and the USDA's FSIS have established a "zero tolerance" for *L. monocytogenes* in RTE foods: no detectable level is permitted in 25 g of the food sample (Thompson et al., 1990). However, in the European Union (EU) and Switzerland, up to 100 CFU/g is currently permitted in RTE foods which do not support the bacterium's growth (e.g., RTE foods with a pH of ≤ 4.4 or a water activity of <0.92). Also, the FDA recently published a proposal to relax its current zero-tolerance policy for *L. monocytogenes* in RTE foods, so that it will be identical to that of the EU and Switzerland (http://www.fda.gov/ICECI/Compliance Manuals/CompliancePolicyGuidanceManual/ucm136694.htm). At the present time, it is unclear whether or not this proposal will pass, but the different tolerance levels in the United States and EU may affect the requirements set forth for *L. monocytogenes* phage preparations in those countries. In that context, although eradication of the targeted bacteria in the foods or on environmental surfaces will be the most desirable outcome, it may be difficult to consistently achieve that type of outcome. However, regulators and food processors may accept a significant reduction in the concentration of *L. monocytogenes* to or below the tolerance levels adopted in their respective countries. From a public safety standpoint, it is likely that any significant reduction in the *L. monocytogenes* levels in foods will be beneficial. In that regard, the FDA, USDA's FSIS, and CDC recently (September 2003) and jointly published the *Quantitative Assessment of the Relative Risk to Public Health from Foodborne* Listeria monocytogenes *among Selected Categories of Ready-to-Eat Foods* (Anonymous, 2003). The assessment analyzed several "what if" scenarios, including one in which a reduction in the levels of contamination was modeled. The model predicted that a 1-log (i.e., 10-fold) preretail reduction in contamination with *L. monocytogenes* would result in an approximately 50% reduction in the annual number of deaths attributable to the consumption of contaminated deli meats by elderly people. *L. monocytogenes*-specific phages have been reported (Leverentz et al., 2003, 2004) to reduce the concentration of *L. monocytogenes* in various experimentally contaminated foods by approximately 2 to 7 logs (i.e., 10^2- to 10^7-fold), compared to that of phage-untreated, contaminated food specimens. This experimentally observed reduction is significantly better than the 10-fold reduction estimated for the above-noted "what if" scenario; thus, if the former degree of efficacy can be reproduced in real-life food processing facilities, phages may have a significantly positive impact on public health. The putative, beneficial public health/food safety impact may not be limited to *L. monocytogenes* phages. In that regard, bacteriophages targeting other food-borne bacterial pathogens have been reported (Abuladze et al., 2008; Bigwood et al., 2008; Goode et al., 2003; Leverentz et al., 2001) to possess a similar degree of potency against their bacterial hosts in various foods and on hard surfaces used to mimic building materials commonly found in food processing facilities.

One very important aspect of using phage preparations to improve food safety is the ability to update the preparations and maintain their efficacy. Although phage resistance is not a very frequent event and it does not typically hamper the efficacy of phage treatment (Bull et al., 2002; Capparelli et al., 2007; Drake et al., 1998; Leverentz et al., 2001), phage-resistant bacterial mutants will eventually emerge (as has been observed for antibiotic-resistant mutants) because of the selective pressure created by using phages in food processing facilities and/or as part of the natural shift in bacterial populations (Lenski, 1984). However, the initial phage preparations may be updated as often as needed to maintain their efficacy (21 CFR part 172.785). The new, updated preparation does not have to be subjected to a new regulatory approval process, as long as the new phages meet the appropriate selection criteria, e.g., the absence of undesirable genes in their genomes, and the manufacturing process has not been substantively altered so that it yields a preparation that is less purified than the original "article of commerce" product. This flexibility enables the manufacturers of the phage product, in collaboration with their customers (i.e., food producers and processors), to continuously monitor the bacterial pathogens being targeted in the latter's facilities, and to update the phage preparations (e.g., by substituting one or more phages in the preparation with more robust phages) if they detect the emergence of mutants or new bacterial strains that are resistant to the original phage preparations. The ability to update phage-based products easily also enables their manufacturers to custom-design phage cocktails for a specific food processor, e.g., for a food processor whose facility is contaminated by a unique strain not specifically targeted by the original, mainstream phage cocktail.

Specific Regulations

OVERVIEW

Substances or products used by food processors generally are in compliance with the requirements of the FFDCA because they (i) are approved as food additives or (ii) are exempt from requiring approval as food additives because they are GRAS by experts who are qualified through training and experience on their properties and intended uses. Phage-based products that may leave residues in or on food are divided among those whose intended antibacterial effects involve food or water that comes into contact with food, and surfaces/substances that come into contact with food. The latter are any substances intended for use as a component of materials used to manufacture, package, transport, or hold/store food, if such use is not intended to have any technical effect on the food. Although phage-based preparations can, under certain conditions, be regulated as food-contact substances, it is likely that most regulatory approvals for phage-based food safety products will be pursued as (i) direct food additives, (ii) GRAS, (iii) dietary supplements, and/or (iv) processing aids. The specific regulations for those four categories are briefly reviewed below.

FOOD ADDITIVES

In general, any substance added to food is a food additive. The legal definition of a food additive, as provided by the FDA, is "any substance the intended use of which results or may reasonably be expected to result—directly or indirectly—in its becoming a component or otherwise affecting the characteristics of any food" (http://www.fda.gov/Food/FoodIngredientsPackaging/ucm094211.htm). This definition includes any substance used in the production, processing, treatment, packaging, transportation, or storage of food. However, it specifically excludes pesticide chemicals and pesticide chemical residues, as a result of the Food Quality Protection Act of 1996, which amended the Federal Insecticide, Fungicide, and Rodenticide Act, and the FFDCA. Food additives are further subdivided into two broad categories: (i) direct food additives and (ii) indirect food additives.

If a substance or product is added to a food to serve a specific purpose in that food, it is referred to as a direct food additive. One example of a direct food additive is aspartame, which is added to various foods as a low-calorie sweetener. Many direct food additives must be listed on the ingredients labels of the foods to which they have been added. Indirect food additives are those that become part of a food, in trace amounts, due to its packaging, storage, or other handling. The FDA's Center for Food Safety and Applied Nutrition maintains a list of indirect food additives, under an ongoing program known as the Priority-Based Assessment of Food Additives. The program contains administrative and chemical information for over 3,000 substances used in food-contact articles, including adhesives and coating components (21 CFR part 175), paper and paperboard components (21 CFR part 176), polymers (21 CFR part 177), and adjuvant and production aids (21 CFR part 178). Additional "indirect" additives that are approved as part of the food-contact substance notification program or are exempt from regulation as food additives (in accordance with 21 CFR part 170.39) are listed in other inventories.

LMP-102™ was the first bacteriophage-based, designed-for-food-safety-enhancement product approved by a U.S. regulatory agency (21 CFR part 172.785). The FDA allowed the product's use "as an antimicrobial agent specific for *Listeria monocytogenes*" and set forth a series of conditions for its safe use. These included the requirement for (i) the product's mean concentration of phages to be at least 1×10^9 PFU/ml; (ii) the absence of viable *L. monocytogenes* in the preparation, as well as any gram-negative and gram-positive bacteria capable of growing in commonly used bacteriological culture media (e.g., Luria-Bertani medium), including *E. coli, Salmonella,* and coagulase-positive staphylococci; (iii) the presence of listeriolysin O (a membrane-damaging toxin produced by *L. monocytogenes*) to be limited to a concentration not to exceed 5 hemolytic units/ml; and (iv) the product's total organic carbon concentration to be $\leq$36 mg/kg. In addition, the FDA specified that the approved LMP-102™ phage cocktail should be composed of six monophages, and it listed specific requirements for the component phages' properties, e.g., the absence of undesirable genes in their genomes. A critically important consideration for maintaining the future efficacy of LMP-102™, as well as that of other current and future phage-based food safety products, was that the FDA allowed its developer to substitute other phages for those in the approved cocktail, as long as the former meet the same stringent safety criteria met by the product's original monophages. The importance of such substitutions is discussed elsewhere in this chapter.

GRAS

An alternative approach for commercializing a phage-based preparation is for it to be considered GRAS, which is, as defined by the FDA under sections 201(s) and 409 of the FFDCA, "any substance that is intentionally added to food as a food additive, that is subject to premarket review and approval by FDA, unless the substance is generally recognized, among qualified experts, as having been adequately shown to be safe under the conditions of its intended use, or unless the use of the substance is otherwise excluded from the definition of a food additive" (http://www.fda.gov/Food/GuidanceCompliance RegulatoryInformation/Guidance Documents/FoodIngredientsandPackaging/ucm061846.htm).

There is no formal requirement for the FDA to approve the GRAS designation of any product; i.e., there is no "GRAS approval" per se. Rather, the GRAS notification program is a voluntary procedure that operates under a proposed rule issued in 1997 (Federal Register, 1997). Each manufacturer determines whether its product meets the GRAS definition and sends an appropriate notification to the FDA. The notification typically includes a short description of the substance (its identity, properties, etc.), the proposed/expected conditions of use, and the rationale for the GRAS determination. The FDA reviews the notification, and if the agency believes that the product meets the definition of GRAS, it typically responds (within 180 days) with a "No Objection" or a "No Questions" letter. Under sections 201(s) and 409 of the FFDCA, in order for a food additive to be considered GRAS, its use should be generally accepted as safe based on the results of published scientific studies [21 CFR part 170.30(b)]—or, for products used in food before 1958, based on a substantial history of safe consumption by a large number of consumers. The rigor of the studies required to obtain a "food additive" designation and a GRAS designation is about the same, although food additive petitions usually contain additional information not commonly found in peer-reviewed publications supporting the GRAS designation. The main differentiating factor is a "common knowledge" criterion; i.e., a GRAS substance is distinguished from a food additive on the basis of common knowledge (e.g., through publication in the scientific literature) concerning the former's safety for its intended use. On the other hand, a food additive designation is obtained after privately held data and information re the substance's use (in addition to whatever publicly available information the petitioner chooses to submit) are sent to the FDA, which then evaluates them and determines whether the data establish that the substance is safe under the conditions of its intended use (21 CFR part 171.1).

Lytic phages are very safe. Therefore, the GRAS designation is applicable to bacteriophages, and it may be used as a strategy for commercializing phage-based products designed for food safety applications. The first company that successfully petitioned the FDA for a GRAS designation for a phage-based product was EBI Food Safety (Table 1). In October 2006, the FDA issued a "No Questions" letter (Center for Food Safety and Applied Nutrition, 2006) regarding EBI's request for a GRAS designation for LISTEX™, a P100 phage-based product designed to be used for direct application on cheeses. This was the first recognition of a GRAS designation for a phage-based product, and it was soon extended (in July 2007) to include using LISTEX™ for other foods commonly associated with the risk of *L. monocytogenes* contamination, e.g., beef, poultry, and fish (Center for Food Safety and Applied Nutrition, 2007). Given the relative simplicity of the GRAS notification process, it is expected that other companies that have phage-based products also will pursue that strategy. Thus, with the possible exception of phage preparations marketed as processing aids, the number of GRAS-designated, food safety-oriented phage products probably will increase more rapidly than will the number of other food safety-oriented phage preparations with other regulatory designations.

DIETARY SUPPLEMENTS

Another potential strategy for obtaining regulatory approval of phage-based food safety products is by designating them "dietary supplements" or "food ingredients." A dietary supplement is a product taken by mouth that contains a "dietary ingredient" intended to supplement the diet. If promoted as a traditional food or beverage, the special rules applicable to a dietary ingredient would not apply and the ingredient would need to be the subject of a food additive petition or a GRAS petition. Many dietary ingredients are simply concentrated forms of components that are otherwise in the diet, so it is likely that humans already have been extensively exposed to them. Thus, such substances typically are considered safe, and there is no requirement to perform the formal types of toxicology studies that would be required to establish safety for a food additive or a GRAS substance. On the other hand, the fact that a substance is naturally present in food does not guarantee that it will be safe when extracted and concentrated. Therefore, under the Dietary Supplement Health and Education Act of 1994 (DSHEA), the dietary supplement manufacturer is responsible for ensuring that a supplement is safe before it is marketed, and the FDA is responsible for taking action against any unsafe dietary supplement after it reaches the marketplace. Generally, manufacturers do not need to register their products with the FDA nor get FDA approval before producing or selling dietary supplements.

The DSHEA amended the FFDCA to provide a less demanding path to market for di-

TABLE 1 Companies involved with therapeutic phage research and commercialization for food safety applications[a]

Company	Web address	Location	Products and regulatory strategy
Biophage Pharma, Inc.	www.biophagepharma.net	Montreal, Quebec, Canada	The company utilizes the in vivo intervention strategy to develop food safety products. Its two phage-based products, Coli-Pro™ and Salmo-Pro™, are for treating *E. coli* and *Salmonella* infections in swine and poultry, respectively.
EBI Food Safety, Inc.	www.ebifoodsafety.com	Wageningen, The Netherlands	A spinoff from Exponential Biotherapies, Inc., it developed LISTEX™ P100, the first phage-based product to receive a GRAS designation.
Exponential Biotherapies, Inc.	www.expobio.com	McLean, VA	One of the first phage therapy companies founded (during the early 1990s) in the United States. It is no longer involved with therapeutic phage research and production, which is currently the main function of its spinoff company EBI Food Safety, Inc.
Gangagen, Inc.	www.gangagen.com	United States and India	Founded in India during 2000, it became a Delaware Corporation in 2001, with its main office located in Palo Alto, CA. Most, and perhaps all, in-house research is done at Gangagen Biotechnologies PVT, Ltd. (Bangalore, India). It utilizes the in vivo intervention strategy to develop food safety products.
InnoPhage, Ltd.	www.innophage.com	Porto, Portugal	A spinoff company from the Catholic University in Porto, it is currently oriented toward developing products for environmental and cosmetic applications and for treating bacterial infections. At the present time, it does not have phage-based products or regulatory approvals.
Intralytix, Inc.	www.intralytix.com	Baltimore, MD	Founded in 1998, it was the first company to receive a food additive approval from the FDA, for LMP-102™, in 2006. It also received EPA approval for the same product in 2008. It has an FDA petition pending for ECP-100™, an *E. coli* O157:H7-specific phage-based product for direct food applications.
Novolytics, Ltd.	www.novolytics.co.uk	Coventry, United Kingdom	A spinout (formed during 2002) from the University of Warwick, which specializes in clinical applications but also is interested in phage products for "disinfection of food and drink industry equipment." At the present time, it does not have food safety products or regulatory approvals.

(Continued)

TABLE 1 *Continued*

Company	Web address	Location	Products and regulatory strategy
OmniLytics, Inc.	www.omnilytics.com	Salt Lake City, UT	Formerly known as AgriPhi, Inc., it developed AgriPhage™ to control bacterial spot/bacterial speck. AgriPhage™ was the first phage-based product to receive EPA approval (in 2005).
Phage Biotech, Ltd.	www.phage-biotech.com	Rehovot, Israel	Incorporated in March 2000. Mostly focuses on developing phage-based products against ear and eye infections; however, recently it has become interested in developing food safety applications for phages. At the present time, it does not have food safety products or regulatory approvals.

[a]The table includes most known "phage therapy" companies involved with commercializing phage-based products for food safety applications. It is not meant to be comprehensive, i.e., to include all companies that utilize phages or phage-derived proteins for human therapeutic, diagnostic, or other non-food-safety applications.

etary ingredients. The manufacturer or distributor must conclude that the dietary ingredient will reasonably be expected to be safe. If the substance was not used in food in the same form prior to enactment of the DSHEA, a "new dietary ingredient notification" (NDIN), which describes the basis for the conclusion that the ingredient is reasonably expected to be safe, must be submitted to the FDA at least 75 days before marketing commences. Usually, the FDA makes a decision concerning an NDIN petition within 180 days of the date the petition is filed with the agency [21 CFR part 350(b)], which may potentially lead to rapid commercialization.

Dietary supplements are subject to labeling requirements. Information that must be included on a dietary supplement's label includes: (i) a descriptive name for the product, which indicates that it is a "supplement"; (ii) the name and place of business of the manufacturer, packer, or distributor; (iii) a complete list of ingredients; (iv) the net contents of the product; and (v) nutritional data in a "Supplement Facts" panel (except for some small-volume products or those produced by eligible small businesses). The dietary supplement route is an attractive option for marketing phage-based food safety products because of the relative ease and rapidity of the NDIN application process. On the other hand, designating a phage-based product as a dietary supplement imposes limitations on how it can be promoted and what claims may be included on its label. For example, the product's label is not permitted to promote it as being effective for the treatment, prevention, or cure of a specific disease or condition—which often are relevant claims for phage-based food safety products. In general, the following three types of claims are permissible for products approved as dietary supplements: (i) health claims, (ii) structure/function claims, and (iii) nutrient content claims. Of these, health claims are likely to be most pertinent for phage-based preparations marketed as dietary supplements. The FDA's 2003 Consumer Health Information for Better Nutrition Initiative (http://www.fda.gov/Food/LabelingNutrition/LabelClaims/ucm111447.htm) stipulates that health claims may be made when there is emerging evidence for a relationship between using a dietary supplement and the reduced risk of a disease or health-related condition. Therefore, marketing phage-based preparations as a dietary supplement (or as components of probiotic preparations) to be regularly consumed by humans in order to improve their overall health status may be a permissible and effective approach. However, the gap between allowable claims (i.e., health claims) and not-allowable claims (i.e., treatment, prevention, or cure of a specific disease) is somewhat fuzzy; therefore, companies wishing to market their phage-based products as dietary supplements must carefully navigate the pertinent regulations to ensure that their products' labeling is consistent with the provisions of the DSHEA of 1994. They may also avoid potential labeling issues altogether if their phage-based products qualify as "processing aids," as described below.

PROCESSING AIDS

The Code of Federal Regulations [21 CFR part 101.100(a)(3)] defines "processing aids" as ingredients that are present during the processing of meat and/or poultry products, but which are present in the finished food at insignificant levels and do not have any technical or functional effect in that food. A processing aid does not have to be declared in the ingredients statement on the label of the meat or poultry products in which it is used. However, it must meet at least one of several criteria in order to qualify as a processing aid. For example, (i) it is added to the food during processing, but is removed before the food is packaged in its final form; (ii) it is added to the food during processing, but is converted into constituents normally present in the food *and* does not significantly increase the amount of those constituents naturally found in the food; (iii) it is added to the food because of its technical or functional effect during the food's processing, but the finished food product contains only insignificant levels that do not have a continued technical or functional effect; or (iv) it is transferred to the food from the processing equipment or from the final product's packaging, but does not have a technical effect on the final product. Processing aids either are not considered to be food additives [as defined in section 201(s) of the FFDCA] or may be food additives used in conformity with regulations established pursuant to section 409 of the FFDCA.

Bacteriophage-based products for food safety applications are added to foods precisely because of their "technical or functional effect," i.e., their ability to eliminate or significantly reduce the levels of their targeted bacterial pathogens in those foods. Thus, the determination as to whether or not they can be considered processing aids will be based on whether they "are present in the finished food at insignificant levels and do not have any technical or functional effect in that food" [21 CFR part 101.100(a)(3)(c)]. For example, bacteriophages may be considered to be processing aids if they are used to treat raw foods that are then processed, packaged, and shown to be completely free of residual phages. Also, the processing aid designation may be applicable even if the packaged food contains a residual amount of phages, provided it does not contain enough phages to protect it from possible recontamination with the targeted bacterial pathogen. One example of such a scenario would be treating beef with an *E. coli* O157:H7-specific phage preparation, which has been reported (Abuladze et al., 2008) to eliminate or significantly reduce the concentration of viable *E. coli* O157:H7 in experimentally contaminated beef before it is ground. If the residual level of phages present in the ground beef is not sufficient to protect it from recontamination (i.e., a residual effect or continued protection is not observed), the phage preparation could warrant a processing aid designation.

Irrespective of their regulatory designation, phage products are likely to be equally safe and effective; thus, the proposed designation may be primarily a marketing issue that influences the (i) ease or difficulty of the regulatory approval process and (ii) the products' labeling and the claims made for them. On the other hand, having phage-based food safety preparations approved as processing aids has the advantage of not having to label the treated food products as "treated with bacteriophages." Although treatment with phages may make the foods significantly safer to eat than untreated foods, customers may not initially realize the potential benefits and therefore may be wary about buying and consuming foods treated with "viruses." Hence, not having to label foods as treated with bacteriophages may simplify marketing efforts, at least during the early years of marketing phage-based products for food safety interventions. However, manufacturers cannot determine whether or not their phage-based products qualify as processing aids; i.e., such determinations are made, on a case-by-case basis, by the FSIS of the USDA, which considers the products' proposed uses. Thus, phage product manufacturers must submit data to the FSIS's Labeling and Program Delivery Division, in which they must demonstrate that their products' proposed uses are consistent with the FDA's definition of a processing aid.

The definitions of a processing aid are fairly similar in the United States and Europe. For example, Article 1(3)(a) of the EU Member States Directive on Additives #89/107/EEC defines a processing aid as "any substance not consumed as a food ingredient by itself, intentionally used in the processing of raw materials, foods or their ingredients, to fulfill a certain technological purpose during treatment or processing and which may result in the unintentional but technically unavoidable presence of residues of the substance or its derivatives in the final product, provided that these residues do not present any health risk and do not have any technological effect on the finished product." Their labeling requirements are also similar; e.g., according to Article 6(4)(c)(iv) of EU Member States Directive #2000/13/EC, processing aids are not regarded as ingredients and therefore do not have to be listed on a product's label. A review of EU regulations pertaining to various applications of bacteriophage-based products for food safety intervention was recently published (von Jagow and Teufer, 2007). Based on that review, it should be possible to designate phage-based food safety products as processing aids in Europe, at least for some types of food safety interventions.

ADDITIONAL APPROVALS AND DESIGNATIONS

Although not required for commercial sales, some companies involved in the commercialization of bacteriophages for various food safety applications have had, or are in the process of having, their phage-based products certified as "kosher," "halal," and/or "organic." Brief descriptions of those designations and of their certification requirements are presented below.

Kosher Designation

"Kosher" is English for the Hebrew term *kashrut* (also *kashruth* or *kashrus;* כַּשְׁרוּת; meaning "fit"), which is a set of dietary laws derived from the Torah/Old Testament, and is used to indicate foodstuffs that meet the dietary requirements of Jewish law. The global kosher market is very significant and it is expected to continue to grow. For example, according to data compiled in 2003 by Integrated Marketing Communications, Inc., ca. 10,000 companies produce products for the kosher market (ca. 3,000 new products are introduced into the kosher market annually), with consumers spending ca. $165 billion annually for kosher products versus ca. $250 million 25 years ago (http://www.star-k.org/ind-advantages-market.htm). Thus, it is anticipated that many companies marketing phage products for food safety applications will pursue the kosher certification for their products, especially since many foods typically accepted as kosher are good candidates for phage-based food safety interventions.

Bacteriophages grown in animal product-free media, using equipment that meets the kosher designation, are clearly eligible for that designation. Thus, if those phages are applied/added to kosher foods, it should not change the latter's kosher status. However, advertising regulations in many jurisdictions prohibit the use of the word "kosher" on a food product's label unless the product has been certified as conforming to Jewish dietary laws. Such certification can be obtained from various organizations or companies that offer kosher certification programs. One of the most commonly used is the Union of Orthodox Congregations (http://www.ou.org/), which uses a "U" inside a circle to symbolize the initials of its organization. The symbol may be followed by various letters of the alphabet to convey additional information about the certified product. For example, (i) "U-D" indicates that the product is kosher (but not necessarily kosher for Passover) and contains a dairy ingredient or a dairy derivative; (ii) "U-M" indicates that the product is kosher (but not necessarily kosher for Passover) and contains meat, a product with meat ingredients, or a derivative of meat; and (iii) "U-P" indicates that the product is *pareve* (foods that do not have milk or meat in them) and that it is kosher for Passover as well as for the rest of the year. Also, a single "K" is used as a symbol for kosher by some kosher-certifying companies, e.g., STAR-K Certification, Inc. (Baltimore, MD; http://www.star-k.org/). For example, the STAR-K certification was granted to all of Intralytix's food safety phage preparations (including LMP-102™) in 2009. The certification process typically involves a detailed review of the manufacturing process and all raw ingredients (including the names of their suppliers), and the equipment used during the manufacturing process (including whether or not the same equipment is used to produce nonkosher products). If all requirements are met, and after all required fees are paid, the company issues a certificate authorizing use of the STAR-K symbol on the approved products. Also, the certificate usually specifies whether the product is *pareve,* kosher for Passover, etc. The fees required for the certification process vary among various certifying organizations and companies, but they are usually a few thousand dollars per year.

Halal Designation

Halal (حلال, halāl, Halaal), an Arabic term meaning "lawful" or "permitted," is used to indicate any object or action that is permissible to use or engage in according to Islamic law and custom. The term is frequently used for food products, to designate foods that are permissible according to Islamic Shariah law. The Islamic Food and Nutrition Council of America (IFANCA) (http://www.ifanca.org/index.php) has compiled a list of ingredients considered to be *halal* (permitted), *haram* (forbidden), and *mashbooh* (questionable or suspect). Currently, bacteriophages are not listed in any of those categories. However, given the size of the market (ca. 70% of Muslims worldwide follow halal food standards, and the global halal market is currently estimated to be a $580 billion industry [Al-Harran and Low, 2008]), it is likely that phage manufacturers will explore halal certification for their products (http://drhalimahali.wordpress.com/2008/04/23/halal-products-marketing-which-way/). As with a kosher certification, the process is relatively straightforward: the manufacturer must complete an application for Halal Certification and Supervision (which can be done online at the IFANCA's website), and provide the IFANCA with a detailed description of the phage manufacturing process, including a list of the raw ingredients and cleaning procedures used. The IFANCA will review the information and arrange for an onsite inspection of the manufacturing facilities. If all requirements are met, and after all required fees are paid, a halal certificate will be issued for either 1 year or for each production lot, depending on the type of product.

Organic Designation

At the present time, the organic food market in the United States is ca. $14 billion. The market has slowed down during the recent economic downturn; however, according to the Organic Consumers Association (http://www.organicconsumers.org), it is expected to reach $30 billion within the next few years. Thus, producers of bacteriophage preparations are likely to explore the "organic" designation for their products, so that the organic foods treated with them retain their status. The organic designation is somewhat of a loose term, and requirements may vary in different countries; however, they are, in general, considered to be foods made according to certain production standards. For example, they are (i) grown without the use of conventional pesticides and artificial fertilizers, (ii) not contaminated with human and industrial waste, and

(iii) processed without ionizing radiation and "artificial" additives. In addition, if livestock are involved, they must be reared without the routine use of antibiotics and growth hormones and they must receive an organic diet. Also, in most countries, organic produce must not be genetically modified.

The National Organic Program (NOP) of the USDA develops, implements, and administers national production, handling, and labeling standards for organic agricultural products. NOP's regulations establish four types of "organic" and similar claims, which are allowed in product labeling according to the percentage of organic ingredients in the product (7 CFR parts 205.300 and 205.301). The four types of claims and the types of foods that may bear these claims are summarized as follows: (i) food labeled "100% organic" contains only organic ingredients; (ii) food labeled "organic" contains at least 95% organic ingredients; (iii) food labeled "made with organic" contains at least 70% organic ingredients; and (iv) the food label contains identification of the organic ingredients when it contains less than 70% organic ingredients in the ingredients list. A product's organic composition is determined by calculating the percentage of organic ingredients in the product (by weight or fluid volume), exclusive of water and salt. Bacteriophages, on their own, clearly fit into the "100% organic" category. However, in order to determine whether foods treated with phage preparations retain their original organic designations, several additional factors must be considered, e.g., the composition and concentration of (i) the bacteriologic culture medium used to grow the host bacteria required for phage propagation, (ii) antifoaming agents (if any were used during phage propagation), and (iii) the carrier solution in which the purified phages were suspended. The synthetic substances that may be used, and the nonsynthetic substances that cannot be used, in organic production and handling operations are contained in the National List of Allowed and Prohibited Substances, which is available on the USDA Agricultural Marketing Service's website (http://www.ams.usda.gov/AMSv1.0).

Companies that want to have their phage-based products placed in one of the four categories described above may seek organic certification. The NOP accredits foreign and domestic agents who inspect organic production and handling operations in order to certify that they meet USDA standards. One such foreign certification organization is Control Union Certifications (formerly International Inspection & Certification Organization, or SKAL) in The Netherlands. In 2007 SKAL issued a letter declaring that LISTEX™ P100 (a phage-based preparation produced by EBI Food Safety) is a food processing aid that may be used for organic foods (confirming EU regulation 2092/91, annex VI, section B). The Organic Materials Review Institute (OMRI) in the United States maintains a list (designated the OMRI Products List) of all products appropriate to use for organic food production, processing, and handling. The OMRI's criteria are based on the NOP regulations, and the products that pass its review are allowed to display the "OMRI® Listed" seal on their labels. At the present time (March 2009), the OMRI Products List contains only one phage-based product: AgriPhage™, developed by OmniLytics. However, because of the popularity and commercial importance of the organic food market, it is likely that many more phage marketing companies will seek to obtain the organic designation for their products during the not-too-distant future.

PHAGE-RELATED PATENTS

Introduction

Bacteriophages have been prominent in the scientific literature since their independent discovery by Félix d'Hérelle and Frederick Twort in the early 20th century (Duckworth, 1976; Summers, 1999); however, they were nearly absent from the patent literature until the ascendance of molecular biology in the early 1980s (Fig. 1). From 1920 to 2008, 37,857 U.S. patents containing the word

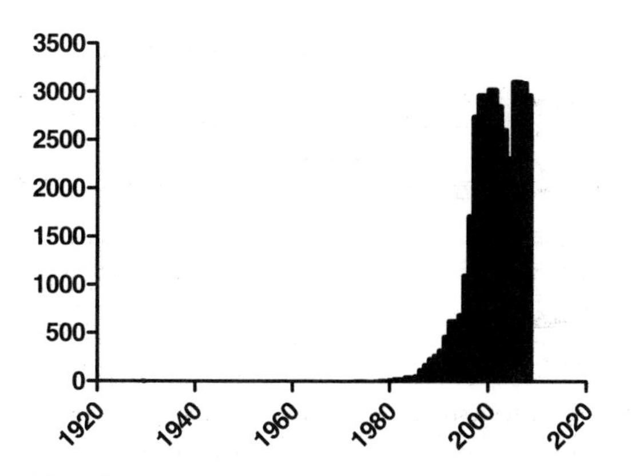

FIGURE 1 Patents containing the word "phage" or "bacteriophage" from 1920 to 2008. Patents containing the indicated search words were identified by Google Patent for the period 1920 to 1975, and by the U.S. Patent and Trademark Office database and search engine for patents issued between 1976 and 2008 (http://patft.uspto.gov/netahtml/PTO/search-adv.htm). With each database, patents were searched year by year over the indicated overall time span.

"phage" or "bacteriophage" were issued, but only 278 (ca. 0.73%) of them were issued between 1920 and 1980. Among various applications of bacteriophages, those developed for animal and human therapeutic uses were fairly common, and several large pharmaceutical companies (including Eli Lilly, E. R. Squibb & Sons, and Swan-Myers) produced commercial phage products for various phage therapy applications (reviewed throughout this book and also in Alisky et al., 1998; Gill et al., 2007; Sulakvelidze and Barrow, 2005; Sulakvelidze and Kutter, 2005; and Summers, 2001). Surprisingly, the early bacteriophage patent literature did not focus on therapy, and until relatively recently it has focused primarily on nontherapeutic applications of bacteriophages. Below, we briefly review various types of phage-related patents. Although the focus of this chapter is on food applications of bacteriophages, patents for all applications of bacteriophages are discussed because often the same patented bacteriophage may be used to improve both food safety and human health. Also, some patented phage-relevant methods are universal irrespective of the type of application.

Overview

The first phage patent was issued approximately 10 years after the discovery of bacteriophages in 1915 to 1917 (Duckworth, 1976; Summers, 1999). In 1926, U.S. patent 1,668,814 was issued to David Legg of Terre Haute, IN (the patent was assigned to the Commercial Solvents Corporation of Maryland). The patent was directed toward preventing the lytic activity of phages against bacteria used in industrial fermentation processes (see chapter 10). Years prior to a general understanding of lysogeny, the patent described a method for heat shocking bacteria in order to eliminate those that might otherwise yield "sluggish" cultures infected by bacteriophages (sometimes referred to in the patent as "ultravirus"). The second patent (U.S. patent 1,734,011), which was issued in 1929 and continued the industrial theme, was awarded to Edward Harrison and assigned to the Chemomechanical Water Company of Philadelphia. It described a procedure for using bacteriophages and their bacterial cell wall-lytic enzymes to purify polluted water, and it presaged some of the sanitation and public health applications of bacteriophage technology that are being revisited today. The next issued patent (U.S. patent 1,949,375, awarded in 1931) did not make any specific claims regarding bacteriophages, but it utilized them to validate the filtration materials that were the true subject of the patent.

As noted above, despite many worldwide studies oriented toward developing therapeutic bacteriophages during the 1930s and 1940s, patents for therapeutic phages were not issued until 1979. The reason for this situation is not clear. However, perhaps it reflected the fact that most of the studies of therapeutic phage technology during that era occurred in academic settings where patents were largely considered anathema. However, several companies did produce and sell therapeutic phage products during the same time period, although they did not seem to pursue patent protection for their products as rigorously as one might expect. One exception might be Eli Lilly, which had a patent (U.S. patent 1,949,375, issued to Frank Jones) describing a superior method for preparing bacterial vaccines, in which the well-known and apparently well-accepted use of bacteriophages to create a bacterial lysate suitable for vaccine use was discussed.

Phage-related patent activity continued to be relatively calm during the 1930s and 1940s. Two additional industrial patents concerning protocols for preventing deleterious phage activity in bacterial fermentations were issued in 1937 and 1946. However, although the field of molecular biology had begun in earnest with the phage studies of Max Delbrück, Salvador Luria, and others during the late 1940s and 1950s (Summers, 1993), none of their academic studies involving bacteriophages resulted in issued patents. Instead, patent activity focusing on industrial fermentations broadened and increased during the 1950s and early 1960s; e.g., patents were issued for eliminating phage-induced problems during cheese production (U.S. patents 2,982,654, 2,982,654, and 2,997,395) and during the production of sourdough pancake starter (U.S. patent 2,857,280).

The first patent for a truly therapeutic phage application was issued in 1958, when Welton Taylor and John Silliker, who found that inoculating eggs with *Salmonella*-specific phages resulted in increased hatch rates, were granted a patent (U.S. patent 2,851,006) entitled "Hatching of eggs." A subsequent patent (U.S. patent 2,876,108, issued in 1959) extended their invention to a food safety application, i.e., treating foodstuffs with phages to mitigate their potential contamination with pathogenic bacteria. That patent's coverage was particularly broad; e.g., its first claim read: "A method of controlling bacterial growth in a food material which comprises adding to the food material containing an objectionable bacterium a bacteriophage having predetermined properties of destroying said bacterium and maintaining said phage in contact with said bacterium under suitable conditions whereby said bacterium are substantially reduced." Thus, in view of their broad coverage, it is not surprising that the two patents were assigned to Swift, Inc., a large food company. The patent is directly relevant to some of the food safety applications of bacteriophages that are currently being pursued by various companies, as discussed throughout this chapter.

The marked annual growth in issued patents that began during the 1980s (Fig. 1) reflected two major trends. The first was the wide dissemination of molecular biological techniques utilizing phages. The second trend was the increased use of phage display technology, which began in earnest during the mid-1990s. The marked growth in phage-based patents may be attributed to the three 1969 patents (U.S. patents 3,444,402, 3,444,403, and 3,444,404) issued to Solomon Spiegelman, another pioneer in the field of molecular biology. They utilized phages as a source of enzymes, and as a means of studying central replicative and biosynthetic processes. Despite the apparent exponential growth of phage-based patents during approximately the last 2 decades (Fig. 1), the true level of activity in issued patents focusing principally upon bacteriophages has seen only modest and more consistent growth. Also although only 5 to 10 bacteriophage patents were issued per year during the1960s, the pace increased during the 1970s, with 10 to nearly 30 patents issuing per year by the end of the decade. This pace of issued patents whose primary subjects are bacteriophages themselves continues to the present time. Thus, much of the growth is attributable to the emergence of phage therapy-oriented patents, even though the issuance of such patents is limited by the extensive prior art.

In recent years, several important categories of issued patents directed at human therapeutic applications have emerged. In 1990, Alan Norris was awarded patents (U.S. patents 4,891,210 and 4,957,686) that utilized intact bacteriophages in oral hygiene preparations such as mouthwash and toothpaste. These appear to be the first issued patents truly directed at human therapeutics. A subsequent patent (U.S. patent 6,635,238) obtained by Allan DeLisle represented an important advance since it also proposed using bacterial cell wall-lytic phage enzymes to improve oral hygiene. DeLisle also extended his approach to food applications, proposing (in U.S. patent 6,955,893) to employ phage enzymes against food spoilage bacteria. In addition, U.S. patent 5,763,251 (issued to John Gasson in 1998) utilized a similar phage enzyme-based approach to control pathogenic *Listeria* in or on food products. Furthermore, Vincent Fischetti and his colleagues at Rockefeller University obtained 31 patents (beginning with U.S. patent 6,056,954 in 2000) involving therapeutic applications for purified bacteriolytic enzymes produced by bacteriophages against various bacterial pathogens. Although the DeLisle patent was issued several years later (in 2003), its priority extended back to June 1997, whereas the first Fischetti patent's priority date was September 1997. Nonetheless, the 31 patents resulting from the detailed studies performed by the Rockefeller University group cover virtually every conceivable formulation and application for purified bacteriolytic enzymes produced by phages, ranging from throat lozenges (U.S. patent 7,232,576, issued in 2007) to vaginal suppositories (U.S. patent 6,428,784, issued in 2002).

Carl Merril's work on prolonging the half-life of systemically administered bacteriophages to treat sepsis and other serious infections comprises another category of therapeutic patents. They (U.S. patents 5,660,812, 5,688,501, 5,766,892, and 5,811,903) formed the basis for an effort to address the issue of rapid elimination of bacteriophages from mammals after parenteral administration. Recently, Ramachandran and his colleagues at Gangagen Inc. (Table 1) have taken a related approach (U.S. patents 6,898,882, 6,913,753, and 7,087,226) by using lysin-deficient bacteriophages to reduce the immune response to phage proteins and to augment an antibacterial immune response. However, the seven patents mentioned in this paragraph do not seem to be relevant to food safety applications (with the exception of, perhaps, in vivo-related applications), and the technical value of at least some of them is uncertain. For example, using multiple dosing regimens is likely to be simpler and less expensive than developing a long-circulating phage preparation by serial, in vivo passaging of the monophages comprising the preparation (Sulakvelidze et al., 2001; Sulakvelidze and Kutter, 2005).

Interestingly, U.S. patent 6,121,036 (issued to Ghanbari and Averback in 2000) seems to "turn the clock back" to the early 1920s, with its extremely broad, general first claim of "A purified, host-specific, non-toxic, wide host range, and virulent bacteriophage preparation, wherein the bacteriophage preparation can be safely administered to patients or mammals in need, wherein (a) the bacteriophage preparation consists essentially of two or more bacteriophage, wherein each bacteriophage is selected against one of the group consisting of staphylococci, hemophilii, helicobacter, mycobacterium, mycoplasmi, streptococci, neisserii, klebsiella, enterobacter, proteus, bacteriodes, pseudomonas, borrelii, citrobacter, escherichia, salmonella, propionibacterium, treponema, shigella, enterococci, and leptospirex; (b) at least two of the bacteriophage are isolated against different strains of bacterial organisms; and (c) each bacteriophage is effective in killing, in vitro, bacteria from at least about 50% of bacterial isolates, wherein the isolates are from the same strain of bacterial organism as that from which the bacteriophage is isolated." Additional claims (a total of 15) cover other aspects of preparing and therapy with phages, including a list of various bacterial pathogens covered by the patent, various carrier solutions for bacteriophages, and methods for propagating and purifying phages. Thus, this patent appears to be an attempt to cover all aspects of bacteriophage therapy, except for a few minor considerations. On the other hand, since the patent's filing date is preceded by numerous other authors' publications providing abundant prior art, it is not clear why a patent with such a broad content range was issued, and defending it is likely to be quite problematic if it ever comes to litigation.

Recent Patenting Trends

The phage therapy literature is extensive, spanning the 1920s to the present. Consequently, intellectual property is difficult to obtain in that area of research. For this reason, one recent trend is for companies and other interested parties to protect the specific phages they are using in their applications without protecting the use or application itself. Such patents, which are of inherently limited scope, generally involve publishing phage nucleotide sequences and/or creating reference patent deposits at an acceptable institution, e.g., the American Type Culture Collection. Patent deposits become available to the general public upon the granting of the associated patent, in order to comply with the obligation that the patent fully enables the practice of the invention. One example of such patents is the recently issued (March 2009) U.S. patent 7,507,571 for *Listeria monocytogenes*-specific bacteriophages and uses thereof. The patent protects specific bacteriophages included in LMP-102™ (ListShield™), the first phage-based preparation approved by the FDA as a direct food additive. Thus, although other *L. monocytogenes*-specific phages may be used by competitors for the same types of applications as LMP-102™, the patent protects against reverse engineering of the original LMP-102™ preparation. Similar protection of subject matter (i.e., the component monophages comprising commercial phage preparations) is likely to be pursued by all companies involved in developing and commercializing phage preparations, whether they are for food safety, agricultural, or clinical applications.

Another interesting approach providing patent protection for therapeutic phage preparations has emerged recently. U.S. patent 7,459,272 (issued in 2008 to scientists at Intralytix) focuses on using phages to eliminate or significantly reduce colonization by various pathogenic bacteria, rather than on the more traditional treatment approach. In this context, virtually all of the literature on therapeutic bacteriophages deals with the *treatment of disease*. However, it is *colonization,* an early first step in the pathogenesis of many bacterial infections, that underlies the transmission of many virulent bacteria and their subsequent ability to cause disease. Therefore, U.S. patent 7,459,272 may stimulate the development of bacteriophage-based therapeutics by providing a proprietary position. It is clearly relevant to some clinical applications (e.g., those instances where colonization often results in disease). In addition, the same approach—if expanded to cover agriculturally important animal species—may be used to control bacterial diseases of domesticated livestock, which would have direct implications for the in vivo type of food safety applications of bacteriophages discussed previously in this chapter and also in chapter 4 of this volume.

In summary, in many ways the shifting patterns of bacteriophage patents reflect not only the evolution that is inherent in scientific progress, but trends indicative of social mores as well. The early academic investigators eschewed patents in favor of the public domain. As a consequence, most of the early patents are from industry. In contrast, the majority of the later patents, beginning with Spiegelman's molecular biology patents, were submitted by academics through their home institutions. As such, the trend in the world of bacteriophages appears to mirror the trend in the world at large. Finally, although many phage-based applications appear to be covered by existing patents, there is no doubt that entrepreneurs and scientists will continue to find novel bacteriophage-based approaches to manage bacterial pathogens, including those responsible for various food-borne illnesses.

ACKNOWLEDGMENTS

John Dubeck is gratefully acknowledged for sharing his expertise on regulatory aspects of phage-based food safety preparations. We thank Chandi Carter and

Bradley Anderson for their help with selected sections of this chapter.

REFERENCES

Abuladze, T., M. Li, M. Y. Menetrez, T. Dean, A. Senecal, and A. Sulakvelidze. 2008. Bacteriophages reduce experimental contamination of hard surfaces, tomato, spinach, broccoli, and ground beef by *Escherichia coli* O157:H7. *Appl. Environ. Microbiol.* **74:**6230–6238.

Adams, M. H. 1959. *Bacteriophages,* p. 365–380. Interscience Publishers, New York, NY.

Al-Harran, S., and P. Low. Mar./Apr. 2008. Marketing of halal products: the way forward. *Halal J.*

Alisky, J., K. Iczkowski, A. Rapoport, and N. Troitsky. 1998. Bacteriophages show promise as antimicrobial agents. *J. Infect.* **36:**5–15.

Anonymous. 2003. *Quantitative Assessment of the Relative Risk to Public Health from Foodborne* Listeria monocytogenes *among Selected Categories of Ready-to-Eat Foods.* Food and Drug Administration, Rockville, MD; Environmental Protection Agency, Washington, DC; and Centers for Disease Control and Prevention, Atlanta, GA. http://www.fda.gov/downloads/Food/ScienceResearch/ResearchAreas/RiskAssessmentSafetyAssessment/UCM197330.pdf.

Araujo, R. M., A. Puig, J. Lasobras, F. Lucena, and J. Jofre. 1997. Phages of enteric bacteria in fresh water with different levels of faecal pollution. *J. Appl. Microbiol.* **82:**281–286.

Armon, R., R. Araujo, Y. Kott, F. Lucena, and J. Jofre. 1997. Bacteriophages of enteric bacteria in drinking water: comparison of their distribution in two countries. *J. Appl. Microbiol.* **83:**627–633.

Armon, R., and Y. Kott. 1993. A simple, rapid and sensitive presence/absence detection test for bacteriophage in drinking water. *J. Appl. Bacteriol.* **74:** 490–496.

Atterbury, R. J., P. L. Connerton, C. E. Dodd, C. E. Rees, and I. F. Connerton. 2003. Isolation and characterization of *Campylobacter* bacteriophages from retail poultry. *Appl. Environ. Microbiol.* **69:**4511–4518.

Berchieri, A., Jr., M. A. Lovell, and P. A. Barrow. 1991. The activity in the chicken alimentary tract of bacteriophages lytic for *Salmonella typhimurium. Res. Microbiol.* **142:**541–549.

Bergh, O., K. Y. Borsheim, G. Bratbak, and M. Heldal. 1989. High abundance of viruses found in aquatic environments. *Nature* **340:**467–468.

Bigwood, T., J. A. Hudson, C. Billington, G. V. Carey-Smith, and J. A. Heinemann. 2008. Phage inactivation of foodborne pathogens on cooked and raw meat. *Food Microbiol.* **25:**400–406.

Bogovazova, G. G., N. N. Voroshilova, and V. M. Bondarenko. 1991. The efficacy of *Klebsiella pneumoniae* bacteriophage in the therapy of experimental *Klebsiella* infection. *Zh. Mikrobiol. Epidemiol. Immunobiol.* **1991:**5–8.

Boyd, F. 2005. Bacteriophages and bacterial virulence, p. 223–265. *In* E. Kutter and A. Sulakvelidze (ed.), *Bacteriophages: Biology and Applications.* CRC Press, Boca Raton, FL.

Brabban, A., T. Callaway, G. Dutta, M. Dyen, T. Edrington, E. Kutter, R. Raya, and P. Varey. 2003. Characterization of a new T-even bacteriophage with potential for reducing *E. coli* O157:H7 levels in livestock, abstr. M-019. *Abstr. 103rd Gen. Meet. Am. Soc. Microbiol.* American Society for Microbiology, Washington, DC.

Breitbart, M., I. Hewson, B. Felts, J. M. Mahaffy, J. Nulton, P. Salamon, and F. Rohwer. 2003. Metagenomic analyses of an uncultured viral community from human feces. *J. Bacteriol.* **185:** 6220–6223.

Brüssow, H., A. Bruttin, F. Desiere, S. Lucchini, and S. Foley. 1998. Molecular ecology and evolution of *Streptococcus thermophilus* bacteriophages—a review. *Virus Genes* **16:**95–109.

Brüssow, H., C. Canchaya, and W. D. Hardt. 2004. Phages and the evolution of bacterial pathogens: from genomic rearrangements to lysogenic conversion. *Microbiol. Mol. Biol. Rev.* **68:**560–602.

Brüssow, H., and R. W. Hendrix. 2002. Phage genomics: small is beautiful. *Cell* **108:**13–16.

Bruttin, A., and H. Brüssow. 2005. Human volunteers receiving *Escherichia coli* phage T4 orally: a safety test of phage therapy. *Antimicrob. Agents Chemother.* **49:**2874–2878.

Bull, J. J., B. R. Levin, T. DeRouin, N. Walker, and C. A. Bloch. 2002. Dynamics of success and failure in phage and antibiotic therapy in experimental infections. *BMC Microbiol.* **2:**35.

Calci, K. R., W. Burkhardt III, W. D. Watkins, and S. R. Rippey. 1998. Occurrence of male-specific bacteriophage in feral and domestic animal wastes, human feces, and human-associated wastewaters. *Appl. Environ. Microbiol.* **64:**5027–5029.

Canadian Food Inspection Agency. 2009. *Meat Hygiene Directive: Control Measures for Listeria in Ready-to-Eat (RTE) Meat Products,* chapter 5, section 5.3.11 and annex I. Canadian Food Inspection Agency, Ottawa, Ontario, Canada.

Capparelli, R., M. Parlato, G. Borriello, P. Salvatore, and D. Iannelli. 2007. Experimental phage therapy against *Staphylococcus aureus* in mice. *Antimicrob. Agents Chemother.* **51:**2765–2773.

Center for Food Safety and Applied Nutrition. 2006. GRAS notice no. GRN 000198. Center for Food Safety and Applied Nutrition, U.S. Food and Drug Administration, Rockville, MD.

Center for Food Safety and Applied Nutrition. 2007. GRAS notice no. GRN 000218. Center for

Food Safety and Applied Nutrition, U.S. Food and Drug Administration, Rockville, MD.

Center for Veterinary Biologics. 2005. Center for Veterinary Biologics notice no. 05-07. Biologics for reduction of colonization and/or shedding of animals. Center for Veterinary Biologics, Animal and Plant Health Inspection Service, U.S. Department of Agriculture, Ames, IA. http://www.aphis.usda.gov / animal_health / vet_biologics / publications/notice_05_07.pdf.

Chen, J., and R. P. Novick. 2009. Phage-mediated intergeneric transfer of toxin genes. *Science* **323:** 139–141.

Civerolo, E. L. 1973. Relationship of *Xanthomonas pruni* bacteriophages to bacterial spot disease in *Prunus*. *Phytopathology* **63:**1279–1284.

Civerolo, E. L., and H. L. Keil. 1969. Inhibition of bacterial spot of peach foliage by *Xanthomonas pruni* bacteriophage. *Phytopathology* **59:**1966–1967.

Davidson, P. M., and M. A. Harrison. 2002. Resistance and adaptation to food antimicrobials, sanitizers, and other process controls. *Food Technol.* **56:**69–78.

Drake, J. W., B. Charlesworth, D. Charlesworth, and J. F. Crow. 1998. Rates of spontaneous mutation. *Genetics* **148:**1667–1686.

Duckworth, D. H. 1976. "Who discovered bacteriophage?" *Bacteriol. Rev.* **40:**793–802.

Federal Register. 1997. Substances generally recognized as safe, proposed rule. *Fed. Regist.* **62:** 18938–18964.

Flaherty, J. E., J. B. Jones, B. K. Harbaugh, G. C. Somodi, and L. E. Jackson. 2000. Control of bacterial spot on tomato in the greenhouse and field with H-mutant bacteriophages. *HortScience* **35:**882–884.

Furuse, K., S. Osawa, J. Kawashiro, R. Tanaka, A. Ozawa, S. Sawamura, Y. Yanagawa, T. Nagao, and I. Watanabe. 1983. Bacteriophage distribution in human faeces: continuous survey of healthy subjects and patients with internal and leukaemic diseases. *J. Gen. Virol.* **64:**2039–2043.

Gautier, M., A. Rouault, P. Sommer, and R. Briandet. 1995. Occurrence of *Propionibacterium freudenreichii* bacteriophages in swiss cheese. *Appl. Environ. Microbiol.* **61:**2572–2576.

Gill, J. J., T. Hollyer, and P. M. Sabour. 2007. Bacteriophages and phage-derived products as antibacterial therapeutics. *Expert Opin. Ther. Pat.* **17:** 1341–1350.

Goode, D., V. M. Allen, and P. A. Barrow. 2003. Reduction of experimental *Salmonella* and *Campylobacter* contamination of chicken skin by application of lytic bacteriophages. *Appl. Environ. Microbiol.* **69:**5032–5036.

Goodridge, L. D. 2004. Bacteriophage biocontrol of plant pathogens: fact or fiction? *Trends Biotechnol.* **22:**384–385.

Grabow, W. O., and P. Coubrough. 1986. Practical direct plaque assay for coliphages in 100-ml samples of drinking water. *Appl. Environ. Microbiol.* **52:**430–433.

Greer, G. G. 2005. Bacteriophage control of foodborne bacteria. *J. Food Prot.* **68:**1102–1111.

Havelaar, A. H., K. Furuse, and W. M. Hogeboom. 1986. Bacteriophages and indicator bacteria in human and animal faeces. *J. Appl. Bacteriol.* **60:**255–262.

Holt, M. E., M. R. Enright, and T. J. Alexander. 1990. Immunisation of pigs with killed cultures of *Streptococcus suis* type 2. *Res. Vet. Sci.* **48:**23–27.

Hsu, F. C., Y. S. Shieh, and M. D. Sobsey. 2002. Enteric bacteriophages as potential fecal indicators in ground beef and poultry meat. *J. Food Prot.* **65:** 93–99.

Hurley, A., J. J. Maurer, and M. D. Lee. 2008. Using bacteriophages to modulate *Salmonella* colonization of the chicken's gastrointestinal tract: lessons learned from *in silico* and *in vivo* modeling. *Avian Dis.* **52:**599–607.

Jalava, K., F. O. Eko, E. Riedmann, and W. Lubitz. 2003. Bacterial ghosts as carrier and targeting systems for mucosal antigen delivery. *Expert Rev. Vaccines* **2:**45–51.

Jalava, K., A. Hensel, M. Szostak, S. Resch, and W. Lubitz. 2002. Bacterial ghosts as vaccine candidates for veterinary applications. *J. Control. Release* **85:**17–25.

Jones, J. B., L. E. Jackson, B. Balogh, A. Obradovic, F. B. Iriarte, and M. T. Momol. 2007. Bacteriophages for plant disease control. *Annu. Rev. Phytopathol.* **45:**245–262.

Kennedy, J. E., Jr., J. L. Oblinger, and G. Bitton. 1984. Recovery of coliphages from chicken, pork sasuage, and delicatessen meats. *J. Food Prot.* **47:**623–626.

Kennedy, J. E., Jr., C. I. Wei, and J. L. Oblinger. 1986. Methodology for enumeration of coliphages in foods. *Appl. Environ. Microbiol.* **51:**956–962.

Krueger, A. P., and E. J. Scribner. 1941. Bacteriophage therapy II. The bacteriophage: its nature and its therapeutic use. *JAMA* **19:**2160–2277.

Kudva, I. T., S. Jelacic, P. I. Tarr, P. Youderian, and C. J. Hovde. 1999. Biocontrol of *Escherichia coli* O157 with O157-specific bacteriophages. *Appl. Environ. Microbiol.* **65:**3767–3773.

Lasobras, J., J. Dellunde, J. Jofre, and F. Lucena. 1999. Occurrence and levels of phages proposed as surrogate indicators of enteric viruses in different types of sludges. *J. Appl. Microbiol.* **86:**723–729.

Lauvau, G., S. Vijh, P. Kong, T. Horng, K. Kerksiek, N. Serbina, R. A. Tuma, and E. G. Pamer. 2001. Priming of memory but not effector CD8 T cells by a killed bacterial vaccine. *Science* **294:**1735–1739.

Lenski, R. E. 1984. Coevolution of bacteria and phage: are there endless cycles of bacterial defenses and phage counterdefenses? *J. Theor. Biol.* **108:** 319–325.

Leverentz, B., W. S. Conway, Z. Alavidze, W. J. Janisiewicz, Y. Fuchs, M. J. Camp, E. Chighladze, and A. Sulakvelidze. 2001. Examination of bacteriophage as a biocontrol method for *Salmonella* on fresh-cut fruit: a model study. *J. Food Prot.* **64:**1116–1121.

Leverentz, B., W. S. Conway, M. J. Camp, W. J. Janisiewicz, T. Abuladze, M. Yang, R. Saftner, and A. Sulakvelidze. 2003. Biocontrol of *Listeria monocytogenes* on fresh-cut produce by treatment with lytic bacteriophages and a bacteriocin. *Appl. Environ. Microbiol.* **69:**4519–4526.

Leverentz, B., W. S. Conway, W. Janisiewicz, and M. J. Camp. 2004. Optimizing concentration and timing of a phage spray application to reduce *Listeria monocytogenes* on honeydew melon tissue. *J. Food Prot.* **67:**1682–1686.

Lopez, V., H. D. Ochs, H. C. Thuline, S. D. Davis, and R. J. Wedgwood. 1975. Defective antibody response to bacteriophage phichi 174 in Down syndrome. *J. Pediatr.* **86:**207–211.

Lucena, F., M. Muniesa, A. Puig, R. Araujo, and J. Jofre. 1995. Simple concentration method for bacteriophages of *Bacteroides fragilis* in drinking water. *J. Virol. Methods* **54:**121–130.

Maciorowski, K. G., S. D. Pillai, and S. C. Ricke. 2001. Presence of bacteriophages in animal feed as indicators of fecal contamination. *J. Environ. Sci. Health B* **36:**699–708.

Mathews, C. K. 1994. An overview of the T4 developmental program, p. 1–8. *In* J. D. Karam (ed.), *Molecular Biology of Bacteriophage T4.* ASM Press, Washington, DC.

Melamed, D., G. Leitner, and E. D. Heller. 1991. A vaccine against avian colibacillosis based on ultrasonic inactivation of *Escherichia coli. Avian Dis.* **35:**17–22.

Merril, C. R., T. B. Friedman, A. F. Attallah, M. R. Geier, K. Krell, and R. Yarkin. 1972. Isolation of bacteriophages from commercial sera. *In Vitro* **8:**91–93.

Merril, C. R., D. Scholl, and S. L. Adhya. 2003. The prospect for bacteriophage therapy in Western medicine. *Nat. Rev. Drug Discov.* **2:**489–497.

Milch, H., and F. Fornosi. 1975. Bacteriophage contamination in live poliovirus vaccine. *J. Biol. Stand.* **3:**307–310.

Miller, E. S., J. F. Heidelberg, J. A. Eisen, W. C. Nelson, A. S. Durkin, A. Ciecko, T. V. Feldblyum, O. White, I. T. Paulsen, W. C. Nierman, J. Lee, B. Szczypinski, and C. M. Fraser. 2003. Complete genome sequence of the broad-host-range vibriophage KVP40: comparative genomics of a T4-related bacteriophage. *J. Bacteriol.* **185:**5220–5233.

Milstien, J. B., J. R. Walker, and J. C. Petricciani. 1977. Bacteriophages in live virus vaccines: lack of evidence for effects on the genome of rhesus monkeys. *Science* **197:**469–470.

Moody, E. E., M. D. Trousdale, J. H. Jorgensen, and A. Shelokov. 1975. Bacteriophages and endotoxin in licensed live-virus vaccines. *J. Infect. Dis.* **131:**588–591.

Moore, E. S. 1926. D'Herelle's bacteriophage in relation to plant parasites. *S. Afr. J. Sci.* **23:**306.

Niu, Y. D., T. A. McAllister, Y. Xu, R. P. Johnson, T. P. Stephens, and K. Stanford. 2009. Prevalence and impact of bacteriophages on the presence of *E. coli* O157:H7 in feedlot cattle and their environment. *Appl. Environ. Microbiol.* **75:** 1271–1278.

Ochs, H. D., R. H. Buckley, R. H. Kobayashi, A. L. Kobayashi, R. U. Sorensen, S. D. Douglas, B. L. Hamilton, and M. S. Hershfield. 1992. Antibody responses to bacteriophage ϕX174 in patients with adenosine deaminase deficiency. *Blood* **80:**1163–1171.

Ochs, H. D., L. G. Lum, F. L. Johnson, G. Schiffman, R. J. Wedgwood, and R. Storb. 1982. Bone marrow transplantation in the Wiskott-Aldrich syndrome. Complete hematological and immunological reconstitution. *Transplantation* **34:**284–288.

Ochs, H. D., S. Nonoyama, M. L. Farrington, S. H. Fischer, and A. Aruffo. 1993. The role of adhesion molecules in the regulation of antibody responses. *Semin. Hematol.* **30:**72–79.

Pedulla, M. L., M. E. Ford, J. M. Houtz, T. Karthikeyan, C. Wadsworth, J. A. Lewis, D. Jacobs-Sera, J. Falbo, J. Gross, N. R. Pannunzio, W. Brucker, V. Kumar, J. Kandasamy, L. Keenan, S. Bardarov, J. Kriakov, J. G. Lawrence, W. R. Jacobs, Jr., R. W. Hendrix, and G. F. Hatfull. 2003. Origins of highly mosaic mycobacteriophage genomes. *Cell* **113:**171–182.

Petricciani, J. C., F. C. Chu, J. B. Johnson, and H. M. Meyer, Jr. 1973. Bacteriophages in live virus vaccines. *Proc. Soc. Exp. Biol. Med.* **144:**789–792.

Petricciani, J. C., T. C. Hsu, A. D. Stock, J. H. Turner, S. L. Wenger, and B. L. Elisberg. 1978. Bacteriophages, vaccines, and people: an assessment of risk. *Proc. Soc. Exp. Biol. Med.* **158:** 378–382.

Puig, A., N. Queralt, J. Jofre, and R. Araujo. 1999. Diversity of *Bacteroides fragilis* strains in their capacity to recover phages from human and animal wastes and from fecally polluted wastewater. *Appl. Environ. Microbiol.* **65:**1772–1776.

Raya, R., T. Callaway, T. Edrington, M. Dyen, R. Droleskey, E. Kutter, and A. Brabban. 2003. In vitro and in vivo studies using phages isolated from sheep to reduce population levels of *Escherichia coli* O157:H7 in ruminants, abstr. P-021. *Abstr. 103rd Gen. Meet. Am. Soc. Microbiol.* American Society for Microbiology, Washington, DC.

Rohwer, F. 2003. Global phage diversity. *Cell* **113:** 141.

Roy, B., H. W. Ackermann, S. Pandian, G. Picard, and J. Goulet. 1993. Biological inactivation of adhering *Listeria monocytogenes* by listeriaphages and a quaternary ammonium compound. *Appl. Environ. Microbiol.* **59:**2914–2917.

Sharma, M., J. H. Ryu, and L. R. Beuchat. 2005. Inactivation of *Escherichia coli* O157:H7 in biofilm on stainless steel by treatment with an alkaline cleaner and a bacteriophage. *J. Appl. Microbiol.* **99:** 449–459.

Smith, H. W., and M. B. Huggins. 1983. Effectiveness of phages in treating experimental *Escherichia coli* diarrhoea in calves, piglets and lambs. *J. Gen. Microbiol.* **129:**2659–2675.

Smith, H. W., M. B. Huggins, and K. M. Shaw. 1987a. Factors influencing the survival and multiplication of bacteriophages in calves and in their environment. *J. Gen. Microbiol.* **133:**1127–1135.

Smith, H. W., M. B. Huggins, and K. M. Shaw. 1987b. The control of experimental *Escherichia coli* diarrhoea in calves by means of bacteriophages. *J. Gen. Microbiol.* **133:**1111–1126.

Stone, R. 2002. Bacteriophage therapy. Stalin's forgotten cure. *Science* **298:**728–731.

Sulakvelidze, A., Z. Alavidze, and J. G. Morris, Jr. 2001. Bacteriophage therapy. *Antimicrob. Agents Chemother.* **45:**649–659.

Sulakvelidze, A., and P. Barrow. 2005. Phage therapy in animals and agribusiness, p. 335–380. *In* E. Kutter and A. Sulakvelidze (ed.), *Bacteriophages: Biology and Applications.* CRC Press, Boca Raton, FL.

Sulakvelidze, A., and E. Kutter. 2005. Bacteriophage therapy in humans, p. 381–436. *In* E. Kutter and A. Sulakvelidze (ed.), *Bacteriophages: Biology and Applications.* CRC Press, Boca Raton, FL.

Summers, W. C. 2001. Bacteriophage therapy. *Annu. Rev. Microbiol.* **55:**437–451.

Summers, W. C. 1999. *Felix d'Herelle and the Origins of Molecular Biology.* Yale University Press, New Haven, CT.

Summers, W. C. 1993. How bacteriophage came to be used by the Phage Group. *J. Hist. Biol.* **26:** 255–267.

Szostak, M. P., A. Hensel, F. O. Eko, R. Klein, T. Auer, H. Mader, A. Haslberger, S. Bunka, G. Wanner, and W. Lubitz. 1996. Bacterial ghosts: non-living candidate vaccines. *J. Biotechnol.* **44:**161–170.

Thiel, K. 2004. Old dogma, new tricks—21st century phage therapy. *Nat. Biotechnol.* **22:**31–36.

Thomas, R. C. 1935. A bacteriophage in relation to Stewart's disease of corn. *Phytopathology* **25:**371–372.

Thompson, P., P. A. Salsbury, C. Adams, and D. L. Archer. 1990. US food legislation. *Lancet* **336:**1557–1559.

von Jagow, C., and T. Teufer. 2007. Which path to go? *Eur. Food Feed Law Rev.* **3:**136–145.

Whitman, P. A., and R. T. Marshall. 1971. Isolation of psychrophilic bacteriophage-host systems from refrigerated food products. *Appl. Microbiol.* **22:**220–223.

INDEX